Smart Innovation, Systems and Technologies

Volume 44

Series editors

Robert J. Howlett, KES International, Shoreham-by-Sea, UK
e-mail: rjhowlett@kesinternational.org

Lakhmi C. Jain, University of Canberra, Canberra, Australia, and
University of South Australia, Adelaide, Australia
e-mail: Lakhmi.jain@unisa.edu.au

About this Series

The Smart Innovation, Systems and Technologies book series encompasses the topics of knowledge, intelligence, innovation and sustainability. The aim of the series is to make available a platform for the publication of books on all aspects of single and multi-disciplinary research on these themes in order to make the latest results available in a readily-accessible form. Volumes on interdisciplinary research combining two or more of these areas is particularly sought.

The series covers systems and paradigms that employ knowledge and intelligence in a broad sense. Its scope is systems having embedded knowledge and intelligence, which may be applied to the solution of world problems in industry, the environment and the community. It also focusses on the knowledge-transfer methodologies and innovation strategies employed to make this happen effectively. The combination of intelligent systems tools and a broad range of applications introduces a need for a synergy of disciplines from science, technology, business and the humanities. The series will include conference proceedings, edited collections, monographs, handbooks, reference books, and other relevant types of book in areas of science and technology where smart systems and technologies can offer innovative solutions.

High quality content is an essential feature for all book proposals accepted for the series. It is expected that editors of all accepted volumes will ensure that contributions are subjected to an appropriate level of reviewing process and adhere to KES quality principles.

More information about this series at http://www.springer.com/series/8767

Atulya Nagar · Durga Prasad Mohapatra
Nabendu Chaki
Editors

Proceedings of 3rd International Conference on Advanced Computing, Networking and Informatics

ICACNI 2015, Volume 2

 Springer

Editors
Atulya Nagar
Department of Computer Science
Liverpool Hope University
Liverpool
UK

Nabendu Chaki
Department of Computer Science
and Engineering
University of Calcutta
Kolkata, West Bengal
India

Durga Prasad Mohapatra
Department of Computer Science
and Engineering
National Institute of Technology Rourkela
Rourkela
India

ISSN 2190-3018   ISSN 2190-3026   (electronic)
Smart Innovation, Systems and Technologies
ISBN 978-81-322-3460-9   ISBN 978-81-322-2529-4   (eBook)
DOI 10.1007/978-81-322-2529-4

Springer New Delhi Heidelberg New York Dordrecht London
© Springer India 2016
Softcover re-print of the Hardcover 1st edition 2016

Printed on acid-free paper

Springer (India) Pvt. Ltd. is part of Springer Science+Business Media (www.springer.com)

Preface

It is indeed a pleasure to receive overwhelming response from researchers of premier institutes of the country and abroad for participating in the 3rd International Conference on Advanced Computing, Networking, and Informatics (ICACNI 2015), which makes our endeavor successful. The conference organized by School of Computer Engineering, KIIT University, India during June 23–25, 2015 certainly marks a success toward bringing the researchers, academicians, and practitioners to the same platform. We received more than 550 articles and very stringently selected through peer review 132 best articles for presentation and publication. We could not accommodate many promising works as we tried to ensure the quality. We are thankful to have the advice of dedicated academicians and experts from industry to organize the conference in good shape. We thank all people participating and submitting their works and having continued interest in our conference for the third year. The articles presented in the two volumes of the proceedings discuss the cutting edge technologies and recent advances in the domain of the conference.

We conclude with our heartiest thanks to everyone associated with the conference and seeking their support to organize the 4th ICACNI 2016.

Organization Committee

Chief Patron

Achyuta Samanta, Founder, Kalinga Institute of Industrial Technology and Kalinga Institute of Social Sciences, India

Patrons

N.L. Mitra, Chancellor, Kalinga Institute of Industrial Technology, India
Premendu P. Mathur, Vice Chancellor, Kalinga Institute of Industrial Technology, India

General Chairs

Amulya Ratna Swain, Kalinga Institute of Industrial Technology, India
Rajib Sarkar, Central Institute of Technology Raipur, India

Program Chairs

Manmath N. Sahoo, National Institute of Technology Rourkela, India
Samaresh Mishra, Kalinga Institute of Industrial Technology, India

Program Co-chairs

Ashok Kumar Turuk, National Institute of Technology Rourkela, India
Umesh Ashok Deshpande, Visvesvaraya National Institute of Technology, India
Biswajit Sahoo, Kalinga Institute of Industrial Technology, India

Organizing Chairs

Bibhudutta Sahoo, National Institute of Technology Rourkela, India
Madhabananda Das, Kalinga Institute of Industrial Technology, India
Sambit Bakshi, National Institute of Technology Jamshedpur, India

Technical Track Chairs

Computing
Umesh Chandra Pati, SMIEEE, National Institute of Technology Rourkela, India
Debasish Jena, International Institute of Information Technology Bhubaneswar, India
Networking
Shrishailayya Mallikarjunayya Hiremath, National Institute of Technology Rourkela, India
Daya K. Lobiyal, Jawaharlal Nehru University, India
Informatics
Korra Sathya Babu, National Institute of Technology Rourkela, India
Hrudaya Kumar Tripathy, Kalinga Institute of Industrial Technology, India

Industrial Track Chairs

Debabrata Mahapatra, Honeywell Technology Solutions, India
Binod Mishra, Tata Consultancy Services, India

Session and Workshop Chairs

Bhabani Shankar Prasad Mishra, Kalinga Institute of Industrial Technology, India
Ramesh K. Mohapatra, National Institute of Technology Rourkela, India

Special Session Chair

Ajit Kumar Sahoo, National Institute of Technology Rourkela, India

Web Chairs

Soubhagya Sankar Barpanda, National Institute of Technology Rourkela, India
Bhaskar Mondal, National Institute of Technology Jamshedpur, India

Publication Chair

Savita Gupta, Vedanta Aluminum Limited

Publicity Chairs

Ram Shringar Rao, Ambedkar Institute of Advanced Communication Technologies and Research, India
Himansu Das, Kalinga Institute of Industrial Technology, India

Registration Chairs

Sachi Nandan Mohanty, Kalinga Institute of Industrial Technology, India
Sital Dash, Kalinga Institute of Industrial Technology, India

Technical Program Committee

Adam Schmidt, Poznan University of Technology, Poland
Akbar Sheikh Akbari, University of Gloucestershire, UK
Al-Sakib Khan Pathan, SMIEEE, International Islamic University Malaysia (IIUM), Malaysia
Andrey V. Savchenko, National Research University Higher School of Economics, Russia
Annappa, SMIEEE, National Institute of Technology Karnataka, Surathkal, India
B. Narendra Kumar Rao, Sree Vidyanikethan Engineering College, India
Biju Issac, SMIEEE, FHEA, Teesside University, UK
C.M. Ananda, National Aerospace Laboratories, India
Ch. Aswani Kumar, Vellore Institute of Technology, India
Dhiya Al-Jumeily, Liverpool John Moores University, UK
Dilip Singh Sisodia, National Institute of Technology Raipur, India
Dinabandhu Bhandari, Heritage Institute of Technology, Kolkata, India
Ediz Saykol, Beykent University, Turkey
Jamuna Kanta Sing, SMIEEE, Jadavpur University, India
Jerzy Pejas, Technical University of Szczecin, Poland
Joel J.P.C. Rodrigues, Instituto de Telecomunicacoes, University of Beira Interior, Portugal
Koushik Majumder, SMIEEE, West Bengal University of Technology, India
Krishnan Nallaperumal, SMIEEE, Sundaranar University, India
Kun Ma, University of Jinan, China
Laszlo T. Koczy, Széchenyi István University, Hungary
M. Murugappan, University of Malaysia, Malaysia
M.V.N.K. Prasad, Institute for Development and Research in Banking Technology (IDRBT), India
Maria Virvou, University of Piraeus, Greece
Mihir Chakraborty, Jadavpur University, India
Natarajan Meghanathan, Jackson State University, USA
Patrick Siarry, SMIEEE, Université de Paris, Paris
Pradeep Singh, National Institute of Technology Raipur, India

Rajarshi Pal, Institute for Development and Research in Banking Technology (IDRBT), India
Soumen Bag, Indian School of Mines Dhanbad, India
Takuya Asaka, Tokyo Metropolitan University, Japan
Tuhina Samanta, Bengal Engineering and Science University, India
Umesh Hodeghatta Rao, SMIEEE, Xavier Institute of Management, India
Vivek Kumar Singh, SMIEEE, MACM, South Asian University, India
Yogesh H. Dandawate, SMIEEE, Vishwakarma Institute of Information Technology, India
Zahoor Khan, SMIEEE, Dalhouise University, Canada

Organizing Committee

Arup Abhinna Acharya, Kalinga Institute of Industrial Technology, India
C.R. Pradhan, Kalinga Institute of Industrial Technology, India
Debabala Swain, Kalinga Institute of Industrial Technology, India
Harish Kumar Patnaik, Kalinga Institute of Industrial Technology, India
Mahendra Kumar Gourisaria, Kalinga Institute of Industrial Technology, India
Manas Ranjan Lenka, Kalinga Institute of Industrial Technology, India
Manjusha Pandey, Kalinga Institute of Industrial Technology, India
Manoj Kumar Mishra, Kalinga Institute of Industrial Technology, India
Pinaki Sankar Chatterjee, Kalinga Institute of Industrial Technology, India
R.N. Ramakant Parida, Kalinga Institute of Industrial Technology, India
Santwana Sagnika, Kalinga Institute of Industrial Technology, India
Sharmistha Roy, Kalinga Institute of Industrial Technology, India
Siddharth Swarup Rautaray, Kalinga Institute of Industrial Technology, India
Subhasis Dash, Kalinga Institute of Industrial Technology, India
Sujoy Datta, Kalinga Institute of Industrial Technology, India
Tanmaya Swain, Kalinga Institute of Industrial Technology, India

External Reviewer Board

Ahmed Farouk Metwaly, Mansoura University, Egypt
Ankit Thakkar, Nirma University, India
Arun Jana, Centre for Development of Advanced Computing, India
Bharat Singh, Indian Institute of Information Technology Allahabad, India
Biswapratap Singh Sahoo, National Taiwan University, Taiwan
Chiheb-Eddine Ben N'Cir, University of Tunis, Tunisia
Gaurav Sharma, Galgotias University, India
Gudikandhula Narasimha Rao, Andhra University, India
Gunter Fahrnberger, University of Hagen, Germany
Hamidreza Khataee, Griffith University, Australia
Jitendra Agrawal, Rajiv Gandhi Proudyogiki Vishwavidyalaya, India
Manuj Darbari, Babu Banarasi Das University, India

Contents

Part I Distributed Systems, Social Networks, and Applications

**An Algorithm for Partitioning Community Graph
into Sub-community Graphs Using Graph Mining Techniques** 3
Bapuji Rao and Anirban Mitra

Traceback: A Forensic Tool for Distributed Systems 17
Sushant Dinesh, Sriram Rao and K. Chandrasekaran

**Deterministic Transport Protocol Verified by a Real-Time
Actuator and Sensor Network Simulation for Distributed Active
Turbulent Flow Control** 29
Marcel Dueck, Mario Schloesser, Stefan van Waasen
and Michael Schiek

Improved Resource Provisioning in Hadoop 39
M. Divya and B. Annappa

Cuckoo Search for Influence Maximization in Social Networks 51
Nikita Sinha and B. Annappa

**Sociopedia: An Interactive System for Event Detection
and Trend Analysis for Twitter Data** 63
R. Kaushik, S. Apoorva Chandra, Dilip Mallya,
J.N.V.K. Chaitanya and S. Sowmya Kamath

**Design and Implementation of a Hierarchical Content Delivery
Network Interconnection Model** 71
Sayan Sen Sarma and S.K. Setua

Part II Networking Systems and Architectures

**Solving Reliability Problems in Complex Networks
with Approximated Cuts and Paths** 85
Baijnath Kaushik and Haider Banka

**Maximal Clique Size Versus Centrality: A Correlation Analysis
for Complex Real-World Network Graphs** 95
Natarajan Meghanathan

On Solving the Multi-depot Vehicle Routing Problem 103
Takwa Tlili, Saoussen Krichen, Ghofrane Drira and Sami Faiz

Prediction of Crop and Intrusions Using WSN 109
S. Sangeetha, M.K. Dharani, B. Gayathri Devi, R. Dhivya
and P. Sathya

**Ethernet MAC Verification by Efficient Verification
Methodology for SOC Performance Improvement** 117
Sridevi Chitti, P. Chandrasekhar, M. Asharani
and G. Krishnamurthy

**MCDRR Packet Scheduling Algorithm for Multi-channel
Wireless Networks** 125
Mithileysh Sathiyanarayanan and Babangida Abubakar

**A Random Access Registration and Scheduling Based MAC
Protocol with Directional Antennas for Improving Energy
Efficiency** 133
Alisha and P.G. Poonacha

**A Key Agreement Algorithm Based on ECDSA for Wireless
Sensor Network** 143
Akansha Singh, Amit K. Awasthi and Karan Singh

**Frame Converter for Cooperative Coexistence Between IEEE
802.15.4 Wireless Sensor Networks and Wi-Fi** 151
Rambabu A. Vatti and Arun N. Gaikwad

Part III   Research on Wireless Sensor Networks, VANETs, and MANETs

Secured Time Stable Geocast (S-TSG) Routing for VANETs   161
Durga Prasada Dora, Sushil Kumar, Omprakash Kaiwartya
and Shiv Prakash

Reduction in Resource Consumption to Enhance Cooperation
in MANET Using Compressive Sensing   169
Md. Amir Khusru Akhtar and G. Sahoo

AODV and ZRP Protocols Performance Study Using OPNET
Simulator and EXata Emulator   185
Virendra Singh Kushwah, Rishi Soni and Priusha Narwariya

Optimal Probabilistic Cluster Head Selection for Energy
Efficiency in WSN   197
Madhukar Deshmukh and Dnyaneshwar Gawali

Firefly Algorithm Approach for Localization in Wireless Sensor
Networks.   209
R. Harikrishnan, V. Jawahar Senthil Kumar and P. Sridevi Ponmalar

Performance Analysis of Location and Distance Based Routing
Protocols in VANET with IEEE802.11p   215
Akhtar Husain and S.C. Sharma

Fuzzy Based Analysis of Energy Efficient Protocols
in Heterogeneous Wireless Sensor Networks   223
Rajeev Arya and S.C. Sharma

Node Application and Time Based Fairness in Ad-Hoc Networks:
An Integrated Approach   231
Tapas Kumar Mishra and Sachin Tripathi

Bee Colony Optimization for Data Aggregation in Wireless
Sensor Networks   239
Sujit Kumar and Sushil Kumar

Part IV   Cryptography and Security Analysis

Extended Visual Cryptography Scheme for Multi-secret Sharing   249
L. Siva Reddy and Munaga V.N.K. Prasad

Identifying HTTP DDoS Attacks Using Self Organizing Map
and Fuzzy Logic in Internet Based Environments   259
T. Raja Sree and S. Mary Saira Bhanu

Improving the Visual Quality of (2, n) Random Grid
Based Visual Secret Sharing Schemes   271
Lingaraj Meher and Munaga V.N.K. Prasad

An Efficient Lossless Modulus Function Based Data Hiding
Method   281
Nadeem Akhtar

On the Use of Gaussian Integers in Public Key Cryptosystems   289
Aakash Paul, Somjit Datta, Saransh Sharma
and Subhashis Majumder

Stack Overflow Based Defense for IPv6 Router Advertisement
Flooding (DoS) Attack   299
Jai Narayan Goel and B.M. Mehtre

Ideal Contrast Visual Cryptography for General Access
Structures with AND Operation   309
Kanakkath Praveen and M. Sethumadhavan

Extending Attack Graph-Based Metrics for Enterprise Network
Security Management   315
Ghanshyam S. Bopche and Babu M. Mehtre

A Graph-Based Chameleon Signature Scheme   327
P. Thanalakshmi and R. Anitha

Design of ECSEPP: Elliptic Curve Based Secure E-cash Payment
Protocol   337
Aditya Bhattacharyya and S.K. Setua

Security Improvement of One-Time Password
Using Crypto-Biometric Model   347
Dindayal Mahto and Dilip Kumar Yadav

An Image Encryption Technique Using Orthonormal Matrices
and Chaotic Maps   355
Animesh Chhotaray, Soumojit Biswas, Sukant Kumar Chhotaray
and Girija Sankar Rath

Learning Probe Attack Patterns with Honeypots 363
Kanchan Shendre, Santosh Kumar Sahu, Ratnakar Dash
and Sanjay Kumar Jena

Part V   Operating System and Software Analysis

Fuzzy Based Multilevel Feedback Queue Scheduler 373
Supriya Raheja, Reena Dadhich and Smita Rajpal

Dynamic Slicing of Feature-Oriented Programs 381
Madhusmita Sahu and Durga Prasad Mohapatra

**Mathematical Model to Predict IO Performance
Based on Drive Workload Parameters** 389
Taranisen Mohanta, Leena Muddi, Narendra Chirumamilla
and Aravinda Babu Revuri

**Measure of Complexity for Object-Oriented Programs:
A Cognitive Approach** 397
Amit Kumar Jakhar and Kumar Rajnish

**Measurement of Semantic Similarity: A Concept Hierarchy
Based Approach** 407
Shrutilipi Bhattacharjee and Soumya K. Ghosh

**Evaluating the Effectiveness of Conventional Fixes
for SQL Injection Vulnerability** 417
Swathy Joseph and K.P. Jevitha

**Salt and Pepper Noise Reduction Schemes Using Cellular
Automata** 427
Deepak Ranjan Nayak, Ratnakar Dash and Banshidhar Majhi

Part VI   Internet, Web Technology, and Web Security

**Optimizing the Performance in Web Browsers Through Data
Compression: A Study** 439
S. Aswini, G. ShanmugaSundaram and P. Iyappan

**Architectural Characterization of Web Service Interaction
Verification** 447
Gopal N. Rai and G.R. Gangadharan

Revamping Optimal Cloud Storage System 457
Shraddha Ghogare, Ambika Pawar and Ajay Dani

**Performance Improvement of MapReduce Framework
by Identifying Slow TaskTrackers in Heterogeneous Hadoop
Cluster** 465
Nenavath Srinivas Naik, Atul Negi and V.N. Sastry

**Router Framework for Secured Network Virtualization
in Data Center of IaaS Cloud** 475
Anant V. Nimkar and Soumya K. Ghosh

**Analysis of Machine Learning Techniques Based Intrusion
Detection Systems** 485
Rupam Kr. Sharma, Hemanta Kumar Kalita and Parashjyoti Borah

Compression and Optimization of Web-Contents 495
Suraj Thapar, Sunil Kumar Chowdhary and Devashish Bahri

**Harnessing Twitter for Automatic Sentiment Identification
Using Machine Learning Techniques** 507
Amiya Kumar Dash, Jitendra Kumar Rout and Sanjay Kumar Jena

Part VII   Big Data and Recommendation Systems

**Analysis and Synthesis for Archaeological Database
Development in Nakhon Si Thammarat** 517
Kanitsorn Suriyapaiboonwattana

**Evaluation of Data Mining Strategies Using Fuzzy Clustering
in Dynamic Environment** 529
Chatti Subbalakshmi, G. Ramakrishna and S. Krishna Mohan Rao

**A Survey of Different Technologies and Recent Challenges
of Big Data** 537
Dipayan Dev and Ripon Patgiri

**Privacy Preserving Association Rule Mining in Horizontally
Partitioned Databases Without Involving Trusted Third Party
(TTP)** 549
Chirag N. Modi and Ashwini R. Patil

**Performance Efficiency and Effectiveness of Clustering Methods
for Microarray Datasets**.   557
Smita Chormunge and Sudarson Jena

**Data Anonymization Through Slicing Based on Graph-Based
Vertical Partitioning**   569
Kushagra Sharma, Aditi Jayashankar, K. Sharmila Banu
and B.K. Tripathy

**An Efficient Approach for the Prediction of G-Protein Coupled
Receptors and Their Subfamilies**   577
Arvind Kumar Tiwari, Rajeev Srivastava, Subodh Srivastava
and Shailendra Tiwari

Part VIII   Fault and Delay Tolerant Systems

**Max-Util: A Utility-Based Routing Algorithm for a Vehicular
Delay Tolerant Network Using Historical Information**.   587
Milind R. Penurkar and Umesh A. Deshpande

**A Novel Approach for Real-Time Data Management in Wireless
Sensor Networks**   599
Joy Lal Sarkar, Chhabi Rani Panigrahi, Bibudhendu Pati
and Himansu Das

**Contact Frequency and Contact Duration Based Relay Selection
Approach Inside the Local Community in Social Delay Tolerant
Network**   609
Nikhil N. Gondaliya, Dhaval Kathiriya and Mehul Shah

**Part IX   Satellite Communication, Antenna Research,
and Cognitive Radio**

**Optimal Structure Determination of Microstrip Patch Antenna
for Satellite Communication**.   621
A. Sahaya Anselin Nisha

**Multi-metric Routing Protocol for Multi-radio Wireless Mesh
Networks**.   631
D.G. Narayan, Jyoti Amboji, T. Umadevi
and Uma Mudenagudi

A Novel Dual Band Antenna for RADAR Application 643
Kousik Roy, Debika Chaudhuri, Sukanta Bose and Atanu Nag

Author Index 651

About the Editors

Prof. Atulya Nagar holds the Foundation Chair as Professor of Mathematical Sciences at Liverpool Hope University where he is the Dean of Faculty of Science. He has been the Head of Department of Mathematics and Computer Science since December 2007. A mathematician by training he is an internationally recognized scholar working at the cutting edge of applied nonlinear mathematical analysis, theoretical computer science, operations research, and systems engineering and his work is underpinned by strong complexity-theoretic foundations. He has an extensive background and experience of working in universities in the UK and India. He has edited volumes on Intelligent Systems and Applied Mathematics; he is the Editor-in-Chief of the International Journal of Artificial Intelligence and Soft Computing (IJAISC) and serves on the editorial boards of a number of prestigious journals such as the Journal of Universal Computer Science (JUCS). Professor Nagar received a prestigious Commonwealth Fellowship for pursuing his Doctorate (D.Phil.) in Applied Non-linear Mathematics, which he earned from the University of York in 1996. He holds a B.Sc. (Hons.), M.Sc., and M.Phil. (with Distinction) from the MDS University of Ajmer, India.

Dr. Durga Prasad Mohapatra received his Ph.D. from Indian Institute of Technology Kharagpur. He joined the Department of Computer Science and Engineering at the National Institute of Technology, Rourkela, India in 1996, where he is presently serving as Associate Professor. His research interests include software engineering, real-time systems, discrete mathematics, and distributed computing. He has published more than thirty research papers in these fields in various international journals and conferences. He has received several project grants from DST and UGC, Government of India. He received the Young Scientist Award for the year 2006 by Orissa Bigyan Academy. He also received the Prof. K. Arumugam National Award and the Maharashtra State National Award for outstanding research work in Software Engineering for the years 2009 and 2010, respectively, from the Indian Society for Technical Education (ISTE), New Delhi. He is nominated to receive the Bharat Sikshya Ratan Award for significant contribution to academics awarded by the Global Society for Health and Educational Growth, Delhi.

Prof. Nabendu Chaki is Professor in the Department of Computer Science and Engineering, University of Calcutta, Kolkata, India. Dr. Chaki did his first graduation in Physics at the legendary Presidency College in Kolkata and then in Computer Science and Engineering, University of Calcutta. He completed Ph.D. in 2000 from Jadavpur University, India. He shares two US patents and one patent in Japan with his students. Prof. Chaki is active in developing international standards for Software Engineering. He represents the country in the Global Directory (GD) for ISO-IEC. Besides editing more than 20 book volumes in different Springer series including LNCS, Nabendu has authored five text and research books and about 130 peer-reviewed research papers in journals and international conferences. His areas of research include distributed computing, image processing, and software engineering. Dr. Chaki served as a Research Assistant Professor in the Ph.D. program in Software Engineering in U.S. Naval Postgraduate School, Monterey, CA. He has strong and active collaborations in US, Europe, Australia and other institutes and industries in India. He is a visiting faculty member for many universities in India and abroad. Dr. Chaki has been the Knowledge Area Editor in Mathematical Foundation for the SWEBOK project of the IEEE Computer Society. Besides being on the editorial board for several international journals, he has also served in the committees of more than 50 international conferences. Prof. Chaki is the founder Chair of the ACM Professional Chapter in Kolkata.

Part I
Distributed Systems, Social Networks, and Applications

An Algorithm for Partitioning Community Graph into Sub-community Graphs Using Graph Mining Techniques

Bapuji Rao and Anirban Mitra

Abstract Using graph mining techniques, knowledge extraction is possible from the community graph. In our work, we started with the discussion on related definitions of graph partition both mathematical as well as computational aspects. The derived knowledge can be extracted from a particular sub-graph by way of partitioning a large community graph into smaller sub-community graphs. Thus, the knowledge extraction from the sub-community graph becomes easier and faster. The partition is aiming at the edges among the community members of different communities. We have initiated our work by studying techniques followed by different researchers, thus proposing a new and simple algorithm for partitioning the community graph in a social network using graph techniques. An example verifies about the strength and easiness of the proposed algorithm.

Keywords Adjacency matrix · Cluster · Community · Graph partition · Sub-Graph

1 Introduction

We use graph theory's some important techniques to solve the problem of partitioning a community graph to minimize the number of edges or links that connect different community [1]. The aim of partitioning a community graph to sub-graphs is to detect similar vertices which form a graph and such sub-graphs can be formed. For example, considering Facebook is a very large social graph. It can be

B. Rao (✉) · A. Mitra
Department of CSE and IT, V.I.T.A.M., Berhampur, Odisha, India
e-mail: rao.bapuji@gmail.com

A. Mitra
e-mail: mitra.anirban@gmail.com

© Springer India 2016
A. Nagar et al. (eds.), *Proceedings of 3rd International Conference on Advanced Computing, Networking and Informatics*, Smart Innovation, Systems and Technologies 44, DOI 10.1007/978-81-322-2529-4_1

partitioned into sub-graphs, and each sub-group should belong to a particular characteristics. Such cases we require graph partitions. In this partition, it is not mandatory that each sub-group contain similar number of members. A partition of a community graph is to divide into clusters, such that each similar vertex belongs to one cluster. Here a cluster means a particular community. Based on this technique, we partition a community graph into various sub-graphs after detecting various vertices belonging to a particular community or cluster.

2 Basics in Graph Theory

Social network, its actors and the relationship between them can be represented using vertices and edges [2]. The most important parameter of a network (i.e., a digraph) is the number of vertices and arcs. Here we denote n for number of vertices and m for number of arcs. When an arc is created by using two vertices u and v, which is denoted by uv. Then the initial vertex is the u and the terminal vertex is the v in the arc uv.

2.1 Digraph

A digraph or directed graph $G = (V, A)$ with $V = \{V_1, V_2, \ldots\ldots, V_n\}$ can be represented as adjacency matrix A. The matrix A is of order nXn where A_{ij} is 1 or 0 depending on V_iV_j is an edge or not. Note that $A_{ii} = 0$ for all i.

2.2 Sub-digraph

A sub-digraph of G to be (V_1, A_1) where $V_1 \subseteq V, A_1 \subseteq A$ and if uv is an element of A_1 then u and v belong to V_1.

2.3 Adjacency Matrix

Let a graph G with n nodes or vertices $V_1, V_2, \ldots, V_n$ having one row and one column for each node or vertex. Then the adjacency matrix A_{ij} of graph G is an nXn square matrix, which shows one (1) in A_{ij} if there is an edge from V_i to V_j; otherwise zero (0).

2.4 Good Partition

When a graph is divided into two sets of nodes by removing the edges that connect nodes in different sets should be minimized. While cutting the graph into two sets of nodes so that both the sets contain approximately equal number of nodes or vertices [1].

In Fig. 1 graph G_1 has seven nodes $\{V_1, V_2, V_3, V_4, V_5, V_6, V_7\}$. After cutting into two parts approximately equal in size, the first partition has nodes $\{V_1, V_2, V_3, V_4\}$ and the second partition has nodes $\{V_5, V_6, V_7\}$. The cut consists of only the edge (V_3, V_5) and the size of edge is 1.

In Fig. 2 graph G_2 has eight nodes $\{V_1, V_2, V_3, V_4, V_5, V_6, V_7, V_8\}$. Here two edges, (V_3, V_7) and (V_2, V_6) are used to cut the graph into two parts of equal size rather than cutting at the edge (V_5, V_8). The partition at the edge (V_5, V_8) is too small. So we reject the cut and choose the best one for cut consisting of edges (V_2, V_6) and (V_3, V_7), which partitions the graph into two equal sets of nodes $\{V_1, V_2, V_3, V_4\}$ and $\{V_5, V_6, V_7, V_8\}$.

2.5 Normalized Cuts

A good cut always balance the size of cut itself against the sizes of the sets of created cut [1]. For this normalized cut method is being used. First it has to define the volume of set of nodes or vertices V which is denoted as Vol (V) is the number of edges with at least one end in the set of nodes or vertices V.

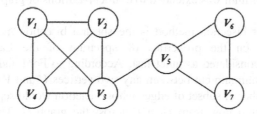

Fig. 1 Graph G_1 with seven nodes

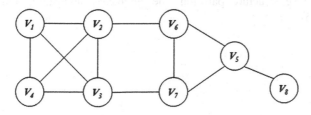

Fig. 2 Graph G_2 with eight nodes

Let us partition the nodes of a graph into two disjoint sets say A and B. So the Cut (A, B) is the number of edges from the disjoint set A to connect a node in the disjoint set B. The formula for normalized cut values for disjoint sets A and B = Cut (A, B)/Vol (A) + Cut (A, B)/Vol (B).

2.6 Graph Partitions

Partition of graph means a division in clusters, such that similar kinds of vertices belong to a particular cluster [1]. In a real world vertices may share among different communities. When a graph is divided into overlapping communities then it is called a cover.

A graph with K-clusters and N-vertices, the possible number of Stirling number of the second kind is denoted as $S(N, K)$. So the total number of possible partitions is said to be the Nth Bell number is given with the formula $B_N = \sum_{K=0}^{N} S(N, K)$ [3]. When the value of N is large then B_n becomes asymptotic [4].

While partitioning a graph having different levels of structure at different scales [5, 6], the partitions can be ordered hierarchically. So in this situation cluster plays an important role. Each cluster displays the community structure independently, which consists of set of smaller communities.

Partitioning of graph means dividing the vertices in a group of predefined size. So that the frequently used vertices are often combined together to form a cluster by using some techniques. Many algorithms perform a partition of graph by means of bisecting the graph. Iterative bisection method is employed to partition a graph into more than two clusters and this algorithm is called as Kernighan-Lin [7]. The Kernighan-Lin algorithm was extended to extract partitions of graph in any number of clusters [8].

Another popular bisection method is the spectral bisection method [9, 10], is completely based on the properties of spectrum of the Laplacian matrix. This algorithm is considered as quiet fast. According to Ford and Fulkerson [11] theorem that the minimum cut between any two vertices U and V of a graph G, is any minimum number of subset of edges whose deletion would separate U from V, and carries maximum flow from U to V across the graph G. The algorithms of Goldberg and Tarjan [12] and Flake et al. [13, 14] are used to compute maximum flows in graphs during cut operation. Some other popular methods for graph partition are level-structure partition, the geometric algorithm, and multilevel algorithms [15].

3 Proposed Algorithms and Analysis

```
Algorithm  Community_Graph_Partition()
// Global Declarations
n     : Number of Communities.
NCM [1:n,1:2]: Holds community number and number of
community members of each community.
tcm      : To count total number of community members.
CMM [1:tcm+1, 1:tcm+1]: Adjacency matrix of Community
Members of order tcmXtcm.

i.[Read Community Data]
   Call Read_Community_Data().
ii.[Generate and assign every members code]
   CallAssign_Community_Member_Codes()
iii.[Creation of adjacency matrix of all the members]
   Call Community_Member_Matrix()
iv. [Partition of Community Graph]
   Call Graph_Partition()
v. (a) Set s:=0.
   (b) Repeat For I:=1, 2,..........,n:
       (1) s:= s+NCM[I][1].
    [Show the 'I'th sub-community graph after partition]
       (2) Call Sub_Community_Matrix_Display(s).
       End For
vi. Exit.

Procedure-I.Read_Community_Data()
i. Set tcm:=0.
ii. Read Number of communities as 'n'.
iii.Read Community details such as community code and
number of members of each community, and assign to the
matrix NCM[][].
iv. Repeat For I:=1, 2, ......., n:
       tcm := tcm + NCM[I][2].
    End For
v. Return.

Procedure-II.Assign_Community_Member_Codes()
i.    Set K := 1.
ii.   Set Pro := 1.
iii.  Repeat For I :=1, 2,........, n:
        (1) If NCM[I][1]>=1 AND NCM[I][1]<=9,
            Then (a) Pro := 10.
            Else If NCM[I][1]>=10 AND NCM[I][1]<=99,
            Then (b)Pro:= 100.
```

```
              Else If NCM[I][1]>=100 AND NCM[I][1]<=999,
              Then(c)Pro:=1000.
              Else
                    (d)Break.
              End If
          (2)Repeat For J := 1, 2,......, NCM[I][2]:
              (a) Set CMM[1][K+1] := (NCM[I][1]*10) + J.
              (b) Set CMM[K+1][1] := (NCM[I][1]*10) + J.
              (c) K := K + 1.
              End For
       End For
iv. Return.
```

Procedure-III.Community_Member_Matrix()
```
i. Get the edge data of all the community members.
ii. Store the above data in the matrix CMM[ ][ ].
iii. Return.
```

Procedure-IV.Graph_Partition()
```
i. Repeat For I := 1, 2, ........, tcm+1:
ii. Repeat For J := 1, 2, ........, tcm+1:
        If CMM[1][J+1]/Pro ≠ CMM[I+1][1]/Pro,
        Then
              [Cut off edge between communities of Different
               group of communities]
                    Set CMM[I+1][J+1] := 0.
        End If
      End For
    End For
iii. Return.
```

Procedure-V.Sub_Community_Matrix_Display(size)
```
size: Size of each community.
i.    Set  x:=0.
ii.   Set count := x.
iii. Repeat For i:=count, count+1,....,size:
iv.      Repeat For j:=count, count+1,....,size:
             (a)If i=count And j=count, Then: Display "C".
                Else If i=count Andj≠count,
                Then: Display CMM[1][j].
                Else If I≠countAnd j=count,
                Then: Display CMM[i][1].
                Else: Display CMM[i][j].
                End If
          End For
          (b)x:=x+1.
      End For
v. x:=x-1.
vi. Return.
```

3.1 Explanation

The proposed algorithm consists of five procedures. Procedure-I allows to read the details about number of communities and number of community members of all the communities. In this example the output has been derived after implemented using C++ programming language. The data related to community and their edges are read from two data files namely "*commun1.txt*" and "*graph.dat*". Procedure-II which generates and assigns community member codes. Procedure-III creates the community adjacency matrix. Procedure-IV allows us to partition the community adjacency matrix by assigning '0' over '1' which indicates the edge between the community members of dissimilar communities. Finally Procedure-V displays every community's adjacency matrix. From the adjacency matrices we can draw the community sub-graphs.

3.2 Example

We propose a community graph [16, 17] with 22 individual communities from four different communities $\{C_1, C_2, C_3, C_4\}$ which is shown in Fig. 3. We try to partition

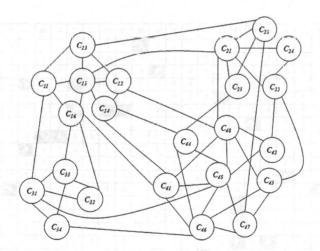

Fig. 3 Community graph of communities $\{C_{11}\ldots C_{16}, C_{21}\ldots C_{25}, C_{31},\ldots C_{34}, C_{41}\ldots C_{48}\}$

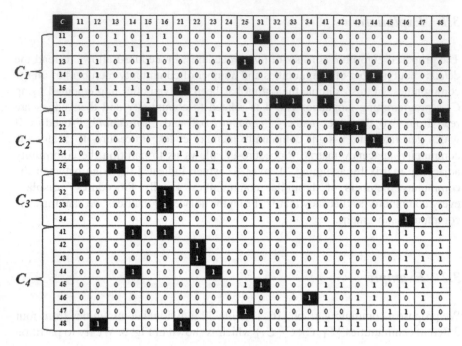

Fig. 4 Adjacency matrix of community graph in Fig. 3

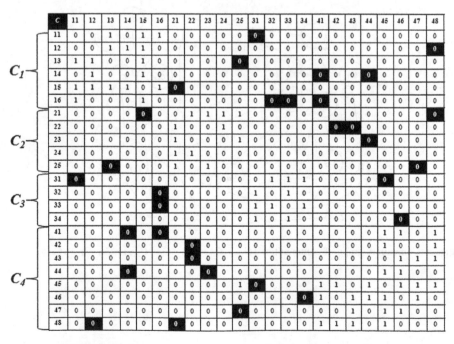

Fig. 5 Adjacency matrix of community graph after cut-off edges between community members of dissimilar communities

C	11	12	13	14	15	16
11	0	0	1	0	1	1
12	0	0	1	1	1	0
13	1	1	0	0	1	0
14	0	1	0	0	1	0
15	1	1	1	1		1
16	1	0	0	0	1	0

i) C_1 Adjacency Matrix

C	21	22	23	24	25
21	0	1	1	1	1
22	1	0	0	1	0
23	1	0	0	0	1
24	1	1	0	0	0
25	1	0	1	0	0

ii) C_2 Adjacency Matrix

C	41	42	43	44	45	46	47	48
41	0	0	0	0	1	1	0	1
42	0	0	0	0	1	0	0	1
43	0	0	0	0	0	1	1	1
44	0	0	0	0	1	1	0	0
45	1	1	0	1	0	1	1	1
46	1	0	1	1	1	0	1	0
47	0	0	1	0	1	1	0	0
48	1	1	1	0	1	0	0	0

C	31	32	33	34
31	0	1	1	1
32	1	0	1	0
33	1	1	0	1
34	1	0	1	0

iii) C_3 Adjacency Matrix

iv) C_4 Adjacency Matrix

Fig. 6 Adjacency matrices of communities C_1, C_2, C_3, and C_4

this graph into four sub-graphs of communities $\{C_1, C_2, C_3, C_4\}$. We try to represent this graph in memory in an adjacency matrix form by following graph techniques which is shown in Fig. 4. Then we try to locate edges between communities members formed from two different communities.

The black filled boxes indicate the edge between the community members of dissimilar communities which is indicated in Fig. 5. These edges are considered as edges between dissimilar communities. So these edges must be cut. Once such edges are cut, then the original graph can be partitioned into so many sub-graphs. And we can say that the graph has been partitioned across edges of community

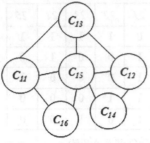

i) Community C_1's Sub-Graph

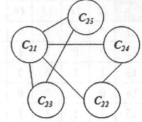

ii) Community C_2's Sub-Graph

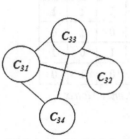

iii) Community C_2's Sub-Graph

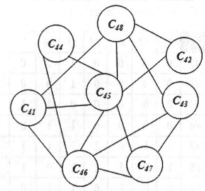

iv) Community C_4's Sub-Graph

Fig. 7 Communities $\{C_1, C_2, C_3, C_4\}$'s sub-graphs

members of dissimilar communities. To do the edge cut operation, we assign 0 over 1 in the black filled boxes of adjacency matrix in Fig. 5. So that we can say there is no physical edge between those community members across the different communities. From the adjacency matrix of Fig. 5, we can construct four different adjacency matrices for the communities C_1, C_2, C_3, and C_4 which is shown in Fig. 6. For C_1 the community members are $\{11, 12, 13, 14, 15, 16\}$. Similarly for C_2, C_3, and C_4 the community members are $\{21, 22, 23, 24, 25\}$, $\{31, 32, 33, 34\}$, and $\{41, 42, 43, 44, 45, 46, 47, 48\}$ respectively. From these four adjacency matrices, now we can construct the sub-graphs which are shown in Fig. 7.

3.3 *Output*

```
Enter the Community Data File Name : commun1.txt

Enter the Edge Data File Name : graph.dat

The Community Matrix Before Partition

C  11 12 13 14 15 16 21 22 23 24 25 31 32 33 34 41 42 43 44 45 46 47 48
11  0  0  1  0  1  1  0  0  0  0  0  1  0  0  0  0  0  0  0  0  0  0  0
12  0  0  1  1  1  0  0  0  0  0  0  0  0  0  0  0  0  0  0  0  0  0  1
13  1  1  0  0  1  0  0  0  0  0  1  0  0  0  0  0  0  0  0  0  0  0  0
14  0  1  0  0  1  0  0  0  0  0  0  0  0  0  1  0  0  1  0  0  0  0  0
15  1  1  1  1  0  1  1  0  0  0  0  0  0  0  0  0  0  0  0  0  0  0  0
16  1  0  0  0  1  0  0  0  0  0  0  0  1  1  0  1  0  0  0  0  0  0  0
21  0  0  0  0  1  0  0  1  1  1  1  0  0  0  0  0  0  0  0  0  0  0  1
22  0  0  0  0  0  0  1  0  0  1  0  0  0  0  0  0  1  1  0  0  0  0  0
23  0  0  0  0  0  0  1  0  0  1  0  0  0  0  0  0  0  1  0  0  0  0  0
24  0  0  0  0  0  0  1  1  0  0  0  0  0  0  0  0  0  0  0  0  0  0  0
25  0  0  1  0  0  0  0  1  0  1  0  0  0  0  0  0  0  0  0  0  0  1  0
31  1  0  0  0  0  0  0  0  0  0  0  0  1  1  0  0  0  0  1  0  0  0  0
32  0  0  0  0  0  1  0  0  0  0  0  1  0  1  0  0  0  0  0  0  0  0  0
33  0  0  0  0  0  1  0  0  0  0  0  1  1  0  1  0  0  0  0  0  0  0  0
34  0  0  0  0  0  0  0  0  0  0  0  1  0  1  0  0  0  0  0  0  1  0  0
41  0  0  0  1  0  1  0  0  0  0  0  0  0  0  0  0  0  0  1  0  0  0  1
42  0  0  0  0  0  0  0  1  0  0  0  0  0  0  0  0  0  0  1  0  0  0  1
43  0  0  0  0  0  0  0  1  0  0  0  0  0  0  0  0  0  0  0  1  1  1
44  0  0  0  1  0  0  0  0  1  0  0  0  0  0  0  0  0  0  0  1  1  0  0
45  0  0  0  0  0  0  0  0  0  0  0  0  1  0  0  1  1  0  1  0  1  1  1
46  0  0  0  0  0  0  0  0  0  0  0  0  0  0  1  1  0  1  1  0  1  0
47  0  0  0  0  0  0  0  0  0  0  1  0  0  0  0  0  0  1  0  1  1  0  0
48  0  1  0  0  0  0  1  0  0  0  0  0  0  0  0  1  1  1  0  1  0  0  0
```

```
The Community Matrix After Partition

C  11 12 13 14 15 16 21 22 23 24 25 31 32 33 34 41 42 43 44 45 46 47 48
11  0  0  1  0  1  1  0  0  0  0  0  0  0  0  0  0  0  0  0  0  0  0  0
12  0  0  1  1  1  0  0  0  0  0  0  0  0  0  0  0  0  0  0  0  0  0  0
13  1  1  0  0  1  0  0  0  0  0  0  0  0  0  0  0  0  0  0  0  0  0  0
14  0  1  0  0  1  0  0  0  0  0  0  0  0  0  0  0  0  0  0  0  0  0  0
15  1  1  1  1  0  1  0  0  0  0  0  0  0  0  0  0  0  0  0  0  0  0  0
16  1  0  0  0  1  0  0  0  0  0  0  0  0  0  0  0  0  0  0  0  0  0  0
21  0  0  0  0  0  0  0  1  1  1  1  0  0  0  0  0  0  0  0  0  0  0  0
22  0  0  0  0  0  0  1  0  0  1  0  0  0  0  0  0  0  0  0  0  0  0  0
23  0  0  0  0  0  0  1  0  0  1  0  0  0  0  0  0  0  0  0  0  0  0  0
24  0  0  0  0  0  0  1  1  0  0  0  0  0  0  0  0  0  0  0  0  0  0  0
25  0  0  0  0  0  0  1  0  1  0  0  0  0  0  0  0  0  0  0  0  0  0  0
31  0  0  0  0  0  0  0  0  0  0  0  0  1  1  1  0  0  0  0  0  0  0  0
32  0  0  0  0  0  0  0  0  0  0  0  1  0  1  0  0  0  0  0  0  0  0  0
33  0  0  0  0  0  0  0  0  0  0  0  1  1  0  1  0  0  0  0  0  0  0  0
34  0  0  0  0  0  0  0  0  0  0  0  1  0  1  0  0  0  0  0  0  0  0  0
41  0  0  0  0  0  0  0  0  0  0  0  0  0  0  0  0  0  0  1  1  0  1
42  0  0  0  0  0  0  0  0  0  0  0  0  0  0  0  0  0  0  1  0  0  1
43  0  0  0  0  0  0  0  0  0  0  0  0  0  0  0  0  0  0  0  1  1  1
44  0  0  0  0  0  0  0  0  0  0  0  0  0  0  0  0  0  0  0  1  1  0
45  0  0  0  0  0  0  0  0  0  0  0  0  0  0  0  1  1  0  1  0  1  1  1
46  0  0  0  0  0  0  0  0  0  0  0  0  0  0  0  1  0  1  1  0  1  0
47  0  0  0  0  0  0  0  0  0  0  0  0  0  0  0  0  1  0  1  1  0  0
48  0  0  0  0  0  0  0  0  0  0  0  0  0  0  0  1  1  1  0  1  0  0  0
```

```
        Community - C1's Adjacency Matrix...

 C  11 12 13 14 15 16
11   0  0  1  0  1  1
12   0  0  1  1  1  0
13   1  1  0  0  1  0
14   0  1  0  0  1  0
15   1  1  1  1  0  1
16   1  0  0  0  1  0

        Press Any Key.....

        Community - C2's Adjacency Matrix...

 C  21 22 23 24 25
21   0  1  1  1  1
22   1  0  0  1  0
23   1  0  0  0  1
24   1  1  0  0  0
25   1  0  1  0  0

        Press Any Key.....
```

```
        Community - C3's Adjacency Matrix...

 C  31 32 33 34
31   0  1  1  1
32   1  0  1  0
33   1  1  0  1
34   1  0  1  0

        Press Any Key.....

        Community - C4's Adjacency Matrix...

 C  41 42 43 44 45 46 47 48
41   0  0  0  0  1  1  0  1
42   0  0  0  0  1  0  0  1
43   0  0  0  0  0  1  1  1
44   0  0  0  0  1  1  0  0
45   1  1  0  1  0  1  1  1
46   1  0  1  1  1  0  1  0
47   0  0  1  0  1  1  0  0
48   1  1  1  0  1  0  0  0

        Press Any Key.....
```

4   Conclusions

We have partitioned our large community graph into sub-community graphs using the concepts of graph technique, especially by detecting an edge between the nodes of different communities. Initial portion of the work is a brief review of the literature on graph partition related to mathematical formulae as well as graph mining techniques. A simple graph technique for partition of a large community graph has been proposed. An appropriate example from social community network background has been represented using the graph theoretic concepts. The paper concludes with focusing on process of partitioning a community graph. There after the various sub-community graphs are to be shown in its adjacency matrix format. Hence extracting knowledge from a particular sub-community graph becomes easier and faster.

References

1. Rajaraman, A., Leskovec, J., Ullman, J.D.: Mining of Massive Datasets. Copyright © 2010, 2011, 2012, 2013, 2014
2. Mitra, A., Satpathy, S.R., Paul, S.: Clustering analysis in social network using covering based rough set. In: 2013 IEEE 3rd International Advance Computing Conference (IACC), India, 22 Feb 2013, pp. 476–481, 2013
3. Andrews, G.E.: The Theory of Partitions. Addison-Wesley, Boston, USA (1976)
4. Lovasz, L.: Combinatorial Problems and Exercises. North-Holland, Amsterdam, The etherlands (1993)
5. Ravasz, E., Barabasi, A.L.: Phys. Rev. E **67**(2), 026112 (2003)
6. Ravasz, E., Somera, A.L., Mongru, D.A., Oltvai, Z.N., Barabasi, A.L.: Science **297**(5586), 1551 (2002)
7. Kernighan, B.W., Lin, S.: Bell Syst. Tech. J. **49**, 291 (1970)
8. Suaris, P.R., Kedem, G.: IEEE Trans. Circuits Syst. **35**, 294 (1988)
9. Barnes, E.R.: SIAM J. Alg. Discr. Meth. **3**, 541 (1982)
10. Scholtz, R.A.: The spread spectrum concept. In: Abramson, N. (ed) Multiple Access, Piscataway, NJ: IEEE Press, ch. 3, pp. 121–123 (1993)
11. Ford, L.R., Fulkerson, D.R.: Canadian J. Math. **8**, 399 (1956)
12. Goldberg, A.V., Tarjan, R.E.: J. ACM **35**, 921 (1988)
13. Flake, G.W., Lawrence, S., Giles, C.L.: In: Sixth ACM SIGKDD International Conference on Knowledge Discovery and Data Mining (ACM Press, Boston, USA), pp. 150–160 (2000)
14. Flake, G.W., Lawrence, S., Lee Giles, C., Coetzee, F.M.: IEEE Comput. **35**, 66 (2002)
15. Pothen, A.: Graph Partitioning Algorithms with Applications to Scientific Computing. Technical Report, Norfolk, VA, USA (1997)
16. Rao, B., Mitra, A.: A new approach for detection of common communities in a social network using graph mining techniques. In: 2014 International Conference on High Performance Computing and Applications (ICHPCA), pp. 1–6, 22–24 Dec 2014. doi: 10.1109/ICHPCA. 2014.7045335
17. Rao, B., Mitra, A.: An approach to merging of two community sub-graphs to form a community graph using graph mining techniques. In: 2014 IEEE International Conference on Computational Intelligence and Computing Research (ICCIC-2014), 978-1-4799-3972-5/14/ $31.00 @2014, pp. 460–466, Coimbatore, India, Dec 2014

Traceback: A Forensic Tool for Distributed Systems

Sushant Dinesh, Sriram Rao and K. Chandrasekaran

Abstract In spite of stringent security measures on the components of a distributed system and well-defined communication procedures between the nodes of the system, an exploit may be found that compromises a node, and may be propagated to other nodes. This paper describes an incident-response method to analyse an attack. The analysis is required to patch the vulnerabilities and may be helpful in finding and removing backdoors installed by the attacker. This analysis is done by logging all relevant information of each node in the system at regular intervals at a centralised store. The logs are compressed and sent in order to reduce network traffic and use lesser storage space. The state of the system is also stored at regular intervals. This information is presented by a replay tool in a lucid, comprehensible manner using a timeline. The timeline shows the saved system states (of each node in the distributed system) as something similar to checkpoints. The events and actions stored in the logs act on these states and this shows a replay of the events to the analyser. A time interval during which an attack that took place is suspected to have occurred can be analysed thoroughly using this tool.

Keywords Forensics · Incident response · Distributed system

S. Dinesh (✉) · S. Rao · K. Chandrasekaran
Department of Computer Science and Engineering, National Institute of Technology
Karnataka, Surathkal, Mangalore, India
e-mail: sushantdinesh94@gmail.com

S. Rao
e-mail: sriram.rao@ieee.org

K. Chandrasekaran
e-mail: kchnitk@ieee.org

© Springer India 2016
A. Nagar et al. (eds.), *Proceedings of 3rd International Conference
on Advanced Computing, Networking and Informatics*, Smart Innovation,
Systems and Technologies 44, DOI 10.1007/978-81-322-2529-4_2

17

1 Introduction

Computer Forensics refers to the field involving inspection and analysis of digital storage to extract information about the computer system and any changes that it may have undergone through processes running on the system. Forensic methods are largely used to gather information about a system after it suffers an illegal attack that may have resulted in compromised data or denial of service, or both, among other consequences. While this information is helpful as legal evidence, it can also be used to study the loopholes in the network and find methods to fix them.

In a distributed system, communication between components is secured using cryptographic methods. A client node (that is making the request) needs to be authenticated by the receiver node before processing the request. On a node, un-trusted code from an external source that has to be executed is run with limitations, prohibiting access to most resources without authorisation. In spite of these mea-sures, the system may be attacked by a malicious agent. In such cases, forensic tools are useful in analysing the method and extent of exploitation, and the information accumulated can be used to patch the system. Any backdoors included in the network by the attacker can also be found and removed. When regular system checkpoints are made, these can be used to restore the compromised system to a known uncompromised state. Currently, analysis involves going through millions of lines of logs from different components of the system manually. This is tedious, time-consuming and there are chances that investigators might miss a crucial part of the attack. A clever attacker might also delete the logs during his attack thereby making them useless in the analysis. With several missing pieces of information, analysing the attack and estimating damage would be very difficult. Due to lack of information about the type of attack, developers must manually go through the whole code to find the vulnerability used. All this suggests that we must have a more robust forensic analysis tool designed to operate on distributed system.

In this paper we conceptualise a forensic tool, Traceback, that helps analyse attacks by simulating events that occur starting from a chosen system state. This makes use of the logs that are kept by various processes that run on each system, and by the systems themselves. These logs are retrieved from storage and a timeline of events is constructed to understand the sequence of events. This timeline also includes regular system states that are stored as checkpoints. When an analyser wants to check a particular time interval for suspicious behaviour, he can choose an initial state and have the logged events perform transitions to this state up to a required point of time.

The paper has been categorised as follows. Section 2 describes related work. Section 3 gives an overview and a brief description of the proposed work. Section 4 details the work and responsibilities of the logging agent. Section 5 discusses the choices available for the central store. Section 6 outlines the tool and useful fea-tures. Section 7 concludes the paper.

2 Related Work

Topics relevant to this paper that are well-studied include efficient compression of files, specifically log files, and good forensic tools to inspect networks. Effective and thorough tools have been made which monitor the network packets at a particular computer system. In a distributed system, a more suited tool would allow detailed monitoring of packets exchanged within the system and with external sources. With respect to log files, there are papers which detail methods proposed to compress log files generated by particular tasks and discuss their compression ratio and speed.

2.1 Compression of Logs

Compression of logs a very well studied topic. Here we consider two papers which talk about two different compression mechanisms to improve the compression ratio and compression speed when dealing with logs. In "Lossless compression for large scale cluster logs" [1], Balakrishnan and Sahoo develop a compression algorithm to compress logs generated by Blue Gene/L supercomputer. Their compression algorithm has limited scope and produces best results only on logs generated by this supercomputer. In "Fast and efficient log file compression" [2], Skibiski and Swacha develop a generic compression scheme for logs with five different variants of it, each varying in degree and speed of compression. Both the papers have a significant improvement in compression ratio and compression speed as compared to general purpose compression schemes like gzip, bzip etc.

2.2 Forensic Tools Used in Network Security

Extensive work has been done in making effective methods to provide network security. Many of these methods, such as those outlined in "Distributed agent-based real time network intrusion forensics system architecture design" [3] are for intrusion detection at the time of attack—to detect real-time attacks. "Achieving Critical Infrastructure Protection through the Interaction of Computer Security and Network Forensics" [4] also concentrates on acquisition of forensic data for real-time security purposes. The Incident Response Support System outlined in [5] is a tool for automatic response to attacks on a system. The attack being made is compared to an information base of known exploits and the response used in the known case is adapted to respond to the current attack. All these procedures are aimed at a general computer system connected to a network. Our paper concentrates on analysis of the attack to gain complete knowledge of the vulnerability in the environment of a distributed system.

3   System Overview

Figure 1 describes an overview of the final Traceback tool. The Traceback tool has
a timeline which allows forensic investigators to go back to the system state during
the time of the attack and analyse the actions of the attacker in detail. Different logs
from various components of the distributed system are put together by the tool to
simulate the situation during the time of attack. This helps the investigators
understand the exact vulnerability in the system and take the required measures to
patch it up. Additionally, as the alteration in the state of the system due to the attack
is logged, any backdoors, worms and residual files left behind by the attacker can be
cleaned up and also used as evidence for further forensic analysis.

Agents are used to "checkpoint" the system states, aggregate logs from the
component, compress and then send them to the central log storage. This allows us
to have a small local log storage and helps us maintain all logs in a central place.

A system state may include information like number of processes running, CPU
usage, Memory usage, network usage etc. Information regarding each process is
stored in the initial state. Each successive checkpoint only stores the changes from
this initial state, in a Git inspired "diff"-like manner. Similar process is followed for
important directories in the file-system. This allows us to think of logs as trans-
formations on a system state.

Logs and checkpoints collected by the agents are compressed before sending to
the central log storage. Compression is done by the agents to avoid excessive
network usage as log files tend to get very big. We can achieve a high compression
ratio as log files tend to have a lot of repeated information. The central log storage
collects logs from all the sources and stores it for future retrieval.

We proceed with a detailed description of the following points:

- Logging agent
- Compression of logs
- Central log storage
- Traceback application

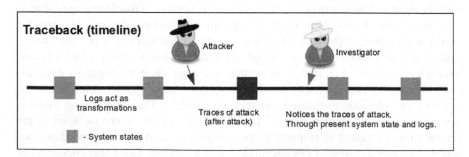

Fig. 1 Overview

4 The Logging Agent

Each component of a distributed system that is open to external communication has an instance of the logging agent running. Briefly, the responsibilities of the logging agent are to save the system state at regular intervals of time, and to log all relevant local events at the system. If there is a centralised storage of all logs, this agent compresses the logs and periodically sends them to the central storage.

4.1 Access to the Agent and Logs

The agent process must run at a higher privilege than other processes so that it is not terminated by other (possibly user-level) processes. Also, the agent must have enough access to collect all relevant logs from the system, and other system-related information such as a list of running processes and their details. The log files written by the agent should be accessible only by the agent and the root of the system so that user-level processes do not interfere or contaminate the logging process. Processes running untrusted code must also not have access to these logs. These reasons point to elevating the agent process to a high privilege.

4.2 Description of a System State

To analyse an attack on the system, investigators must look at the state of the system before the attack and analyse the changes made to this state during and after the attack interval. The changes made to the system are studied to single out those initiated by the attacker as part of the exploit and accurately define the attack interval and method. To aid this procedure, it is beneficial to save the states of each system in the distributed network at regular intervals. System snapshots have large storage size, so it is not practical to intermittently take snapshots of each system. Instead, specific data such as a list of the details of all running processes and the general file system information can be stored. The information about the file system tells us what files have been added or deleted, and when a file was last modified. Process-relevant information that can be stored includes CPU usage, Memory usage, Process ID and Parent Process ID, and Process running time.

After an initial system state containing complete information as outlined, newer save states can be stored as a difference from this initial state. This again reduces the storage space occupied. It also helps visualise the changes made to the initial system state as state transitions, making the changes easier to understand. The differences that need to be considered are changes in the file system, and changes in the processes that are running. Every change in the details of running processes need not be considered as the CPU and memory usage are not strictly constant, and the process running time is different at different instants of time. Only a significant difference in usage of resources (above a configurable threshold) can initiate noting this in the system save state.

This information about system states and the logs made can be effectively represented in the form of a timeline. The events in the log files can be simulated on the saved system states to show a replay of events that makes it simpler for the investigators to understand the sequence of developments during a suspected attack interval.

4.3   Information Collection from Different Logs

To adequately represent the developments on a system state, all significant events must be logged by the agent. These logs are then compressed before storage at the central store. The information that needs to be collected for the logs includes:

Network Traffic When an attack is made on a node, traffic from external agents to a node in the distributed system holds information about the attack. To capture information about the traffic from and to a node, tools such as libpcap or snort can be used. These tools also allow filtering based on protocol and IP addresses since all packets are not required to be stored. Packets of some protocols may have low chances of carrying malicious code.

Internal Communication Internal communication between nodes of the distributed system is generally logged by the middleware that specifies communication procedures for the system. These logs can be used for forensic purposes, in cases where a compromised node may communicate with other nodes of the system to try and spread the attack.

Running Processes At regular intervals, the agent makes a check on the running processes in order to log changes. The process-related information stored is elucidated in Sect. 4.2. This information is needed as part of the system state.

System Logs and Authentication Logs The logs that are kept by the system can be vital in the analysis and search of information about an attack that the system suffered. The event logs of each node have information about all operations carried out and events that occurred, among other useful information. These contain the details of the malicious operations carried out, except in cases where a cautious attacker erases relevant information from these logs.

Authentication logs contain information about the Authentication Provider and the time of login of all users on a system. It follows that remote login procedures used, such as SSH or Telnet, will also be logged in this file. Thus, the auth logs collected from each node can reveal information about login by an unfamiliar user or at an odd time.

Database Logs Any change made to a database existing on the node is stored in the database logs. These can hold information about back-doors installed in the database by the intruder. When the distributed system uses a database to store necessary information, it is necessary to ensure its security so as not to allow unauthorised changes to the information, or a loss of database entries.

4.4 Compression of Logs

Compression of the logs generated is a very important responsibility of the agent. Better compression ensures efficient storage of logs and better network utilisation as the data to be transferred to the central store would be smaller.

Table 1 compares various general purpose compression mechanisms. It is clear that gzip the fastest among the four and achieves a compression of 93.51 %. The other algorithms are considerably slower and offer only a marginal difference in size of the logs after compression.

Compression of logs is a well studied subject and several journals have been released about the same. In [1] Balakrishnan and Sahoo talk about a specialized lossless compression for large scale cluster logs. In the paper the authors talk about exploiting the redundancy in the log files to allow a large amount of compression. Compression is done in multiple stages to exploit redundancy in information at various stages and maximize the compression. The authors were able to achieve a 28.3 % better compression ratio and 43.4 % of improvement in compression time as compared to standalone compression utilities. However, the algorithm is very specific to the logs their systems generate.

In [2] Skibiski and Swacha talk about a custom compression scheme and five different variants of it. Using the faster variant, the log files were transformed to be 36.6 % shorter than the original files compressed with gzip. Using the slower variant, the log files were transformed to be 62 % shorter than the original files compressed with gzip, and 41 % shorter than the original files compressed with bzip2. The advantage of this scheme is that there is no pre defined format for the log and the compression will work for all generic logs.

We could also develop a custom compression method based on our logs to achieve a better compression. However, this is not a subject of this paper and more of a scope for future work. We will use a general purpose algorithm like gzip for our application.

Table 1 Comparison of popular compression algorithms. *Source* [6]

Compression algorithm	Compression ratio	Bytes/s
GZIP	93.51	0.5191
BZIP	96.52	0.2785
7-Zip	97.50	0.2002
LZXQ 0.4	95.84	0.3325

5   Central Storage

The central store is responsible for aggregating logs from various components and store it for further analysis. The central store should be highly secure as it stores all vital information regarding the behaviour of the distributed system. An attack on the central store could lead to loss of large amounts of information. Ideally, the central store should not be a part of the distributed system and must not be accessible from anywhere outside the organization's network. The only means of communication between the distributed system and the central store must be through the agents monitoring them. Confidentiality and integrity of the data sent over the network by the agents can be ensured by using a secure transfer mechanism such as SCP or SFTP. SCP and SFTP rely on security provided by SSH to transfer. SSH is very secure as long as we use non-trivial passwords or authentication is allowed only through SSH keys. Hence we will use this in our application. There are two ways we can go about storing the logs: a database like MongoDB [7, 8] to store logs, or in the file-system like normal files.

5.1   Using a Database—MongoDB

MongoDB is an open-source document database, and the leading No-SQL database. Written in C++, MongoDB offers features like:

- Full indexing—Index any attribute
- Replication support—Mirror across LAN's and WAN's
- Auto sharding—Horizontal scaling
- Querying—Feature rich query support
- Map/Reduce—Flexible aggregation and data processing

Using a database like MongoDB to store the logs would be easier for using the information in applications. MongoDB has adapters for various languages and hence it would be convenient to use while building the Traceback application. MongoDB is a full fledged database management system, this means the search and retrieval of documents is highly optimized and can provide a better performance than a traditional file-system without any such optimizations. MongoDB is fast, secure and also supports sharding which can be leveraged when the data set becomes very large. Having built-in support for replication allows us to maintain multiple replicas of the database. This would be useful in case of an attack on the central store.

5.2   Using a File-System

For a smaller system with a relatively smaller amount of logs a file-system maybe a reasonable way to store these logs. File-system access is easier and can be used by anyone without having knowledge of a particular database. Hence if manual, regular checks of logs are needed, then a file-system would suit this need better. A custom implementation for the specific application allows developers to have a finer control over its performance and storage. As the number of request for retrievals for logs is expected to be small (only in case of attack or suspected breach) using a database might not be necessary. As the size of log store grows we may have to consider a distributed file-system such as Hadoop to store the logs in a distributed manner.

MongoDB ensures integrity of data and also has built-in security mechanisms and fault tolerance. Additionally, it is easier to build applications to use MongoDB rather than a normal file-system. Due to the advantages of storing logs in MongoDB over a file-system we decided to go with using MongoDB to store logs for this particular application.

6 Traceback Tool

Traceback can be used to obtain a replay of events that occurred in the distributed system during some interval of time. It uses the information stored in the logs, along with the system states that were saved, to provide an analytical visualisation of events. The logs obtained from the store are decompressed and then used for replay purposes.

6.1 Timeline Format

To make the visualisation easy, the tool presents the system states and changes on a timeline. The system states are visualised as points on the timeline, and all events that change the state of the system can be represented as transitions on this state. The analyser can choose to view the state of the distributed system at a particular point of time on the timeline, and simulate the events that occurred from this instant to another chosen instant. Important instants of time can be bookmarked for easy reference, and any comments that follow from analysis can be added to the timeline to help future study.

6.2 Modules

Traceback also allows forensic investigators to write and include modules. Modules are well known attack vectors which can be used by other investigators to quickly search if a particular attack was used against the system. This will allow investigators to fly through their analysis and know exactly what was exploited on the system. The modules are stored centrally on Traceback's repository which makes it easy for analysers to download and include them for their analysis.

6.3 Playback Mechanism

When the starting state is chosen and the replay is begun, all events occurring are shown to the analyser. All the information about events and system states is obtained from the logs at the central store. The simulation details packets being sent and received from a node in the distributed system to an external node, as well as packets interchanged between nodes within the system. Processes changing the states of each node are also shown. The simulation on all nodes in the system is shown together. On finding suspicious action in any node, the analysers can concentrate on—or "zoom" into—the behaviour of that node.

Filters can be used to observe specific events at a time. If only network interactions are to be analysed, or only events from a particular log (like the authorisation log) the appropriate filter is applied and the playback is observed.

7 Conclusion and Future Work

Avoiding attack is of first and foremost importance when we design a distributed system. But as the system grows and becomes more complicated, security vulnerabilities may creep in. Malicious users may exploit these flaws in the system to gain access to sensitive information or to cause damage. For such situations, we need a robust tool to analyse the extent of damage caused by the attacker, remove all back-doors and understand the actual vulnerability and the system state that allowed such an attack. This tool helps investigators as manually sifting through logs to find the source of attack is tedious and will lead to delay in patches. To summarise, this paper discusses the implementation of a forensic analysis tool in a distributed system to investigate the system post attack. It uses agents to collect information about the status of various components of the system and stores it centrally. This information is retrieved by the tool, decompressed and displayed to the investigator in a convenient manner so as to make the analysis fast and effective. We have also discussed about some of the issues which are faced during the design of such a tool and presented a few ways to overcome these issues.

Future work that can improve this tool can involve the following details. Sharding the central log and storing the primary log in a distributed manner is an option to resolve the issue of efficient storage on large distributed systems. The tool can be made more intuitive in the sense that it analyses the information provided and marks suspicious locations on the timeline for the analysers to scrutinise. The security of the central log store and the agent processes is also a matter that needs to be looked into thoroughly.

References

1. Balakrishnan, R., Sahoo, R.K.: Lossless compression for large scale cluster logs. In: 20th International Parallel and Distributed Processing Symposium (IPDPS 2006), IEEE (2006)
2. Skibiski, P., Swacha, J.: Fast and efficient log file compression. In: CEUR Workshop Proceedings of 11th East-European Conference on Advances in Databases and Information Systems (ADBIS 2007) (2007)
3. Ren, W., Jin, H.: Distributed agent-based real time network intrusion forensics system architecture design. In: 19th International Conference on Advanced Information Networking and Applications (AINA 2005), vol. 1, IEEE (2005)
4. Hunt, R., Slay, J.: Achieving critical infrastructure protection through the interaction of computer security and network forensics. In: 2010 Eighth Annual International Conference on Privacy Security and Trust (PST), IEEE (2010)
5. Capuzzi, G., Spalazzi, L., Pagliarecci, F.: IRSS: Incident response support system. In: International Symposium on Collaborative Technologies and Systems (CTS 2006), IEEE (2006)
6. Benchmarks for popular compression algorithms. http://www.maximumcompression.com/data/log.php
7. MongoDB Official Documentation. http://www.mongodb.org/
8. Using MongoDB to store logs. http://docs.mongodb.org/ecosystem/use-cases/storing-log-data/

Deterministic Transport Protocol Verified by a Real-Time Actuator and Sensor Network Simulation for Distributed Active Turbulent Flow Control

Marcel Dueck, Mario Schloesser, Stefan van Waasen
and Michael Schiek

Abstract Total drag of common transport systems such as aircrafts or railways is primarily determined by friction drag. Reducing this drag at high Reynolds numbers ($<10^4$) is currently investigated using flow control based on transversal surface waves. For application in transportation systems with large surfaces a distributed real-time actuator and sensor network is in demand. To fulfill the requirement of real-time capability a deterministic transport protocol with a master slave strategy is introduced. With our network model implemented in Simulink using TrueTime toolbox the deterministic transport protocol could be verified. In the model the *Master-Token-Slave* (*MTS*) protocol is implemented between the application layer following the IEEE 1451.1 smart transducer interface standards and the Ethernet medium access protocol. The model obeys interfaces to the flow control and the DAQ-hardware allowing additional testing in model in the loop simulations.

Keywords TrueTime · Real-Time transport protocol · Distributed actuator and sensor network · Network model

M. Dueck (✉) · M. Schloesser · S. van Waasen · M. Schiek
Central Institute of Engineering, Electronics and Analytics ZEA-2: Electronic Systems,
Forschungszentrum Juelich GmbH, Wilhelm-Johnen-Straße, 52428 Juelich, Germany
e-mail: m.dueck@fz-juelich.de
URL: http://www.fz-juelich.de

M. Schloesser
e-mail: m.schloesser@fz-juelich.de

S. van Waasen
e-mail: s.van.waasen@fz-juelich.de
URL: https://www.uni-due.de/

M. Schiek
e-mail: m.schiek@fz-juelich.de

S. van Waasen
University of Duisburg-Essen, Faculty of Engineering, 47057 Duisburg, Germany

© Springer India 2016   29
A. Nagar et al. (eds.), *Proceedings of 3rd International Conference
on Advanced Computing, Networking and Informatics*, Smart Innovation,
Systems and Technologies 44, DOI 10.1007/978-81-322-2529-4_3

1 Introduction

In the FOR1779 research group "Active drag reduction by transversal surface waves" [1] fundamental research in the field of active drag reduction in high Reynolds numbers is done based on both wind tunnel experiments and numerical studies. The final aim is to realize turbulent flow control based on transversal surface waves by a distributed real-time actuator and sensor network.

The flow control will be embedded into a cascade control loop. Where the task of the outer loop is to optimize drag reduction by adapting wave parameters such as amplitude, frequency and wavelength. The inner loop is responsible for the wave control and thus minimizes deviations of the surface from the desired wave form.

The development of the distributed actuator and sensor network is based on modelling the network using Simulink and the TrueTime toolbox [2]. The model is used to investigate the most efficient topology and communication flow for the distributed turbulent flow control. The required real-time behavior will be assured by deterministic network communication. The network simulation is designed as a model in the loop approach to be used either as stand-alone network simulation or to utilize flow control experiments in the wind tunnel. For this the model includes interfaces to the flow control and the DAQ-hardware driving the actuators and sensors [3].

In Sect. 2 the model based on IEEE 1451.1 smart transducer interface standards is described. In Sect. 3 the three layer model and its origin is explained. Section 4 deals with the developed real-time *Master-Token-Slave* protocol to cover the missing features in the communication model. In Sect. 5 evaluation of the network behavior is presented as the verification of the proposed transport protocol. In Sect. 6 the work is summarized and future steps are presented.

2 Model

Data exchange in the actuator and sensor network is based on the IEEE 1451.1 smart transducer interface standards [4] and the node names are also chosen referring to this standard as network capable application processor (NCAP) and smart transducer module (STM).

The different node types are representing the interface node ($NCAP_I$), controller nodes ($NCAP_C$), nodes for actuation and local control (STM_A) and sensor nodes (STM_S) to measure data for the flow control (see Table 1). The strategy for a standard communication can be described as follows:

1. Flow control provides wave parameters as actuating variables
2. $NCAP_I$ takes data and forwards it to $NCAP_C$
3. $NCAP_C$ distributes data to STM_A
4. STM_A generates closed loop controlled surface wave
5. STM_S measures friction coefficient and supplies it to $NCAP_C$

Table 1 Description of the modelled nodes

Name	Behavior
NCAP$_I$	Interface for external flow control
NCAP$_C$	Controlling, distributing and gathering data
STM$_A$	Actuator node with internal closed loop wave control
STM$_S$	Sensor node for flow control

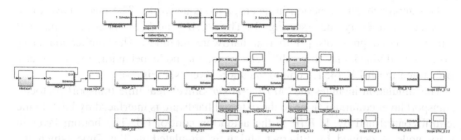

Fig. 1 Simulink network model consisting of eleven TrueTime network nodes and splitted into three different networks. One to connect $NCAP_I$ and two $NCAP_C$ nodes. Each $NCAP_C$ is connected to its assigned STM_A and STM_S nodes in a separate network. There are also interfaces to common Simulink subsystems on $NCAP_I$ and STM_A for providing model in the loop behavior

6. $NCAP_C$ gathers drag reduction information and forwards it to $NCAP_I$
7. $NCAP_I$ provides drag reduction information to flow control

To set up a communication between the nodes the TrueTime toolbox provides a low level network communication which is used here. The parameters of the low level Ethernet network have been verified using a Raspberry Pi testbed [5]. The described approach additionally needs to cover the real-time capability for a predictable message arrival time on the destination node. This *Master-Token-Slave* protocol represents the transport layer between the particular IEEE 1451.1 communication and Ethernet representing the physical layer. The current TrueTime simulation chart is shown in Fig. 1.

3   Three Layer Communication Concept

Referring to the ISO-OSI communication model [6], common real-time protocols are settled either above the transport layer such as RTP [7] or below the transport layer. This is realized either within medium access control protocols including hardware components [8] or industrial bus technology, e.g. fieldbusses [9]. Inside the layer architecture there is a logical communication between equal layers on each communication partner. Physical communication only takes place on the bottom layer.

Following a simplified OSI-model, a three layer architecture is used in this paper as outlined by Deleuze [10]. It consists of application layer, transport layer and network access layer (see Fig. 2). For the presented purpose every layer uses defined interfaces to communicate with top and bottom layer. No cross-layer communication is allowed. One advantage using a layer strategy is to achieve replaceable protocols.

This approach can be nested in heterogeneous networks. The layers have to be implemented in every node (see Fig. 3). The $NCAP_I$ and STM nodes are connected with predefined protocols [3] to real hardware modules. The application layer protocol is divided into three different parts, the node behavior, the parameter format and the IEEE 1451.1 communication layer (see Fig. 3).

The simulation includes the "*Master-Token-Slave*" real-time transport protocol. Transport layer communication to the external hardware is interfaced by UDP to the central control and by TCP/IP to the actuation control [3]. The bottom layer in every node is defined by Ethernet and it is simulated in TrueTime which we recently investigated using a hardware testbed [5].

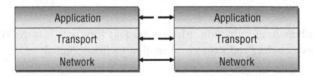

Fig. 2 Simplified protocol stack based on internet protocol architecture. This simplification enables easier modelling of network behavior

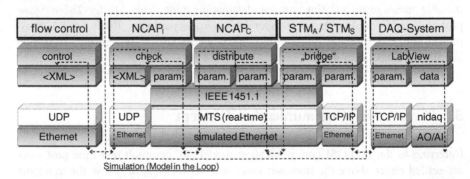

Fig. 3 Package delivery through communication layers from central control algorithm to actuation hardware for analog in- and output. The nodes in the *middle* are parts of the Simulation with UDP and TCP/IP interfaces to the hardware

4 Real-Time *Master-Token-Slave* Protocol

As proposed by Verissimo [11], it is possible to gain real-time behavior in networks by deterministic package distribution. Because application layer and medium access layer have already been defined by IEEE 1451.1 and Ethernet the transport protocol is apparently designated to deliver the real-time capability. The described network model is sealed to the outside and has not to be secured against external errors.

The defined transport protocol is named *Master-Token-Slave*, short *MTS*. The general behavior of interactions is similar to common protocols. An interface is provided to the upper layer using queues to buffer data. The layer below is used by the send- and receive interfaces available in TrueTime. Usually packages are embedded into frames running through network layers. There are two *C*-structures defined in this model: the mts_msg structure provides necessary information as outer frame added by the *MTS* transport layer. I1451_msg represents the IEEE 1451.1 message and is a dummy structure for the data to send (e.g. commands, functions, parameters, values) by the application itself. So an I1451_msg is handed over to the *MTS* interface and can also be expected by receiving messages through the *MTS* transport layer. An example for message exchange can be found in Fig. 4.

The communication is strictly ordered to assure determinism. It is initiated and driven by one master. Slave nodes just have to react on incoming messages. In the

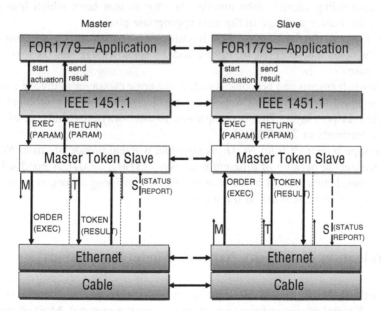

Fig. 4 Example to clarify the *MTS* interactions with the *upper* and *lower* layers. According to the IEEE 1451.1 standard, messages are sent to invoke actions on the other nodes and wait for return values. The *MTS*-protocol sorts it into two phases to send the order and carry the answer. The third phase is optionally provided to the slave node for expressing any errors or giving status reports

described example (see Fig. 1) the communication flow is divided into three different communication phases, nested in two separated network sections. This ensures that every $NCAP_I$ is master for all $NCAP_C$ and every $NCAP_C$ controls his own set of STM_A and STM_S nodes separately. In contrast to a homogeneous network our heterogeneous approach allows higher efficiency in communication since data exchange takes place in different network sections in parallel. To identify messages, every message is marked with a message identification number. Requests, answers and acknowledgements carry the same identification number as corresponding messages.

As shown in Fig. 4 every phase is initiated with a beacon-broadcast by the master. The beacon is marked with a phase number, so if any node misses a beacon it will catch up with the next one and the communication has not to be reset completely. In the first phase, the *Master*-phase, the master node (see Fig. 4, left side) can send information to his slaves. Sent messages are saved for potentially resends.

The second phase is the *Token*-phase. The master node sends a token to the slaves which received a message in the first phase. This is a short message which indicates some free time for the slave node to talk. The token is addressed to a specific slave node, it is not possible to use broadcasts here. If a message is enqueued the slave node responds to the received identification number with a corresponding message. If the queue for answers contains no message, the queue for incoming messages is investigated. In case there is no unprocessed message with corresponding identification number an error is sent back which initiates a resend of the master message in the next appropriate phase.

In the last part, the *Slave*-phase there is a time slot for status messages from slave nodes. Messages are accepted by master and an acknowledgement is sent in the *Master*-phase.

Acknowledgements and beacons as well as the error messages to invoke a resend on *Token*-phase are internal messages and are not noticed in any protocol layer above the transport layer. This is comparable to three-way-handshake or message acknowledgements in TCP/IP [12].

A message broadcast from $NCAP_I$ is possible, whereat broadcasts from $NCAP_C$ are not reasonable because of different node types (STM_A, STM_S). A good solution to save network time would be to send to groups consisting of one type by multicast. This has to be realized in the future.

5   Evaluation of *MTS* by Network Model Simulations

The described *MTS*-protocol is implemented in the model as TrueTime code functions. Several processes have to be started to run the protocol. Most of the time the processes sleep or wait to execute new communications. Interface functions can be used to send and receive messages from the main code function of a node.

The behavior can be analyzed by Simulink scopes which record datasets from the TrueTime kernel blocks. For evaluation and analysis of timing and behavior an example has been implemented using the *MTS*-protocol in TrueTime. The deterministic simulation working on application layer is structured as follows:

1. $NCAP_I$ sends 16 messages to each connected $NCAP_C$
2. $NCAP_I$ waits for answers
3. $NCAP_{C1}$ and $NCAP_{C2}$ answer to every message
4. $NCAP_{C1}$ sends one status messages for every incoming message

The network model used is shown in a Simulink chart in Fig. 1. The messages are sent from and to *MTS*-layer by predefined interfaces. In this experiment there are no additional disturbances like interrupts, dying nodes or lost packages introduced. The message size is in general 1400 Byte which is below the maximum transfer unit of Ethernet at about 1500 Byte. Beacons and token have the size of Ethernet minimum frame size of 64 Byte. The timing windows for the communication phases are equally spaced by $\frac{100}{3}$ ms. The experiment takes 0.3 s which equals three full *MTS*-cycles.

Figure 5 shows the result of this experiment. The network events are initiated by the transport layer protocol in the simulated Ethernet layer. The events are indicated by a peak on the scope-output. Also the limitation in time for the different communication states occurs in the right moment.

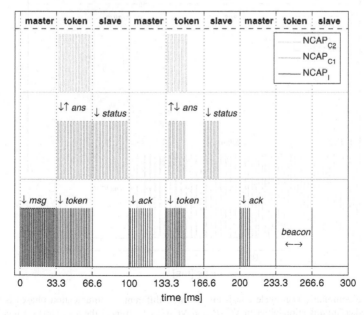

Fig. 5 The result of an experiment for communication between one master ($NCAP_I$) and two slave nodes ($NCAP_{C1}$, $NCAP_{C2}$) shows a full communication which is initiated once by enqueueing some messages on master node and then waiting for reactions on slave nodes

The master permanently controls all slaves. At the start a beacon is sent to indicate the start of *Master*-timeslot. 16 messages are enqueued by $NCAP_I$. The messages are broadcasted to both $NCAP_{C1}$ and $NCAP_{C2}$ (see Fig. 5) within the *Master*-phase. An answer is enqueued by the controller nodes after receiving a message. Then the *Token*-phase is initiated by a second beacon. Tokens are sent to all nodes which should have received a message in the first timeslot. The corresponding answer is read from the queue and sent directly. The end of the *Token*-phase is indicated by another beacon. The following *Slave*-phase is open for status messages by the slave nodes. It is proposed to be used for registration to the network at start of a slave node or for error notifications to the master node. The detailed behavior of the previous described communication is depicted in Fig. 6.

In this example the whole communication was not finished during the first cycle and it is continued within the next cycle. Hereby the master starts with acknowledging the status messages from $NCAP_{C1}$, then the token is sent to both slaves and requested answers are received. Again status messages are sent by $NCAP_{C1}$ and acknowledged by the master. After no further messages occur, the flow is just continued by sending beacons for every new phase.

The result of this experiment proofs the correctness of the deterministic transport layer protocol embedded in the network model. The reaction time in the *Token*-phase from token send until the message is fully received is 1.25 ms. The time of

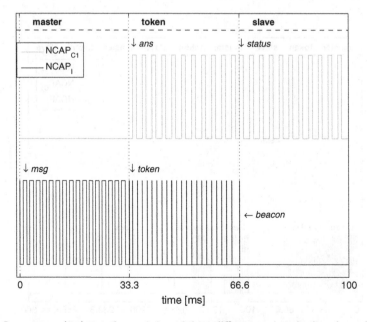

Fig. 6 One communication cycle consisting of three different communication phases is shown. The detailed communication between $NCAP_I$ and $NCAP_{C1}$ highlights the token answer mechanism and the status message part

network occupation and buffer time by beacon, token and acknowledgements is below 1 ms. Messages with the size of 1400 Byte like master messages or token answers need a propagation time of 1.14 ms.

6 Conclusion and Outlook

In this paper we presented a deterministic transport protocol for a real-time actuator and sensor network based on a three layer network architecture. The *Master-Token-Slave* (*MTS*) protocol is implemented between the application layer defined by IEEE 1451.1 smart transducer interface standards and the Ethernet medium access protocol delivered by the TrueTime toolbox. Real-time capability of the protocol is checked using a reduced network model where the response time depends on the length of the three different communication windows. The in-house development allows to adapt the phase-lengths to the sample rate of the flow control.

Error influences like packet loss, node malfunction, restart or blackout have to be covered in a future version of the model. Also collisions which can occur in the last phase of a communication cycle only have to be investigated. To cover e.g. a $NCAP_C$ blackout the network has to be changed from heterogeneous to a full-mesh topology. Multicasts from $NCAP_C$ to STM_A and STM_S for *Master*-phase should be considered to save time supplying the same data to a specific node type.

To summarize a flexible real-time transport protocol has been developed which can now be used to investigate model in the loop simulations to develop distributed real-time actuator and sensor networks for turbulent flow control.

Acknowledgments The authors would like to thank all partners in the research group FOR1779 and acknowledge the funding by the DFG (German Research Foundation).

References

1. DFG. GEPRIS. [12.06.2014]. http://gepris.dfg.de/gepris/projekt/202175528
2. Cervin, A., Henriksson, D., Lincoln, B., Eker, J., Årzén, K.-E.: How does control timing affect performance? Analysis and simulation of timing using Jitterbug and TrueTime. IEEE Control Syst. Mag. **23**(3), 16–30 (2003)
3. Dueck, M., Kaparaki, M., Srivastava, S., van Waasen, S., Schiek ,M.: Development of a real time actuation control in a network-simulation framework for active drag reduction in turbulent flow. In: Automatic Control Conference (CACS), 2013 CACS International, pp. 256–261, Dec 2013
4. IEEE std 1451.1-1999, standard for a smart transducer interface for sensors and actuators— network capable application processor (ncap) information model, June 1999. The Institute of Electrical and Electronics Engineers, Inc. 3 Park Avenue, New York, NY 10016-5997, USA
5. Dueck, M., Schloesser, M., Kaparaki, M., Srivastava, S., van Waasen, S., Schiek ,M.: Raspberry pi based testbed verifying truetime network model parameters for application in distributed active turbulent flow control. In: Proceedings of the SICE Annual Conference (SICE), pp. 1970–1975, Sept 2014

6. Day, J.D., Zimmermann, H.: The OSI reference model. Proc. IEEE **71**(12), 1334–1340 (1983)
7. Schulzrinne, H., Casner, S., Frederick, R., Jacobson, V.: RFC 3550: RTP: A Transport Protocol for Real-Time Applications. Technical report, IETF, 2003
8. Carvajal, G., Wu, C.W., Fischmeister, S.: Evaluation of communication architectures for switched real-time ethernet. IEEE Trans. Comput. **63**(1), 218–229 (2014)
9. Tovar, E., Vasques, F.: Real-time fieldbus communications using profibus networks. IEEE Trans. Ind. Electron. **46**(6), 1241–1251 (1999)
10. Deleuze, C.: Content networks. Internet Protoc. J. **7**(2), 2–11 (2004)
11. Verissimom P.: Real-time communication. In: Mullender, S. (ed.) Distributed Systems, pp. 447–490. Addison-Wesley (1993)
12. Cerf, V., Dalal, Y., Sunshine, C.: Specification of Internet Transmission Control Program. RFC 675, December 1974

Improved Resource Provisioning in Hadoop

M. Divya and B. Annappa

Abstract Extensive use of the Internet is generating large amount of data. The mechanism to handle and analyze these data is becoming complicated day by day. The Hadoop platform provides a solution to process huge data on large clusters of nodes. Scheduler play a vital role in improving the performance of Hadoop. In this paper, MRPPR: MapReduce Performance Parameter based Resource aware Hadoop Scheduler is proposed. In MRPPR, performance parameters of Map task such as the time required for parsing the data, map, sort and merge the result, and of Reduce task, such as the time to merge, parse and reduce is considered to categorize the job as CPU bound, Disk I/O bound or Network I/O bound. Based on the node status obtained from the TaskTracker's response, nodes in the cluster are classified as CPU busy, Disk I/O busy or Network I/O busy. A cost model is proposed to schedule a job to the node based on the classification to minimize the makespan and to attain effective resource utilization. A performance improvement of 25–30 % is achieved with our proposed scheduler.

Keywords Hadoop · Job scheduling · Resource awareness

1 Introduction

The digital data generating today is not just the relational data which is stored in the form of rows and columns. The data is unstructured because of the extensive use of Internet for social network, email, videos, images and online business. The effect of extensive use of Internet is the generation of huge bytes of data everyday which in

M. Divya (✉) · B. Annappa
Department of Computer Science and Engineering, National Institute of Technology
Karnataka, Surathkal 575025, Karnataka, India
e-mail: divyakeerthi.m@gmail.com

B. Annappa
e-mail: annappa@ieee.org

© Springer India 2016
A. Nagar et al. (eds.), *Proceedings of 3rd International Conference on Advanced Computing, Networking and Informatics*, Smart Innovation, Systems and Technologies 44, DOI 10.1007/978-81-322-2529-4_4

turn makes the handling of data and analyzing more complicated. Rise of cloud computing and cloud data stores have been a boon to the emergence of big data [1]. The concept of virtualization and distributed system technology of cloud computing is empowering organization to deal with large scale processing of data because of the elasticity of cloud resources.

The need to deal with network communication, replicate data and failure of components increases programming overhead associated with distributed computing. Big data need mechanism to analyze and distribute the data, parallelize the computation, fault tolerance and balance of load in the system automatically without any human intervention. Hadoop [2] is a platform which is being extensively used nowadays to handle data intensive applications. It could provide a solution with which most of the programming overhead associated with the distributed computing can be skipped and it can be scalable to a large cluster.

Schedulers play a critical role for the efficient performance of the system because real-time applications are deadline bound. Deciding the best match between the job and the resources is critical for the optimal performance of the system. Richard et al. [3] has observed that the most optimal network path for both computation and storage should be identified and the scheduling system should calculate and incorporate network measurement while planning for the submission of the job in a distributed environment. Experiments in [4–6] have shown that the characteristics of jobs plays a critical role in optimizing the performance of hadoop. There is a need for scheduler, that consider the type of workload and the characteristic of nodes in the cluster with network awareness to reduce the bottleneck in the whole system.

In this work, the parameters considered are proposed in [7] such as the time to parse the data, map, sort and merge for map task and time to merge, parse and reduce, for reduce task. The node status is calculated based on the TaskTrackers response. Based on these parameters, a cost model is proposed. The resources in the cluster are classified as CPU busy, Disk I/O busy or Network I/O busy and tasks as CPU bound, Disk I/O bound or Network I/O bound and accordingly schedule the tasks based on the cost model. We have identified that the loss of packets in the network and available bandwidth plays a critical role in making scheduling decisions for large scale data intensive application. These metrics are used in this work for calculating the network cost which has never been used in any of the earlier work of job scheduling in Hadoop.

2 Related Works

The schedulers proposed earlier for Hadoop are based on data locality, resource management and speculative execution. JobTracker decides which job to be assigned to which TaskTracker node and TaskTracker node does the actual job execution. FIFO [2] assigns jobs to the nodes based on first come first serve basis. FIFO suffers from head of queue blocking and starvation issues. Fair Scheduler [8] allocates jobs to the 'pools' and then assures specific number of slots to each pool.

Capacity scheduler [9] is based on capacity of the resources. In capacity scheduling, each queue is given a fraction of the guaranteed capacity of the resource. These schedulers does not take into account the resource utilization of the applications.

LATE [10] tries to find stragglers by computing the remaining time of all tasks in a multiuser environment. If the remaining time is greater than some threshold defined, then such tasks are identified as stragglers and the scheduler does the speculative execution. Ganesh et al. [11] alleviated the effect of stragglers through the 'kill and restart' of tasks that has been recognized as probable stragglers. The main drawback of these techniques is that killing and duplicating tasks results in wasted resources.

The basic notion of Hadoop is moving computation is cheaper than moving data. Zaharia et al. [12] proposed a scheduler to address the conflict between the fairness and locality. They have addressed a technique called delay scheduling that employ a delay for non-local tasks. Zhang et al. [13] proposed a method called Next K node Scheduling (NKS) to improve data locality by calculating the probability of each task, and schedule the one with the highest probability to these nodes. Such strong assumptions cannot be held in heterogeneous environment.

3 Overview of Scheduling in Hadoop

A hadoop system can be described based on three factors: cluster, workload and user [14]. These factors can be either homogeneous or heterogeneous, that exhibit the level of diversity of the hadoop system. Hadoop scheduler works by dividing the job into two phases, map phase and reduce phase. In each phase, the job is divided into a set of tasks based on the available number of slots. Map tasks should pass through the steps like parsing the data, execution of mapper function, shuffle and sort the result locally based on the key as shown in Fig. 1.

In the reduce phase, tasks has to pass through steps like parsing the data, merging the data based on key and to execute the reduce function as shown in Fig. 2. Then the output of the reduce function is written back to HDFS or the location desired by the user. The design of HDFS, where the file is replicated from

Fig. 1 Phases of execution of map task

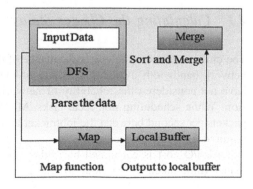

Fig. 2 Phases of execution of reduce task

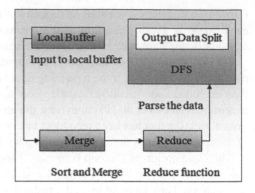

node to node as a protection against node failure, depends completely on the speed of the network infrastructure, which is responsible for distribution of replicas. The network plays a critical role even during the shuffle phase of MapReduce. If there is multiple data reduction, then the network performance impacts the performance of hadoop cluster.

4 Proposed Work

In MRPPR, the status of the node is classified as CPU busy, CPU idle, Disk I/O busy and Disk I/O idle. Two separate queues of nodes are maintained, i. CPU busy node and ii. Disk I/O busy node. If the task is not local, the Network cost of transferring the data is considered to make a decision of which node to choose for task execution. The MapReduce performance parameters are considered to characterize the Hadoop jobs as CPU bound, I/O bound or Network I/O bound job. CPU bound and Disk I/O bound jobs are assigned to nodes based on the node status. If the task is not local or if it is Network I/O bound then the Network cost is considered and based on the proposed cost model, the task is assigned to the node.

4.1 Calculating the Cluster Status

The cluster resources include performance of disk, CPU utilization of the node and network bandwidth availability. Many of the previous research work in hadoop have not considered the reliability of network while making the scheduling decision. While scheduling non local tasks, MRPPR scheduler considers the dropped packets per second between the jobtracker and the tasktracker, to avoid loss of data required for the job execution.

Disk I/O cost is calculated by finding the read performance and write performance (time taken by the node to read and write the data) and the access time of hard disk [7] of tasktracker node. Disk I/O cost of the node is calculated as in Eq. (1).

$$Disk\_I/0 = Avg_{read} + Avg_{write} + Avg_{access} \tag{1}$$

Scheduling in distributed environment is affected mostly by the network cost. The load on the network and availability of bandwidth play a critical role in job scheduling. While calculating the network I/O of the link, MRPPR considers the loss of packets/second in the link between the jobtracker and the tasktracker. This has never been considered in any of the schedulers before to find the quality of service in the link between the master and the slaves. The network having less packet loss is more efficient than a network having more packet loss. This cost is high if the loss is high and the bandwidth is low. The network cost is calculated as shown in Eq. (2).

$$Network\_I/O = \frac{(CPU_{current} - CPU_{last}) * loss}{Bandwidth} \tag{2}$$

CPU performance of a node is calculated based on the number of CPU cycles used to run the task, the heartbeat time and the number of processors in the node. If the number of processors is more, then the overall performance of CPU bound task is higher. CPU cost of the node is calculated as shown in Eq. (3).

$$CPU = \frac{CPU_{no\_of\_cycles} * (CPU_{current} - CPU_{last})}{Number\ of\ processors} \tag{3}$$

Once the resource status is found, the nodes in the cluster are assigned to one of the two queues: CPU busy queue and Disk I/O busy queue as explained in Algorithm 1.

Algorithm 1: Node Classification
---
if $Disk\_I/0 > CPU$ then
| classify the node as I/O busy node;
end
else
| classify the node as CPU busy node;
end
if $Network\_I/O$ considered for non-local tasks then
| find the link which has less $Network\_I/O$ in each queue;
end

4.2 Job Classification Based on Its Characteristics

Map reduce programming model has a set of map tasks and reduce tasks. Feature of the job has to be evaluated based on the cost of execution of set of map tasks and reduce tasks. From the literature [7], it is evident that hadoop job scheduler has to consider the features of map tasks as in Fig. 1 and reduce tasks as in Fig. 2. The cost of Disk I/O of map task is calculated as in Eq. (4).

$$D\_I/O_{map} = \frac{map\_i/p\_size}{read/sec} + \frac{map\_o/p\_size}{write/sec} \tag{4}$$

The time to parse the data for maptask T_map_{parse}, time to execute the maptask T_map_{map}, time to sort and merge the result T_map_{sort} and T_map_{merge} add to the cpu cost, calculated in Eq. (6) of the maptask. The CPU time for the execution of each steps is calculated as in Eq. (5).

$$CPU\_time = CPU_{current} - CPU_{last} \tag{5}$$

$$C_{map} = T\_map_{parse} + T\_map_{map} + T\_map_{sort} + T\_map_{merge} \tag{6}$$

The network cost of map task, calculated as in Eq. (7) includes the size of the map task input and the bandwidth of the link as in 7.

$$N\_I/O_{map} = \frac{map\_i/p\_size}{bandwidth} \tag{7}$$

The overall cost of map function is given in Eq. (8)

$$Map\_cost = d_m * D\_I/O_{map} + c_m * C_{map} + n_m * N\_I/O_{map} \tag{8}$$

d_m, c_m and n_m are the weights given based on the values of the parameters considered.

Similarly, for the reduce task, time to merge map output data T_reduce_{merge}, time to sort T_reduce_{sort}, time to execute the reduce function T_reduce_{reduce} and parse the data T_red_{parse} is considered. Execution time of each step is calculated as explained in Eq. (5) and the overall CPU cost of reduce task is given as in Eq. (9).

$$C_{red} = T\_red_{merge} + T\_reduce_{sort} + T\_red_{reduce} + T\_red_{parse} \tag{9}$$

The cost of Disk I/O of reduce task is calculated as in Eq. (10).

$$D\_I/O_{red} = \frac{map\_o/p\_size}{read/sec} + \frac{reduce\_o/p\_size}{write/sec} \tag{10}$$

The network cost of reduce task includes the size of map task input and bandwidth of the link as in Eq. (11).

Table 1 Cluster node information

No. of nodes	CPU	Disk	RAM
3	Core i7	320	8
1	Core i3	320	4
1	Core i7	250	4

$$N\_I/O_{red} = \frac{map\_o/p\_size}{bandwidth} \tag{11}$$

The overall cost of reduce task is given in Eq. (12).

$$Reduce\_cost = d_r * D\_I/O_{red} + c_r * C_{red} + n_r * N\_I/O_{red} \tag{12}$$

d_r, c_r and n_r are the weights given based on values of the parameters considered as explained for map tasks. Our objective is to minimize *Map_cost* and *Reduce_cost* and to make efficient resource utilization of the computing resources. Allocation of tasks to the nodes with respect to cost model is explained in Algorithm 2.

5 Experimental Setup

The proposed scheduler is implemented on Hadoop 0.20.0. Five nodes with specifications used in our experiment is given in Table 1. Status of the node is obtained by the heartbeat message sent by the TaskTracker to the JobTracker. Heartbeat message is modified to provide node status like CPU utilization and disk read/write performance, so that it will piggyback these data along with heartbeat. Information about the network bandwidth utilization and dropped packets in the heartbeat message is considered in this work. Modification for the heartbeat message is done in /src/mapred/org/apache/mapred/TaskTracker.java and /src/mapred/org/apache/mapred/TaskTrackerStatus.java [15].

Sample tasks are extracted from each job in job queue and assigned to the available free slots in the cluster. Scheduler then calculates the time of completion of each step in map tasks and reduce tasks. Once it get these data, the next step is to use this data in our proposed mathematical model and decide the characteristics of tasks. The scheduler uses the data of already run sample tasks to give weight to the parameters. MRPPR is designed in such a way that none of the compute node in the cluster is overloaded.

6 Evaluation

The experiment is evaluated with several benchmark programs in hadoop like i. word count, ii. terasort, iii. grep and iv. MultiFetch program, which fetches the title of web page for the URLs given as input. It is found from the experiments that

Terasort and grep are I/O bound tasks, word count is CPU bound task and MultiFetch is network I/O bound task. Experiment is conducted with 1, 5 and 10 GB of data for each examples.

Algorithm 2: MRPPR Scheduler

Data: Run sample tasks form set of jobs in $J = J_1, J_2, \ldots, J_n$;
Extract the job features;
Extarct the feature of the node;
Output 1: Minimize Map_cost for Map Tasks;
Output 2: Minimize $Reduce\_cost$ for Reduce Tasks;
for each map task t_{mi} of J_i
if t_{mi} *is local* **then**
    if $Disk\_I/O_{map} < CPU_{map}$ **then**
        | Allocate t_{mi} to I/O busy node;
    end
    else
        | Allocate t_{mi} to CPU busy nodes;
    end
end
else
    if $Disk\_I/O_{map} < CPU_{map}$ **then**
        | Allocate t_{mi} to I/O busy node which has less $Network\_I/O$;
    end
    else
        | Allocate t_{mi} to CPU busy node which has less $Network\_I/O$;
    end
end
else
    | Allocate t_{mi} to a node which has less $Network\_I/O$ and less overhead;
end
for each reduce task t_{ri} of J_i
if t_{ri} *is local* **then**
    if $Disk\_I/O_{reduce} < CPU_{reduce}$ **then**
        | Allocate t_{ri} to I/O busy node;
    end
    else
        | Allocate t_{ri} to CPU busy node;
    end
end
else
    if $Disk\_I/O_{reduce} < CPU_{reduce}$ **then**
        | Allocate t_{ri} to I/O busy nodes with less $Network\_I/O$;
    end
    else
        | Allocate t_{ri} to CPU busy nodes with less $Network\_I/O$;
    end
end
else
    | Allocate t_{mi} to a node which has less $Network\_I/O$ and less overhead;
end

Table 2 Makespan of workloads on different schedulers

Workload	FIFO	Fair	Capacity	MRPPR
Word count	10	8	7	6
TeraSort	9	7	6	4
Grep	13	12	12	9
Multifetch	18	15	15	13

Table 3 Resource utilization of workloads

Workload	CPU	Disk I/O	Network
Word count	9	7	6
TeraSort	7	5	6
Grep	8	7	6
Multifetch	6	5	9

Table 2 shows the completion time of each job with respect to FIFO, Fair, Capacity and our scheduler. We could observe that the completion time of jobs with FIFO, Fair and Capacity scheduler is comparatively less. These schedulers doesn't take into account type of workload before allocating the tasks into compute node.

Table 3 shows the resource utilization of different workloads with the proposed scheduler. The usage of resource for each job is created from the sample tasks. These approximations are passed to our scheduler. The scheduler then assigns different weights to each parameters based on values obtained from the sample tasks.

Figures 3, 4 and 5 shows the performance of hadoop with MRPPR for different size of data sets for each job. The makespan of the jobs with our scheduler does not

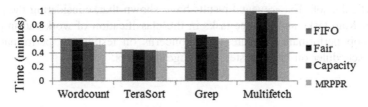

Fig. 3 Execution time (Y-axis) of 1 GB data for each job (X-axis)

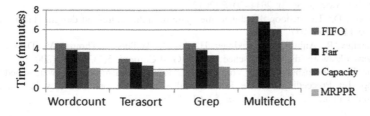

Fig. 4 Execution time (Y-axis) of 5 GB data for each job (X-axis)

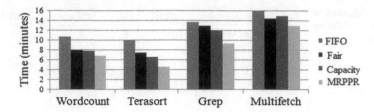

Fig. 5 Execution time (Y-axis) of 10 GB data for each job (X-axis)

differ much from the other three schedulers when the size of input data is small as shown in Fig. 3. When the size of input data increases our scheduler makes significant performance improvements that can be seen from Figs. 4 and 5 compared to other three schedulers. This is because, when the size of the data is high, all the nodes are busy in processing the data and computing. In such cases, when the new job arrives the scheduler should consider the type of the job and the node status before scheduling tasks of the node. With our proposed scheduler there is a performance improvement of 25–30 % compared to Hadoop's FIFO scheduler.

7 Conclusion

In this paper, a new Hadoop job scheduler is proposed which considers different parameters of Map Task and Reduce Task. The heartbeat message is propagated with data such as utilization of resources in cluster like CPU, disk I/O along with bandwidth utilization and packet loss that helps in deciding best match of resource for a particular job. Experimental results have shown that, considering the performance of Map tasks and Reduce tasks in the classification of workload improves the Hadoop performance. Future work concerns on testing the performance of MRPPR scheduler on large cluster with huge data sets.

References

1. Dittrich, J., Quiané-Ruiz, J.A.: Efficient big data processing in hadoop mapreduce. Proc. VLDB Endowment **5**(12), 2014–2015 (2012)
2. Borthakur, D.: The hadoop distributed file system: architecture and design. Hadoop Project Website **11**, 21 (2007)
3. McClatchey, R., Anjum, A., Stockinger, H., Ali, A., Willers, I., Thomas, M.: Data intensive and network aware (diana) grid scheduling. J. Grid Comput. **5**(1), 43–64 (2007)
4. Kumar, K.A., Konishetty, V.K., Voruganti, K., Rao, G.: Cash: context aware scheduler for hadoop. In: Proceedings of the International Conference on Advances in Computing, Communications and Informatics, pp. 52–61. ACM (2012)

5. Mude, R.G., Betta, A., Debbarma, A.: Capturing node resource status and classifying workload for map reduce resource aware scheduler. In: Intelligent Computing, Communication and Devices, pp. 247–257. Springer (2015)
6. Ren, Z., Xu, X., Wan, J., Shi, W., Zhou, M.: Workload characterization on a production hadoop cluster: a case study on taobao. In: 2012 IEEE International Symposium on Workload Characterization (IISWC), pp. 3–13. IEEE (2012)
7. Lin, X., Meng, Z., Xu, C., Wang, M.: A practical performance model for hadoop mapreduce. In: Cluster Computing Workshops (CLUSTER WORKSHOPS), 2012 IEEE International Conference on. pp. 231–239. IEEE (2012)
8. Fair Scheduler: http://hadoop.apache.org/mapreduce/docs/r0.21.0/fairscheduler.html
9. Capacity Scheduler: http://hadoop.apache.org/mapreduce/docs/r0.21.0/Capacityscheduler.html
10. Zaharia, M., Konwinski, A., Joseph, A.D., Katz, R.H., Stoica, I.: Improving mapreduce performance in heterogeneous environments. OSDI **8**, 7 (2008)
11. Ananthanarayanan, G., Kandula, S., Greenberg, A.G., Stoica, I., Lu, Y., Saha, B., Harris, E.: Reining in the outliers in map-reduce clusters using mantri. OSDI **10**, 24 (2010)
12. Zaharia, M., Borthakur, D., Sen Sarma, J., Elmeleegy, K., Shenker, S., Stoica, I.: Delay scheduling: a simple technique for achieving locality and fairness in cluster scheduling. In: Proceedings of the 5th European Conference on Computer systems, pp. 265–278. ACM (2010)
13. Zhang, X., Zhong, Z., Feng, S., Tu, B., Fan, J.: Improving data locality of mapreduce by scheduling in homogeneous computing environments. In: 2011 IEEE 9th International Symposium on Parallel and Distributed Processing with Applications (ISPA), pp. 120–126. IEEE (2011)
14. Song, G., Yu, L., Meng, Z., Lin, X.: A game theory based mapreduce scheduling algorithm. In: Emerging Technologies for Information Systems, Computing, and Management, pp. 287–296. Springer (2013)
15. Hadoop mapreduce: http://hadoop.apache.org/

Cuckoo Search for Influence Maximization in Social Networks

Nikita Sinha and B. Annappa

Abstract In a social network, the influence maximization is to find out the optimal set of seeds, by which influence can be maximized at the end of diffusion process. The approaches which are already existing are greedy approaches, genetic algorithm and ant colony optimization. Eventhough these existing algorithms take more time for diffusion, they are not able to generate a good number of influenced nodes. In this paper, a Cuckoo Search Diffusion Model (CSDM) is proposed which is based on a metaheuristic approach known as the Cuckoo Search Algorithm. It uses fewer parameters than any other metaheuristic approaches. Therefore parameter tuning is an easy task for this algorithm which is the main advantage of the Cuckoo Search algorithm. Experimental results show that this model gives better results than previous works.

Keywords Social network · Influence maximization · Cuckoo search

1 Introduction

A social network is the interconnection and interaction of a set of entities. The social network is the best way to spread the information as compared to any other medium such as newspapers and advertisements, etc. In the present era, online social networking sites have become popular since they connect people very effectively. One of the applications of influence maximization is viral marketing in social networks [1], to target a small number of potential customers by which information may be spread very fast [2, 3].

N. Sinha (✉) · B. Annappa
Department of Computer Science and Engineering, National Institute of Technology
Karnataka, Surathkal 575025, Karnataka, India
e-mail: nikitasinha27@gmail.com

B. Annappa
e-mail: annappa@ieee.org

© Springer India 2016   51
A. Nagar et al. (eds.), *Proceedings of 3rd International Conference
on Advanced Computing, Networking and Informatics*, Smart Innovation,
Systems and Technologies 44, DOI 10.1007/978-81-322-2529-4_5

Suppose a company wants to launch a new product in the market, but it has budget constraint to inform a large population. So, it selects some people to spread the information in the network at the earliest, enabling them to convince people to use their product and spread information regarding the product amongst their friends, each of whom in turn spread it further to their friends and so on. In this way, information will spread in the whole network. Hence the main objective of the company is to select the most influencing nodes by which the maximum part of the network will be influenced. It is also beneficial from the customer's point of view since people are more influenced by their friends as compared to other mediums. Social network itself has a very complex and large structure so to find out a feasible solution to get the set of most influential people in the network is a big challenge.

2 Literature Survey

Richardson took initiative to solve the influence maximization problem [4] in social networks. Kempe et al. [5] took the initiative to solve the discrete optimization problem. Propagation of influence is based on a stochastic cascade model. The drawback of this algorithm is its efficiency. They proved that finding a small seed set, by which the influence will be maximized is an NP-Hard problem. Monte-Carlo simulation was run a good number of times in the influence cascade model. This gives an estimate of the influence which is spread by the seed set. It took days to complete one run of this algorithm.

Leskovec presented an optimization scheme which is known as Cost-Effective Lazy Forward (CELF) scheme [6]. This approach was better than the previous approach because the number of evolutions is less than that of Kempe's algorithm but it still took a number of days for a few hundreds or thousands of vertices. Hence it is not very efficient for large social networks like Facebook and Twitter.

The scheme proposed by Chen consists of two algorithms, DegreeDiscount [7] and Prefix excluding Maximum Influence Arborescence model (PMIA) [8]. In DegreeDiscount algorithm, the node added to the seed set is the one with the highest degree and degrees of nodes adjacent to that node are discounted. The PMIA occupies a large amount of memory so it is not scalable for large social graphs.

Marginal Discount of Influence Spread Path (MISP) algorithm, which is proposed by Liu, works faster than Monte-Carlo [9] which is a greedy algorithm based on simulation. An ACO algorithm is a heuristic algorithm based on a parameterized probabilistic model [10]. By using artificial ants we can find high quality solutions in fewer iterations. Advanced research says that metaheuristic approaches give better results than greedy approach, deterministic approach and stochastic approach.

3 Cuckoo Breeding Behaviour and L'evy Flight

The Cuckoo is famous for its aggressive reproductive system. The cuckoo has a quality that it lays its eggs in the nests of another community. Some species of cuckoo like Guira and Ani lay their eggs in communal nests [11]. If a surrogate parent knows that the eggs are not their own, then they either destroy those eggs or the whole nest and make a new nest.

Tapera cuckoos lay their eggs in host nests whose pattern of eggs and color is similar. In this way, the probability of discovering the eggs is reduced ensuring that reproductivity is not reduced.

Parasitic cuckoos lay their eggs where the surrogate parent just laid their eggs. Cuckoo's egg has a property that it hatches slightly earlier than the surrogate's egg. Once the first cuckoo egg hatches, surrogate parents take the first action instinctively to destroy the host eggs by blindly throwing out the eggs from the nest. In this way, the cuckoo chicks share of food supplied by the surrogate bird would increase. It can be simply said that the cuckoo has a tendency to move towards the survival of fitness.

L'evy flights are a random walk whose step length is distributed according to the levy distribution. Levy distribution [12–16] follows the power of law formula. The Levy distribution formula Eq. (1) is as follows:

$$Levy \sim u = t^{-\lambda} \tag{1}$$

To generate random numbers, L'evy flights consist of two steps. First, the step length, which is taken from L'evy distribution is determined. Second, the direction of the step is determined according to a uniform distribution.

4 Cuckoo Search Algorithm

Cuckoo search algorithm is a search algorithm which is totally based on the breeding pattern of cuckoos. There are three basic assumptions [17] to implement cuckoo search which is described as follows:

(1) At any point of time, a cuckoo can lay only one egg in a nest which is selected randomly.
(2) The best nest, which has good quality of eggs will go to the next generation.
(3) Number of host nests should be fixed and the probability of discovering an egg is Pa. Pa specifies the fraction of n nests that is replaced by good quality of new nests.

Each nest represents a solution. In each iteration the best nest is preserved and new nests which has a potentially better solution take place of worse nests. To generate a new solution, L'evy flight is performed according to Eq. (2):

$$x_i^{(t+1)} = x_i^t + \alpha \oplus L'evy(\lambda) \tag{2}$$

L'evy flight by random walk gives more efficient results. Because of randomization and not being trapped in local search, Cuckoo Search gives a globally optimal solution.

5 Proposed Method

Traditionally, a social network can be represented by a graph. In a graph, each node represents people and each edge represents the connectivity between them. We consider each node either in active mode or inactive mode. A node has a tendency to become active from inactive mode, but not vice versa. In active mode, a person knows about the information and in inactive mode person is not aware of it.

Our proposed model is based on discrete timestamp. Suppose one person knows the information at a certain timestamp, then this information will spread to his friend by the next timestamp, that is, within the time period between these two timestamps.

Yang [18, 19] proved that cuckoo search is efficient for finding the global optimal in an uncertain environment. Now, the challenge is how to design the cuckoo search for the influence maximization problem. The Solution for this problem is designed in the following manner:

5.1 Nest Representation

In influence maximization, nest represents the solution in the form of randomly chosen K nodes of the graph.

5.2 Initial Solution, Step Size and Stop Criteria

To implement Cuckoo Search, first we have to initialize the nest on the basis of some heuristic. To implement this problem three heuristic approaches are considered which is described as follows:

(1) Degree centrality heuristic approach—Degree centrality is defined as the in-degree of a vertex. But in this influence maximization, out-degree of a node is considered to spread the information to its neighbors.
(2) Distance centrality heuristic approach—The distance centrality of a node is the average distance to all other nodes from that node.

(3) Simulated influence heuristic approach—In this approach, random number of nodes are taken without any consideration and the nest is initialized with these nodes.

Step size (α) is related to the scale of the problem. Mostly, $\alpha > 0$. For influence maximization problem, step size for cuckoo can be defined as the distance between the nest and the best nest at its generation. To calculate α, Eq. (3) is used.

$$\alpha = x_i - x_{best} \tag{3}$$

The distance between two nests i and j is the difference of influence between these nodes. The step size is calculated in a biased manner. In case of the best nest, no change occurs. If $x_i = x_{best}$ then alpha will be zero.

Stop criteria for this problem is the difference between the current best solution and a previous best solution. If this difference is less, it means that influence is not varying much from the current best solution to the previous best solution so that the unnecessary time taken to calculate influence in further iterations is saved. This criterion is considered in order to reduce the computation time of the algorithm. Stopping criteria values will vary according to the scale of the data set and the timestamp. The greater the time stamp, higher the value of the stopping criteria.

5.3 Pseudo Code for Influence Maximization in Social Networks

Cuckoo Search Diffusion Model works according to the following pseudo code.

```
begin
Objective function is influence(I)
Initialize stopping criteria m according to scale of problem
Initialize host nests according to degree heuristic/ distance heuristic/ simulated
heuristic approach
while difference of current best nest and best nest in previous iteration is
greater than m do
    Find cuckoo by L'evy Flight
    Evaluate it's influence(Ic) for timestamp t
    Select a random nest(j)
    Evaluate it's influence(Ij) for timestamp t
    if (Ic > Ij) then
    |   nest j is replaced by the new solution
    end
    A fraction Pa of worse nests are destroyed and new nests are built
    Store the solution which has maximum influence
    Rank the solution based on influence
    end
end
Display the best nest which has k number of nodes and its influence
end
```

6 Experimental Results and Evaluation

For the experiments, a real world data set is used. We evaluated the efficiency of the proposed approach by doing experiments on a real world data set which is known as Wiki-Vote. This data consists of 7115 nodes and 1,03,689 edges. A network is constructed as a directed graph in which each node represents Wikipedia users and an edge between node i and node j denotes that user i votes for user j. Pa is equal to 0.25 is considered because for this value cuckoo search gives better results [17] than other values. Three heuristic approaches are used for initializing the nest. Initially, each node is in inactive mode.

6.1 Case 1: Simulated Heuristic Approach

In a simulated heuristic model, each nest takes K number of random nodes and the nest is initialized. At this time, these K nodes become in active mode. After that, cuckoo search is applied for finding the seed set to maximize the influence. Figure 1a, b shows the running time plotted against seed set size at different

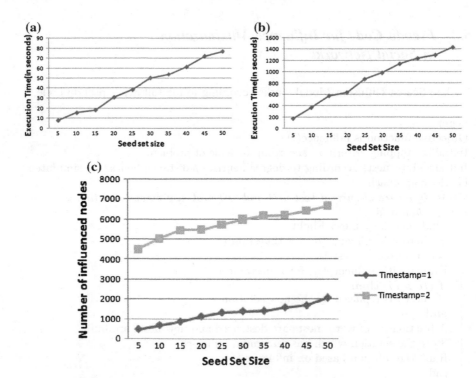

Fig. 1 Simulated heuristic approach. **a** Timestamp = 1. **b** Timestamp = 2. **c** Total number of influenced nodes

timestamps. Figure 1c shows number of influenced nodes plotted against seed set size at different timestamps after the diffusion process. As shown in Figs. 1c, 2c and 3c, when seed set size is increased, number of influenced nodes increases because the number of friends of seed set member will increase. For timestamp = 2 numbers of influenced nodes will be more than when timestamp = 1 because the number of friends of friends is more than the number of friends.

6.2 Case 2: Degree Centrality Heuristic Approach

In the degree heuristic model, the degree of each node in the network is calculated and sorted in decreasing order of degree of the node. Some nodes (in our experiment we took 500 nodes) which have maximum degree are considered and each nest is initialized with a K number of random nodes among the selected nodes. After that, cuckoo search is applied for finding the seed set to maximize the influence. Figure 2a, b shows the running time plotted against seed set size at different timestamps according to this model.

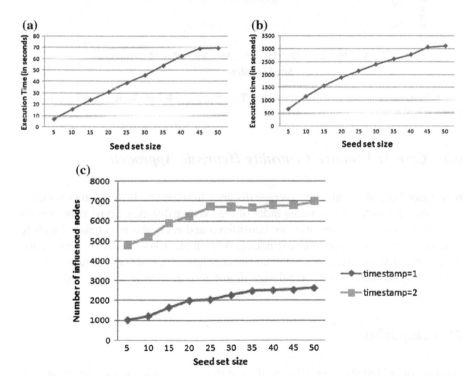

Fig. 2 Degree centrality heuristic approach. **a** Timestamp = 1. **b** Timestamp = 2. **c** Total number of influenced nodes

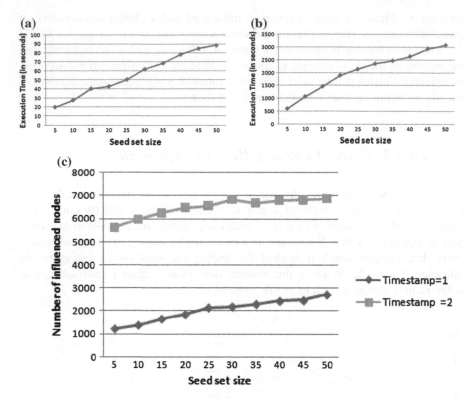

Fig. 3 Distance centrality heuristic approach. **a** Timestamp = 1. **b** Timestamp = 2. **c** Total number of influenced nodes

6.3 Case 3: Distance Centrality Heuristic Approach

In distance heuristic model, the average distance from one node to all other nodes is calculated and sorted in ascending order of the average distance. Some nodes which have minimum average distance are considered and each nest is initialized with K random nodes among the selected nodes. After that,cuckoo search is applied for finding the seed set to maximize the influence. Figure 3a, b shows the running time plotted against seed set size at different timestamps according to this model.

7   Comparison

On the basis of the above results, the three different heuristic models are compared. The maximum degree approach and the minimum distance approach are considered for baseline comparison. In maximum degree approach, K nodes which have maximum degree are selected and nodes influenced by these nodes are found. Similarly, in minimum distance approach, K nodes with minimum distance to other nodes are selected and after selecting such nodes, the nodes which are influenced by these selected nodes at different timestamp values are found.

Consider an example to understand the issue with the maximum degree approach. Suppose there is a network where around half of the nodes create a complete graph between them, but the remaining nodes are connected to each other sparsely and hence have low out-degree. If we take nodes which have a maximum out-degree, then nodes which make the complete graph would be considered, but by taking these nodes influence will not be maximized because the remaining nodes which are sparsely connected will not get influenced by them.

Consider an example to understand the issue with the minimum distance centrality approach. Suppose there is a network where some nodes are in the center, but degree of these nodes is small. As distance increases from the center, degree of

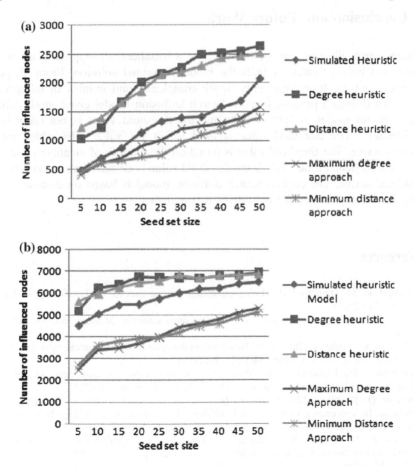

Fig. 4 Comparison between three heuristic approaches. **a** Timestamp = 1. **b** Timestamp = 2

nodes increases. In this case also by taking the center nodes as a seed set, we will not get a good number of influenced nodes.

Thus we can say that there is no specific criteria by which we can always get good results since it depends upon the structure of the network, which is uncertain. To solve these problems, randomization approach gives better results than a certain fixed criteria. Figure 4a, b shows that the three heuristic models give better results than the maximum degree approach and minimum distance approach. Hence it also proves that the cuckoo search diffusion model gives better results. In case of space requirement, CSDM takes memory, mainly to store the network and does not require much memory for running the model. Because of the above reasons, we conclude that Cuckoo Search Diffusion Model works well as compared to other methods.

8 Conclusion and Future Work

Cuckoo search diffusion model is based on a metaheuristic approach which is known as Cuckoo Search. It finds the globally optimal solution. From all perspectives, such as, time complexity, space complexity and number of influenced nodes after diffusion process, Cuckoo Search Diffusion Model gives good results. In the present model, weight of links and the threshold value is not taken into consideration. The threshold value is used to set the value after which node becomes active. The threshold value is based on the number of neighbors that are active at a particular time. Only if the threshold value is crossed then a node can be considered active. The cuckoo Search diffusion model is based on discrete time-stamp hence our future work is to make it continuous.

References

1. Richardson, M., Domingos, P.: Mining knowledge-sharing sites for viral marketing. In: KDD, pp. 61–70 (2002)
2. Bass, F.: A new product growth model for customer durables. Manage. Sci. **15**, 215–227 (1969)
3. Mahajan, V., Muller, E., Bass, F.M.: New product diffusion models in marketing: a review and directions for research. J. Mark. **54**(1), 1–26 (1990)
4. Domingos, P., Richardson, M.: Mining the network value of customers. In: Seventh International Conference on Knowledge Discovery and Data Mining (2001)
5. Kempe, D., Kleinberg, J.M., Tardos, E.: Maximizing the spread of influence through a social network. In: Proceedings of the 9th ACM SIGKDD Conference on Knowledge Discovery and Data Mining, pp. 137–146 (2003)
6. Leskovec, J., Krause, A., Guestrin, C., Faloutsos, C., VanBriesen, J., Glance, N.: Cost-effective outbreak detection in networks. In: KDD, pp. 420–429 (2007)
7. Chen, W., Wang, Y., Yang, S.: Efficient influence maximization in social networks. In: KDD, pp. 199–208 (2009)

8. Chen, W., Wang, Y., Yang, S.: Scalable influence maximization for prevalent viral marketing in large-scale social networks. In: KDD, pp. 1029–1038 (2010)
9. Liu, B., Cong, G., Xu, D., Zeng, Y.: Time constrained influence maximization in social networks. In: Proceedings ICDM, Washington, DC, USA, pp. 439–448 (2012)
10. Yang, W.S., Weng, S.X., Guestrin, C., Faloutsos, C., VanBriesen, J., Glance, N.: Application of the ant colony optimization algorithm to the influence-maximization problem. Int. J. Swarm Intell. Evolut. Comput. $1(1)$, 1–8 (2012)
11. Payne, R.B., Sorenson, M.D., Klitz, K.: The Cuckoos. Oxford University Press, Oxford (2005)
12. Brown, C., Liebovitch, L.S., Glendon, R.: Levy flights in Dobe Ju/hoansi foraging patterns. Human Ecol. 35, 129–138 (2007)
13. Pavlyukevich, I.: Levy flights, non-local search and simulated annealing. J. Comput. Phys. 226, 1830–1844 (2007)
14. Pavlyukevich, I.: Cooling down Levy flights. J. Phys. A: Math. Theory 40, 12299–12313 (2007)
15. Barthelemy, P., Bertolotti, J., Wiersma, D.S.: A Levy flight for light. Nature 453, 495–498 (2008)
16. Reynolds, A.M., Frye, M.A.: Free-flight odor tracking in Drosophila is consistent with an optimal intermittent scale-free search. PLoS ONE 2, e354 (2007)
17. Yang, X.-S., Deb, S.: Cuckoo search via Levy flights. In: Proceedings of the Nabic—World Congress on Nature and Biologically Inspire Computing, pp. 210–214 (2009)
18. Yang, X.S.: Nature-Inspired Metaheuristic Algorithms, 2nd edn. University of Cambrige, Luniver Press, United Kingdom (2010)
19. Yang, X.S., Deb, S.: A engineering optimisation by cuckoo search. Int. J. Math. Modell. Numer. Optim. $1(4)$, 330343 (2010)

Sociopedia: An Interactive System for Event Detection and Trend Analysis for Twitter Data

R. Kaushik, S. Apoorva Chandra, Dilip Mallya,
J.N.V.K. Chaitanya and S. Sowmya Kamath

Abstract The emergence of social media has resulted in the generation of highly versatile and high volume data. Most web search engines return a set of links or web documents as a result of a query, without any interpretation of the results to identify relations in a social sense. In the work presented in this paper, we attempt to create a search engine for social media datastreams, that can interpret inherent relations within tweets, using an ontology built from the tweet dataset itself. The main aim is to analyze evolving social media trends and providing analytics regarding certain real world events, that being new product launches, in our case. Once the tweet dataset is pre-processed to extract relevant entities, Wiki data about these entities is also extracted. It is semantically parsed to retrieve relations between the entities and their properties. Further, we perform various experiments for event detection and trend analysis in terms of representative tweets, key entities and tweet volume, that also provide additional insight into the domain.

Keywords Social media analysis · Ontology · NLP · Semantics · Knowledge discovery

R. Kaushik (✉) · S. Apoorva Chandra · D. Mallya · J.N.V.K. Chaitanya ·
S. Sowmya Kamath
Department of Information Technology, National Institute of Technology Karnataka,
Surathkal, Mangalore 575025, India
e-mail: kaushik1603@gmail.com

S. Apoorva Chandra
e-mail: apoorvachandras@gmail.com

D. Mallya
e-mail: dmallya93@gmail.com

J.N.V.K. Chaitanya
e-mail: chaitanya.jnvk@gmail.com

S. Sowmya Kamath
e-mail: sowmyakamath@nitk.ac.in

© Springer India 2016   63
A. Nagar et al. (eds.), *Proceedings of 3rd International Conference
on Advanced Computing, Networking and Informatics*, Smart Innovation,
Systems and Technologies 44, DOI 10.1007/978-81-322-2529-4_6

1 Introduction

Modern day social media are already a valuable source of information that exhibit the inherent qualities of volume, variety, veracity and velocity, making social media analysis a big data analytics problem. Corporations use social media to map out their demographics and analyze customer responses. People generally rely on the opinion of their peers on social media to shape their own opinions. Social media, then, can be considered to be a strong biasing factor in the era of Web 2.0 [1]. Various social networking sites also act as a source of information for its users and for analytics applications alike. Dissemination of news and trends has been proven to be faster on social media than by conventional media or even news websites. Twitter and Facebook tend to be the most commonly used Web 2.0 services on the Web and a big part of the fabric of web savvy population's daily life.

The widespread acceptance of social media as a source of useful data is based on the fact that, the interactions made by the users are of great social significance and are highly dependent on both the user's sentiments and also on peer consensus. Taking micro-blogging as an example, Twitter currently has over 280 million active users and over 500 million tweets are sent per day. Using such a high volume and velocity data to provide useful information has become a tedious task bordering on the impossible. Automating this process through intelligent, semantic-based information extraction and knowledge discovery methods is therefore increasingly needed [2]. This area of research merges methods from several fields, in addition to semantic web technologies, for example, Natural Language Processing (NLP), behavioral and social sciences, data mining, machine learning, personalization, and information retrieval [3].

The problem of making sense of the data we get from various websites like Twitter is currently an area of huge research interest. It is near impossible to perform manually and we need to utilize various big data and semantic web techniques to gather usable information from the Twitter feed. The proposed solution discussed in this paper, relies on the concept of an ontology to understand the relation between various terms used in the tweets and for enabling automatic analysis of the data. The ontology is constructed from the tweet dataset and the relations between terms are mapped. We then use the constructed ontology to provide statistics and usable information from the data to serve an user inputted search query.

The rest of the paper is organized as follows—In Sect. 2, we discuss various related work in the literature and also some relevant ontologies that currently exist for social media representation. Section 3 discusses the proposed methodology and the various components of system and the development process. In Sect. 4, we present experimental results observed and discussion, followed by conclusion and future work.

2 Related Work

The different kinds of social media coupled with their complex characteristics, make semantic interpretation extremely challenging. State-of-the-art automatic semantic annotation, browsing, and search algorithms have been developed primarily on news articles and other carefully written, long web content. In contrast, most social media streams (e.g. tweets, Facebook messages) are strongly inter-related (due to the streams being expressions of individual thought and experiences), highly temporal, noisy, bursty, short, and full of colloquial references and slang, thus making it difficult to generate accurate analysis, without semantic and linguistic parsing.

Iwanaga et al. [4] presented a ontology based system for crisis management during crucial time periods like during an earthquake. The requirement for an ontology in that scenario is to ensure the proper evacuation of the victims. Their dataset consisted of tweets on the earthquake in Japan's Tohoku region in Japan. Several researchers have concentrated on the issue of learning semantic relationships between entities in Twitter. Celik et al. [5] discuss the process of inferring relationships between the various entities present in a tweet such as timestamp, unique id and # tags and @ tags. There exist many methods for inferring relationships [6, 7] such as using already existing ontologies and web documents crawling, using a bag of words approach based on the term frequency and inverse document frequency of the terms involved and using term co-occurrence based strategies. They concluded that co-occurrences based approaches result in the highest precision.

Bottari et al. [8] proposed an ontology, which has been developed specifically to model relationships in Twitter, especially linking tweets, locations, and user sentiment (positive, negative, neutral), as extensions to the Socially-Interlinked Online Communities (SOIC) ontology. Unlike SIOCT, Bottari identifies the differences between retweets and replies. Geo-locations (points-of-interest) are represented using the W3C Geo vocabulary, and hence can be used to develop location-based logic and relations. For interlinking Social Media, Social Networks, and Online Sharing Practices, DLPO (LivePost Ontology) is used, which provides a comprehensive model of social media posts, going beyond Twitter. It is strongly grounded in fundamental ontologies, such as FOAF, SOIC, and the Simple Knowledge Organization System (SKOS). DLPO models personal and social knowledge discovered from social media, as well as linking posts across personal social networks. For modeling tag semantics, Meaning-Of-A-Tag (MOAT) ontology can be used allowing users to define the semantic meaning of a tag by linking Open Data and thus, manually creating semantic annotations within the social media data itself. The ontology defines two types of tags: global (i.e. across all content) and local (referring to a particular tag on a given resource). MOAT can be combined with SIOCT to tag microblog posts.

3   System Development

3.1   Data Cleaning

Figure 1 depicts the various processes of the proposed system. Since it is intended
for the use of company officials for gathering the social media users' opinion about
their brands, we need social media data, for which we chose Twitter. Tweets
containing particular keywords were extracted from Twitter using a tool called
Sysomos [9], which also provides the URL, associated timestamp, location, sentiment and the user who tweeted. We obtained as average of about 40,000 tweets on
each of the datasets—namely Apple's *iphone6*, Chinese Manufacturer, *Xiaomi's*
smartphones and Motorola's *Moto* smartphone.

As the first step to cleaning, we filtered the tweets to retain only English language tweets, as natural language processing is supported for English only, currently. This was done by analyzing the stopwords found in the tweet and assigning
the tweet a ratio for each language that would indicate its chance of belonging to
that language. After the ratio was assigned, the tweets in English obtained the
maximum scores and all tweets with scores above a threshold were selected. Tweets
with too many # and @ are considered to be spammy tweets and are automatically
eliminated during stopword detection technique. The remaining English tweets are
collected along with timestamp and stored.

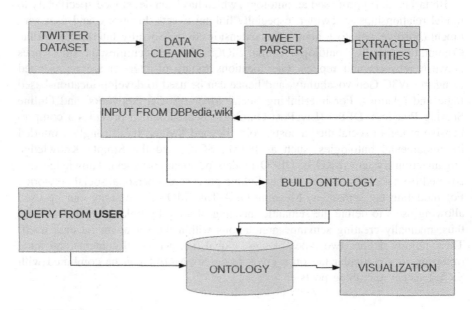

Fig. 1 Workflow of the proposed system

3.2   Data Preprocessing

After tweet cleaning, tweets are sorted according to timestamp and are counted on a daily basis. The primary requirement of this application is that the company executives must know about the current social media buzz about their products. This can be best indicated by computing the volume of tweets generated during the day and plotting the volume versus time. The granularity level was maintained as a day because lesser than that would result in the appearance of many peaks which would be difficult for the marketer to analyze and many of them may not be interesting events as well.

The system as to provide a search mechanism that can help the marketer to research and analyze the topics that were being discussed in the social media. In order to build such a search engine, we used an novel approach using an ontology constructed from the tweets to capture the trends and relationships between events. An ontology is a collection of entities and relationships between them. The keywords or entities extracted from the tweets are to be represented in the constructed ontology. These entities or keywords were extracted from the tweets using the CMU tweet parser [10, 11].

The CMU tweet parser tokenizes tweets and returns the parts of speech tagging for each of the tweets that are given as an input. The confidence associated with each part of speech was also returned after the tagging [12]. The output of the parser was stored in a comma separated file. A separate parser was required for this purpose because a POS tagger that works on formal sentences cannot be used with tweets, where many local abbreviations are used due to the limitation in tweets (only 140 characters can be present at the maximum in a tweet).

After POS tagging, the entities that were used in the further analysis are *named entities, hashtags* and *nouns* as these conveyed the maximum information about the topic. Figure 2 shows the percentage of each in the resultant processed dataset. These entities were extracted from the tweets and a frequency distribution was generated for all extracted entities. From this frequency distribution, the top 100 keywords were chosen as final entities based on which the ontology would be built. The frequency of occurrence of these entities were also stored.

After selecting the top 100 keywords as entities for constructing the ontology, the next phase is to capture any relationships between these entities. The ontology can be complete only if relationships exists between the entities. Since, the tweet data retrieved from Twitter cannot be considered as a reliable published source of information in the standard sense, it is imperative to establish relationships between documents using other techniques. We chose Wikipedia as a source of reliable published information, available as a structured dataset, to recognize the relationships between entities. We systematically retrieved documents from Wikipedia, which are then stored in a directory, for later access during the ontology building process.

Fig. 2 Entities extracted
using CMU Tweet Parser

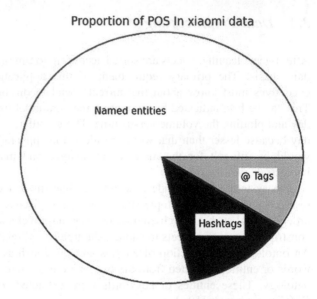

3.3 *Build Ontology Module*

The ontology is a graph or a network of nodes (entities) that are connected by edges (relationships identified by relationship names). The relationships between the extracted entities are inferred using Wikipedia, DBpedia and web documents for the extracted entities. The ontology is built using Web Ontology Language (OWL), as it is a standardized ontology language based on description logic. The generated ontology is saved in an ontology base which is one of the main underlying components for the search engine to be built.

Currently, we provided a basic search interface, that works using the constructed ontology saved in the ontology base. The user can enter a query which will be processed to retrieve documents regarding the topic on which the ontology was developed, corresponding to most recent events and trends. The results of the search engine will depend upon the type of the dataset.

4   Results and Analysis

We considered two different tweet datasets pertaining to the launch of two popular products, Motorola Moto series of mobile phones and Apple's iPhone 6. The *Moto Dataset* contains 40,000 tweets collected during the period Sep 1st to Sep 9th 2014 and the *iPhone6 dataset* contains about 35,000 tweets collected during the period Sep 6th to Sep 12th 2014. As seen from Fig. 3, spikes were detected in the tweet volume versus time graph on Sep 2nd to 5th, i.e., during the time when Moto

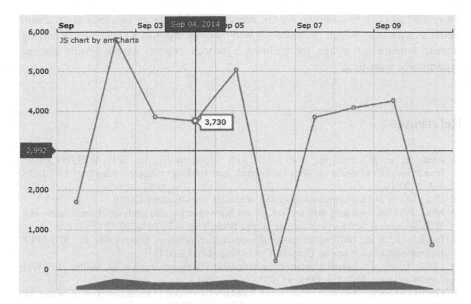

Fig. 3 Tweet volume versus time for the Motorola Dataset

released the next generation of the Moto X and the Moto G smartphones and Moto 360, a wearable smartwatch Android device. This is an indication that the system was able to detect events that pertain to certain events interesting to the target user, that is release of a product and the hype surrounding it during the initial days.

Certain other interesting observations could be made from the work that was carried out. Out of all the entities generated by the CMU tweet parser, those corresponding to the nouns, hashtags and the named entities were the most useful. Verbs, even though may be useful in establishing relationships between the entities can be obtained from Wikipedia and other the web documents that were used to infer relationships. Hashtags generally form a lesser proportion of the entities after parsing the tweets, so some weightage can be given to hashtags, as information gain from hashtags can be significant.

5 Conclusion and Future Work

In this paper, we proposed an interactive system for event detection and trend analysis of Twitter data. It is intended to automatically detect trends in social media regarding certain real world events like launch of new products, advertising campaigns etc., to aid companies to analyze the effectiveness of their marketing approach or the mood of the consumer through their tweets on Twitter. Currently, available ontologies are mostly based on web documents or direct information from sources as a result of which, they cannot be contemporary to any event/topic they are related to [13]. In contrast, we are attempting to build an ontology from social media as a result of which the ontology will be contemporary to currently occurring events, a major advantage for building a search engine, whose results change dynamically with time.

References

1. Asur, S., et al.: Predicting the future with social media. In: 2010 IEEE/WIC/ACM International Conference on Web Intelligence and Intelligent Agent Technology (WI-IAT), vol. 1 (2010)
2. Ritter, A., et al.: Unsupervised modeling of twitter conversations (2010)
3. Mika, P.: Flink: semantic web technology for the extraction and analysis of social networks. Web Semantics: Sci. Services Agents World Wide Web 3(2), 211–223 (2005)
4. Iwanaga, I., et al.: Building an earthquake evacuation ontology from twitter. In: 2011 IEEE International Conference on Granular Computing (GrC) (2011)
5. Celik, I., et al.: Learning semantic relationships between entities in twitter. In: Web Engineering, pp. 167–181. Springer (2011)
6. Ozdikis, O., et al.: Semantic expansion of hashtags for enhanced event detection in twitter. In: 1st International Workshop on Online Social Systems (2012)
7. Owoputi, O., et al.: Improved part-of-speech tagging for online conversational text with word clusters. In: HLT-NAACL, pp. 380–390 (2013)
8. Celino, I., et al.: Towards bottari: using stream reasoning to make sense of location-based micro-posts. In: The Semantic Web: ESWC 2011 Workshops, pp. 80–87. Springer (2012)
9. Sysomos.com: Sysomos heartbeat
10. Gimpel, K., et al.: Part-of-speech tagging for twitter: annotation, features, and experiments. In: 49th Annual Meeting of the Association for Computational Linguistics: Human Language Technologies: short papers, vol. 2, pp. 42–47 (2011)
11. Kong, L., et al.: A dependency parser for tweets. In: International Conference on Empirical Methods in Natural Language Processing, Doha, Qatar (2014)
12. Derczynski, L., et al.: Twitter part-of-speech tagging for all: Overcoming sparse and noisy data. In: RANLP, pp. 198–206 (2013)
13. Zavitsanos, E., et al.: Gold standard evaluation of ontology learning methods through ontology transformation and alignment. IEEE Trans. Knowl. Data Eng. 23(11), 1635–1648 (2011)

Design and Implementation of a Hierarchical Content Delivery Network Interconnection Model

Sayan Sen Sarma and S.K. Setua

Abstract Content Management System (CMS) is an infrastructure for efficient distribution, organization, and delivery of digital content. It is desirable that the content must be successfully delivered regardless of the end users location or attachment network. For the end to end delivery of content, a virtual open content delivery infrastructure is formed by interconnecting several CDNs. In this paper, we focus on Content Delivery Network Interconnection. An efficient Hierarchical CDNI Architecture, named as HCDNI, is proposed to reduce the limitations of CDNIs. Next, a content distribution and redistribution scheme is proposed so that the searching time and the round trip time for the content delivery can be minimized. Finally, analysis and simulation studies show that proposed algorithm results significant improvement in terms of data routing, path selection, content distribution and redistribution.

Keywords Content management · Content delivery network · Content delivery network interconnection · Hierarchical CDNI architecture

1 Introduction

Content Management System (CMS) is an infrastructure for efficient organization, distribution and delivery of digital content [1] with associated management. Content Delivery Networks (CDN) are overlay networks designed to deliver content to web users with high availability and performance. Web users demand faster and higher quality of services from their media hosting companies. CDN is

S.S. Sarma (✉)
Advanced Computing and Microelectronics Unit, Indian Statistical Institute, Kolkata, India
e-mail: sayansensarma@gmail.com

S.K. Setua
Department of Computer Science & Engineering, University of Calcutta, Kolkata, India
e-mail: sksetua@gmail.com

© Springer India 2016
A. Nagar et al. (eds.), *Proceedings of 3rd International Conference on Advanced Computing, Networking and Informatics*, Smart Innovation, Systems and Technologies 44, DOI 10.1007/978-81-322-2529-4_7

71

basically a large distributed system consisting of multiple servers deployed in multiple data centers across the Internet. Now-a-days a large fraction of the Internet content is served by CDN. It is generally desirable that delivery of content must be successful regardless of the end users location or attachment network. This is the main inspiration to interconnect the stand alone CDNs to have a virtual open content delivery infrastructure for the end to end delivery of the content. The CDNI Working Group (CDNIWG) [1] allows the interconnection of separately adminis- tered CDNs to support end-to-end delivery of content from CSPs through multiple CDNs and ultimately to end users (via their respective User Agents).

In this paper we go through the existing CDNI model and their various features and functionalities as proposed in IETF [1–4], which introduce the concept of some interfaces for efficient interconnection of several CDNs. We have also identified the limitations and the less specified areas of the interfaces of these existing models. Next, we focus on Content Delivery Network Interconnection and an efficient Hierarchical CDNI Architecture, named as HCDNI, is proposed to reduce the limitations of CDNIs. In HCDNI, the end to end delivery of content is done through a virtual open content delivery infrastructure, formed by efficient interconnection of several CDNs. A content distribution and redistribution scheme is also proposed so that the searching time and the round trip time for the content delivery can be minimized. Finally, analysis and simulation studies show that proposed algorithms cause significant improvement of system performance in terms of data routing, path selection, content distribution and redistribution.

The rest of the paper is organized as follows. Section 2 defines the existing model and introduces the basic scheme of interconnection between different CDNs. Section 3 presents the Hierarchical model for CDNI. Redistribution of Content to reduce the round trip time is defined in Sect. 4. Finally, Sect. 5 concludes the paper.

2 Related Work

2.1 Existing CDNI Model

To constitute the problem space of CDN Interconnection the concept of CDNI interfaces are introduced along with the required functionality of each interface in RFC6707 [2]. These interfaces are named as CDNI Control interface, CDNI Metadata interface, CDNI Request Routing interface, CDNI Logging interface. Basic functionality of the above mentioned CDNI interfaces are as follows:

1. CDNI Control interface: This interface allows establishment, updation or ter- mination of CDNI interconnection as well as bootstrapping and configuration of the other CDNI interfaces etc. Some up-gradation of the existing CDNI model is suggested in the active Internet draft proposed by CDNI working group [4], where the Control Interface is divided to introduce another interface named

Trigger Interface. It performs request specific actions, communications to be undertaken.

2. CDNI Request Routing interface: This interface allows several interconnected CDNs to communicate through two modules CDNI Request Routing interface, CDNI Footprint and Capabilities advertisement interface.

3. CDNI Metadata interface: This interface ensures that CDNI Metadata can be exchanged across the interconnected CDNs. It enables a Downstream CDN to obtain CDNI Metadata from an Upstream CDN so that the Downstream CDN can properly process and respond.

4. CDNI Logging interface: Logging data consists of potentially sensitive information (which End User accessed which media resource, IP addresses of End Users, potential names and subscriber account information, etc.). These log records are moved between CDNs through logging interface.

2.2　The Constraints of the CDNI Model

In this paper as a part of the development of the CDNI interfaces we have identified some desired features which were missing in the existing model as described in IETF [1–4]. The features are as following:

1. Routing: Efficient selection of path for end to end data delivery or selection of proper surrogates is very important for reducing the round trip time of the content delivery, which leads to a better system performance [5].

2. Content distribution and redistribution: To minimize the end to end delivery cost and betterment of the system performance [6] by means of both reduced searching time and round trip time for the content we need to find efficient distribution of the contents. It is also the idea behind replication of contents [7, 8]. Some times the content need to be redistributed also for the improvement of system performance. It is more important to identify when we need content redistribution because it involves a cost.

3　Proposed Model

In this paper we propose an efficient architectural model, named as Hierarchical CDNI model (HCDNI). The proposed model is an advancement of the existing CDNI frameworks [2–4] where we focus on Data Routing, Path Selection, Content Distribution and Content Redistribution.

3.1 HCDNI Model

The basic idea of the HCDNI is to cover the whole geographical area of the CDN users with minimum hazards and minimum service cost. In this model, the tasks of identification of CDNs and interconnection among them is done using hierarchical structure [9, 10] as shown in Fig. 1. The hierarchy contains

- A CDN controller for each CDN.
- Clusters of CDNs.
- A Cluster Controller for each cluster.
- A Master Cluster Controller.

Now two types of awareness [11] is to be provided as Content Awareness and Network Awareness. These awarenesses play a crucial role in case of load balancing, path selection and data delivery, content distribution and redistribution etc. They are implemented through several strategies like Meta Data Exchange, Content registration.

3.2 Terminologies

1. Content Awareness: In our proposed model Content Awareness is achieved by assigning a location independent content ID for each and every item present in the network. In contrast to other existing methods [5, 12], we have used a 128 bit content ID to solve the purpose which is created by MD5 hash function.

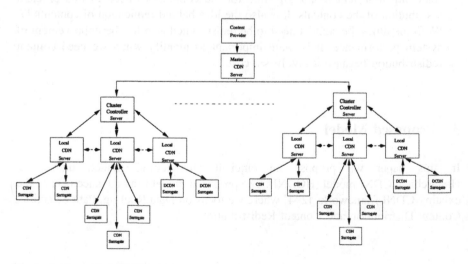

Fig. 1 A hierarchical CDNI model

2. Content Registry: It keeps track of every data item present in the network by storing the content ID along with the content location and Content Meta Data i.e. other additional properties that are useful while a data being searched or processed [12]. It is implemented by data base partitioning and the partition is done according to the hierarchical distribution of the CDNI structure.

3. Conventional CDN architectures: There are mainly two standard approaches [9] as

   (a) Commercial Architecture: It is basically client-server architecture.
   (b) Academic Architecture: It is peer-to-peer architectures and for sharing of computer resources (content, storage, CPU cycles) by direct exchange.

4. Content Resolution and delivery: It is the selection of the data item among its various replicas for the best possible delivery. It depends on various parameters like network bandwidth and availability, congestion, load factor.

5. Conversion of content: The user request may come for different types or formats of the same content. So we need to replicate all the available types of a particular content. It will result better user experience but not always a cost effective solution. Hence, a content converter module is introduced in the controller nodes to convert the content from one format to another. Content conversion is done depending on factors like navigation cost, delivery cost, conversion cost etc.

6. Data base Partitioning: It is splitting of a database table into a number of overlapping tables with smaller cardinalities [13]. Each partition is stored in a different physical location. Partitioning enhances the performance, manage-ability, and availability of a wide variety of applications and helps to reduce the total cost of ownership for storing large amounts of data.

7. Meta Data: Two types of Meta Data are introduced in HCDNI model as

   (a) Content Distribution Meta Data—It is mainly responsible for conveying the message about the distribution of data items to the upstream CDN and downstream CDNs.
   (b) Content Description Meta Data—It is mainly used to describe the data item i.e. its genre, type, rating etc.

8. Intra domain and Inter domain Routing: In the clustered architecture routing is of two types Intra domain routing and Inter domain routing. Intra domain routing path is calculated by the topology and the availability of Servers. But Inter domain routing is mainly done by peering of CDNs [14].

3.3 Procedure to Create Hierarchical CDNI Structure

Step 1: Depending on available set of CDNs an initial set of minimal CDNs are identified using well accepted statistical distribution pattern and commercial architecture to cover the whole region of interest. Uncovered region is countered by academic architecture. Each CDN has its own controller named as CDN controller.

Step 2: Using standard clustering technique CDNs are grouped into several clusters having geographic references. Each cluster is associated with one cluster controller and for synchronization among clusters there is one master controller at the top of hierarchy. All controllers communicate among themselves using CDNI meta data. The necessary security association among clusters is made for requisite transfer and/or exchange of information of any type and is done by exchanging CDNI meta data.

Step 3: Content service providers are assumed to be under direct control of master controller. Initially contents are statistically distributed all over the network using timeliness of data, location wise usage/request pattern, cost of storage and delivery.

Step 4: When a new content is introduced in the network, it is registered with master controller and stored according to request pattern or on load distribution policy in absence of requests. Content registry is updated accordingly using content distribution meta data.

Step 5: The records of the contents, available in the network are stored in content registry of master controller and distributed along the hierarchy using database partitioning. The cluster controller stores items, kept in its cluster and the CDN controller keeps track of its own items.

Step 6: Any kind of redistribution of data item will take place from the lowest level of CDNI structure and it will be reported to the next higher levels of the hierarchy and the needful modifications will be made. It is also done by exchanging content distribution meta data.

3.4 Routing Mechanism in Hierarchical CDNI Architecture

Step 1: Any request for content is processed by the local CDN controller using its content registry for absence or presence of the desired content.

Step 2: On negative search result, the request is forwarded to cluster controller for successful search. But again on unsuccessful search result, the request is forwarded to master controller for necessary search. All of these done through the exchange of content description meta data.

Step 3: On successful search of content in desired format, the respective controller serves the request immediately. But, if the available content is not in desired format then the controller optimizes between the conversion cost plus delivery costs from identified CDN and the navigation cost plus delivery cost from the CDN where the content is found in desirable format.

Step 4:   In case of requests, forwarded to next higher level of CDN and for multiple copies of the content in the network, the respective controller applies the content resolution technique to serve in request. It will consider the following facts

   (a)   Server Load: that is the numerical representation of the server status.
   (b)   Path length: It can be calculated by using average round trip time of data item as per previous transactions.
   (c)   Band Width: Maximum average supported band width of the path. Etc.

This is how the content is served by the Hierarchical CDNI structure.

4   Redistribution of Content

In our proposed model, initial distribution of content is done by statistical analysis. But the content will loose its relevance with time and usage pattern. As a result the desired content may have to be retrieved from the remotest site, network may be flooded with type-C contents over the times and finally the system performance will be degraded. Hence with time the content need to be redistributed in the network. For a fruitful redistribution policy we have introduced the concept of rank associated with content. It will reflect the relevance of the content. Based on the rank, unused content can be identified and may be removed from the network making place for newer contents. Relevant content and optimal number of copies in the network will improve the performance in terms of searching time, distribution and maintenance costs of the proposed system.

4.1   Ranking of Content

Relevance of content depends on multiple number of factors. For the sake of simplicity in simulating our proposed model, we have considered the most relevant factors for the generation of rank of content. They are- User request pattern, Size factor, Timeliness of the data, Validation of security policy agreement between the CDNs and Component compatibility. We defined the factors as follows:

- Size factor = Data size/Total space available.
- Usage pattern = Number of request/Total request.
- Timeliness factor = Current version/Latest version available.
- Policy factor = Current Year/Year of expire of policy agreement.
- Component compatibility factor = According to the type of content it is decided by statistical analysis. The value is something between 0 and 1.

In our policy Ranks do not have any global existence. It is calculated in the lowest level of the hierarchy and do not have any existence in the higher levels. The weight of the factors in case of rank generation may vary depending on the type of content. Hence, we propose a ranking scheme where software generated easily modifiable weights are used for rank determination.

$$\text{Rank} = (\text{Size factor} * W_1) + (\text{Usage pattern} * W_2) + (\text{Timeliness factor} * W_3)$$
$$+ (\text{Policy factor} * W_4) + (\text{Component compatibility factor} * W_5).$$

where, $\sum_i W_i = 100$.

4.2 Simulation Study and Observations

We have simulated the performance of our proposed methods using C++. Behavior and performance of CDNI infrastructure totally depends on the type of content and type of users. Hence all the statistics we have provided is as per our simulation results. Here, the weight factors for calculation of rank of a content are set to, $W_1 = 25$, $W_2 = 45$, $W_3 = 15$, $W_4 = 10$ and $W_5 = 5$. For the simplicity of simulation we have used 8 bit content ID.

- Observation1: After the initial distribution of data, new data should replace the less relevant data based on their ranks. Simulating the scenario accordingly we have the performance graph as shown in Fig. 2. It shows that filtering the content with increasing rank will consistently decrease the number of hits. It seems to be an unusual behavior but actually it is not because for the new data

Fig. 2 Variation of total number of hits w.r.t. rank

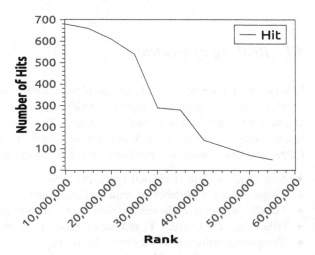

Fig. 3 Variation of total number of hits w.r.t cut-off frequency

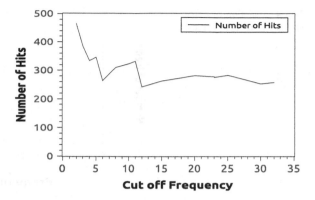

requests we do not have proper usage pattern and the factor determining the usage pattern in the rank is low for them. As a consequence they usually do not score high rank. Hence we need to calculate the usage pattern for the elements currently existing in the system as well the newly requested data items for a window of time T.

- Observation2: Now to identify the potential contents to be stored we need to set a cut-off frequency or frequency threshold value. It is done by the analysis of the request pattern. From the results of simulation studies as shown in Fig. 3 we can observe that the system performance is quite consistent for a frequency threshold value between 12 and 30.

Remark 1 From observation 1 and observation-2 we can conclude that before redistribution of the data items we need to watch them for time T, that is our window size. All the contents including the existing ones and the newly requested ones which passes a threshold limit are the potential candidates to be stored in the end nodes of the CDNI hierarchy.

- Observation3: After selecting the threshold value by statistical analysis the next challenge is to determine a frame size i.e., number of items to be compared with the existing items. By simulating the scenario we can have an estimated frame size as shown in Fig. 4. From Fig. 4 frame size is found to be 5.
- Observation4: After having the frame size and the frequency threshold we have our potential data items which can replace the existing elements. But before that we need to determine the ranks of the elements. Simulation is done by setting the threshold value at 15 and 20 consecutively and setting the frame size to 5 we can have the hit ratio of the system as shown in Figs. 5 and 6.

Remark 2 From Figs. 5 and 6 we observe that percentage of hit of the system is constant and it is above 65 % which is very consistent for the CDNI system. Hence we can conclude that selecting a suitable window size and frame size and then a proper replacement policy can yield a consistent system performance.

Fig. 4 Variation of frame size w.r.t. frequency threshold

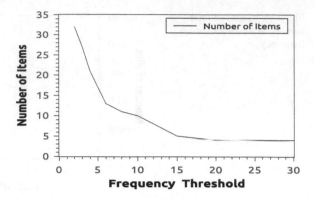

Fig. 5 Variation of number of hits w.r.t. total number of user requests

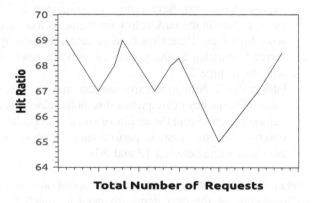

Fig. 6 Variation of total number of hits w.r.t. total requests

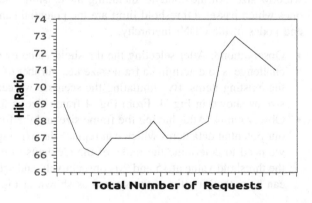

Remark 3 For every end node we need to estimate the frame size, window size and cut off frequency individually. And it will differ for every individual implementation platform and environment.

4.3 Content Redistribution Procedure

From the above simulation study of the system behavior and performance we can propose an efficient content redistribution procedure as follows:

Step 1: First we need to determine a suitable Window size of time (T), frame size (F) and the frequency threshold (f).

Step 2: We need to keep track of usage pattern for the items that belong to the CDN as well the newly requested items.

Step 3: After time T we need to identify the items those cross the frequency threshold value (f). Among those we need to select the F most frequent items.

Step 4: After that we need to calculate rank for all the existing items as well the items, belonging to the frame.

Step 5: According to the allowed storage size the higher ranked items are stored in the CDN. This algorithm needs to be executed in every end CDN of the hierarchical CDNI structure.

5 Conclusions

In this paper, we address the problem on Content Delivery Network Interconnection (CDNI) and proposed an efficient Hierarchical CDNI Architecture, named as HCDNI. In our model the probability of having content in the lowest level of hierarchy is very high and this results faster delivery of the content. Here, unique identification number of the content results content awareness and faster retrieval of content is achieved by the implementation of content registry through database partitioning. Contents format conversion is feasible at CDN level, if necessary. Next, a content distribution and redistribution scheme is proposed so that the searching time and the round trip time for the content delivery can be minimized. Finally, by some analysis and simulation studies we go through some important observations and then proposed a procedure for content redistribution. We also simulated the system with the newly proposed procedure which showed a satisfactory system performance.

References

1. https://datatracker.ietf.org/wg/cdni/character
2. Niven-Jenkins, B., Le Faucheur, F., Bitar, N.: Content Distribution Network Interconnection (CDNI) Problem Statement Internet Engineering Task Force (IETF) Request for Comments: 6707 Category: Informational, Sept 2012. ISSN:2070-1721

3. Bertrand, G., Stephan, E., Burbridge, T., Eardley, P., Ma, K., Watson, G.: Use Cases for Content Delivery Network Interconnection Internet Engineering Task Force (IETF) Request for Comments: 6770 Obsoletes: 3570 Category: Informational, Nov 2012. ISSN: 2070-1721
4. https://datatracker.ietf.org/doc/draft–ietf–cdni–control-tiggers
5. Gritter, M., Cheriton, D.R.: An Architecture for Content Routing Support in the Internet fm gritter, cheritong
6. Wissingh, B.: Content delivery network interconnection footprint versus capability information and exchange. Institute of Informatics at University of Amsterdam, Netherlands
7. Cooper, I., Melve, I., Tomlinson, G.: Internet Web Replication and Caching Taxonomy Network Working Group Request for Comments: 3040 Category: Informational, Jan 2001
8. Day, M., Cain, B., Tomlinson, G., Rzewski, P.: A Model for Content Internetworking (CDI) Network Working Group Request for Comments: 3466 Category: Informational
9. Mulerikkal, J.P., Khalil, I.: An architecture for distributed content delivery network distributed systems and networking school of computer science. RMIT University, Melbourne 3000, Australia
10. Menal, M.F., Fieau, F. (Orange Labs), Souk, A. (USTL), Jaworski, S. (TelecomBretagne): Demonstration of Standard IPTV Content Delivery Network Architecture Interfaces
11. Garcia, G., Bben, A., Ramon, F.J., Maeso, A., Psaras, I., Pavlou, G., Wang, N., Liwiski, J., Spirou, S., Soursos, S., Hadjioannou, E.: COMET: content mediator architecture for content-Web Replication and Caching Taxonomy Network Working aware networks IIMC International Information Management Corporation (2011). ISBN:978-1-905824-25-0
12. Beben, A., Wisniewski, P., Krawiec, P., Nowak, M., Pecka, P., Mongay Batalla, J., Bialon, P., Olender, P., Gutkowski, J., Belter, B., Lopatowski, L.: Content aware network based on virtual infrastructure. In: 2012 13th ACIS International Conference on Software Engineering, Artificial Intelligence, Networking and Parallel/Distributed Computing
13. https://docs.oracle.com/cd/B2835901/server.111/b32024/partition.htm
14. https://datatracker.ietf.org/doc/draft–ietf–cdni-redirection

Part II
Networking Systems and Architectures

Solving Reliability Problems in Complex Networks with Approximated Cuts and Paths

Baijnath Kaushik and Haider Banka

Abstract The paper solves reliability problems for the design of complex network without failure. The traditional approaches based on minimal paths and cuts require significant computational effort equals to NP-hard. The major difficulty lies in calculating the minimal cuts and paths to improve exact reliability bounds. Therefore, a neural network algorithm based on approximated paths and cuts to deal with this difficulty. The proposed approach is divided into two parts. The first part is used to approximate the computed minimal cuts and paths from two networks. The approach in other part improves the reliability bounds. The proposed approach has been tested on mesh network of 256 nodes and hyper-tree of 496 nodes. The performance of proposed approach is compared with PSO for the reliability bounds improvement in low failure probability.

Keywords Approximated paths · Approximated cuts · Combinatorial spectrum · Approximated combinatorial spectrum · Neural network · PSO

1 Introduction

Managing the design of complex network without failure based on solving the reliability problem. Therefore, system reliability computation becomes an important component for solving reliability problem. Optimizing the reliability parameters have gains much attention in the past years and received significant contributions. The fundamental concepts of network reliability, network and algebraic structures and combinatory of network reliability theory has been studied from following has

B. Kaushik (✉) · H. Banka
Department of Computer Science & Engineering, Indian School of Mines,
Dhanbad 826 004, India
e-mail: bkaushik99@gmail.com

H. Banka
e-mail: hbanka2002@yahoo.com

© Springer India 2016
A. Nagar et al. (eds.), *Proceedings of 3rd International Conference on Advanced Computing, Networking and Informatics*, Smart Innovation, Systems and Technologies 44, DOI 10.1007/978-81-322-2529-4_8

been reported in the literature [1–3]. Solving the exact reliability problem in complex network is NP-Hard problems, that is computational effort required growing exponentially with growth of network size [1, 2]; therefore, estimation by simulation and other approaches often becomes an alternative choice.

Monte-Carlo Simulation (MCS) is a straightforward simulation method for solving reliability problems for complex systems [4–7]. Most of the authors have claimed that MCS is an optimal algorithm to compute system reliability. However; these methods require simulation to be repeated numerous times in order to ensure good estimates which require significant computational effort and time. The approach stated in [8] is based on simple cut-sets for reliability measure in comparison to spanning tree approach. If there are r paths and s cuts in the network, then calculating the exact overall reliability using the paths involve $2^r - 1$, using the cuts will involve $2^s - 1$ term. Therefore, the methods using minimal paths should be used if and only if $r \leq s$. Moreover, it is easier to find minimal paths than minimal cuts in the network. Therefore, methods using paths have to be used because finding all cuts may be infeasible and intractable. It is quite evident that calculating the exact overall reliability in a complex network is quite difficult. Therefore, the method proposed in this paper focus on obtaining the approximated lower and upper bounds are desirable, so that, the calculations be substantially reduced.

Shpungin [9–11] proposed a combinatorial approach for Monte Carlo reliability estimation of a network with unreliable nodes and unreliable edges. The results shows that the MCS techniques for evaluating reliability based on the minimal-cuts and paths requires significant computational effort required is very high and the significant decrease in the lifetime reliability evaluation for complex layered network structures. MCS require significant computational effort and time while solving the network reliability problems.

The approach stated in [12] iteratively calculates root mean squared error (RMSE) for minimizing the reliability errors. Ratana et al. [12] proposed ANN approach to overcome the difficulties in improving the reliability bounds. The approach stated in [12] works on small set of topologies consisting of 10 nodes only and with fixed upper-bound reliability value. The approach requires significant time and computational effort when networks have complex size due its NP-Hard nature. Our proposed approach uses comparably large and complex networks such as mesh of 256 nodes and hyper-tree of 496 nodes with reasonable time. The novelty of the proposed approach in improving the reliability bound (lower and upper bound) by incorporating failure reliability to the exact reliability.

In this paper, we have proposed a neural network algorithm to evaluate the performance of complex network for improving reliability and minimizing cost. The neural network approach has been compared with PSO. The proposed approach incorporates approximated minimal paths and cuts for the improvement of reliability bounds for complex networks. The PSO compute the reliabilities (for random and heuristic links) from complex networks. Our approach achieves improvement for reliabilities computed by PSO and also reduces the cost C(X) incurs by the same.

The remaining part of the paper is organized as follows. The preliminaries of reliability computation introduced in Sect. 2.1. The proposed neural network algorithm discussed in Sect. 2.2. The framework to perform computational study discussed in Sect. 3. The simulation results are discussed in Sect. 4 and finally, Sect. 5 concludes the paper.

2 Reliability Prelims

The major difficulty lies in calculating the minimal cuts and paths to improve exact reliability bounds. This section introduces the reliability evaluation approach to be incorporated into the proposed approach.

2.1 Reliability Evaluation Approach

Reliability in a network represented as a graph defined as the probability that all nodes must communicate with each other using some arbitrary paths. The minimal path (MP) or cut is defined as the minimal number of components whose failure or functioning maintains the operational status of the system. The MC or MP of a system is computed by computing the path from the input to output component. System may have number of components connected in series or parallel.

A series system requires that all components must be operational, and, a parallel system requires that at least one of the components must be operational. If a system has $X = (x1, x2, .., xn)$, and, the reliability function $R(X) = \prod_{i=1}^{n} x_i = \min_{1 \leq i \leq n} x_i$ and, $R(X) = 1 - \prod_{i=1}^{n} (1 - x_i) = \max_{1 \leq i \leq n} x_i$ for series and parallel system. A k-out-of-n system is operational if at least k components are functional, the function $R(X) = 1$, if, $\sum_{i=1}^{n} x_i \geq k$ of the system. If a system has $P1, P2, \ldots, PS$ and $C1, C2, ..,$ and CK be the minimal paths and cuts, then the reliability function is defined as: $R(X) = 1 - \prod_{j=1}^{s} [1 - \prod_{i \in P_j} x_i]$ and $R(X) = \prod_{j=1}^{k} [1 - \prod_{i \in C_j} (1 - x_i)]$.

These two formulations can be verified as follows: let there be at least one minimal path which is operational, say $P1$. The reliability function corresponds to the path $\prod_{i \in P_1} x_1 = 1$ is also 1, if and only if there is one minimal path in which all components are operational. Similarly, the other can also be proved. The proposed neural approach uses approximation of minimal paths and cuts from two networks input to the neural network so that the results obtained at the output is improved. The minimal paths and cuts are approximated by computing the combinatorial spectrum of the calculated paths and cuts.

Let us define the minimal paths and cuts to compute the reliability bounds of the system. Let $A_1, A_2, .., A_i$ and $C1, C2, .., Cm$ be the paths and cuts. Then, the exact reliability $R(X)$ $R(X) = \prod_{k=1}^{m} [1 - \prod_{j \in C_k} (1 - p_j)] \leq R_0 \leq 1 - \prod_{r=1}^{i} [1 - \prod_{j \in A_i} p_j]$ bounded between the expected reliability R_O. In other words, the expected reliability is bounded between the lower and upper bound computed using the minimal paths and cuts. Then, we understand the approximation minimal paths and cuts. Let us define combinatorial spectrum S given by $S = \{\{x_{i,j}\}, 1 \leq i \leq n, 1 \leq j \leq m\}$ of a network of N nodes. The set S is the number of all permutation π such that $N(\pi(i,j))$ is an anchor of π. The combinatorial spectrum $x_{i,j}$ is a set of all possible operational permutations nodes and links of a network. Let us define the approximated spectrum $A = \{(i,j) | i \geq r, j \leq s\}$ from combinatorial spectrum for acquiring possible cuts and paths from the network. An approximated combinatorial spectrum is possible permutations of paths and cuts. A possible permutation is a Cartesian product $\Pi = \Pi_{V/K} \times \Pi_E$ defined such that every permutation $\pi \in \Pi$ is a pair $(\pi_{V/K}, \pi_E)$, where $\pi_{V/K} \in \Pi_{V/K}$, $\pi_E \in \Pi_E$.

The approach proposed in this paper use the approximated spectrum for minimal cuts and paths. Here, we introduce how to acquire the computations for approximated spectrum from graph shown in Fig. 1.

For the shown graph, the permutation will be computed along the edges $(B, E), \Pi = ((\Pi_V = B, E), \Pi_E = (1, 2, 3, 4, 5, 6, 7, 8))$, and then index r is found so that $N(\Pi(r, 8))$ with $r = 1$. Therefore, the first anchor of permutation is $N(\Pi(1, 4))$ that is node B and edges 1, 2, 3, 4, 5, 6, 7, 8 are operational and rest will be down. The combinatorial spectrum $S = \{X_{1,2} = 12, X_{1,3} = 42, X_{1,4} = 48, X_{1,5} = 68, X_{1,6} = 84, X_{1,7} = 98, X_{1,8} = 128, X_{2,2} = 36, X_{2,3} = 48, X_{2,4} = 72, X_{2,5} = 96, X_{2,8} = 144\}$ can be easily acquired, and for rest of other (i, j) the $X_{i, j} = 0$. The computed combinatorial spectrum is used by the proposed approach.

The algorithmic framework to perform the computation discussed in the next Section.

Fig. 1 Graph used for computing approximated combinatorial spectrum

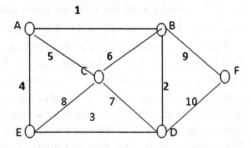

2.2 Proposed Approach

This section gives a brief idea to the proposed neural network algorithm for solving the reliability problem. The objective of the proposed algorithm to approximate reliability values (lower and upper bounds) obtained from the paths and cuts of the given network. The detail of the proposed approach is discussed as follows:

Neural Network Algorithm
Input: Adjacency matrix for reliability values (R_O) for mesh network of 256 nodes and hyper-tree network of 496 nodes.

Output: a. Reliability (lower and upper bound) improvement, b. Reliability (random and heuristic links) improvement, c. cost minimization.

Step 1: Compute the minimal cuts and paths from the adjacency matrix for mesh and hyper-tree network.

$$R(X) = \prod_{k=1}^{m} [1 - \prod_{j \in C_k} (1 - p_j)] \le R_0 \le 1 - \prod_{r=1}^{i} [1 - \prod_{j \in A_i} p_j]$$

Step 2: Compute the approximated combinatorial spectrum from the minimal cuts and paths acquired in step 1. This is obtained from the permutations of the minimal paths and cuts.

Step 3: Compute the net input Z_{inj} for the hidden unit Z_j.

$$Z_{inj} = V_{oj} + \sum_{i=1}^{n} X_i V_{ij}$$

And output at Z_j unit

$$Z_j = f(Z_{inj})$$

Step 4: The error terms δ_j and δ_k used for updating weight and bias so that computed output match with the desired (target) output from the problem. The error updation term δ_k will be computed as follows:

$$\delta_k = [(t_k - Y_k) . f'(Y_{ink})]$$

Similarly, δ_j will be calculated by computing δ_{inj} first as follows:

$$\delta_{inj} = \sum_{k=1}^{m} \delta_j W_{jk}$$

And δ_j will be computed as

$$\delta_j = \delta inj . f'(Z_{inj})$$

Step 5:   The output layer Y_K compute the calculated output from neural network.
The output Y_K computed by computing the net input Y_{ink} at the output
layer. The net input is given by:

$$Y_{ink} = W_{ok} + \sum_{j=1}^{p} Z_j W_{jk}$$

The output is given by:

$$Y_k = f(Y_{ink})$$

The proposed approach achieves improvement in the network reliability (lower
and upper bound when the failure reliability $f(r,s)$ is added to them. The
improvement in the lower bound $R(X)$ is given by:

$$R(X) \leq 1 - \sum_{i=1}^{N_{nm}} q_i^d \prod_{k=1}^{n_i} (1 - q_{k-1}^d) \prod_{k=n_i-1}^{i=1} (1 - q_k^d) + f(r,s) = \frac{1}{r!(n-r)!} p_v^r q_v^{n-r} \sum_{j=s}^{m} \frac{1}{j!(m-j)!} p_e^j q_e^{m-j}$$

Similarly, the improvement for the lower bound is achieved with following
computation:

$$R(X) \leq 1 - [\sum_{i=1}^{N_{nm}} [\prod_{(k,i)\in E_i} (1 - p_{ki}) \prod_{j=1}^{N_{nm}} [1 - \frac{\prod_{(k,j)\in E_j} (1 - p_{kj})}{(1 - p_{ij})}]] + f(r,s) = \frac{1}{r!(n-r)!} p_v^r q_v^{n-r} \sum_{j=s}^{m} \frac{1}{j!(m-j)!} p_e^j q_e^{m-j}$$

The next Section gives the computational framework to solve the reliability
problems using neural network and PSO.

3   Computational Framework

We have used the proposed approach and the PSO for solving the reliability
problem on two networks. A mesh network of $K^q = 256$ nodes, where $K = 4$ is
degree and $q = 4$ is dimension, and 4-*Ary* hyper-tree network of depth $d = 4$ has
$2^d(2d+1) - 1) = 496$ nodes. In mesh network, each internal node has $2q = 8$ edges,
i.e., $256 * 8 = 2048$ edges. The hyper-tree network has 496 nodes organized such
that the root has 16 nodes, level 1 has 32 nodes, level 2 has 64 nodes, level 3 has
128 node, and 256 nodes in the leaves. Each internal node in the hyper-tree network
has 6 edges, therefore, $496 * 6 = 2976$ edges. We have used an adjacency matrix for
mesh and hyper-tree network to solve the reliability problem. The neural network
and PSO consider segments of network from mesh and hyper-tree network. The
approach considers 4 node segment of mesh, i.e., $256/4 = 64$ segments of mesh

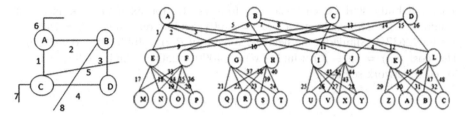

Fig. 2 Mesh and hyper-tree networks used for reliability computation

network. Similarly, a 6 node segment of hyper-tree network, i.e., 496/6 = 84 segments of hyper-tree network.

The neural network computes the exact reliability bounded by the lower and upper bound. The exact reliability $R(X)$ requires the computation of minimal paths and cuts from the mesh and hyper-tree segment. We have considered a segment of mesh and hyper-tree network to perform the reliability computation shown in the Fig. 2.

Following steps will be used in the computation:

1. The reliability (upper and lower bound) from minimal paths and cuts computed for $R_O = 0.93$ as follows:

$$0.9850339 = [1 - (0.07)^2]^3[1 - (0.07)^3] \le R(0.93, 0.93, 0.93, 0.93, 0.93) \le 1$$
$$- [1 - (0.93)^2]^2[1 - (0.93)^3]$$
$$= 0.99643$$

2. The exact reliability from minimal paths and cuts will be obtained for $R_O = 0.93$ as follows:

$$R(0.93, 0.93, 0.93, 0.93, 0.93, 0.93) = 2 \times (0.93)^2 + (0.93)^3 - 3(0.93)^4$$
$$+ (0.93)^5$$
$$= 0.985689$$

3. Approximated reliability is obtained by dividing cost of operational links by number of links. If link cost for 5 operational links are $X_1X_2X_3X_4X_5 = 112$, therefore, the approximated reliability will be $\frac{112}{5!} = 0.9333$.

Next, we give the computation flow for PSO algorithm as follows:

Step 1: Particles are assigned with reliability lower bound $R(X)$ computed cuts and path. For example, if a 5 component mesh system, if $R_O = 0.95$, exact reliability from $R(0.95, 0.95, 0.95, 0.95, 0.95) = 2 \times (0.95)^2 + (0.95)^3 - 3(0.95)^4 + (0.95)^5 = 0.992637$ minimal paths and cuts. Thus, for $k = 4$ cuts or paths, the PSO reliability $R(k) = 0.992637/4 = 0.2444$, therefore, $R(k) \le R(X)$. Now, the random and heuristic links reliability will be calculated as follows: If, for example link has cost (random $(\beta_i) = 2500$ and heuristic $(\alpha_i) = 4000$), random reliability $R(k) = 2500$ * 5 * 0.992637/4. Similarly, the heuristic reliability $R(k) = 4000$ * 5 * 0.992637/4. The cost $c(x) = (\alpha_i - \beta_i) \times \ln(1 - xi) = (4000 - 2500) \times \ln(1 - 0.95) = 1146.$ for each network segment.

4   Simulation Results

The neural network approach improves the reliability in comparison to the PSO for both random and heuristic links. The neural network approach minimizes the link cost in comparison to PSO approach. The simulation has been performed over 10^4 iterations. The simulation result in Fig. 3 shows reliability (random and heuristic) improvement by proposed approach comparably to PSO.

The proposed approach significantly reduces link cost comparably to the PSO. The results in Fig. 4 show comparison between the cost and reliability.

Next, we discuss the conclusions and future scope of the proposed work.

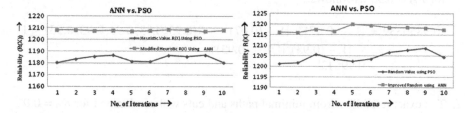

Fig. 3 Plot for reliability (random and heuristic) links

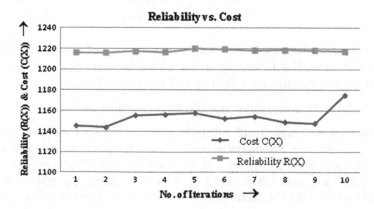

Fig. 4 Plot for reliability versus cost

5 Conclusions and Future Scope

The approach proposed in this paper uses reliability (exact and bounded) and approximated reliability as inputs for improving the reliability bounds (lower and upper bound) in output. The novelty of the proposed approach in improving the reliability bound (lower and upper bound) by incorporating failure reliability to the exact reliability. This has been applied on comparably large and complex networks such as mesh network of 256 nodes and hyper-tree of 496 nodes with reasonable time. The proposed approach achieves significant reliability bounds (lower and upper bound) comparably to PSO. We hope that the paper may open up new research directions for calculating reliability bounds in large and complex networks.

References

1. Gertsbakh, I.: Reliability Theory. Springer-Verlag, Heidelberg (2000)
2. Shier, D.R.: Network Reliability and Algebraic Structure. Oxford, New York (1991)
3. Ahuja, R.K., Magnanti, T.L., Orlin, J.B.: Network Flows: Theory, Algorithms, and Applications. Prentice Hall, Englewood Cliffs (1993)
4. Aggarwal, K.K., Rai, S.: Reliability evaluation in computer-communication networks. IEEE Trans. Reliab. 30(1), 32–35 (1981)
5. Aggarwal, K., Gupta, J., Misra, K.: A simple method for reliability evaluation of a communication system. IEEE Trans. Comm. 23(5), 563–566 (1975)
6. Misra, K.B., Prasad, P.: Comments on: reliability evaluation of flow networks. IEEE Trans. Reliab. 31(2), 174–176 (1982)
7. Aggarwal, K.K., Chopra, Y.C., Bajwa, J.S.: Capacity consideration in reliability analysis of communication systems. IEEE Trans. Reliab. 31(2), 177–181 (1982)
8. Rai, S.: A cutset approach to reliability evaluation in communication networks. IEEE Trans. Reliab. 31(5), 428–431 (1982)
9. Shpungin, Y.: Combinatorial approach to reliability evaluation of network with unreliable nodes and unreliable edges. Int. J. Comp. Sci. 1(3), 177–183 (2006)
10. Shpungin, Y.: Combinatorial approaches to Monte Carlo estimation of dynamic systems reliability parameters. Comm. Depend. Quality Manage.: An Int. J. 9(1), 69–75 (2006)
11. Shpungin, Y.: Networks with unreliable nodes and edges: Monte Carlo lifetime estimation. World Acad. Sci. Eng. Tech. 27, 349–354 ((2007))
12. Ratana, C.S., Konak, A., Smith, A.E.: Estimation of all-terminal network reliability using an artificial neural network. Comp. Oper. Res 29(7), 849–868 (2002)

Maximal Clique Size Versus Centrality: A Correlation Analysis for Complex Real-World Network Graphs

Natarajan Meghanathan

Abstract The paper presents the results of correlation analysis between node centrality (a computationally lightweight metric) and the maximal clique size (a computationally hard metric) that each node is part of in complex real-world network graphs, ranging from regular random graphs to scale-free graphs. The maximal clique size for a node is the size of the largest clique (number of constituent nodes) the node is part of. The correlation coefficient between the centrality value and the maximal clique size for a node is observed to increase with increase in the spectral radius ratio for node degree (a measure of the variation of node degree in the network). As the real-world networks get increasingly scale-free, the correlation between the centrality value and the maximal clique size increases. The degree-based centrality metrics are observed to be relatively better correlated with the maximal clique size compared to the shortest path-based centrality metrics.

Keywords Correlation · Centrality · Maximal clique size · Complex network graphs

1 Introduction

Network Science is a fast-growing discipline in academics and industry. It is the science of analyzing and visualizing complex real-world networks using graph theoretic principles. Several metrics are used to analyze the characteristics of the real-world network graphs; among them "centrality" is a commonly used metric. The centrality of a node is a measure of the topological importance of the node with respect to the other nodes in the network [1]. It is purely a link-statistics based measure and not based on any offline information (such as reputation of the node,

N. Meghanathan (✉)
Department of Computer Science, Jackson State University, 18839,
Jackson 39217, MS, USA
e-mail: natarajan.meghanathan@jsums.edu

© Springer India 2016
A. Nagar et al. (eds.), *Proceedings of 3rd International Conference on Advanced Computing, Networking and Informatics*, Smart Innovation, Systems and Technologies 44, DOI 10.1007/978-81-322-2529-4_9

cost of the node, etc.). The commonly used centrality metrics are degree centrality, eigenvector centrality, closeness centrality and betweenness centrality. Degree centrality (DegC) of a node is simply the number of immediate neighbors for the node in the network. The eigenvector centrality (EVC) of a node is a measure of the degree of the node as well as the degree of its neighbor nodes. DegC and EVC are hereafter referred to as degree-based centrality metrics. Closeness centrality (ClC) of a node is the inverse of the sum of the shortest path distances of the node to every other node in the network. Betweenness centrality (BWC) of a node is the ratio of the number of shortest paths the node is part of for any source-destination node pair in the network, summed over all possible source-destination pairs that do not involve the particular node. ClC and BWC are hereafter referred to as shortest path-based centrality metrics. Computationally efficient polynomial-time algorithms have been proposed in the literature [1–4] to determine exact values for each of the above centrality metrics; hence, centrality is categorized in this paper as a computationally lightweight metric.

A "clique" is a complete sub graph of a graph (i.e., all the nodes that are part of the sub graph are directly connected to each other). Cliques are used as the basis to identify closely-knit communities in a network as part of studies on homophily and diffusion. Unfortunately, the problem of finding the maximum-sized clique in a graph is an NP-hard problem [3], prompting several exact algorithms and heuristics to be proposed in the literature [5–9]. In this paper, a recently proposed exact algorithm [5] has been chosen to determine the size of the maximum clique for large-scale complex network graphs and extended to determine the size of the maximal clique that a particular node is part of. The maximal clique size for a node is defined as the size of the largest clique (in terms of the number of constituent nodes) the node is part of. Note that the maximal clique for a node need not be the maximum clique for the entire network graph; but, the maximum clique for the entire graph could be the maximal clique for one or more nodes in the network.

Since the maximal clique size problem is a computationally hard problem and exact algorithms run significantly slower on large network graphs, the paper explores whether the maximal clique size correlates well to one of the commonly studied computationally lightweight metrics, viz., centrality of the vertices, for complex real-world network graphs: if a high positive correlation is observed between maximal clique size and one or more centrality metrics, one could then infer the corresponding centrality values of the vertices as a measure of the maximal clique size of the vertices in real-world network graphs. The work available in the literature so far considers these two metrics separately. This will be the first paper to conduct a correlation study between centrality and maximal clique size for real-world network graphs. To the best of the author's knowledge, there is no other work that has done correlation study between these two metrics (and in general, a computationally hard metric vis-a-vis a computationally lightweight metric) for real-world network graphs.

The rest of the paper is organized as follows: Sect. 2 describes the six real-world network graphs that are used in this paper and presents an analysis of the degree distribution of the vertices in these graphs. Section 3 presents the results of the correlation studies between centrality and maximal clique size at the node level for each of the real-world network graphs. Section 4 concludes the paper. Throughout the paper, the terms 'node' and 'vertex' and 'link' and 'edge' are used interchangeably.

2 Real-World Networks and Their Degree Distribution

The network graphs analyzed are briefly described as follows (in the increasing order of the number of vertices): (i) *Zachary's Karate Club*: Social network of friendships (78 edges) between 34 members of a karate club at a US university in the 1970s; (ii) *Dolphins' Social Network*: An undirected social network of frequent associations (159 edges) between 62 dolphins in a community living off Doubtful Sound, New Zealand; (iii) *US Politics Books Network*: Nodes represent a total of 105 books about US politics sold by the online bookseller Amazon.com. A total of 441 edges represent frequent co-purchasing of books by the same buyers, as indicated by the "customers who bought this book also bought these other books" feature on Amazon; (iv) *Word Adjacencies Network*: This is a word co-appearance network representing adjacencies of common adjective and noun in the novel "David Copperfield" by Charles Dickens. A total of 112 nodes represent the most commonly occurring adjectives and nouns in the book. A total of 425 edges connect any pair of words that occur in adjacent position in the text of the book; (v) *American College Football Network*: Network represents the teams that played in the Fall 2000 season of the American Football games and their previous rivalry—nodes (115 nodes) are college teams and there is an edge (613 edges) between two nodes if and only if the corresponding teams have competed against each other earlier; (vi) *US Airports 1997 Network*: A network of 332 airports in the United States (as of year 1997) wherein the vertices are the airports and two airports are connected with an edge (a total of 2126 edges) if there is at least one direct flight between them in both the directions. Data for networks (i) through (v) and (vi) can be obtained from http://www-personal.umich.edu/ ∼ mejn/netdata/ and http://vlado. fmf.uni-lj.si/pub/networks/pajek/data/gphs.htm respectively.

Figure 1 presents the degree distribution of the vertices in the six network graphs in the form of both the Probability Mass Function (the fraction of the vertices with a particular degree) and the Cumulative Distribution Function (the sum of the fractions of the vertices with degrees less than or equal to a certain value). The average node degree and the spectral radius degree ratio (ratio of the spectral radius and the average node degree) have been also computed; the spectral radius (bounded below by the average node degree and bounded above by the maximum node degree) is the largest Eigenvalue of the adjacency matrix of the network graph, obtained as a result of computing the Eigenvector Centrality of the network graphs. The spectral radius degree ratio is a measure of the variation in the node degree with respect to the average node degree; the closer the ratio is to 1, the smaller the variations in the

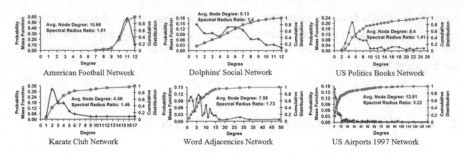

Fig. 1 Node degree: probability mass function and cumulative distribution

node degree and the degrees of the vertices are closer to the average node degree (characteristic of random graph networks). The farther is the ratio from 1, the larger the variations in the node degree (characteristic of scale-free networks).

3 Correlation Analysis: Centrality Versus Maximal Clique Size

This section presents the results of correlation coefficient analysis conducted between the centrality values observed for the vertices vis-a-vis the maximal size clique that each vertex is part of. The analysis has been conducted on the six real-world network graphs with respect to centrality and the maximal clique size measured for the vertices in these graphs. The algorithms implemented include those to determine each of the four centrality metrics (Degree, Eigenvector, Betweenness and Closeness) and the exact algorithm to determine the maximal clique size for each vertex in a graph.

Table 1 presents results of the correlation coefficient analysis of the four centrality metrics and the maximal clique size observed for the vertices in each of the six real-world network graphs studied in this paper. Values of correlation coefficient greater than or equal to 0.8 (high correlation) have been indicated in bold; values below 0.5 (low correlation) are indicated in italics; and values between 0.5 and 0.8 (moderate correlation) are indicated in roman. If $\overline{X}$ and $\overline{Y}$ are the average values of the two metrics (say X and Y) observed for the vertices (IDs 1 to n, where n is the number of vertices) in the network, the formula used to compute the Correlation Coefficient between two metrics X and Y is as follows:

$$CorrCoeff(X, Y) = \frac{\sum\limits_{ID=1}^{n} (X[ID] - \overline{X}) * (Y[ID] - \overline{Y})}{\sqrt{\sum\limits_{ID=1}^{N} (X[ID] - \overline{X})^2} \sqrt{\sum\limits_{ID=1}^{N} (Y[ID] - \overline{Y})^2}} \tag{1}$$

Table 1 Correlation coefficients: centrality metrics and maximal clique size for the nodes

Network index	Network name (increasing order of spectral radius ratio)	Degree versus clique	Eigenvector versus clique	Closeness versus clique	Betweenness versus clique
(v)	American College Football Network	0.32	0.35	−0.03	−0.17
(ii)	Dolphins' Social Network	0.78	0.56	0.42	0.28
(iii)	US Politics Books Network	0.70	0.75	0.32	0.37
(i)	Zachary's Karate Club Network	0.64	0.77	0.62	0.46
(iv)	Word Adjacencies Network	0.71	**0.82**	**0.84**	0.48
(vi)	US Airports 1997 Network	**0.87**	**0.95**	**0.84**	0.40

As one can see in Table 1, in general, the correlation between the centrality metrics and the maximal clique size increases as the spectral radius ratio for node degree increases. This implies, the more scale-free a real-world network is, the higher the correlation between the centrality value and the maximal clique size observed for a node. With several of the real-world networks being mostly scale-free, one could expect these networks to exhibit a similar correlation to that observed in this paper.

The degree-based centrality metrics (degree centrality and eigenvector centrality) have been observed to be very positively and highly correlated with the maximal clique size observed for the nodes. Between the two degree-based centrality metrics, the eigenvector centrality metric shows higher positive correlations to the maximal clique size. This could be attributed to the eigenvector centrality of a node being a measure of both the degree of the node as well as the degrees of its neighbors. That is, a high degree node located in a neighborhood of high degree vertices is more likely to be part of a maximal clique of larger size. In addition, as the networks get increasingly scale-free, nodes with high degree are more likely connected to other similar nodes with high degree (to facilitate an average path length that is almost independent of network size: characteristic of scale-free networks [1] contributing to a positive correlation between degree-based centrality metrics and maximal clique size.

With respect to the two shortest-path based centrality metrics, the betweenness centrality metric is observed to exhibit a low correlation with maximal clique size for all the six real-world network graphs; the correlation coefficient increases as the network becomes increasingly scale-free. In networks with minimal variation in node degree (like the American College Football network that is more closer to a random network), nodes that facilitate shortest-path communication between several node pairs in the network are not part of a larger size clique; on the other hand, nodes that are part of larger size cliques in such random networks exhibit a relatively lower betweenness centrality. Since the degrees of the vertices in random networks are quite comparable to the average node degree, there is no clear ranking of the vertices based on the degree-based centrality metrics and maximal size cliques that they are part of. Also, if at all a vertex ends up being in a larger sized clique in random network graphs, it is more likely not to facilitate shortest path communication between the majority of the vertices (contributing to a negative/zero correlation or at best a low correlation with betweenness centrality). As the network becomes increasingly scale-free, the hubs that facilitate shortest-path communication between any two nodes in the network exhibit higher betweenness and closeness centralities as well as form a clique with other high-degree hubs—exhibiting the ultra small-world property (the average path length is $\ln(\ln N)$, where N is the number of nodes in the network) [1]. The correlation of the closeness centrality values and the maximal clique size values observed for the vertices in real-world network graphs is significantly higher (i.e., positive correlation) for networks that are increasingly scale-free.

Overall, the degree-based centrality metrics exhibit a relatively better correlation with the maximal clique size compared to that of the shortest-path based centrality metrics (especially in networks with low-moderate variation in node degree). For real-world networks that exhibit moderate-high variation in node degree, the shortest-path based centrality metrics (especially closeness centrality) fast catch up with that of the degree-based centrality metrics and exhibit higher levels of positive correlation with maximal clique size. As the networks become increasingly scale-free, the hubs (that facilitate shortest-path communication between any two nodes) are more likely to form the maximum clique for the entire network graph—contributing to higher levels of positive correlation between node centrality and maximal clique size.

4 Conclusions

The correlation coefficient analysis studies between the centrality metrics and the maximal clique size for the vertices in the real-world network graphs unravel several significant findings that have been so far not reported in the literature: (i) the degree-based centrality metrics (especially the eigenvector centrality) exhibit a significantly high positive correlation to the maximal clique size as the networks get increasingly scale-free; (ii) the betweenness centrality of the vertices exhibits a low correlation with that of the maximal size cliques the vertices can be part of; (iii) in real-world networks that are close to random network graphs, the centrality metrics exhibit a low correlation to maximal clique size (especially in the case of shortest-path based closeness and betweenness centrality metrics); (iv) for all the four centrality metrics, the extent of positive correlation with maximal clique size increases as the real-world networks become increasingly scale-free.

With the problem of determining maximal clique sizes for individual vertices being computationally time consuming, the approach taken in this paper to study the correlation between maximal clique sizes and centrality can be the first step in identifying positive correlation between cliques/clique size in real-world network graphs to one or more network metrics (like centrality) that can be quickly determined and thereby appropriate inferences can be made about the maximal size cliques of the individual vertices. The degree-based centrality metrics (especially the eigenvector centrality) have been observed to show promising positive correlations to that of maximal clique sizes of the individual vertices, especially as the networks get increasingly scale-free; this observation could form the basis of future research for centrality-clique analysis for complex real-world networks.

References

1. Newman, M.: Networks: An Introduction, 1st edn. Oxford University Press, Oxford (2010)
2. Strang, G.: Linear Algebra and its Applications, 1st edn. Cengage Learning, Boston (2005)
3. Cormen, T.H., Leiserson, C.E., Rivest, R.L., Stein, C.: Introduction to Algorithms, 3rd edn. MIT Press, Cambridge (2009)
4. Brandes, U.: A faster algorithm for betweenness centrality. J. Math. Sociol. **25**, 163–177 (2001)
5. Pattabiraman, B., Patwary, M.A., Gebremedhin, A.H., Liao, W.-K., Choudhary, A.: Fast algorithms for the maximum clique problem on massive sparse graphs. In: Bonato, A., Mitzenmacher, M., Pralat, P. (eds.): 10th International Workshop on Algorithms and Models for the Web Graph. Lecture Notes in Computer Science, vol. 8305, pp. 156–169. Springer-Verlag, Berlin Heidelberg New York (2013)
6. Fortunato, S.: Community detection in graphs. Phys. Rep. **486**, 75–174 (2010)
7. Palla, G., Derenyi, I., Farkas, I., Vicsek, T.: Uncovering the overlapping community structure of complex networks in nature and society. Nature **435**, 814–818 (2005)
8. Sadi, S., Oguducu, S., Uyar, A.S.: An efficient community detection method using parallel clique-finding ants. In: Proceedings of IEEE Congress on Evolutionary Computation, pp. 1–7. IEEE, Piscataway NJ (2010)
9. Tomita, E., Kameda, T.: An efficient branch-and-bound algorithm for finding a maximum clique with computational experiments. J. Global Optim. **37**, 95–11 (2007)

On Solving the Multi-depot Vehicle Routing Problem

Takwa Tlili, Saoussen Krichen, Ghofrane Drira and Sami Faiz

Abstract Problems associated with seeking the lowest cost vehicle routes to deliver demand from a set of depots to a set of customers are called Multi-depot Vehicle Routing Problems (MDVRP). The MDVRP is a generalization of the standard vehicle routing problem which involves more than one depot. In MDVRP each vehicle leaves a depot and should return to the same depot they started with. In this paper, the MDVRP is tackled using a iterated local search metaheuristic. Experiments are run on a number of benchmark instances of varying depots and customer sizes. The numerical results show that the proposed algorithm is competitive against state-of-the-art methods.

Keywords Multi-depot vehicle routing problem · Cplex · Iterated local search

1 Introduction

Nowadays, companies try to manage efficiently their resources and propose beneficial strategies, in order to improve their services and satisfy customers. Hence, it is important to focus on the logistic sector considered as a key factor in business competitiveness, especially in the delivery of goods. In this respect, interested researchers addressed such routing problems and used optimization tools.

This paper handles strategic routing problem that is a general variant of the basic VRP including multiple depot, namely Multi-Depot Vehicle Routing Problem (MDVRP). It models perfectly the routing process from several depots to customers

T. Tlili (✉) · S. Krichen · G. Drira
LARODEC, Institut Supérieur de Gestion Tunis, Université de Tunis,
Tunis, Tunisia
e-mail: takwa.tlili@gmail.com

S. Faiz
LTSIRS, Ecole Nationale dIngenieurs de Tunis, Université de Tunis El Manar,
Tunis, Tunisia

© Springer India 2016
A. Nagar et al. (eds.), *Proceedings of 3rd International Conference on Advanced Computing, Networking and Informatics*, Smart Innovation, Systems and Technologies 44, DOI 10.1007/978-81-322-2529-4_10

taking into account some constraints, in order to reduce routing cost and satisfy customers.

The MDVRP is a generalization of the well known Vehicle Routing Problem (VRP), thus it is a NP-hard combinatorial problem. Laporte et al. [1], Contardo and Martinelli [2], and Escobar et al. [3] developed exact methods for solving the MDVRP. In the literature, there are numerous approximate approaches for solving large MDVRP instances. Metaheuristic examples include hybrid genetic algorithm [4], variable neighborhood search [5] and ant colony optimization [6].

In this paper, we evoke the multi-depot vehicle routing problem and give an illustrative example solved by Cplex solver. Then, we adopt an Iterated local search approach to handle more complex Benchmark instances.

The remaining of the paper is organized as follows: Sect. 2 describes mathematically the problem. Section 3 presents a detailed example solved by Cplex solver. Section 4 shows the solution methodology based on the Iterated local search. Section 5 details the experimental results.

2 Problem Statement

The MDVRP can be defined as follows. Let G = (V, E) be a graph where V is the vertex set and E is the edge set. The set V contains two subsets V1 and V2 that are the set of customers and the set of depots, respectively. The depots number is known in advance. Each depot is large enough to store all customer demands. Each vehicle, with maximum capacity C_{max}, begins and ends at the same depot. The vehicles are required to visit all corresponding customers and return to the same starting depot. Each customer is visited by a vehicle exactly once. The MDVRP is to decide the tours of each vehicle so that the total traveled distance is minimized.

2.1 Illustrative Example

We propose an illustrative example and we solve it through the Branch and Cut algorithm by using IBM ILOG CPLEX 12.1. Minimum requirements to run this solver are: Intel Core Duo processor, 3 GO RAM, operating system windows XP or higher.

Our example is described as follows: Lets consider a delivery planning for 2 depots and 9 customers. For each depot, a maximum number of vehicles is fixed. Each vehicle is characterized by a traveling distance restriction (expressed in meter) and a maximum loading capacity (expressed in kilogram). Two vehicles are assigned to each one. The maximum capacity of each vehicle is limited to 250 kg and the distance that can be traveled by each one is fixed to 160 km.

Those noted parameters are summarized in what follows: This example is illustrated in Fig. 1, where customers are expressed with circles and depots are

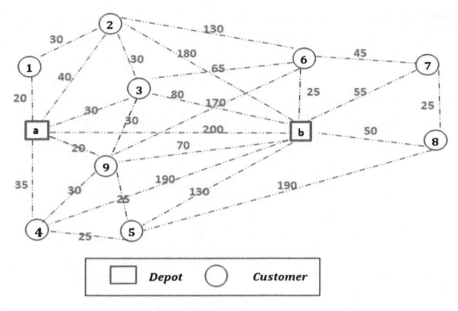

Fig. 1 Example of MDCVRP

Table 1 Description of
example parameters

Table 1 Description of example parameters

Parameters	Values
Number of vehicles for each depot	2
Vehicle capacity (kg)	250
Maximum distance traveled by each vehicle (km)	160

mentioned using squares. In order to check the formulation model previously stated, we denote a matrix resuming distances between depots and customers (Table 1).

$$M = \begin{array}{c} \\ a \\ b \\ C_1 \\ C_2 \\ C_3 \\ C_4 \\ C_5 \\ C_6 \\ C_7 \\ C_8 \\ C_9 \end{array} \begin{pmatrix} a & b & C_1 & C_2 & C_3 & C_4 & C_5 & C_6 & C_7 & C_8 & C_9 \\ 0 & 200 & 20 & 40 & 30 & 35 & 50 & 180 & 240 & 280 & 20 \\ & 0 & 220 & 180 & 80 & 190 & 130 & 25 & 55 & 50 & 70 \\ & & 0 & 30 & 40 & 65 & 75 & 75 & 95 & 125 & 30 \\ & & & 0 & 30 & 100 & 125 & 130 & 140 & 150 & 65 \\ & & & & 0 & 70 & 65 & 65 & 110 & 120 & 30 \\ & & & & & 0 & 25 & 210 & 225 & 215 & 30 \\ & & & & & & 0 & 180 & 200 & 190 & 25 \\ & & & & & & & 0 & 45 & 70 & 170 \\ & & & & & & & & 0 & 25 & 190 \\ & & & & & & & & & 0 & 190 \\ & & & & & & & & & & 0 \end{pmatrix}$$

Table 2 Customers' demands

Customers	C_1	C_2	C_3	C_4	C_5	C_6	C_7	C_8	C_9
Demands	80	60	70	20	60	80	60	70	20

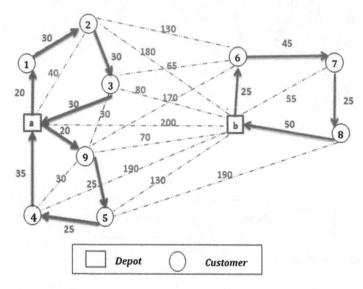

Fig. 2 Vehicles itineraries related to depots a and b

The following table resumes the customers' demands (Table 2):

The obtained solution after 10 iterations of Branch and Cut algorithm leads to: $Z(X) = 360$.

The outputs of this example is noticed in Fig. 2.

Figure 2 illustrates the solution of precedent example showing that for depot B just one vehicle is used to serve customers C_6, C_7 then C_8. Concerning depot A, 2 vehicles are required to serve assigned customers. The first one, supplies C_1, C_2 and C_3, the second serves C_9, C_4 and C_5.

3 Solution Approach

As the DCVRP is NP-hard, we propose to solve it using a metaheuristic approach. Indeed, based on the fact that the Iterated Local Search (ILS) is efficient for generating high quality solutions, we propose to adapt it for the MD-DCVRP.

ILS, proposed by Glover and Laguna [7], is a simple stochastic local search method based on the proximate optimality principle. It is a neighborhood searching paradigm that consists in exploring the space of local optima, with respect to a given neighborhood. ILS requires (i) a constructive heuristic for the initial solution, (ii) an improvement procedure, and (iii) a random perturbation procedure.

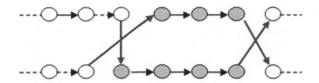

Fig. 3 Cross-exchange move

① **Constructive heuristic**: The initial solution is generated as follows. Customers are inserted one by one into a vehicle route, and when the vehicle capacity is reached, a new empty route is started.

② **Improvement procedure (Local search)**: The local search operator designed for the routing problem consider simple arc exchange moves to transit from one solution to the other. The best insertion is performed, and the process is repeated until no improvement is possible.

③ **Perturbation procedure**: The perturbation is carried out by applying a randomly cross-exchange move on the current solution [8]. For the exchange, we choose randomly two routes and two starting points as presented in Fig. 3.

The search is initiated from a good feasible solution, computed by the constructive heuristic. ILS applies local search to this solution in order to obtain a first local optimum. At each iteration a cross-exchange move is applied to the current solution until improvement.

4 Numerical Results

In order to show the effectiveness of our approach, we adapt the existing tabu search (TS) metaheuristic from this framework to the MDVRP and compare its results with those of our approach. In this respect, we compare also our results with those of Genclust heuristic (Thangiah and Salhi 2001). This is shown in Table 3. Table 3 shows that the ILS out performs clearly the TS results and it competes with Genclust. We can mention also that our approach is able to improve upon the Genclust results for a considerable number of instance.

5 Conclusion

In this paper the Multi-Depot Vehicle Routing Problem (MDVRP) is evoked and a metaheuristic to solve it is proposed. The proposed algorithm integrates an adaptive neighborhood selection into Iterated local search (ILS). The MDVRP is about

Table 3 Experimental results

I	TS	Genclust	ILS	I	TS	Genclust	ILS
P01	630	591	606.11	P12	1464	1421.94	1390.04
P02	504	468	496.45	P13	1474	1318.95	1400.2
P03	685	694	675.32	P14	1742	1360.12	1554.25
P04	1108	1062	1062.6	P15	2912	3059.15	2766.3
P05	810	754	782.34	P16	3248	2719.98	2885.45
P06	925	976.02	910.13	P17	2231	2894.69	2969.45
P07	945	976.48	904.44	P18	4634	5462.90	4202.3
P08	5036	4812.52	4784.2	P19	4460	3956.61	4137.03
P09	4300	4284.62	4102.0	P20	4807	4344.81	4466.56
P10	4180	4291.45	3960.01	P21	6179	6872.11	5770.05
P11	4249	4092.68	4036.55	P22	7183	5985.32	6623.85

constructing a set of vehicle routes over a subset of available customers with the minimum routing costs. Computational experiments on the DCVRP benchmark instances show that the SA metaheuristic is effective.

References

1. Laporte, G., Nobert, Y., Arpin, D.: Optimal solutions to capacitated multidepot vehicle routing problems. Congr. Numer. **44**, 283–292 (1984)
2. Contardo, C., Martinelli, R.: A new exact algorithm for the multi-depot vehicle routing problem under capacity and route length constraints. Discret. Optim. **12**, 129–146 (2014)
3. Escobar, J.W., Linfati, R., Toth, P., Baldoquin, M.G.: A hybrid granular tabu search algorithm for the multi-depot vehicle routing problem. J. Heuristics **20**, 483–509 (2014)
4. Ghoseiri, K., Ghannadpour, S.: A hybrid genetic algorithm for multi-depot homogenous locomotive assignment with time windows. Appl. Soft Comput. **10**, 53–65 (2010)
5. Polacek, M., Hartl, R., Doerner, K., Reimann, M.: A variable neighborhood search for the multi depot vehicle routing problem with time windows. J. Heuristics **10**, 613–627 (2004)
6. Yu, B., Yang, Z., Xie, J.: A parallel improved ant colony optimization for multi-depot vehicle routing problem. J. Oper. Res. Soc. **62**, 183–188 (2010)
7. Glover, F., Laguna, M.: Tabu search. Kluwer Academic Publishers, Boston (1997)
8. Chen, P., Kuan Huang, H., Dong, X.-Y.: Iterated variable neighborhood descent algorithm for the capacitated vehicle routing problem. Expert Syst. Appl. **27**, 1620–1627 (2010)

Prediction of Crop and Intrusions Using WSN

S. Sangeetha, M.K. Dharani, B. Gayathri Devi,
R. Dhivya and P. Sathya

Abstract Nowadays, the major problem in the agriculture sector is stumpy crop production due to less number of workers in the farm and animal intrusion. The main objective is to improve the sustainable agriculture by enhancing the technology using wireless sensor technology. It uses Micro Electro Magnetic System which is used to measure temperature, humidity and moisture. The characteristic data obtained from the Wireless Sensor Network will be compared with the pre-defined data set in the Knowledge Base where historical data's are stored. The corresponding decisions from the Knowledge Base are sent to the respective land owner's mobile through SMS using radio frequency which has less power consumption. The sensors are co-ordinated using the GPS and are connected to the base station in an ad hoc network using WLAN. Another common issue is animal intrusion, especially in the places like Mettupalayam, Coimbatore, and Pollachi where elephants are destroying the crops. To protect the crops and common people, Seismic sensors are used to detect the footfalls of elephants in hilly areas. This sensor uses geophone to record the footfalls of elephants and immediately alert message is sent to the people.

Keywords WSN-Wireless sensor network · MEMS-Micro electro magnetic system · SMS-Short message service · WLAN-Wireless local area network · GPS-Global positioning system

S. Sangeetha (✉) · M.K. Dharani · B. Gayathri Devi ·
R. Dhivya · P. Sathya
Department of Computer Science and Engineering, Avinashilingam
University for Women, Coimbatore, Tamil Nadu, India
e-mail: visual.sangi@gmail.com

© Springer India 2016
A. Nagar et al. (eds.), *Proceedings of 3rd International Conference on Advanced Computing, Networking and Informatics*, Smart Innovation, Systems and Technologies 44, DOI 10.1007/978-81-322-2529-4_11

1 Introduction

Recently the modern agriculture uses advanced technology such as Wireless Sensor network to enhance the crop cultivation [1]. The crop prediction and animal prediction increases the efficiency of crop production. The ontology based crop prediction understands and analyzes the knowledge of agriculture. It establishes a semantic network to predict the crop to be grown [2].

The crop monitoring system makes farmers to be more profitable and sustainable, since it provides better water management [3, 4]. If rainfall comes, land owners no need to irrigate the land because humidity gets changed. So, water can be saved which in turn consumes power. Also if the temperature or humidity or pH goes beyond the threshold level, then it generates alert to the corresponding land owners. An early warning system is used to minimize the elephant intrusion. To protect the crops and common people around, Seismic sensors are used to detect the footfalls of elephants in hilly areas [5, 6]. This sensor uses geophone to record the footfalls of elephants and immediately alert message is sent to the people and to the forest authorities as a advanced information to take necessary actions [7].

2 Ontology Based Crop Cultivation

The Ontology classifies the crop cultivation based on knowledge base. It consists of 3 attributes such as Soil, Climate, and life span. Climate is classified into temperature and humidity. Humidity sensors are used to measure the amount of water vapour in air. Temperature sensors are used to sense the temperature level of air from radiation and moisture. pH sensors measures the pH value and is used to analyze the acid level of the soil, by which the fertilizer usage can be reduced. Based on this pH value of the soil, the crop to be grown can be identified. The attribute and its related variables of crop using Agriculture ontology are represented [8, 9] in Tables 1 and 2. Based on this Classification using knowledgebase, the type of crop yielding best can be identified.

Table 1 Soil related variables

Attribute	Types	Crops to be grown
Soil	Alluvial soil	Rice, wheat, sugarcane, cotton, jute
	Black soil	Rice, wheat, sugarcane, cotton, groundnut, millet
	Red soil	
	Laterite soil	Tropical crops, cashew, rubber, coconut, tea, coffee
	Mountain soil	Tea, coffee, spices, tropical fruits
	Dessert soil	Barley, millet

Table 2 Climate related variables

Attribute	Types	Relative temperature	Relative humidity	Crops to be grown
Climate	Summer	Very hot (32–40 °C)	Very high to moderate	Millets, paddy, maize, groundnut, red chillies, cotton, sugarcane, turmeric,
	Autumn	Warm days (<30 °C) cool nights (21–29 °C)	Low	Maize, oats
	Spring	Warm days (<30 °C) cool nights (25–29 °C)	Low to moderate	Wheat, barley, mustard, peas
	Winter	Cold (10–15 °C)	High	Oats

3 Knowledge Representation for Crop Prediction

Ontology knowledge based crop prediction system is used to estimate the cultivation of crop based on weather and soil conditions. Figure 1 shows functional flow of crop prediction. The function takes soil, Climate, pH of the soil as input variables. Fuzzy rules are constructed to predict the climate based on temperature and

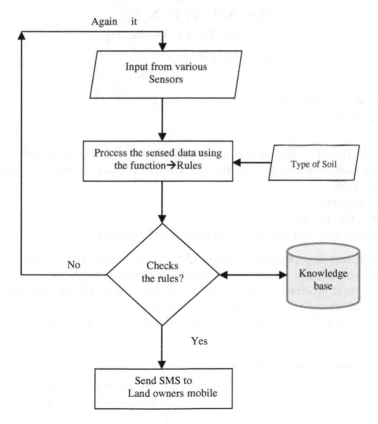

Fig. 1 Functional flow of crop prediction

Table 3 Climate related variables

h	t		
	High	Moderate	Low
High	Summer	Spring	Rainy
Moderate	Summer	Spring	Winter
Low	Autumn	Autumn	Autumn

humidity, since it keeps on changing over time [10, 11]. Fuzzy Associate Memory (FAM) is used to map fuzzy rules in the form of a matrix which is shown in Table 3. These rules take two variables (temperature and humidity) as input and map them into a two dimensional matrix. The rules in the FAM follow a simple if-then-else format. Samples rules for predicting the climate is shown in Table 4. Fuzzy Associative Memory reduces the rate of false negatives.

3.1 Attribute Representation

$$Si = \{A, \ B, \ C, \ D, \ E, \ F\}$$
$$X = \{Ai, \ Bi, \ Ci, \ Di, \ Ei, \ Fi\}$$
$$f(\chi) = f(S_i, \ pH, \ Z)$$

where
x crop to be grown and $x \subseteq X$.
Si   set of Soils

A, B, C, D, E, F—Alluvial Soil, Black Soil, Red Soil, Dessert Soil, Laterite Soil, Mountain Soil respectively. Ai, Bi, Ci, Di, Ei, Fi—set of crop to grown in respective soils.
X   Set of crops
pH   pH value of soil
z   Climate predicted based on temperature and humidity

The function $f(\chi)$ predicts the crop to be grown. It takes type of soil, Climate, pH of the soil as input variables. It calculates the plant to be grown on each input variable then it predicts the crop to be grown by intersecting the matched inputs with the knowledge base. The sample rule for predicting the crop is shown below.
   If Si = A AND Z = Summer AND pH > 7.0 then Ai ∩ Zi ∩ pHi

Table 4 Sample rules for predicting the climate

Rule 1	If (t==Low) AND (h==*Moderate*) THEN (Season==Winter);
Rule 2	If (t==Moderate) AND (h==*Moderate*) THEN (Season==Spring);
Rule 3	If (t==Moderate) AND (h==*Low*) THEN (Season==Winter);

4 Crop Monitoring System

Due to scarcity of workers, automatic monitoring system has been established to measure temperature, humidity, pH at different time and from different locations by deploying various sensors [12, 13] which is shown in Fig. 2. The measured value will be converted to Digital signal using Analog to Digital Converter (ADC). Micro Electro Magnetic System which acts as a processing Unit transmits the data to Wireless Sensor Network. The sensors are co-ordinated using the GPS and they are connected to the base station in an ad hoc network using WLAN [14]. These values are compared with the pre-defined data set in the Knowledge Base where historical data's are stored in ontology knowledgebase. The corresponding decisions from the Knowledge Base are sent to the respective land owner's mobile through SMS. If rainfall comes, land owners no need to irrigate the land because humidity gets changed. So, water can be saved which in turn consumes power. Also if the temperature or humidity or pH goes beyond the threshold level, then it generates alert to the corresponding land owners.

5 Elephant Intrusion Prediction

Elephants are destroying the crops in the places like Mettupalayam, Coimbatore, and Pollachi. The elephant intrusions are predicted in order to resist the crop cultivated [15]. Seismic sensors are deployed around the hills which are used to detect the elephant and to protect the crops and people. It uses geophone to record

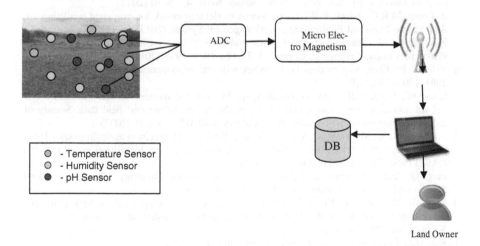

Fig. 2 Crop monitoring system

the footfalls of elephants. The exact measurement of each sensor identifies the exact place of elephant intrusion. After detecting, the alert message is sent to the land owners through RF and to the forest department immediately to protect the people.

6 Conclusion

This paper uses WSN technology for Crop cultivation, it provides low cost, consume less power. The ontology based crop cultivation predicts the crop to be cultivated with high yield. Fuzzy Associative Memory is used to predict the climate condition which is mainly to reduce the rate of false negatives. Automatic crop monitoring system has been established to avoid paucity of workers. Seismic sensors are used to detect the footfalls of elephants in hilly areas. This sensor uses geophone to record the footfalls of elephants and immediately alert message is sent to the people. The crop prediction and elephant intrusion prediction makes the farmer to have high yield.

References

1. Othman, M.F., Shazali, K.: Wireless sensor network applications: a study in environment monitoring system. International Symposium on Robotics and Intelligent Sensors, Procedia Engineering **41**, 1204–1210 (2012)
2. Nengfu, X., Wensheng, W.: Ontology and acquiring of agriculture knowledge. Agric. Netw. Inf. **8**, 13–14 (2007)
3. Keshtgari, M., Deljoo, A.: A wireless sensor network solution for precision agriculture based on zigbee technology. Sci. Res. J. Wirel. Sensor Netw. **4**, 25–30 (2012)
4. de Leona, M.R.C., Jalaob, E.R.L.: A prediction model framework for crop yield prediction. In: Asia Pacific Industrial Engineering and Management System (2013)
5. Kays, R., et al.: Tracking animal location and activity with an automated radio telemetry system in a tropical rainforest. Comput. J. **54**(12), 1931–1948 (2011)
6. Zviedris, R., Elsts, A., Strazdins, G.: LynxNet: wild animal monitoring using sensor. Networks **2009**, 170–173 (2010)
7. Hons, M., Stewart, R., Lawton, D., Bertram, M.: Ground motion through geophones and MEMS accelerometers: sensor comparison in theory, modelling and field data. Society of Exploration Geophysicists, University of Calgary, CREWES Project (2007)
8. Song, G., Wang, M., Ying, X., Yang, R., Zhang, B.: Study on precision agriculture knowledge presentation with ontology. In: AASRI Conference on Modelling, Identification and Control, AASRI Procedia, vol. 3, pp. 732–738 (2012)
9. Ping, Q., Yelu, Z.: Study and Application of Agricultural Ontology. China Agricultural Science and Technology Publishing House, Beijing (2006)
10. Stathakis, D., Savin, I., Nègre T.: Neuro-fuzzy modeling for crop yield prediction. In: The International Archives of the Photogrammetry, Remote Sensing and Spatial Information Sciences, vol. 34, Part XXX
11. Klir, G.J.: Fuzzy sets and fuzzy logic theory and applications
12. El-kader, S.M.A., El-Basioni, B.M.M.: Precision farming solution in Egypt using the wireless sensor network technology. Egypt. Inform. J. **14**, 221–233 (2013)

13. Jiang, X., Zhou, G., Liu, Y., Wang, Y.: Wireless sensor networks for forest environmental monitoring, pp. 2–5 (2010)
14. Majone, B., Viani, F., Filippi, E., Bellin, A., Massa, A., Toller, G., Robol, F., Salucci, M.: Wireless sensor network deployment for monitoring soil moisture dynamics at the field scale. Procedia Environ. Sci. **19**, 426–435 (2013)
15. Wood, J.D., O'Connell-Rodwell, C.E., Klemperer, S.: Using seismic sensors to detect elephants and other large mammals: a potential census technique. J. Appl. Ecol. **42**, 587–594 (2005)

Ethernet MAC Verification by Efficient Verification Methodology for SOC Performance Improvement

Sridevi Chitti, P. Chandrasekhar, M. Asharani
and G. Krishnamurthy

Abstract Verification of Gigabit Ethernet Media Access Control (MAC), part of most of the networking SOC is accomplished by using the most advanced verification methodology i.e. Universal Verification Methodology (UVM) has been presented in this paper. The main function of MAC is to forward Ethernet frames to PHY through interface and vice versa. With the use of UVM factory and configuration mechanism, coverage driven verification of MAC Characteristics such as frame transmission, frame reception etc. is achieved in best possible way. Coverage metrics and self-checking which reduces the time spent on verifying design. By using UVM methodology, a reusable test bench is developed which has been used to run different test scenarios on same TB environment.

Keywords UVM · MAC · Verification IP · Gigabit ethernet · TB environment

1 Introduction

In general, for verifying a SoC firstly, we need to verify the standard bus interconnecting IP Cores present in the system [1]. The whole verification process of SoC consumes approximately 70 % of total design time. In this research work, the problems taken care of are as follows:

S. Chitti (✉) · P. Chandrasekhar · M. Asharani · G. Krishnamurthy
Department of Electronics and Communication Engineering, SRITW,
Warangal, Telangana, India
e-mail: Sridevireddy.arram@gmail.com

P. Chandrasekhar
e-mail: sekharpaidimarry@gmail.com

M. Asharani
e-mail: ashajntu1@yahoo.com

© Springer India 2016
A. Nagar et al. (eds.), *Proceedings of 3rd International Conference
on Advanced Computing, Networking and Informatics*, Smart Innovation,
Systems and Technologies 44, DOI 10.1007/978-81-322-2529-4_12

117

1. Verification of Ethernet MAC which is an essential part of Ethernet SoC verification.
2. Development of VIP for MAC unit.
3. Using that MAC VIP, Ethernet MAC has been verified and coverage analysis has been performed [2].

2   Universal Verification Methodology (UVM)

UVM uses system Verilog as its base language. UVM methodology is vendor independent which is not the case for rest of the verification methodologies [3]. Need of a common verification methodology which provides the base classes and framework for constructing scalable and reusable verification environment has been achieved by the introduction of UVM [3]. UVM Improves productivity and ensures re-usability. Maintenance of the verification components is much easier because the components are standardized.

3   Proposed System

10 Gigabit Ethernet MAC implements a MAC controller conforming to IEEE 802.3 specification. This proposed system consists of two modules namely transmit module and receive module. Table 1 shows IEEE 802.3 data frame which consists of seven different fields. These fields are put together to form a single data frame which illustrates the seven fields: Preamble, Start-of-Frame delimiter, Destination Address, Source Address, Length, Data, and Frame Check Sequence [4–6].

3.1   Transmit and Receive Module

The transmit and receive engine provides the interface between the client and physical layer. Figure 1 shows a block diagram of the transmit and receive engine with the interfaces to the client and physical layer and vice versa [7].

The Fig. 2 describes the components of ETHERNET MAC verification Architecture, which consists of verification components like agent, driver etc.

Table 1 IEEE 802.3 ethernet frame	PRE	SOF	DA	SA	Length	Data	FCS
	7	1	6	6	2	46–1500	4

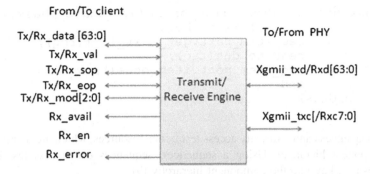

Fig. 1 Block diagram of ETHERNET transmit module

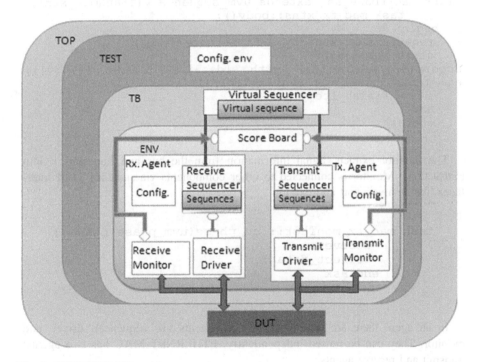

Fig. 2 ETHERNET VIP architecture

Generally data items are generated and transmitted to the DUV in a typical test. A large number of meaningful tests can be created by randomizing data item fields using System Verilog constraints thus maximizing coverage [8].

A driver fetches data repeatedly from sequencer, drives the DUT based on the protocol using the virtual interface.

```
task mac_tx_driver::run_phase(uvm_phase phase);
          forever begin
              seq_item_port.get_next_item(req);
              send_to_dut(req);
              seq_item_port.item_done();
          end
     endtask
```

The sequences cannot directly access test bench resources, which are available in the component hierarchy. Using a sequencer, sequences can access test bench resources as a key into the component hierarchy [9].

```
class mac_tbase_seq extends uvm_sequence #(transmit_xtn);
       task mac_tx_xtns::body();
  begin
   req=transmit_xtn::type_id::create("req");
   strat_item(req);
assert(req.randomize()   with   {data[2]   inside   {[20:0]};
en==1'b1; val==1'b1; avail==1'b1;});
          finish_item(req);
   end
   endtask
```

The monitor extracts signal information from the bus and translates it into transactions. Monitor is connected to other components via standard TLM interfaces like Analysis port and export.

```
task mac_tx_monitor::run_phase(uvm_phase phase);
          forever
          collect_data();
          endtask
     // add logic here
```

In an agent there are three specific components viz: sequencer, driver, and monitor. They can be reused independently. ETHERNET MAC has two agents: transmit and receive agents.

A scoreboard is a crucial element which verifies the proper operation of the design at functional level by comparing the predicted output from reference model with the actual output from receiver [8, 9].

```
function    void    mac_scoreboard::mac_transmit(transmit_xtn
tr);
if(txd_fifo_status ==0 && tr.full==1)
`uvm_info("MAC              transmit              function",
$psprintf("fifo_data=%b",fifo_data), UVM_LOW)
    begin
    fifo_en=tr.val;
     fifo_data = tr.data;
end
// add logic here
// similarly add mac_receive() method
 $cast(ref_xtn , re.clone());
   if(mac_receive(ref_xtn))
   begin
//compare
`uvm_info(get_type_name(), $sformatf("Scoreboard  -  Data
Match successful"), UVM_MEDIUM)
xtns_compared++ ;
```

The environment acts as the top-level component for all the verification components. Interface is a static component that encapsulates communication between the hardware blocks. It provides a mechanism to group together multiple signals into a single unit that can be passed around the design hierarchy thus reducing the amount of code and promotes reuse [8].

Based on the declared directions, Modport restricts the interface access within a module. The function of the clocking block is to identify the clock signals and to capture the synchronization and timing requirements of the modeled blocks. With the help of the clocking block, test bench drives the signals on time. Interface can contain more than one clocking block depending on the environment. Set up and hold time of the DUV can also be modeled [8].

```
interface mac_if(input bit clock);
    //  transmiter driver CB
   clocking tdr_cb @ (posedge clock);
        default input #1  output #1;
        output tx_data, tx_val,tx_sop, tx_eop,tx_mod;
        input tx_full;
    endclocking
modport tdr_mp(clocking tdr_cb);
```

Virtual interface instance is created by using keyword "virtual". By which drivers and monitors can be created and deleted dynamically during run time [9].

A sequencer that is not attached to any driver and does not process items by itself can be used for high-level control of multiple sequencers from a single sequencer. This kind of sequencer is referred to as a virtual sequencer [10].

Table 2 Local instance coverage details

Weighted average				94.61 %
Coverage type	Bins	Hits	Misses	Coverage (%)
Branch	400	357	43	89.23
Assertion attempted	10	10	0	100.00
Assertion failures	10	0	–	0.00
Assertion successes	10	10	0	100.00

Table 3 Coverage details for transmit and receive packet cover group

Coverage group	Goal (%)	% of Goal
Cover group type mac_fcov1	100.00	89.58
Cover group type mac_fcov2	100.00	88.88

3.2 Test Cases

To check the functionality of the ETHERNET according to the specification the scenarios which have been covered are as follows: Receive Enable, Receive available, Valid data (Tx and Rx), Start of packet (Tx and Rx), End of packet (Tx and Rx), Modulus length (Tx and Rx), Packets data (Tx and Rx), Receive error and Transmit full.

4 Coverage Reports and Results

According to Test Plan, the test cases are verified by developing the Verification IP for Ethernet Protocol. The Test Cases are written in the form of sequences in the Sequencer using System Verilog UVM methodology.

By using Questa simulation software, the Verification of Ethernet components such as transmit Agent and receive agent are done and the log files for the test cases are generated with Coverage report.

Table 2 shows the coverage of the whole environment for code and functional coverage. 94.61 % overall coverage has been obtained. Table 3 shows Tx coverage is 89.58 % and Rx coverage is 88.88 %. The cover group coverage is not 100 % as all the registered address is not required to be checked which results in 89.23 % coverage.

5 Conclusion

The specifications of ETHERNET are verified successfully using UVM methodology on QuestaSim simulator. Functional coverage i.e. measure of implementation of design is carried out and 94.61 % of coverage is extracted. The coverage can be improved by modifying the code according to the need. The scoreboard successfully compares the result of every transaction generated.

Acknowledgments Authors would like to express sincere thanks to Department of Science and Technology, New Delhi for their financial support to carry out this work under project Grant No. SR/WOS-A/ET-17/2012(G). Further our sincere feelings and gratitude to management and principal of Sumathi Reddy Institute of Technology for Women, for their support and encouragement to carry out the research work.

References

1. Chauhan, P., Clarke, E.M., Lu, Y., Wang, D.: Verifying IP core based system-on-chip designs. Carnegie Mellon University Research Showcase
2. Samanta, P., Chauhan, D., Deb, S., Gupta, P.K.: UVM based STBUS Verification IP for Verifying SOC Architectures. In: Proceedings of IEEE VLSI Design and Test, 18th International Symposium, doi:10.1109/ISVDAT.2014.6881037, Coimbatore, July 2014
3. Vaidya, B., Pithadiya, N.: An introduction to universal verification methodology. J. Inf. Knowl. Res. Electron. Commun. Eng. **2**, Nov-12 to Oct-13
4. Assaf, M.H., Arima; Das, S.R., Hernias, W., Petriu, E.M.: Verification of ethernet IP core MAC design using deterministic test methodology. In: IEEE International Instrumentation and Mesurements Technology Conference, victoria, May 2008. doi:10.1109/IMTC.2008.4547312
5. Tonfat, J., Reis, R.: Design and verification of a layer-2 ethernet MAC classification engine for a gGigabit ethernet switch. In: Proceedings of IEEE Electronics, Circuits, and Systems, Athens, Dec 2010. doi:10.1109/ICECS.2010.5724475
6. Frazier, H.: The 802.3z gigabit ethernet standard. In: Proceedings of IEEE Journal, vol. 12, May–June 1998. doi:10.1109/65.690946
7. Lau, M.V., Shieh, S., Wang, P.-F., Smith, B., Lee, D., Chao, J., Shung, B., Shih, C.-C.: Gigabit ethernet switches using a shared buffer architecture. Communications Magazine, IEEE, **41**(12), 76–84 (2003)
8. Bergeron, J.: Writing Test Benches using SystemVerilog. Springer, ISBN-10: 0-387-29221-7, Business Media (2006)
9. www.accellera.org/
10. www.testbench.in/ [online]

MCDRR Packet Scheduling Algorithm for Multi-channel Wireless Networks

Mithileysh Sathiyanarayanan and Babangida Abubakar

Abstract In this paper we considered multi-channel Deficit Round Robin scheduler (MCDRR) for the multi-channel wireless networks to provide better fairness to the users. The scheduler needs to exploit channel availability to achieve higher network performance. The MCDRR scheduling algorithm was first implemented in hybrid TDM/WDM optical networks and this was mainly developed to study multi-channel communication. The results provided nearly perfect fairness with ill-behaved flows for different sets of conditions. Now shifting our focus back to wireless networks, many algorithms proposed for the multi-channel wireless networks have fairness issues. This paper address fairness issue by investigating the existing scheduler for the IEEE 802.11n multi-channel wireless network case to provide efficient fair queueing. We take into account the availability of channels, the availability of data packets and efficiently utilize channels to achieve better fairness. Simulation results show that the MCDRR for multi-channel wireless networks can provide nearly perfect fairness with ill-behaved flows for different sets of conditions. Finally, after comparing our results, we say MCDRR performs better than the existing schedulers Round-robin (RR) and Deficit round-robin (DRR) in terms of fairness and throughput.

Keywords Multi-channel scheduling · Fair queueing · MCDRR · Wireless networks · Quality of service (QoS)

M. Sathiyanarayanan (✉) · B. Abubakar
School of Computing, Engineering and Mathematics,
University of Brighton, Brighton, UK
e-mail: M.Sathiyanarayanan@brighton.ac.uk

B. Abubakar
e-mail: b.abubakar@brighton.ac.uk

© Springer India 2016
A. Nagar et al. (eds.), *Proceedings of 3rd International Conference
on Advanced Computing, Networking and Informatics*, Smart Innovation,
Systems and Technologies 44, DOI 10.1007/978-81-322-2529-4_13

1   Introduction

From years, packet scheduling in wireless networks faces unique challenges compared with the wireline networks (optical networks) due to its bursty errors, time and location dependency [1–3]. The performance of a wireless packet scheduler can be enhanced by integrating channel state information and other protocol and application related information into the scheduling decision. From a user perspective, a scheduler must possess an element of fairness, so that each user can get it's due share of service.

Due to the number of high increase in the demand for wireless services especially real-time transmission such as video streaming, packet scheduling is much more challenging due to the time-sensitivity of the video packets and the time-varying nature of the wireless channel [4]. Large number of users may compete for the wireless channel simultaneously. Therefore, effective multichannel scheduling should be put in place to handle several packet scheduling. In multi-channel scheduling, it is the scheduling algorithm that is key to achieve high performance. In most of the research, major focus is on the delay and throughput performance of the whole system, but there is not much support for fairness and QoS guarantee [5].

From multi-channel wireless point of it, wireless radios operate on a designed radio frequency (RF) channel identified by numbers. The 2.4 and 5 GHz frequency bands have designated channels each. Certain channels may have less interference than others, depending on the network configuration. To optimise the network performance, a choice of channel is paramount. There are possibilities for the adjacent channels in 2.4 GHz to overlap and interfere with each other, whereas The 5 GHz band channels are non-overlapping, thus, all the channels have the capacity to be used in a single wireless system.

In our paper we use IEEE 802.11n, which has 5 GHz frequency band as the 802.11n devices work on both 5 GHz and 2.4 GHz. We consider multi-channel Deficit Round Robin scheduler (MCDRR) scheduler for the wireless networks. The first MCDRR [6] scheduler was implemented in the hybrid hybrid time division multiplexing (TDM)/wavelength division multiplexing (WDM) *optical networks* which used tunable transmitters and fixed receivers for multi-channel communication. This MCDRR scheduler for multi-channel communication is based on the deficit round-robin (DRR) scheduling for single channel case [7].

In a standard wireless scheduling framework, data packets arriving at the base station (BS) are classified into connections which are then classified into service flows. A packet scheduler is employed to decide the service order of the packets from the queues. If scheduling algorithm is properly designed, then the system can provide the desired service guarantees. Baring this in mind, we investigate the performance of a multi-channel deficit round-robin (MCDRR) scheduling algorithm for wireless networks, which can provide better fairness (in terms of throughput). First, the algorithm is explained, followed by discussion of the experimental results based on the simulation. Then we conclude our discussions in this paper (Fig. 1).

```
Initialization;
for i ← 0 to N − 1 do
    DC[i] = 0;
end

Arrival on the arrival of a packet p;
if Enqueue(i, p) is successful then
    if a channel is available then
        (ptr, ch) ← Dequeue();
        if pkt ≠ NULL then
            Send(*ptr, ch);
            if VOQ[i] is empty then DQ[i] ← 0;
        end
    end
end

Dequeue;
startQueueIndex ← (currentQueueIndex + 1)%N;
for i ← 0 to N − 1 do
    idx ← i + startQueueIndex%N;
    if VOQ[idx] is not empty then
        DC[idx] ← DC[idx] + Q[idx];
        if numPktsScheduled[idx] == 0 then
            currentQueueIndex ← idx;
            pos ← 0;
            ptr ← &packet(VOQ[idx], pos);
            repeat
                DC[idx] ← DC[idx] − length(*ptr);
                numPktsScheduled[idx] + +;
                pos + +;
                ptr ← &packet(VOQ[idx], pos);
                if ptr is NULL then Exit the loop;
            until DQ[idx] ≥ length(*ptr);
            Return (&packet(VOQ[idx], 0),      currentQueueIndex);
        end
    end
end
Return NULL;

Departure at the end of transmission on channel i;
numPktsScheduled[i] − −;
if numPktsScheduled[i] > 0 then
    ptr ← &packet(VOQ[i], 0);
    Send(*ptr, i);
    if VOQ[i] is empty then DC[i] ← 0;
end
else
    (ptr, ch) ← Dequeue();
    if ptr ≠ NULL then
        Send(*ptr, ch);
        if VOQ[ch] is empty then DC[ch] ← 0;
    end
end
```

Fig. 1 Pseudocode for the MCDRR algorithm [6]

2 MCDRR Algorithm for Multi-channel Wireless Networks

Multi-Channel Deficit Round-Robin (MCDRR) is the existing scheduler used in the case of multi-channel communication with tunable transmitters and fixed receivers in hybrid TDM/WDM Optical networks. The same MCDRR scheduler is tested on multi-channel wireless network as shown in Fig. 2a which consists of four flows, two channels and three users. We consider a wireless downlink network with multiple channels which consists of a single base station (including the scheduler) and multiple users. The users join the network for the purpose of receiving files from a source which is not modelled in our framework, and leave the network after downloading the complete file. The source transmits the file to the base-station, and then the base-station transmits to the user (wirelessly) using any of the channels in the network.

The MCDRR for wireless networks is implemented based on the MCDRR for optical networks considering: availability of the channels, availability of the data packets and overlaps 'rounds' in scheduling to efficiently utilize channels which is provided in a detailed pseudocode in Fig. 1. The virtual output queues (VOQs) are serviced by the simple round-robin algorithm with a quantum of service assigned to each queue. At each round, enqueue and dequeue process takes place.

Enqueue(i, p) is a standard queue operation to put a packet p into a VOQ for channel i. *Dequeue()* is a key operation of the MCDRR scheduling and returns a pointer to the head-of-line (HOL) packet in the selected VOQ or *NULL* when the scheduler cannot find a proper packet to transmit. *packet(queue, pos)* returns a pointer to the packet at the position of *pos* in the *queue* or *NULL* when there is no such packet. For each *VOQ[i]*, we maintain *DC[i]* which basically contains the byte that *VOQ[i]* did not use in the previous round and *numPktsScheduled[i]* which counts the number of packets scheduled for transmission during the service of *VOQ[i]*.

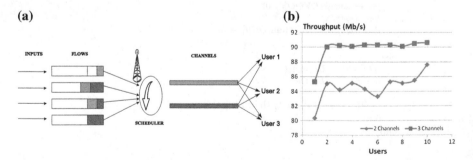

Fig. 2 **a** Block diagram of using MCDRR scheduler in wireless networks. **b** Throughput for the MCDRR scheduler: case 1

3 Simulation Results

To demonstrate the performance of the proposed MCDRR scheduling algorithm for multi-channel wireless networks, we carried simulation experiments with a existing model of flows with the base station along with the multi-channel and users as shown in Fig. 2a. Analogy to the previous paper on MCDRR for the optical networks, we consider each VOQ can hold up to 1000 frames. We measure the throughput of each flow at a receiver for 10 min of simulation time.

The results in Fig. 2b are based on the case 1. For the 10 users, 4 flows and 2 channels used, throughput is slightly lower when compared with the 3 channels used for the same number of users. The results in Fig. 3a are based on the case 2. For the 20 users, 4 flows and 2 channel used, throughput is again slightly lower when compared with the 3 channels used for the same number of users. In both the cases, the inter-frame times are exponentially distributed with the average 48 μs for all the flows, while the frame sizes are uniformly distributed between 64 and 1518 bytes for all the flows. In case 1, Raj Jain's fairness index [8] for 2 channel MCDRR is 0.9997756 and 3 channel MCDRR is 0.9999814. In case 2, Raj Jain's fairness index [8] for 2 channel MCDRR is 0.9998798 and 3 channel MCDRR is 0.9999876.

Later, we tested using the existing scheduling algorithms for single-channel case: round-robin (RR) and deficit round-robin (DRR) algorithms. DRR had a good consistency, fairness and better throughput compared with the RR. Again, as expect MCDRR outperformed DRR and RR as shown in the Fig. 3b. For all the cases: if number of flows and channels are increased, throughput gets better and we can see consistency in the results. Raj Jain's fairness index [8] for the 3 channel MCDRR is 0.9999876, 1 channel DRR is 0.9999813 and 1 channel RR is 0.6798836. So, from the simulation results, we found that the proposed MCDRR scheduling algorithm provides better throughput and fairness for the multi-channel wireless networks.

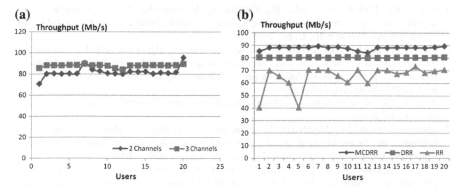

Fig. 3 **a** Throughput for the MCDRR scheduler: case 2. **b** Throughput for the RR, DRR and MCDRR scheduler: case 3

4   Conclusion

In this paper, the original contribution is that we adapted multi-channel Deficit Round Robin (MCDRR) scheduler which was used for the optical networks to work for the wireless network and investigated the scheduler's performance. Many algorithms proposed for the multi-channel wireless networks had fairness issues but MCDRR provided nearly perfect fairness. This algorithm tried to efficiently utilize the network resources (i.e., channels) by overlapping rounds, while maintaining its low complexity (i.e., $O(1)$). The scheduler exploited channel availability to achieve good network performance (i.e., throughput and fairness). We argue, taking into account the availability of channels and efficiently utilizing channels aids to achieve better fairness. Simulation results show that the MCDRR scheduler for the multi-channel wireless network provide better throughput and fairness when compared to the existing single channel schedulers (Round-robin (RR) and Deficit round-robin (DRR)) with ill-behaved flows for different sets of conditions.

There is quite a lot to consider in the future. First, more simulation results can be produced considering delay and latency. Performance test can be implemented considering wide range of parameters in the simulation model. Many multi-channel wireless packet scheduling algorithms like Idealized Wireless Fair Queuing (IWFQ), Channel-Condition independent Packet Fair Queuing (CIF-Q) and Wireless Packet Scheduling (WPS) can be implemented and compared with our MCDRR. Also, mathematical bounds can be established to prove the simulation results. Finally, we can look into the interesting future of testing MCDRR in the integrated optical and wireless networks, some of the existing architectures are GROW-NET (Grid Reconfigurable Optical and Wireless Network) [9] and FiWi (Fiber-Wireless Network) [10].

References

1. Bagaa, M., Derhab, A., Lasla, N., Ouadjaout, A., Badache, N.: Semi-structured and unstructured data aggregation scheduling in wireless sensor networks. In: INFOCOM. 1em plus 0.5em minus 0.4em IEEE, pp. 2671–2675 (2012)
2. Hassan, M., Landolsi, T., El-Tarhuni, M.: A fair scheduling algorithm for video transmission over wireless packet networks. 1em plus 0.5em minus 0.4em IEEE, pp. 941–942 (2008)
3. Ramanathan, P., Agrawal, P.: Adapting packet fair queueing algorithms to wireless networks. In: MOBICOM'98, pp. 1–9 (1998)
4. Zhao, S., Lin, X.: Rate-control and multi-channel scheduling for wireless live streaming with stringent deadlines
5. Sathiyanarayanan, M., Abubhakar, B., Dual mcdrr scheduler for hybrid tdm/wdm optical networks. In: Proceedings of the 1st International Conference on Networks and Soft Computing (ICNSC 2014), pp. 466–470. Andhra Pradesh, India (2014)
6. Sathiyanarayanan, M., Kim, K.S.: Multi-channel deficit round-robin scheduling for hybrid tdm/wdm optical networks. In: Proceedings of the 4th International Congress on Ultra Modern Telecommunications and Control Systems (ICUMT 2012), pp. 552–557. St. Petersburg, Russia (2012)

7. Shreedhar, M., Varghese, G.: Efficient fair queueing using deficit round-robin. IEEE/ACM Trans. Netw. **4**(3), 375–385 (1996)
8. Jain, R., Chiu, D., Hawe, W.: A quantitative measure of fairness and discrimination for resource allocation in shared computer systems. Digital Equipment Corporation, Technical Report DEC-TR-301 (1984)
9. Shaw, W.-T., Gutierrez, D., Kim, K.S., Cheng, N., Wong, S.-W., Yen, S.-H., Kazovsky, L.G.: GROW-net—a new hybrid optical wireless access network architecture. In: Proceedings of JCIS 2006 (invited paper). Kaohsiung, Taiwan (2006)
10. Ghazisaidi, N., Maier, M., Assi, C.M.: Fiber-wireless (fiwi) access networks: a survey. Comm. Mag. **47**(2), 160–167 (2009)

A Random Access Registration and Scheduling Based MAC Protocol with Directional Antennas for Improving Energy Efficiency

Alisha and P.G. Poonacha

Abstract In this paper, we consider a random access registering and scheduling based MAC protocol with directional antennas for improving energy efficiency. In this scheme senders interested in sending packets register with the Access Point (AP) using a low data rate CSMA/CA channel. Based on the available registered users the AP schedules data transfer over the high data rate communication channel in an efficient way using directional antennas for transmission. It also provides flexibility in scheduling stations. We study two scheduling schemes—Priority based scheduling and Earliest Deadline First (EDF). Our simulation results show that the proposed scheme works better than CSMA/CA under high load with less delay and use of directional antennas result in significant energy savings compared to CSMA/CA.

Keywords Wireless LAN · CSMA/CA · Polling · Location table · Data table · Transmission opportunity · Block acknowledgment · Directional antenna

1 Introduction

A channel access mechanism [1–3] is a way to allocate channel efficiently between stations. This is done through different channel access mechanisms defined by Medium Access Control (MAC) protocols. Such protocols provide rules for stations to transmit and receive data over a common access channel. In IEEE 802.11 standards, we have a number of different MAC protocols with each having its own advantages and disadvantages. We do come across a large number of protocols in the literature starting with controlled access protocols such as token passing,

Alisha (✉) · P.G. Poonacha
International Institute of Information Technology (IIIT-B), Bangalore 560100, India
e-mail: alisha.iiitb@gmail.com

P.G. Poonacha
e-mail: poonacha.pg@iiitb.ac.in

© Springer India 2016
A. Nagar et al. (eds.), *Proceedings of 3rd International Conference on Advanced Computing, Networking and Informatics*, Smart Innovation, Systems and Technologies 44, DOI 10.1007/978-81-322-2529-4_14

133

reservation based and polling protocols [4-5]. These protocols avoid collisions and may or may not use priority-based access. Second category consists of channelization protocols using FDMA, TDMA, and CDMA. In such protocols bandwidth, time is divided among stations. It results in wastage of resources if any station is not transmitting and if users come and go out of the network in a random fashion. Dynamic assignment of time slot or bandwidth every time is also used.

The most popular method uses random access protocols such as pure ALOHA, slotted ALOHA, CSMA, CSMA/CA and CSMA/CD with a number of variations. They allow stations to access the channel randomly (type of random access may vary from protocol to protocol). Recently, there has been lot of interest in using directional antennas to save energy and increase reach of transmitters in mobile ad hoc networks [6–9]. In this paper we use directional antennas at the AP and propose a random access registering and scheduling based MAC protocol with directional antennas.

Antenna is an electrical device, which converts electric power into electromagnetic waves or vice versa. In smart antenna's, array pattern can be adjusted according to the received signal to enhance the performance of the device. Smart antenna is categorized into two types: Switched beam antenna consist of fixed predefined patterns which are finite in number and Adaptive array antenna which can be adjust in real time and can have infinite number of patterns. We will use Adaptive antenna as directional antenna. This antenna can be programmed to direct its main beam towards some station based on (X, Y) coordinates. (X, Y) coordinates are the latitude and longitude of that station.

We will use directional antenna at Access Point and suppose that beam width of directional antenna is directly proportional to the power needed by a transmitter. This means that if we use directional antenna with 45° beam width at AP then we will need only 1/8th the total of power needed by an Omni-directional antenna [10].

2 New Channel Access Method

In our proposed model architecture, we use two channels, one for registration and control purposes and the other for data transfer.

- Control Channel: This channel is used for sending registration messages and control messages. It is a low data rate channel (about 100 Kbps). This channel will use an Omni-directional antenna.
- Data communication Channel: This channel will have directional antennas and it is used for sending high-speed data from the senders. The data rate assumed is 20 Mbps.
- The Access Point will have two types of antennas. One of them is an omni-directional antenna to receive signals from all directions and the other will be a directional antenna which can beam in required direction.

Omni and Directional
Antenna both in one AP

AP Contains Data Table
and Location Table

Wireless Stations

Wireless Stations
know their location
with the help of GPS

Fig. 1 Proposed method

In our proposed scheme, we have one Omni and two Directional antennas with an AP. Proposed Method setup is shown in Fig. 1.

The access point contains two tables with it, called Location Table and Data Table.

These tables are updated by the sender using the control channel. AP uses an Omni-directional antenna for receiving registration and control packets. These packets are sent by stations to the AP using CSMA/CA technique. Whereas AP uses directional antenna for receiving and sending data packets from sender to receiver. AP uses Location Table and Data Table for sending and receiving data.

Every wireless AP has some coverage area under it and AP keeps track of the all mobile stations under it in location table. If any new station wants to sending data then before sending data it should get itself registered under that AP then only station can send data.

Location Table in the AP gets updated under the following three conditions:

- If a new station is coming in coverage area of an AP or
- Already registered station wants to leave that particular AP or
- Station wants to change its location within the coverage area of AP.

We assume that each station determines its own geographical location by using Global Positioning System (GPS) and then this location information is use by AP to transmit/receive data (Tables 1 and 2).

Table 1 Location table

IP Address	MAC address	Location

Table 2 Data table

Sender IP	Receiver IP	Priority	Number of packets	TXOP	ACK

IP and MAC Address is the address of the station which is interested in updating LT. Location column represents the latitude and longitude information of that station. It is represented as (X, Y) coordinates.

Data Table is updated when an interested station which is registered under an AP wants to send data to the other station under the same AP. Each interested station sends its priority, number of packets and receiver information to the AP and AP sorts its Data Table based on priority and the number of packets. Different stations may have different type of traffic.

Basic Traffic Priority is of Four Types:

- Voice with Priority: 3
- Video with Priority: 2
- Best effort with Priority: 1
- Background application with Priority: 0

We assume that three (i.e. 3) is the highest priority and zero (i.e. 0) is the lowest priority. Voice has the highest priority and Background Application has lowest priority.

A Transmit Opportunity (TXOP) is a bounded time interval during which a station can send as many frames as possible.

AP will calculate the TXOP based on the number of packets with each station and AP with the help of directional antenna will beam form in senders direction for that much time to receive packets.

Acknowledgement (ACK) is initially set to False. When a particular station is served and AP receives Block ACK from that receiver, then this field is set to True and entry is deleted by AP from the data table.

2.1 Algorithm

- Stations interested in sending data register with the access point using low data rate random access channel. Packets are generated according to Poisson distribution. Each interested Station sends control packet to the AP using CSMA/CA technique. AP will update either Data Table or Location Table depending upon the control packet it receives.
- AP sorts the data table either on priority and number of packets or on earliest deadline first scheme.
- AP calculates the TXOP for the each station in the sorted data table.

$$\text{TXOP} = \text{Total Number of Packets} * (\text{Transmission Time of each Packet}) \quad (1)$$

- AP gets sender and receivers location information from the Location Table.
- AP uses directional antenna to beam form in sender and receivers direction. After beam forming in sender's direction AP sends TXOP time to sender.
- Sender sends its data within that time limit (TXOP) to AP and AP sends this data to receiver using another directional antenna.
- Receiver after receiving data sends block acknowledgement to AP via random access channel. On receiving block acknowledgement, AP deletes that station's entry from data table.

Note: Data Table and Location Table updating will happen in parallel with transmission of data.

3 Simulation Results

The designed algorithm was implemented in Java language using Eclipse. Our model consists of N independent stations. Each station generates packets for transmission. Two separate channels are used. One is control channel for control packets and other is the data channel for transferring data.

All the stations that are interested in sending data have data traffic with different priority and number of packets. The AP capacity for control channel is consider as 100 Kbps where as for data channel its 20 Mbps. We are assuming one packets size to be 2000 bytes and all packets are of same size.

$$M * lambda < R/X \quad (2)$$

where, M = number of stations,
R = bandwidth i.e. 20 Mbps
X = packet length i.e. 2000*8 bits

$$M * lambda < 1250 \quad (3)$$

We assume that almost every station will be able to update Data Table and Location Table. AP will have some extra storage space to store and then transfer data packets. All stations generate packets according to Poisson distribution. We assume that switching time of directional antenna is negligible. We measure delay, throughput and power consumption.

3.1 Energy Consumed

We have assumed that a directional antenna with 45° beam width at AP will require only 1/8th of the total of power needed by an Omni-directional antenna.

We use the known formula,

$$Energy = Power(Watt) * time \qquad (4)$$

In the graph above, we are comparing total energy consumed by proposed scheme versus CSMA/CA model. We are using beam width θ, (theta) of 45° in directional antenna. We can change θ, according to our requirement. As we can see from the graph as number of packets increases the energy used by standard CSMA/CA increases exponentially but in our proposed method, it is increasing very slowly (Fig. 2).

3.2 Delay and Throughput

In Fig. 3, we plot the average delay experienced by all the packets taken together. In Fig. 4, we plot the delay experienced by each station before it transmits data.

From Fig. 3, we can see that Average Delay Faced increases as the number of packets are increases but it is always less than 1 s. As can be seen from the plot in

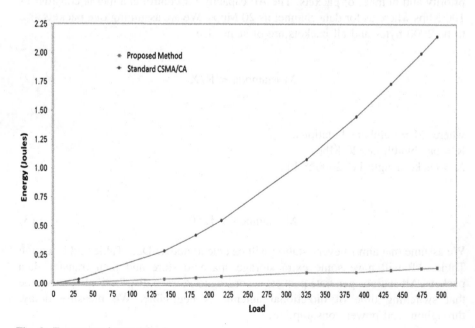

Fig. 2 Energy used versus load

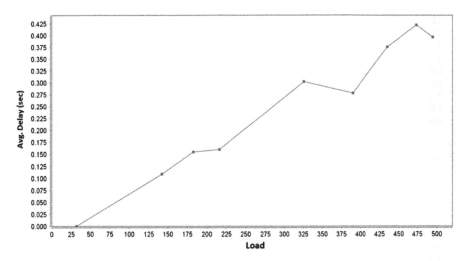

Fig. 3 Average delay (in sec) versus load

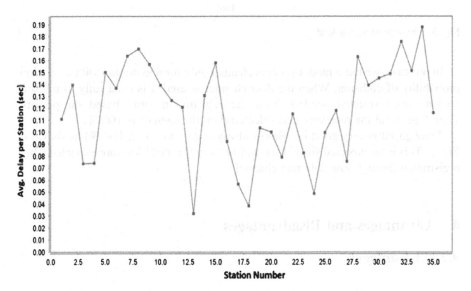

Fig. 4 Average delay (in sec) versus station number

Fig. 4, the delay faced by stations each time is different as sometimes stations have high priority data and sometimes low priority data.

We calculate throughput as the ratio of the total number of packets successfully transmitted and the total number of packets that can be serviced by the channel in a given time duration.

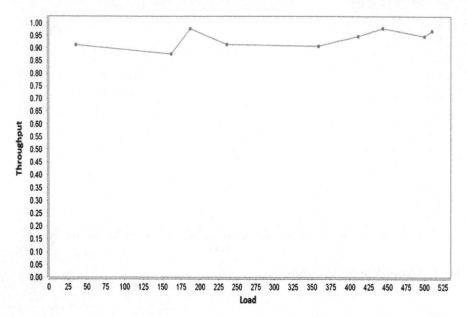

Fig. 5 Throughput versus load

In our case, we use a random access channel only for registration with a very low probability of collision. When the data channel is used it is used fully to transmit packets from selected senders from the registration table based on priority. Therefore, if all the receivers gets selected then throughput is 100 %.

Throughput is very good i.e. almost always equal to 1 (i.e. 100 %) as shown in Fig. 5. This is as expected since there will be very few collisions are expected while registration through low data rate channel.

4 Advantages and Disadvantages

4.1 Advantages

- Less power consumption: Use of directional antennas leads to much less power consumption in proposed scheme.
- Throughput: Much higher throughput can be achieved compared to CSMA/CA. This is due to much less collision of packets in our scheme. We use a low data rate channel for registration on which collision probability is very low. Transmission of data packets that are registered is guaranteed.
- No collision: Transmission of data packets that are generated is guaranteed. As there is no collision because we are using directional antenna and at a particular time AP will serve only one station rest will wait for their turn.

- Reduced ACK overhead: The acknowledgment sent by the receiver is send as block ACK so there is no need to acknowledge for every single frame.
- Reduced polling overhead: Each station will register under one AP. So it avoids polling by another AP and AP will give TXOP to only those stations registered with it.
- Priority Based Scheduling: Stations having high priority can always be serviced first.
- Absence of Hidden and Exposed Terminal Problems: As we are using two separate channels one for registration and control purposes and the other for data transfer. Therefore, there will be very few collisions at AP.
- Absence of Deafness problem at AP: We use a low data rate channel for registration using random access method. Therefore, deafness problem experienced while using directional antenna does not arise in our case.

4.2 Disadvantages

- Computations at AP: A processor at the AP is required to handle all calculations. These computations include sorting of Data Table based on priority and number of packets. After serving a particular station AP will delete its entry from the table.
- Increased hardware at the AP: We need two directional antennas and one Omni-directional antenna at AP in contrast to only one transmit and receive Omni-directional antenna.
- Non-real time traffic may starve because of priority based scheduling. But with EDF we can take care of this problem.
- Switching Speed of directional antennas could affect throughput and delay. We are assuming that with more innovations in the future such effects will be minimized.

5 Conclusions

Implementation of the proposed protocol in WLAN will be advantageous because power consumption by directional antenna is very less and more number of senders can send their data without loss of packets due to collision. In this scheme, priorities can be better handled. In addition hidden terminal and exposed terminal problems are significantly reduced due to directional antennas.

5.1 Future Scope

It is needed to achieve maximum throughput and minimum delay by predicting user traffic and adjusting the parameters in an adaptive manner. Switching speed of directional antenna may play a crucial role in limiting throughput and delay. This aspect needs further investigation.

If misleading station tries to attack AP with wrong/invalid information i.e. with highest priority and more number of packets then it is possible that AP waists its time by assigning TXOP to that misleading station.

We can save more energy of the system by making receiver or sender to sleep for the idle time.

References

1. Anjaly Paul, R.C.: A review of packet scheduling schemes in wireless sensor networks (2014)
2. Forouzan, B.A.: Data Communication and Networking. Tata Mcgraw Hill Education Private Limited (2006)
3. O'Reilly, P.J.P., Hammond, J.L.: Performance Analysis of Local Computer Networks. Addison-Wesley, Reading (1986)
4. Tomar, P., Poonacha, P.: An efficient channel access method using polling and dynamic priority assignment. In: 2013 4th International Conference on, Computer and Communication Technology (ICCCT), pp. 145–150, Sept 2013
5. ElBatt, T., Anderson, T., Ryu, B.: Performance evaluation of multiple access protocols for ad hoc networks using directional antennas. In: IEEE Wireless Communications and Networking Conference, WCNC 2003, no. 1, pp. 982–987, Mar 2003
6. Chang, Y., Liu, Q., Zhang, B., Jia, X., Xie, L.: Minimum delay broadcast scheduling for wireless networks with directional antennas. In: Global Telecommunications Conference (GLOBECOM 2011), 2011, pp. 1–5. IEEE, Dec 2011
7. Gossain, H., Cordeiro, C., Agrawal, D.: Mda: An efficient directional mac scheme for wireless ad hoc networks. In: Global Telecommunications Conference, 2005, GLOBECOM '05, vol. 6, pp. 3633–3637. IEEE, Dec 2005
8. Timm-Giel, A., Subramanian, A., Dhanasekaran, K., Navda, V., Das, S.: Directional antennas for vehicular communication-experimental results. In: Vehicular
9. Technology Conference 2007: VTC2007-Spring. In: IEEE 65th, pp. 357–361, April 2007
10. Nasipuri, A., Li, K., Sappidi, U.: Power consumption and throughput in mobile ad hoc networks using directional antennas. In: Proceedings Eleventh International Conference on, Computer Communications and Networks 2002, pp. 620–626 Oct 2002

A Key Agreement Algorithm Based on ECDSA for Wireless Sensor Network

Akansha Singh, Amit K. Awasthi and Karan Singh

Abstract Today the wireless sensor networks are being used extensively for general purposes and military aviation that results into the fulfillment of the security requirements. These networks are easily susceptible to attack. Many researchers provided security by using symmetric key cryptography but advanced research shows public key cryptography algorithms can also be used in WSN. Major studies and researches in literature nowadays stress on RSA and ECC algorithms but RSA consumes more energy than ECC. We propose a protocol to provide secure data delivery between node and gateway. Our protocol is based on ECDSA and the proposed scheme imposes very light computational and communication overhead with reducing the measure of key storage. Analysis depicts that the scheme has few merits in connectivity of key, consumption of energy and communication.

Keywords Wireless sensor networks · Network security · Key management

1 Introduction

Now a days such type of networks are widely used in industrial and home application [1]. Sensor network are commonly used in home monitoring, agriculture [2], disaster relief operation, railways and many more. Due to wireless nature, security must be carefully carried out in WSN. The secure clustering and key establishment are the challenging issues in WSN. Issues related to security in ad hoc networks

A. Singh (✉) · A.K. Awasthi
Gautam Buddha University, Greater Noida, India
e-mail: singhakansha1@gmail.com

A.K. Awasthi
e-mail: awasthi.amitk@gmail.com

K. Singh
Jawaharlal Nehru University, Delhi, India
e-mail: karancs12@gmail.com

© Springer India 2016   143
A. Nagar et al. (eds.), *Proceedings of 3rd International Conference on Advanced Computing, Networking and Informatics*, Smart Innovation, Systems and Technologies 44, DOI 10.1007/978-81-322-2529-4_15

very much resemble to those in sensor networks but the mechanisms of defense developed for ad hoc networks are not suitable for sensor networks because of the constrained energy. The key management scheme is significant form to maintain security in wireless sensor network. In order to cater a secure communication all messages transmitted between two sensor nodes should be authenticated and encrypted. Key management mechanism includes generation, distribution and maintenance within sensor nodes. According to literature, some of the key distribution mechanisms for WSNs are divided in two categories; symmetric-key protocols and asymmetric-key protocols to ground secure connections among the WSN nodes. In symmetric key protocols, communicating nodes use only one key for encryption and decryption.

Protocol based on using single shared key for the whole WSN, is not a flourishing concept as this key can be easily obtain by the adversary. Another overture is based on secure pair wise communication where, a set of secret keys assign to each sensor node. This solution requires each and every node to store $n-1$ keys in a network which consist n nodes. However this scheme is speculative for WSN as n can be large so does not scale well with memory requirement. Moreover, addition of node is also difficult in existing system. Major protocols depend on symmetric key cryptography but some research work lies on the basis of public key cryptography shows that it may be suitable for WSN. Recent research depicts about the validation of public key schemes for wireless sensor network.

In this research paper, we project a key management scheme for WSN lies on ECDSA for clustered wireless sensor networks. Proposed scheme intent to minimize the storage, calculation and communication overhead of the network to maximize the network lifetime.

The remaining part of research paper is detailed below: In Sect. 2, the existing key management scheme in WSN is discussed. In Sect. 3, proposed protocol is presented. Section 4, presents the analysis of security and working performance evaluation. The conclusion and result is given in Sect. 5.

2 Related Works

Key management and security issues in WSN are analysed in some literature survey [3–5]. Recent research shows that the algorithms of public key cryptography are extensively used in WSN [6, 7]. Major studies and researches emphasis on RSA and ECC algorithms. Wander et al. compared the cost value of key exchange and energy of authentication depends on RSA and ECC cryptography on an Atmel ATmega128 processor [8]. With the use of Elliptic Curve Digital Signature Algorithm (ECDSA), The ECC-based signature is rendered and proved. Further two scheme have been designed, Tiny PK [9] base on RSA and Tiny ECC [10] based on ECC. Some work and study has also been done in analyzing security protocols in WSN [11].

Research shows that ECC achieves the replica of security with smaller key sizes, low memory use, energy and computational cost. Consumption as compared to RSA. Thus, ECC is much more suitable for WSNs. Elliptic curve digital signature algorithm (ECDSA) is used for the verification of identity of cluster heads. The next section describes the system architecture.

3 Proposed Protocol Description

Sensor nodes are the core element of the network. Cluster head are also sensor nodes and the leaders of their groups having more computational power than the other sensor nodes. Gateway is a more powerful machine that performs costly operation and manage the whole network. Gateway is assumed secure and trustworthy.

For easiness and appliance, the notation and symbols used in this paper are summarized in Table 1.

In this research paper, we suggest a new ECDSA based key agreement protocol for WSN. The target of proposed scheme is to establish a secure communication link between gateway to cluster head and then cluster head to nodes using ECC. Our scheme divided into three phase. The first phase is key generation and key pre distribution phase which is performed before deployment of sensor nodes. The second phase consist with two sub phases. One is the key agreement between gateway and cluster head and second is establishing session key between cluster head and sensor nodes which are performed after deploying the sensor nodes. The three phases associated with the proposed scheme are illustrated as follows:

3.1 Key Generation and Pre Distribution Phase

The first phase is usually referred to as key pre-distribution phase. The mentioned phase is carried out offline before deploying sensor nodes. Gateway is fuddled by

Table 1 Notations

Symbol	Definition
E	Elliptic curve
G	Elliptic curve base point, a generator of the elliptic curve
N	Order of point, where n is a prime, $n \times G = O$
O	Point of E at infinity
pu_{GW}, pr_{GW}	Public and private key of gateway
pu_{CH_i}, pr_{CH_i}	Public and private key of ith cluster
T	Time stamp
h(.)	Secure and a lightweight one way hash function
K_m	Network key employed for registration

keys that are public and private based on ECDLP. The gateway generates public as well as private key pairs based on ECDLP. These keys are assigned to all cluster head. Gateway performs following process to generate keys

- Compute the pair, (pr_{CH_i}, pu_{CH_i}) such that $pu_{CH_i} = pr_{CH_i} G$ and $1 \le i \le N$ (N is the number of cluster head). Every cluster head are loaded with its ID_i, K_m, pu_{CH_i}, pr_{CH_i} and public key of gateway. Every node which is sensor loaded with its ID_j, K_m, K_{S_j} (session key) and public key of gateway.

3.2   Key Agreement Phase

Key agreement is divided into two parts: one is in between base station and cluster head, whereas, the next is to establish communication key between cluster and the members of the same cluster.

3.2.1   A Key Agreement Between Cluster Head and Gateway

When gateway wants to send message to cluster head it will take the following procedure:

- Gateway picks a random number r_i and compute

$$R = r_i G = (r_1, r_2)$$
$$K = r_i pu_{CH_i} = (k, l)$$
$$C = E_k(m)$$

And send $< ECDS_{pr_{GW}}\{h(m)\}, C, R, T >$ to the *ith* cluster head.

After receiving signed message at time T' cluster head execute the following steps

- Check validity of time-stamp, $T - T' \le \Delta T$, then cluster node executes the further operation, or else the mutual authentication phase is removed. Here, ΔT depicts the expected time interval for the transmission delay.
- Compute $= pr_{CH_i} R = (k, l)$
- *Compute $m = D_k(m)$*

Using ECDSA with public key of gateway cluster head can verify authenticity of the gateway. Now cluster head accepts the message m and k as a common key.

3.2.2   Key Agreement Between Cluster Head and Sensor Node

After deployment every sensor node should be associated with nearby cluster. Sensor node is charged with network key and public key of gateway. Each sensor node will be the legitimate member of cluster head through the following steps:

- All sensor node generates a random no. x_j, for the identification of this transaction and to shield communication from replay attack and link it with ID_j and encrypt the result with gateway public key and forward the following to nearby cluster head $\{E_{pu_{GW}}(x_j\|ID_j), h_{K_m}(ID_j), ID_j\}$. If more than one cluster are in the range then only one of them will be the cluster head of a node decided by the gateway according to the routing map.
- After receiving this message from the sensor node, cluster head confirm with network registration key (K_m) that this message is from identified sensor.
- Cluster head send $\{E_{pu_{GW}}(x_j\|ID_j), h_k(ID_i), ID_i, T\}$ to gateway. Gateway check the authenticity of the message and position of sensor node and classified every sensor into a cluster head according to the routing map and generate session key K_{S_j} and send to cluster head for communication between cluster and sensor node.

Once the key setup phase completed K_m is deleted from the memory of cluster and the sensor node.

3.3 Key Updating Phase

To increase security session key must be refreshed periodically. To support this functionality sensor node repeat the same phase of key set up with x_j as new session key, and K_{S_j} becomes the network key because K_m is no longer available in the memory.

4 Performance Analysis

Performance of the sensor network depends on Memory usage, computation and communication costs. Sensor node are the device with low memory capacity, limited transmission range and limited battery source. Memory usages should be kept small in WSN. In our proposed protocol, for each node there is no need to store public key of all other sensor node. To enhance the network lifetime, the number of message transmission and the amount of computation must be kept as low as possible. We concentrate on the performance of sensor nodes as the Gateway is usually termed as a powerful nodes. In key establishment phase there is only one message is broadcasted by the sensor node for session key generation so our scheme provides security with minimum communication cost. Computation cost depends on the key generation algorithm. In our proposed protocol, the main computational overhead for each node is to generate a random no., compute one way hash function and one encryption with gateway public key. Random number generation and hashing costs are comparability less energy consuming. So our protocol fulfilled the requirements of sensor networks.

In the case of cluster head, use of ECC-160 cater a reasonable security level with less energy consumption. Cluster head compute one ECDH and one ECDSA. The ECDH and ECDSA protocols desire point multiplication. There were various efforts to pursue ECC on MICA2 platform. The cost of energy for the authentication and exchange of key mainly depends on RSA and ECC which was investigated by Wander et al. on ATmega128 processor. They found that energy cost of digital signature computations with ECDSA consumes less energy than RSA and energy cost of key exchange computations with ECC also consumes less energy than RSA.

5 Conclusion

This paper generates and results a new ECC based key agreement protocol. We suggested in this paper, a periodic authentication scheme. Also, the proposed scheme does not increase keys in number, stored in sensor nodes memory and has a reasonable communication and computation overhead. The proposed protocol can prevent the general security issues. Session keys are updated periodically that increases the network security. Therefore, the suggested protocol is suited to wireless sensor networks.

References

1. Su, J.M., Huang, C.F.: An easy-to-use 3D visualization system for planning context-aware applications in smart buildings. Int. J. Comput. Stand. Interfaces **32**(2), 312–326 (2014)
2. Rehman, A.U., Abbasi, A.Z., Islam, N., Shaikh, Z.A.: A review of wireless sensors and networks' applications in agriculture. Comput. Stand. Interfaces **36**(2), 263–270 (2014)
3. Akyildiz, I.F., Su, W., Sankarasubramaniam, Y., Cayirci, E.: Wireless sensor networks: a survey. Comput. Netw. Int. J. Comput. Telecommun. Netw. **38**(4), 393–422 (2002)
4. Akkaya, K., Younis, M.: A survey on routing protocols for wireless sensor networks. Ad Hoc Netw. **3**(3), 325–349 (2005)
5. Wang, Y., Attebury, G., Ramamurthy, B.: A survey of security issues in wireless sensor networks. IEEE Commun. Surv. Tutorials, pp 223–237. 2nd Quarter (2006)
6. Eldefrawy, M.H., Khan, M.K., Alghathbar, K.: A key agreement algorithm with rekeying for wireless sensor networks using public key cryptography. In: IEEE International Conference on Anti-Counterfeiting Security and Identification in Communication, pp. 1–6. IEEE (2010)
7. Sahingoz, O.K.: Large scale wireless sensor networks with multi-level dynamic key management scheme. J. Syst. Architect. **59**, 801–807 (2013)
8. Wander, A.S., Gura, N., Eberle, H., Gupta, V., Shantz, S.C.: Energy Analysis of public-key cryptography for wireless sensor networks. In: PerCom'05: Proceedings of the 3rd IEEE International Conference Pervasive Computing and Communication, March (2005)
9. Watro, R., Kong, D., Cuti, S.F., Gardiner, C., Lynn, C., Kruus, P.: Tinypk: securing sensor networks with public key technology. In: Proceedings of the 2nd ACM Workshop on Security of Ad Hoc and sensor Networks (SASN 04), pp. 59–64. ACM Press (2004)

10. Liu, A., Ning, P.: Tiny ECC: a configurable library for elliptic curve cryptography in wireless sensor networks. In: IEEE International Conference on Information Processing in Sensor Networks, April 2008, pp. 245–256 (2008)
11. Casola, V., Benedictis, A.D., Drago, A., Mazzocca, N.: Analysis and comparison of security protocols in wireless sensor networks. In: 30th IEEE Symposium on Reliable Distributed Systems Workshops (2011)

Frame Converter for Cooperative Coexistence Between IEEE 802.15.4 Wireless Sensor Networks and Wi-Fi

Rambabu A. Vatti and Arun N. Gaikwad

Abstract With the exponentially increased users of wireless communication, a large numbers of wireless networks coexists in 2400 MHz Industrial, Scientific and Medical (ISM) band. The IEEE 802.15.4 Wireless Sensor Networks with their low power consumption and low cost advantages are widely adopted in low data rate industrial and consumer applications. The Wi-Fi, IEEE 802.11b/g offers high data rate and larger range. Both these technologies co-exist in ISM Band. The Wi-Fi signals having high signal strength interfere with the weak signals of IEEE 802.15.4, which degrades the throughput performance of IEEE 802.15.4 Wireless sensor networks. It results in a coexistence with non-cooperation. The authors developed a Frame Converter System to establish co-operative coexistence between IEEE 802.15.4 and IEEE 802.11b/g networks. The frame converter is implemented at the Media Access Control Layer using ARM7 processor. It converts the frames of IEEE 802.15.4 network to frames WiFi and WiFi to IEEE 802.15.4.

Keywords IEEE 802.15.4 wireless sensor networks · Wifi · IEEE 802.11b/g · Coexistence · Throughput

R.A. Vatti (✉)
Department of Electronics and Telecommunication Engineering,
Sinhgad College of Engineering, Vadagaon (Bk), Pune, Maharashtra, India
e-mail: rambabuvatti.india@gmail.com

A.N. Gaikwad
Dnyanaganga College of Engineering and Research, Narhe,
Pune, Maharashtra, India
e-mail: arungkwd47@gmail.com

© Springer India 2016
A. Nagar et al. (eds.), *Proceedings of 3rd International Conference
on Advanced Computing, Networking and Informatics*, Smart Innovation,
Systems and Technologies 44, DOI 10.1007/978-81-322-2529-4_16

151

1 Introduction

Various Wireless technologies like IEEE 802.15.4 Wireless Sensor Networks (WSN) or Wireless Personal Area networks, IEEE 802.11b/g Wireless Local Area Networks or Wi-Fi, IEEE 802.15.1 Bluetooth, Microwave Ovens, some more communication and non-communication devices are operating in the License free 2400 MHz ISM (Industrial, Scientific and Medical) band. The 2400 MHz band is over crowded with increasingly deployed low power, low cost WSN and high data rate IEEE 802.11b Wireless Local Area Networks in industrial and consumer applications. The data frames of these coexisted networks interfere with each other and result in significant performance degradation of each other. Both of these wireless network technologies operate in 2400 MHz frequency band with similar physical layer characteristics are encountering channel conflicts due to their over-lapped channels. The overlap of the channels of IEEE 802.15.4 networks and Wi-Fi are shown in Fig. 1.

Due to the independent design and development of the IEEE 802.15.4 Wireless Sensor Networks and the Wi-Fi technologies, there exist co-existence problems during their deployment [4]. There are only four Channels 15, 20, 25 and 26 in North America, channels 15, 16, 21 and 22 in Europe, which are non-overlapping with Wi-Fi [1, 5]. The IEEE 802.19 Coexistence technical advisory group develops and maintains policies that define the responsibilities of 802 standards developers in addressing issues of coexistence with existing standards and other standards is under development [6].

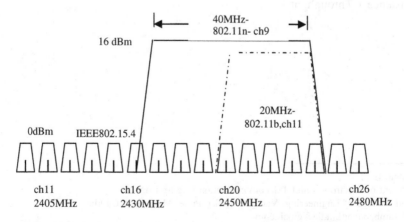

Fig. 1 Channels overlap between coexisted wireless networks (2400 MHz PHY) [1–3]

2 The Coexistence Problems

Two or more different network technologies operating in the same frequency band causes the coexistence issues. When two or more similar or different network technologies operating in the radio range of each other, there exist colocation issues. The two major problems to deal with these two coexisted technologies IEEE 802.15.4 and IEEE802.11b are: (a) Channel collision and (b) Interference.

2.1 Channel Collision Probability Between IEEE 802.15.4 and Wi-Fi

The IEEE 802.15.4 standard employs Direct Sequence Spread Spectrum (DSSS). It operates in four different bands. There are total 27 channels in all four bands. Channel 0 is available in 868 MHz band and is used in Europe. Channels 1–10 are available in 915 MHz band and are used in North America. Channels, 11–26 are available in 2400 MHz band and are used across the globe [1]. 950 MHz band is used in Japan [7]. In this paper, we consider only the 2400 MHz band because the coexistence problem between IEEE 802.15.4 and Wi-Fi occurs in 2400 MHz band only. The center frequencies of the channels in 2400 MHz band are given in Eq. (1) [1].

$$f_{IEEE802.15.4\_2400\,MHz} = 2405 + 5(k-11)\,MHz;\ k = 11, 12, 13, \ldots\ldots, 26 \quad (1)$$

IEEE 802.11b also employs Direct Sequence Spread Spectrum (DSSS) technique, and it defines 14 channels with 22 MHz band width for each one. The channels 1–13 are used in Europe and Singapore. Channels 1–11 are used in U.S and most of the countries of the world. The channel 14 is used in the Japan [8]. The center frequencies of IEEE 802.11 are separated by 5 MHz as shown in the Eq. (2).

$$f_{IEEE802.11b} = 2412 + 5k;\ k = 0, 1, 2, 3, \ldots, 13 \quad (2)$$

The adjacent channels in IEEE 802.11b are partially overlapped and the adjacent channel interference will occur when two IEEE 802.11b are co-located and the throughput performance of both WSN and Wi-Fi will be degraded. if the distance between the center frequencies is at least 25 MHz, it is possible to operate multiple networks simultaneously without interference [8]. Hence, the only three non-overlapping channels (1, 6, and 11) are used in all practical communications [9].

When the IEEE 802.15.4 network coexist with 'n' non-conflicting IEEE 802.11b networks, there exists two possible cases: (a) The IEEE 802.15.4 network transmits on one of the four available non-overlapping channels (channels 15, 20, 25 and 26). (b) The IEEE 802.15.4 network operates in one of the three overlapped channels of IEEE 802.11b/g.

In the first case, the probability of the non-conflicting channel allocation is always 1 regardless of number of coexisted Wi-Fi networks. In the second case, the non-conflicting channel allocation probability:

$$P_{nc} = (3 - n)/3 \tag{3}$$

The non-conflicting channel probability, for a single IEEE 802.15.4 network existing with 'n' IEEE 802.11b/g networks can be estimated by

$$P_{nc} = \begin{cases} \{(4/12) + (12/16) \times (3n)/3\}; & 0 \le n \le 3, \\ \{0; & \text{otherwise.} \end{cases} \tag{4}$$

Four channels are non-overlapped out of available 16 channels in 2400 MHz band.

2.2 The Wi-Fi Interference on IEEE 802.15.4 Wireless Sensor Networks

The Wi-Fi interference occurs as the channels of the IEEE 802.15.4 sensor network and the Wi-Fi network overlap. The normal transmission power of IEEE 802.11 b/g is between 15 and 20 dBm, and bit rates range from 1 to 54 Mbps using variety of modulations. Whereas, the normal transmission power of the IEEE 802.15.4 is only 0 dBm. The strong WiFi signals suppress the weak IEEE 802.15.4 signals [9]. The interference of Wi-Fi on the IEEE 802.15.4 networks can be characterized by energy detection by spectrum sensing process. The Wi-Fi signals are discontinuous and bursty. If the signal is present in N1 samples and absent in remaining samples, N0 = N − N1, The presence rate p1 = N1/N and absence rate P0 = N0/N and the energy distribution can be expressed as [10]:

$$Fw(x) = p_0 f_0(x) + p_1 f_1(x) \tag{5}$$

$f_0(x)$ is the energy distribution when the interfering signal is absent and $f_1(x)$ is the energy distribution when the interfering signal is present.

3 The Proposed Frame Converter System

The NXP Semicoductor-LPC2294 ARM7 TDMI processor is used in the prototype of the frame converter (FC) [11]. The FC has both the IEEE 802.15.4 and Wi-Fi radio interfaces. The FC converts the IEEE 802.15.4 frame into Wi-Fi frame at the one end and coverts back into IEEE 802.15.4 frame at the other end. This allows the IEEE 802.15.4 network to use the infrastructure of the Wi-Fi. Hence the radio range of the IEEE 802.15.4 network is extended to that of Wi-Fi. It establishes cooperative coexistence between the IEEE 802.15.4 Wireless Sensor Networks and Wi-Fi.

3.1 The Frame Conversion Algorithm

3.1.1 Transmission

1. Read DA, SA fields of IEEE 802.15.4 frame
2. Copy DA, SA in the respective Address fields of the Wi-Fi frame
3. Read data from data field of IEEE 802.15.4 frame
4. Copy into the data field of the Wi-Fi
5. Transmit with Wi-Fi signal strength.

3.1.2 Reception

During the data transfer from WSN to Wi-Fi, the data of the IEEE 802.15.4 can be easily accommodated in Wi-Fi frame. During the data transfer from Wi-Fi to WSN, if the Wi-Fi frame has more than 127 bytes of data, it has to be placed in N number of frames of IEEE 802.15.4. each containing maximum of 127 bytes of data.

Data transfer from Wi-Fi to IEEE 802.15.4 wireless senor network

1. Read Addr1, Addr2 from Wi-Fi frame
2. Copy DA, SA into the IEEE 802.15.4 wireless senor network frame
3. Seq. no. = 0
4. Read 127 Bytes of data
5. Copy into the IEEE 802.15.4 wireless senor network frame
6. Perform CSMA/CA and transmit
7. Seq. no = seq. no +1
8. Repeat steps 4–7 till the entire data is copied and transmitted.

4 Results

Experiments are conducted on the prototype Frame Converter (FC) and tested the working of the converter with and without the Wi-Fi interference. The channel and the signal strength are monitored with the InSSider Wi-Fi tracking tool. The number of packets received with and without the FC in the presence of different number of Wi-Fi channels are tabulated in Table 1 (Figs. 2 and 3).

Table 1 No. of packets received under Wi-Fi Interference

No. pkts transmitted	No Wi-Fi signal		1 Wi-Fi signal		2Wi-Fi signals		3Wi-Fi signals	
	Without FC	With FC	Without FC	With FC	Without FC	With FC	Without FC	With FC
10	10	10	7	9	6	9	5	7
15	15	15	12	13	11	13	7	10
20	20	20	17	18	15	17	8	13
25	25	25	21	24	19	23	13	16
30	30	30	24	26	22	25	19	24

Fig. 2 The No. of packets transmitted versus packets received in presence of Wi-Fi signals

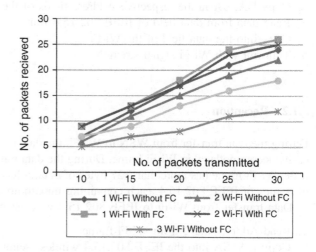

Fig. 3 Cumulative packet loss reduction with frame converter

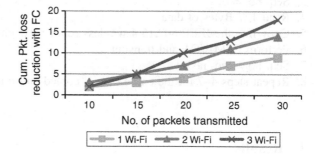

5 Conclusion

The FC is implemented on ARM 7 Processor. It is observed that the IEEE 802.15.4 can transmit to a larger distance with FC. The packet loss is less with the FC. The throughput performance of IEEE 802.15.4 is improved. Hence, the Frame Converter is providing the cooperative coexistence between IEEE 802.15.4

Wireless Sensor Networks and Wi-Fi Networks. The system can be further modified to convert the frames of other wireless networks, which operate in 2400 MHz ISM Band.

References

1. IEEE. 802.15.4., Standard 2006, Part 15.4: Wireless Medium Access Control (MAC) and Physical Layer (PHY) Specifications for Low Rate Wireless Personal area Networks (LR WPANs). IEEE–SA Standard Board (2006)
2. IEEE. 802.11n, Standard 2009, amendment 5 to Part 11: Wireless LAN Medium Access Control (MAC) and Physical Layer (PHY) Specifications. IEEE—SA Standard Board (2009)
3. Petrova, M., Wu, L., Mahonen, P., Rihijarvi, J.: Interference measurements on performance degradation between colocated IEEE 802.11 g/n and IEEE 802.15.4 networks. In: Proceedings of the Sixth International Conference on Networking, IEEE Computer Society, Washington, DC, USA, pp. 93 (2007)
4. Mahalin, N.H, Sharifa, H.S, Yousuf, S.K.S., Fisal, N., Rashid, R.A.: RSSI Measurements for Enabling IEEE 802.15.4. Coexistence with IEEE 802.11b/g. TENCON2009-IEEE, Singapore, pp. 1–4 (2009)
5. Rambabu, A.V., Gaikwad, A.N.: Congestion control in IEEE 802.15.4 wireless personal area networks. Int. J. Adv. Manage. Technol. Eng. Sci. III3(1), pp. 5–8 (2013)
6. IEEE. 802.19.1–2014—IEEE Standard for Information technology–Telecommunications and information exchange between systems—Local and metropolitan area networks—Specific requirements—Part 19: TV White Space Coexistence Methods (2014)
7. IEEE. 802.15.4d., Standard 2009, amendment 3 to Part 15.4: Wireless Medium Access Control (MAC) and Physical Layer (PHY) Specifications for Low Rate Wireless Personal area Networks (WPANs). *IEEE*—SA Standard Board (2009)
8. IEEE. 802.11, Standard 2012, Part 11: Wireless LAN Medium Access Control (MAC) and Physical Layer (PHY) Specifications. *IEEE*—SA Standard Board (2012)
9. Rambabu, A.V., Gaikwad, A.N.: Throughput improvement of randomly deployed wireless personal area networks. In: 2013 International Conference on Applied Computing, Computer Science, and Computer Engineering, Elsevier B.V. pp. 42–48
10. Penna, F., Pastrone, C., Spirito, M.A., Garello, R.: Measurement—based analysis of spectrum sensing in adaptive WSNs under Wi-Fi and bluetooth interference. In: 69th IEEE Conference on Vehicular Technology, Barcelona, pp. 1–5 (2009)
11. NXP Semiconductors LPC2294ARM7 Data Sheetsat: http://pdf1.alldatasheet.com/datasheet-pdf/view/86125/PHILIPS/LPC2294.html5. Accessed 10 Oct 2014

Wireless Sensor Networks and Wi-Fi Networks. The system can be configured both ... to convert the homes of other wireless networks, which operate in 2.4 GHz ... ISM Band.

References

1. ... ... (2004)

Part III
Research on Wireless Sensor Networks, VANETs, and MANETs

Secured Time Stable Geocast (S-TSG) Routing for VANETs

Durga Prasada Dora, Sushil Kumar, Omprakash Kaiwartya and Shiv Prakash

Abstract Enhancement of safety and security on road, and reduction in traffic congestion is the theme of Intelligent Transportation System (ITS). Vehicular Ad hoc Network as an integral part of ITS protects from Denial of Service attack and maintain Security Associations (SA) due to decentralized, open, dynamic, limited bandwidth and control overhead. Dynamic Time Stable Geocast (DTSG) routing protocol unable to protect from attacks of the intruder. Data authentication, Data integrity and non-repudiation are some security features which need to be taken into account for making geocast protocol robust and critically secured. This paper proposes algorithms for achieving the above mentioned goal of securing time stable geocast routing protocol in vehicular traffic environment. The proposed protocol is simulated in Network Simulator (ns2) and the results are compared with DTSG. The comparative analysis of results reveals the usefulness of the proposed protocol under security attack.

Keywords Geocast routing · Security attack · Vehicular ad hoc networks

D.P. Dora (✉) · S. Kumar · O. Kaiwartya
Wireless Communication and Networking Research Lab,
School of Computer and Systems Sciences, Jawaharlal Nehru University,
New Delhi 110067, India
e-mail: doradurga@gmail.com

S. Kumar
e-mail: skdohore@yahoo.com

O. Kaiwartya
e-mail: omokop@gmail.com

S. Prakash
Department of Chemical Engineering, Indian Institute of Technology,
New Delhi 110 016, India
e-mail: shivprakash@chemical.iitd.ac.in

© Springer India 2016
A. Nagar et al. (eds.), *Proceedings of 3rd International Conference on Advanced Computing, Networking and Informatics*, Smart Innovation, Systems and Technologies 44, DOI 10.1007/978-81-322-2529-4_17

1   Introduction

Due to the predictability and preventability of accidents, new techniques are being evolved by vehicle manufactures. Significant investment are being done in research and development in scientific and intelligent vehicular design by incorporating advanced wireless technologies in traffic control system (TCS) [1]. Vehicular Ad Hoc Networks (VANETs) is treated as a subclass of Mobile Ad Hoc Networks (MANETs) with some distinguishing characteristics such as high mobility, distributed and self-organizing network architecture, constrained movement pattern, unlimited battery power, etc. [2].

This paper proposes a Secured Time Stable (STSG) Geocast Routing protocol based on security concepts; namely, confidentiality, integrity, and nonrepudiation in data. We have designed four security algorithms for signing, key generation, voting and verification. In the following sections, the subsequent divisions of this paper are presented. Section 2 reviews the relevant literatures on time stable geocast routing protocol design in vehicular traffic environment. Section 3 introduces the proposed STSG routing protocol, its different phases and security algorithms. Section 4 provides the details regarding simulation and analysis of results. Section 5 concludes this paper.

2   Related Work

In [3], Christian Maihofer has purposed server approach, election approach and neighbor approach for storing and disseminating geocast message and in [4] he used caching concept and controlled dissemination range method in routing. Hamidreza Rahbar proposed the variability of geocast time in vehicular nodes by allowing it to work in two phases. When a node interprets an event it broadcasts warning packets until a helping vehicle is acknowledged [5]. Vehicular node and RSU shares mutual authentication by sharing and disseminating the authentication report obtained from third party [6].

In [7], pseudonyms are used with multiple key pairs storing facility for the preservation of the privacy. Different types of RFID authentication protocols are compared by the authors, use hash chain method [8]. Hash chain function is used to generate authentication keys for broadcasting in TESLA. A random selection is done for the last part of the chain and to generate key chain hash function is used. The time synchronization is manipulated to show the robustness of symmetric cryptography. In [9], Public Key Infrastructure having digital signature as its component is used to achieve data Non repudiation.

3 S-TSG

For warning vehicles, Geocast routing protocols give result with less precision of error which is a characteristics of ITS. For forwarding data packets in geocast routing the spatiotemporal features of nodes are taken into account which minimizes the bandwidth wastage and data packet dissemination overhead. As VANETs give traffic congestion information as well as emergency alerts in real time basis, safety and information security is the foremost important. Only legal nodes can be in the geocast region as well as forwarding region which should be authenticated.

3.1 Sender Side Authentication (SSA)

Setup phase, Key generation phase, Signing phase are executed on SSA. The spatiotemporal data such as latitude, longitude, time stamp values generated from Geographic Position System (GPS) which is embedded in each vehicle are used by setup phase as the parameters. In Key generation phase public key and private key

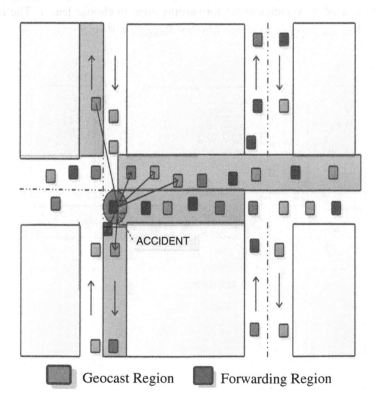

Geocast Region   Forwarding Region

Fig. 1 Sender side authentication

are generated by the vehicles and beaconing only the hashed value of public key, which is the beaconed key to avoid the manipulation of keys by intruder.

As shown in Fig. 1 when the vehicle faced an accident, it forms a geocast region where the vehicles get signal. The opposite lane is the forwarding region.

$K = Xlat^{Xlon}$, $R = g^{pvKlogK}$, where g is very large prime number.

We use Triple Encryption System (TES) by using encryption algorithm and hashing for making the data more secured.

C = Encrypt (p_v, msg)...1st round of TES, M = Encrypt (K, C)...2nd round of TES
E = h (h (h (h..........R times (M)))) = h^R (M).......3rd round of TES
Signature $S = p_vKlog(EK)$

Then the vehicle broadcasts (E, S) which is to be verified by the receiver to authenticate.

3.2 Leader Side Authentication (LSA)

Voting of leader and verification are done on LSA. As shown in Fig. 2, voting algorithm is used by vehicles in the forwarding zone to choose leader. The leader presumed $R_{Verified} = \frac{g^s}{pu^{KlogE}}$ to check whether $R_{Verified}$ = R or not.

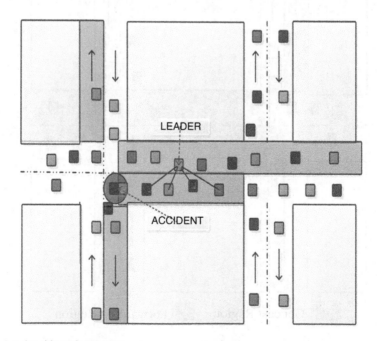

Fig. 2 Leader side authentication

$$R_{\text{Verified}} = \frac{g^s}{pu^{K \log E}} = \frac{g^{pvK \log(EK)}}{g^{PvK \log E}} = g^{pvK \log K} = R$$

If $R_{\text{Verified}} = R$ then the leader checks data integrity and data confidentiality.
Hence, $\text{msg}_{\text{Verified}} = \text{DC1 (DC2 (Revh}^{R\text{verified}}(E), K), p_u) = \text{msg}$

Algorithm 1. Secured Time Stable Geocast(S-TSG) Routing

Notations: R_{Verified}: a random large number chosen by leader; DC: decrypt function;
Encrypt: encrypt function; Hc: hashing function; E: beaconed message;
S: digital signature; g: very large prime number

Signing algorithm

 1. **While** (! Broadcast (E, S))
 2.   Parameters obtained from Setup phase;
 3.   Recursively generate K from Key generation phase;
 4.   **If** (K!=NULL)
 5.   **Then** R=$g^{pvKlogK}$;
 6.   C=Encrypt (private key, message);
 7.   M=Encrypt (K, C);
 8.   **While** (E != NULL);
 9.   Hash the message R times;
10.   S= p_vKlog(EK);
11. Broadcast (E, S);

Key generation algorithm

 1. **While** (! Accident)
 2.   **Calculate** $p_v = Xlat^{Xlon}$, $p_u = g^{Pv}$;
 3.   Hc=H(H(H(p_u))) .... $\alpha = \beta p_v$ times= Hα(p_u);
 4.   Beacon Hc;

Voting algorithm

 1. **Calculate** $\text{Lat}_{mid} = (\text{Lat}_{max}+\text{Lat}_{min})/2$, $\text{Lon}_{mid} = (\text{Lon}_{max}+\text{Lon}_{min})/2$;
 2. **While**(FR$_{axis}$=East-West)
 3.   **If**(ABS(Lon$_{nod}$)<ABS(Lon$_{mid}$))
 4.   **Then** Node=Next(Node);
 5. Lon$_{led}$ = Lon$_{nod}$;
 6. **While**(FR$_{axis}$=North-South)
 7.   **If**(ABS(Lat$_{nod}$)<ABS(Lat$_{mid}$))
 8.   **Then** Node=Next(Node);
 9. Lat$_{led}$ = Lat$_{nod}$;
10. **If**(ABS(TS$_{nod}$)<ABS(TS$_{mid}$))
11.   **Then** Node=Next(Node);
12. TS$_{led}$ = TS$_{nod}$;

Verifying algorithm

 1. The parameters E, S received by receiver
 2. **Calculate** R_{Verified}=g^S/ (p$_u$)$^{Klog E}$
 3. **Calculate** E_{Verified} = h^{Rverified}(M)
 4. **If** (Decrypt(E, p$_u$))
 5.   **Then** Confidentiality achieved
 6. **Else If** (E_{Verified}=E)
 7.   **Then** Integrity achieved
 8. **Else If** (Sign(sender)=Sign(receiver))
 9.   **Then** Nonrepudiation achieved
10. **Else**
11.   Discard Msg and wait

---

4 Simulation Results and Performance Analysis

Network Simulator ns-2 is used in Linux environment to evaluate the performance of S-TSG. The following simulation parameters are used such as a 2X3 double lane road having six junction points perceived as simulation area, 1000–1500 is taken as the range of vehicles, 50–80 km/h is considered as the speed of the vehicles. Transmission range is taken as 300 m and packer size is 512 bytes. CBR is the traffic type, MAC protocol 802.11 DCF.

In Fig. 3, it shows that that in S-TSG when the number of intruder vehicles increases the rate of traffic congestion is increasing but at a slower rate as compared to DTSG due to Triple Encryption System [TES]. The Leader blocks the messages that are being forwarded by the intruder vehicles to the intended vehicles to increase the traffic congestion. But in DTSG no such features available.

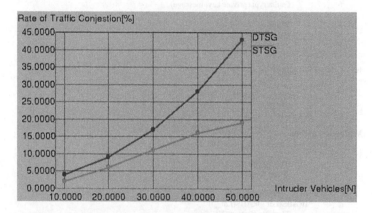

Fig. 3 Rate of traffic congestion (%) verses intruder vehicles

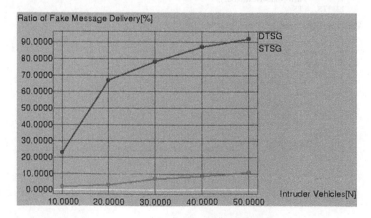

Fig. 4 Ratio of fake message delivery (%) verses intruder vehicles

Figure 4 shows the ratio of fake delivery messages increases marginally with the increment of the number of intruder vehicles in S-TSG. Hence, the graph depicts nearly parallel to the horizontal axis. As the intruder vehicles is not able to bypass the authentication system TES the rate of generation and dissemination of fake messages. But in case of DTSG it increases sharply for small range of intruder vehicles but after that it increases slowly due to the unavailability of any authentication system.

5 Conclusion and Future Work

In this paper, a secured time stable geocast routing approach protocol is proposed to provide security in terms of confidentiality, integrity, authentication and nonrepudiation in traffic data. The performance of STSG has been evaluated in terms of fake message delivery, network load and rate of traffic congestion. The performance analysis shows that STSG has lower fake message delivery rate, negligible network load and lesser rate of traffic congestion as compared to that of DTSG. In future, authors will explore the comparative study of S-TSG with some existing time stable approaches in the literature.

References

1. Li, F., Wang, Y.: Routing in vehicular Ad Hoc networks: a survey. IEEE Veh. Technol. Mag. **2**, 12–22 (2007)
2. Abdalla, G.M.T., Abu-Rgheff, M.A., Senouci, S.M.: Current trends in vehicular ad-hoc networks. In: IEEE Global Information Infrastructure Symposium, Morocco (2007)
3. Maihöfer, C., Leinmüller, T., Schoch, E.: Abiding geocast: time–stable geocast for ad hoc networks. In: Proceedings of the 2nd ACM International Workshop on Vehicular Ad Hoc Networks. Cologne, Germany, 02 Sept 2005
4. Mahofer, C., Eberhard, R.: Geocast in vehicular environment: caching and transmission range control for improved efficiency. In: IEEE Intelligent Vehicles Symposium (2004)
5. Rahbar, H., Naik, K., Nayak, A.: DTSG: dynamic time stable geocast routing in vehicular ad hoc networks. In: 2010 The 9th IFIP Annual Mediterranean Ad Hoc Networking Workshop (Med-Hoc-Net), pp. 1–7, 23–25 June 2010
6. Chaurasia, B.K., Verma, S.: Infrastructure based Authentication Scheme for VANETs. Int. J. Multimedia Ubiquitous Eng. **6**(2), 41–53 (2011)
7. Raya, M., Hubaux, J.P.: The security of vehicular ad hoc networks. In: Proceedings of ACM Workshop on Security of Ad Hoc and Sensor Networks (2005)
8. Syamsuddin, I., et.al.: A survey of RFID authentication protocols based on hash chain method. In: Proceedings of Third International Conference on Convergence and Hybrid Information Technology (2008)
9. Ma, D., Tsudik, G.: Security and privacy in emerging wireless networks. IEEE Wirel. Commun. **17**(5), 12–21 (2010)

Reduction in Resource Consumption to Enhance Cooperation in MANET Using Compressive Sensing

Md. Amir Khusru Akhtar and G. Sahoo

Abstract Energy and bandwidth are the scarce resource in a wireless network. In order to prolong its life nodes drop packets of others to save these resources. These resources are the major cause of selfish misbehavior or noncooperation. To enforce nodes cooperation this paper presents the reduction in resource consumption using Compressive Sensing. Our model compresses the neighborhood sparse data such as routing table updates and other advertisement. We have divided a MANET in terms of the neighborhood called neighborhood group (NG). Sparse data are compressed by neighborhood node and then forwarded to the leader node. The leader node joins all neighborhood data to reconstruct the original data and then broadcasts in its neighborhood. This gives a reduction in resource consumption because major computations are performed at leader end which saves battery power of neighborhood nodes. It compresses sparse data before transmission thus reduces the amount of transmitting data in the network which saves the total energy consumption to prolong life of the network. It also prevents from several attacks because individual nodes do not accept the advertisement and updates directly rather it uses leader node processed information.

Keywords Compressive sensing · Neighborhood group · Leader node · Border node · Regular node · Malicious node · Neighborhood compressive sensing

Md.A.K. Akhtar (✉)
Cambridge Institute of Technology, Tatisilwai 835103, Ranchi, India
e-mail: akru2008@gmail.com

G. Sahoo
Birla Institute of Technology, Mesra 835215, Ranchi, India
e-mail: gsahoo@bitmesra.ac.in

© Springer India 2016
A. Nagar et al. (eds.), *Proceedings of 3rd International Conference on Advanced Computing, Networking and Informatics*, Smart Innovation, Systems and Technologies 44, DOI 10.1007/978-81-322-2529-4_18

169

1 Introduction

A MANET work in a standalone manner in which every node has additional responsibilities of forwarding and routing. When nodes drop packets of others due to its honest or malicious causes it is called misbehavior or packet dropping attack. There are several reasons for the packet dropping attack categorized into honest and malicious [1, 2]. In the wireless environment most of the energy is consumed in the transmit mode and by dropping packets of other nodes want to prolong its life. The other scarce resource in a wireless network is bandwidth. Thus, nodes drop packets of others and save its bandwidth. The other honest reasons are network congestion, jamming and burst channel errors caused by interference, fading etc. These reasons are also responsible for the data dropping attack. While in malicious misbehavior nodes intentionally drop packets of others by deploying wormhole and blackhole attacks. All these reasons are responsible for the data dropping attack which degrades efficiency of packet transfer, maximizes the packet loss rate, increases the packet delivery time and creates network partitioning.

This paper presents the use of Compressive Sensing (CS) [3, 4] in the reduction of resource consumption to minimize battery and bandwidth usage. It also focuses on how attacks and misbehavior can be nullified. In this work a MANET is logically divided into several neighborhood groups (NGs) on the basis of one hop distance. A neighborhood group consists of a set of regular nodes, one or more border node(s) and one leader node. Regular nodes are responsible for compressing and forwarding the sparse data to the leader node. The border node is a regular node with an additional responsibility of passing the traffic to other neighborhood group. We have used the term neighborhood node to denote the regular or border node that belongs to a neighborhood group. The leader nodes are responsible for joining and reconstruction of original data as well as it checks the validity of the advertisement. The original data is broadcasted by the leader node in its neighborhood. This gives a reduction in resource consumption because major computations are performed by the leader node and only limited computations are performed by neighborhood nodes. Thus, it saves battery power of low processing devices because it compresses sparse data before transmission to reduce the amount of transmitting data. This gives a reduction in total energy consumption to prolong the network life. The NCS model prevents a network from several attacks because individual nodes do not accept the advertisement and updates directly rather it uses leader node processed information.

1.1 Overview of Compressive Sensing

Candes [3] and Donoho [4] proposed the new concept of signal sensing and compression called Compressive Sensing (CS). CS gained a wide acceptance in the recent years and applied in signal and image processing, pattern recognition, wireless communication, medical systems and analog-to-digital converters. In compressive sensing sampling and compression is done simultaneously and accurately. It compresses directly without involving intermediate steps of conventional compression techniques.

In the field of ad hoc network a set of sparse data can be compressed and forwarded to the leader node. The leader node uses projection to recover the actual data and forwards in its neighborhood, which reduces the amount of transmitting data in the network. The neighborhood nodes simply use leader node advertisements. Thus, the proposed model prevents networks from several attacks (such as blackhole, replay attacks, etc.) because wrong advertisements are not considered. It also declines the power consumption of neighborhood nodes because major computations are performed at the leader end. When we compare it with conventional routing methods all nodes are responsible for calculations, updates and forwarding. Thus, the battery of neighborhood nodes drains faster and lead to misbehave and noncooperation.

1.2 Motivation

The motivation behind this work is to define some new way to tackle from selfish misbehavior and attacks by reducing the resource consumption and by centralizing the control. In a MANET we have mostly low battery power devices. These nodes have the dual responsibilities of routing and forwarding and in order to save itself noncooperation is genuine. Noncooperation is in terms of misbehavior or packet dropping attacks to reduce the resource consumption such as battery lifetime and other resources. That's why we want to minimize the battery usage so that these devices can survive more and cooperate in network activities. In a battle zone the captain's Laptop can be taken as a leader node because it has high processing power and battery lifetime [5, 6]. We have used the compressive sensing in a neighborhood sparse domain. It compresses the neighborhood sparse data such as routing table updates and other advertisement and trust information. This saves battery power of low processing devices and protects the network from attacks and misbehavior because all advertisements and routing updates are under the care of leader nodes.

The rest of the paper is organized as follows. Section 2, enlightens the literature review in the field of reduction in resource consumption and compressive sensing. Section 3, presents the proposed NCS model. Analytical study is discussed in Sect. 4. Finally, Sect. 5 concludes the paper.

2 Literature Review

MANET is most susceptible to selfishness, that's why it needs some mechanisms to enhance cooperation [2]. Enforcing cooperation using complex techniques protects the network from attacks and misbehavior but consumes more battery power and bandwidth thus, reduces the life of the network. Let's take a look on how reduction enhances cooperation in MANET. After that we have discussed some prior work on Compressive Sensing.

2.1 Reduction Enhances Cooperation

Reduction in resource consumption surely enhances cooperation because it minimizes the real cause of selfish misbehavior [1, 2]. The reduction is in terms of minimizing the routing activities that saves battery life and bandwidth, so that misbehavior can be controlled up to the maximum extent. A lot of work is proposed in the literature to minimize total control traffic overhead using partitioning and subnetting. These works minimizes resource consumption, because they limit unnecessary broadcasts. Hence, nodes have fewer chances for misbehaving and it is true also because they have enough energy to survive.

Chiang et al. [7] proposed a Partition Network Model to minimize the routing overhead using Mobile agents. This model enhances network cooperation by reducing routing overhead. Lopez et al. [8] proposed the subnetting concepts to reduce routing overhead. It divides a network into several subnets to reduce the unwanted packets in a network. By reducing the routing overhead it saves battery power thus enhance cooperation in MANET. This model uses an internet type structure to group nodes into subnets. But, subnetting concept is difficult to apply in MANET due to their dynamic and distributed nature. This paper introduces several open challenges such as subnet formation and address acquisition, mobility of nodes between subnets and the intra-subnet and inter-subnet routing. Lots of efficient virtual subnet model for MANET were proposed in the literature [9–11]. These models enhance cooperation by minimizing overhead. But, these solutions are not appropriate for devices having low computation power. The limitation is that nodes in the subnet are authenticated using certificates which involve lots of computations. Akhtar and Sahoo have proposed a novel approach for securing an ad hoc network using the Friendly Group model [12]. It uses two Network Interface Cards (NICs) in the border node to partition a MANET into several friendly groups. This model reduces battery usage by reducing total control traffic overhead which enhances cooperation in MANET.

2.2 Prior Work on Compressive Sensing

Compressive sensing is a new technique of signal sensing and compression, it samples and compresses a signal simultaneously and reconstructs with high accuracy. Lots of work has been proposed in the field of signal processing, pattern recognition and wireless communication to reduce the energy consumption and maximize the network lifetime. Here, we have discussed some of the prior work on compressive sensing.

Lee et al. [13] have proposed a combined compressed sensing. It incorporates routing design with compressive sensing for energy efficient data gathering in sensor networks. In the Simulation results they have shown the effectiveness of the proposed combined technique in comparison with standard compressed sensing. Chou et al. [14] have proposed an adaptive algorithm based on the compressive sensing. This algorithm is used to gather information from WSNs in an energy efficient way. Feizi et al. [15] demonstrated some applications of compressive sensing over networks. They make connection between compressive sensing and traditional information theoretic techniques for source coding and channel coding. They show the explicit trade-off between the rate and the decoding complexity. Zhang et al. [16] proposed compressed neighbor discovery protocol that allows all nodes to concurrently discover their neighborhoods with a single frame of transmission, which is normally of a few thousand symbol epochs. This scheme is more efficient than conventional random-access discovery, because nodes have to retransmit lots of frames with random delays to effectively discover. For the reduction in energy consumption. Xiong et al. [17] proposed a combined method by taking compressive sensing and network coding inner contact in WSN. This work enhances network lifetime by reducing total energy consumption. Recently, for improving the performance of routing in wireless sensor network an adaptive and efficient technique based on compressive sensing (ECST) is proposed by Aziz et al. [18]. The proposed technique gives better results than the existing protocol in terms of the network life time and energy consumption. This paper inspired us to use compressive sensing in MANET.

3 Proposed NCS Model

3.1 Overview

This section presents the proposed model for the reduction in resource consumption using compressive sensing (CS) technique [3, 4, 19–21]. In a MANET we have heterogeneous devices having different storage, processing capability and battery lifetime. Processing of routing table updates and other advertisement consumes valuable resources (such as energy and bandwidth), which curtail the life of low battery power devices. That's why this model defines Neighborhood Compressive Sensing (NCS), so that the major computations are performed at leader end. Thus, it saves battery usage of low battery power devices. It prevents from attacks and misbehavior because the fake advertisement shows misclassified coefficients with abrupt and dissimilar values. In CS we check the sparse matrix in terms of zero and nonzero elements or in terms of similar and dissimilar coefficients. We have mostly zero elements and some are nonzero elements. Thus, the malicious advertisement can be easily identified because it has dissimilar coefficient or nonzero elements. But, the limitation of this technique is that if the network having a larger number of misbehaving nodes or attackers then it misclassifies the identity because we have large number of malicious data. Thus, we take action on regular or honest nodes because they are less in number. To handle this point we have considered the leader node reputation value as the basis for classification and validity of zero elements or matching coefficients. In this paper we are assuming mostly regular or honest nodes and a lesser number of malicious nodes in the network. The limitation will be considered for future work.

3.2 Mathematical Background

Consider in an ad hoc network in which K number of nodes are cooperating each other for its network operations. This work divides the network into N number of NGs having W number of nodes per NGs defined as $K_1 + K_2 + K_3 + \cdots + K_N = K$

Let $d = [d_1, d_2, \ldots, d_N]^T$ represent the data received by N number of nodes in a neighborhood. The received data (d) is sparse in nature. Here, routing table updates and other advertisement are taken as sparse data. This work assumes a less percentage of selfish nodes than regular nodes. The leader node coefficient is the basis of authenticity and sparsity because regular node coefficient is approximately same in the neighborhood. Hence, the leader node and regular nodes data is sparse in nature. Whereas malicious advertisement contain dissimilar coefficients and can be easily identified by leader node. Suppose that Ψ denotes the domain which is an MXN orthonormal basis represented as

$$\Psi = \begin{bmatrix} \Psi_{11} & \Psi_{12} & \cdots & \Psi_{1N} \\ \Psi_{21} & \Psi_{22} & \cdots & \Psi_{2N} \\ \vdots & \vdots & \ddots & \vdots \\ \Psi_{M1} & \Psi_{M2} & \vdots & \Psi_{MN} \end{bmatrix}$$

Then data **d** can be represented in Ψ domain as

$$\begin{bmatrix} d_1 \\ d_2 \\ \vdots \\ d_M \end{bmatrix} = \begin{bmatrix} \Psi_{11} & \Psi_{12} & \cdots & \Psi_{1N} \\ \Psi_{21} & \Psi_{22} & \cdots & \Psi_2 \\ \vdots & \vdots & \ddots & \vdots \\ \Psi_{M1} & \Psi_{M2} & \vdots & \Psi_{MN} \end{bmatrix} \begin{bmatrix} x_1 \\ x_2 \\ \vdots \\ x_N \end{bmatrix} \tag{1}$$

i.e., $d = \Psi X$

where $X = [x_1, x_2, \ldots, x_N]^T$ are the obtained coefficients of **d** in domain Ψ.

3.3 Phases of the NCS Model

Our model contains three phases i.e. the Neighborhood creation and leader selection phase, the Data compression and forwarding phase and the Data gathering, reconstruction and advertisement phase.

3.3.1 Neighborhood Creation and Leader Selection Phase

In this phase one or more neighborhood group(s) is defined on the basis of single hop neighbors or one hop distance. The one hop distance is computed on the basis of leader node and regular node locations as shown in Fig. 1.

Let $LN = (x_1, y_1)$ and $RN = (x_2, y_2)$ then the Euclidean distance d is computed using

$$d(LN, \ RN) = \sqrt{(x_1 - x_2)^2 + (y_1 - y_2)^2}$$

A node is in the coverage area of a NG if Euclidean distance is less than or equal to communication range. On the other hand when a RN is within the coverage of two or more NGs, then the minimum Euclidean distance is the basis for selecting a RN as a member for the NG. If two or more LNs are at the same distance from the RN, in that case the RN is added to any one of the NGs.

After that a leader node is elected for each neighborhood group on the basis of high computational power and battery lifetime. For example, in a battle zone a leader node could be a captain's laptop [5, 6]. Figure 2a, b shows the proposed diagram of the NCS model. Figure 2a shows a neighborhood group in which nodes are at a distance of one hop and arranged in a grid pattern, where arrows indicate bidirectional flow. The center node acts as a leader node. The leader node could also be in the corner or periphery of the network. But, choosing a node at the center gives large geographical coverage. Here, this work employs only grid pattern, but nodes can be arranged in other patterns on the basis of one hop distance.

The complete network is shown in Fig. 2b in which 4 nodes act as leader nodes having high processing power and battery lifetime. For intergroup routing only relevant information is forwarded by border nodes. The border nodes can also be elected from the periphery of the neighborhood group. The selection of leader node and border node can be defined as per the requirement of the network and mobility. In this paper we are concentrating on data compression using compressive sensing to maximize the network lifetime and to prevent a network from attacks and misbehavior. We have not discussed the selection of leader node in this paper. To select a leader node we can use any of the existing algorithms [22, 23].

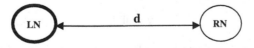

Fig. 1 Euclidean distance (d) between LN and RN

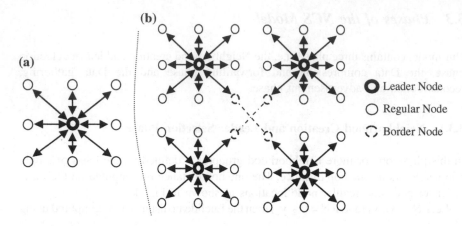

Fig. 2 **a** Neighborhood group. **b** A MANET in NCS model

3.3.2 Data Compression and Forwarding Phase

In this phase, neighborhood sparse data are compressed by neighborhood nodes individually and forwarded to the leader node instead of its direct use. Data means any type of advertisements or routing table updates.

For example, the reception and broadcast of hello message in a neighborhood is used to determine the network connectivity. A network having 4 NGs is shown in Fig. 3. It shows that data within a NG is sparse or of the same coefficient. Table 1a, b shows the routing table of node $NG_{1,1}$ which is sparse because, it has few large or dissimilar coefficients and many small or similar coefficients. The compression is performed individually by the nodes of a neighborhood group. Here, sparse data denotes such type of data in which some of its coordinates having similar value and the rest having another value.

Let $d = [d_1, d_2, \ldots, d_N]^T$ represent the data received by N number of nodes in a neighborhood group. The compressed version y can be defined for the data **d** through the measurement matrix Φ as

$$\begin{bmatrix} y_1 \\ y_2 \\ \vdots \\ y_M \end{bmatrix} = \begin{bmatrix} \Phi_{11} & \Phi_{12} & \cdots & \Phi_{1N} \\ \Phi_{21} & \Phi_{22} & \cdots & \Phi_{2N} \\ \vdots & \vdots & \ddots & \vdots \\ \Phi_{M1} & \Phi_{M2} & \vdots & \Phi_{MN} \end{bmatrix} \begin{bmatrix} d_1 \\ d_2 \\ \vdots \\ d_N \end{bmatrix}$$

or, we can write

$$y = \Phi d \tag{2}$$

where Φ is MXN random Gaussian or Bernoulli matrix with M $\ll$ N.

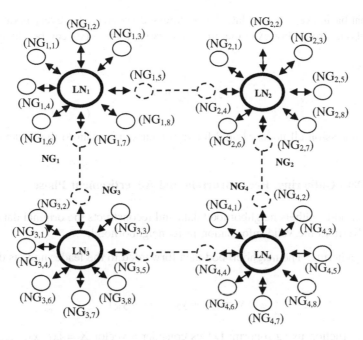

Fig. 3 Hello message broadcast is sparse

Table 1 (a) Hello message within the NG$_1$ for node NG$_{1,1}$, (b) hello message in NG$_2$ for node NG$_{1,1}$

Dest	Next hop	Metric
a		
NG$_{1,2}$	LN$_1$	2
NG$_{1,3}$	LN$_1$	2
NG$_{1,4}$	LN$_1$	2
NG$_{1,5}$	LN$_1$	2
NG$_{1,6}$	LN$_1$	2
NG$_{1,7}$	LN$_1$	2
NG$_{1,8}$	LN$_1$	2
b		
NG$_{2,2}$	LN$_1$	5
NG$_{2,3}$	LN$_1$	5
NG$_{2,4}$	LN$_1$	5
NG$_{2,5}$	LN$_1$	5
NG$_{2,6}$	LN$_1$	5
NG$_{2,7}$	LN$_1$	5
NG$_{2,8}$	LN$_1$	5

On that basis, we can calculate the compressed version y for every nodes of the neighborhood group. For example, the compressed version (y_i) can be obtained for node$_i$ as

$$y_i = \sum_{j=1}^{n} \Phi_{ij}d_i$$

After compression all nodes should forward their sparse data to the leader node.

3.3.3   Data Gathering, Reconstruction and Advertisement Phase

The leader node gathers neighborhood data and reconstructs the original data. After that it advertises the valid information in its neighborhood.

- Data gathering: The neighborhood data forwarded to the leader node is denoted as

$$y = (y_1 + y_2 + \cdots + y_N)$$

- Reconstruction using p-norm: Let us consider a vector $X = [x_1, x_2, \ldots, x_N]^T$ The p-norm of the vector is defined using

$$\|X\|_p = \sqrt[p]{\|x_1\|^p + \|x_2\|^p + \cdots + \|x_N\|^p}$$

Similarly the one norm can be obtained as

$$\|X\|_1 = \|x_1\| + \|x_2\| + \cdots + \|x_N\|$$

From Eqs. (1) and (2) the leader node reconstructs the original data by solving the given L1-minimization problem.

$$\min \|X\|_{L1}$$

subject to

$$y = \Phi\Psi X, \, d = \Psi X$$

We have used the "Signal recovery from random measurements via orthogonal matching pursuit" algorithm [21] at the leader end to reconstruct the original data.

- Advertisement of valid information: The leader node after reconstruction advertises the valid information in its neighborhood.

4 Analytical Study

4.1 Scenario Description

To demonstrate the working of the proposed NCS model, let us consider a network of 9 nodes in which node n_5 is chosen as a leader node and node n_3 is defined as malicious node. The network is arranged in a grid pattern as shown in Fig. 4a. The proposed model protects the MANET from threats and misbehavior as well as it reduces resource consumption. The detection of attacks and misbehavior is presented in Sect. 4.2 and reduction in resource consumption is discussed in Sect. 4.3.

4.2 Detection of Attacks and Misbehavior

The malicious advertisement gives wrong information to the network and thus nodes having false routes in the MANET. To check the validity of the advertisements neighborhood nodes forwards the Advertisement Check Request (ACREQ) compressed message to leader node as shown in Fig. 4b. The ACREQ message is denoted by a pair (node_number, data). The data forwarded by neighborhood nodes is 'i' and malicious node is 'j'. Arrows in the figure indicate transmission of ACREQ message from neighborhood nodes to leader node.

Then the leader node advertises the Advertisement Check Reply (ACREP) message in its neighborhood after processing (joining and reconstructing) of the received compressed data. The ACREQ information is sparse in nature because all neighborhood nodes are at a distance of one hop. The neighborhood node compresses these advertisements and forwards it to the leader node. The leader node gathers all neighborhood data and reconstructs the original data. Finally, it advertises the original data and identity of misbehaving nodes in its neighborhood.

The leader node receives all correlated data including the malicious nodes advertised fake or misclassified data. Thus, it identifies that malicious nodes have broadcasted wrong advertisements in the neighborhood. Now, we can punish the

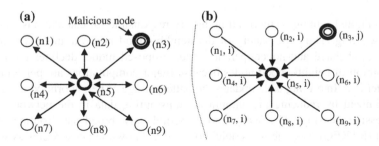

Fig. 4 a Scenario description. b ACREQ message from neighborhood nodes

malicious node by isolating misbehaving nodes from the routing path and discarding its messages by sending an alarm message in the neighborhood. A neighborhood node forwards the node_number and compressed advertised data to leader node for checking the validity of the advertisement and to know the identity of the malicious nodes. The sparse data gathered by leader node can be represented as

$$y = (y_1 + y_2 + \cdots + y_N)$$

The data gathered by leader node is sparse in nature because all neighborhood nodes are at a distance of one hop. Thus malicious advertisements can be easily identified from the given matrix S. The data forwarded by neighborhood node is 'i' and malicious node is 'j' on the basis of its coefficient. This work assumes a less percentage of malicious nodes than regular nodes in a neighborhood. That's why the malicious advertisements contain dissimilar coefficients and can be easily identified by leader node. Matrix 'S' shows that we have very little malicious/fake advertisements. The 'j' value indicates fake advertisements by malicious nodes.

$$S = \begin{bmatrix} i & i & i & i & i & i & i & i & i \\ i & i & i & i & j & i & i & i & i \\ j & j & j & j & j & j & j & j & j \\ i & i & i & i & i & i & i & i & i \\ i & i & i & i & j & i & i & i & i \\ i & i & i & i & i & i & i & i & i \\ i & i & i & i & i & i & i & i & i \\ i & i & i & i & i & i & i & i & i \\ i & i & i & i & i & i & i & i & i \end{bmatrix}$$

Thus, the proposed model protects a network from unwanted advertisement and attacks because neighborhood nodes do not accept the advertisement and updates directly rather it uses its leader node processed information.

4.3 Reduction in Resource Consumption

In conventional routing methods all nodes are responsible for calculations, updates and forwarding. Thus, the battery of the neighborhood nodes drains faster and it leads to misbehave and noncooperation. Our proposed model declines the power consumption of neighborhood nodes because major computations are performed at the leader end. In routing table updates the battery usage of neighborhood nodes can be minimized by reducing the amount of transmitting data in the network using Compressive sensing. The leader node periodically broadcast Routing Table Request (RTREQ) message in its neighborhood to know the latest routes as shown in Fig. 5a, where arrows indicate RTREQ advertisement. After that neighborhood

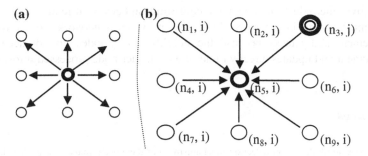

Fig. 5 **a** ACREP advertisement by leader node. **b** RTREP from neighborhood nodes

nodes compress and forward their routing table to the leader node using RTREP message as shown in Fig. 5b, where arrows indicate transmission of RTREP message from neighborhood nodes to leader node.

The leader node finally joins and reconstructs the routing table and broadcast the latest route in the neighborhood. This saves unwanted broadcast and minimizes the energy consumption of low battery power nodes. The leader node gathers all RTREPs and the data is represented as

$$y = (y_1 + y_2 + \cdots + y_N)$$

After that the leader node reconstructs the original data using "Signal recovery from random measurements via orthogonal matching pursuit" algorithm [21]. Finally, the leader node broadcast the original data or latest route in its neighborhood.

5 Conclusion

Compressive Sensing gained a wide acceptance in the recent years and applied in signal and image processing, pattern recognition, wireless communication, medical systems and analog-to-digital convertors. It can be used in MANET to reduce the resource consumption to enhance cooperation and protect a network from several attacks (such as blackhole, replay etc.). The proposed NCS model compresses the neighborhood sparse data such as routing table updates and other advertisement and trust informations. We have divided a MANET in terms of one or more neighborhood groups having a set of regular nodes, one or more border nodes and one leader node. The regular node compresses the received sparse data and forwarded it to the leader node. The leader node joins all neighborhood compressed data and reconstructs the original data. After that it broadcasts the correct message and latest route in its neighborhood. The border nodes act as a gateway for the neighborhood which passes the traffic to other neighborhood group. The use of compressive sensing gives the reduction in resource consumption because it reduces the amount of transmitting data in the network. It also prevents a network from unwanted advertisement and attacks because the neighborhood nodes do not accept the advertisement and updates directly rather it uses leader node processed information.

References

1. Anusas-Amornkul, T.: On detection mechanisms and their performance for packet dropping attack in ad hoc networks. ProQuest (2008)
2. Wang, Y.: Enhancing node cooperation in mobile wireless ad hoc networks with selfish nodes. University of Kentucky, Doctoral Dissertations. Paper 602 (2008). http://uknowledge.uky.edu/gradschool_diss/602
3. Candes, E.: Compressive sampling. In: Proceedings of International Congress of Mathematicians, pp. 1433–1452. Mathematical Society Publishing House (2006)
4. Donoho, D.L.: Compressed sensing. IEEE Trans. Inf. Theory 52(4), 1289–1306 (2006)
5. Saeed, N.H., Abbod, M.F., Al-Raweshidy, H.S.: IMAN: an Intelligent MANET routing system. In: Proceedings of the IEEE 17th International Conference on Telecommunications (ICT), pp. 401–404. IEEE (2010)
6. Saeed, N.: Intelligent MANET optimisation system. Ph.D. Thesis, School of Engineering and Design, Electronic and Computer Engineering Department, Brunel University, Brunel University (2011)
7. Chiang, T.C., Tsai, H.M., Huang, Y.M.: A partition network model for ad hoc networks. In: IEEE International Conference on Wireless and Mobile Computing, Networking and Communications, 2005 (WiMob'2005), vol. 3, pp. 467–472. IEEE (2005)
8. Lopez, J., Barceló, J.M., García-Vidal, J.: Subnet formation and address allocation approach for a routing with subnets scheme in MANETs. In: Wireless Systems and Network Architectures in Next Generation Internet, pp. 62–77. Springer, Berlin (2006)
9. Chowdhury, M.A.H., Ikram, M., Kim, K.H.: Secure and survivable group communication over MANET using CRTDH based on a virtual subnet model. In: Asia-Pacific Services Computing Conference, (APSCC'08), pp. 638–643. IEEE (2008)
10. Chang, C.W., Yeh, C.H., Tsai, C.D.: An efficient authentication protocol for virtual subnets on mobile ad hoc networks. In: International Symposium on Computer Communication Control and Automation (3CA), vol. 2, pp. 67–70. IEEE (2010)
11. Vilhekar, A.A., Jaidhar, C.D.: Modified authentication protocol using elliptic curve cryptosystem for virtual subnets on mobile adhoc networks. In: Wireless Communications and Applications, pp. 426–432. Springer, Berlin Heidelberg (2012)
12. Akhtar, M.A.K., Sahoo, G.: A novel methodology for securing ad hoc network by friendly group model. In: Computer Networks and Communications (NetCom), vol. 131, pp. 23–35. Springer (2013)
13. Lee, S., Pattem, S., Sathiamoorthy, M., Krishnamachari, B., An Ortega, A.: Compressed sensing and routing in multi-hop networks. University of Southern California CENG Technical Report (2009)
14. Chou, C.T., Rana, R., Hu, W.: Energy efficient information collection in wireless sensor networks using adaptive compressive sensing. In: Proceedings of 34th Conference on Local Computer Networks, (LCN 2009), pp. 443–450. IEEE (2009)
15. Feizi, S., Médard, M., Effros, M.: Compressive sensing over networks. In: Proceedings of 48th Annual Allerton Conference on Communication, Control, and Computing (Allerton), pp. 1129–1136. IEEE (2010)
16. Zhang, L., Luo, J., Guo, D.: Neighbor discovery for wireless networks via compressed sensing. Performance Evaluation (2012)

17. Xiong, J., Zhao, J., Xuan, L.: Research on the combining of compressed sensing and network coding in wireless sensor network. J. Theor. Appl. Inform. Technol. **47**(3) (2013)
18. Aziz, A., Salim, A., Osamy, W.: Adaptive and efficient compressive sensing based technique for routing in wireless sensor networks. In: The INTHITEN (INternet of THings and ITs ENablers) Conference, The Bonch-Bruevich State University of Telecommunications (SUT), St Petersburg, Russia (2013)
19. Zheng, H., Wang, X., Tian, X., Xiao, S.: Data gathering with compressive sensing in wireless sensor networks: an in-network computation perspective (2012). http://iwct.sjtu.edu.cn/Personal/xwang8/paper/INFCOM2012_InNetworkComputation_techreport.pdf
20. Zheng, H., Xiao, S., Wang, X., Tian, X.: Energy and latency analysis for in-network computation with compressive sensing in wireless sensor networks. In: Proceedings of the INFOCOM, pp. 2811–2815. IEEE (2012)
21. Tropp, J.A., Gilbert, A.C.: Signal recovery from random measurements via orthogonal matching pursuit. IEEE Trans. Inf. Theory **53**(12), 4655–4666 (2007)
22. Singh, A.K., Sharma, S.: Elite leader finding algorithm for MANETs. In: Proceedings of the 10th International Symposium on Parallel and Distributed Computing (ISPDC), pp. 125–132. IEEE (2011)
23. Mohammed, N., Otrok, H., Wang, L., Debbabi, M., Bhattacharya, P.: Mechanism design-based secure leader election model for intrusion detection in MANET. IEEE Trans. Dependable Secure Comput. **8**(1), 89–103 (2011)

AODV and ZRP Protocols Performance Study Using OPNET Simulator and EXata Emulator

Virendra Singh Kushwah, Rishi Soni and Priusha Narwariya

Abstract There are two important protocols based property in mobile wireless networks and these are ZRP and AODV for understanding concept of wireless networks. Both have own properties and characteristics. As we know that mobile based temporary network is an infrastructure free network and it has ability for self configurability, easy deployment etc. Effective and efficient routing protocols will help to make this network more reliable. The characteristics of self-organization and wireless medium make Mobile Ad hoc Network (MANET) easy to set up and thus attractive to users. Its open and dynamic nature enhance its easy to handle and operate. Motivation of this paper is to determine basic difference ZRP and AODV under two different simulators.

Keywords AODV · ZRP · OPNET · EXATA · Routing

1 Basics of Mobile Based Network

A mobile based network is an integration of lots of mobile nodes or devices and it is normally recognized to a decentralized autonomous system. It works as a client-server system. Nodes in this type of network is either fixed or mobile. There are so many mobile devices such as laptop, cellular phone and personal digital assistant etc. Mobile deviced or nodes can be installed at anywhere say on ships, airplanes or land area. A MANET has self configurability, movability, dynamic nature and self connection to the network as on demand etc. [1, 2] as we can see in the Fig. 1.

V.S. Kushwah (✉) · R. Soni · P. Narwariya
Department of Computer Science and Engineering, ITM, Gwalior 474 001, India
e-mail: kushwah.virendra248@gmail.com

R. Soni
e-mail: rishisoni17@gmail.com

P. Narwariya
e-mail: pri_nar_7933@yahoo.co.in

© Springer India 2016
A. Nagar et al. (eds.), *Proceedings of 3rd International Conference on Advanced Computing, Networking and Informatics*, Smart Innovation, Systems and Technologies 44, DOI 10.1007/978-81-322-2529-4_19

Fig. 1 Ad hoc environment

There are two important parts in a mobile wireless network in terms of wireless moving node, one is source node and other one is destination node. But instead of these, the intermediate nodes are required to channellized the data in proper way because these nodes are known as router and have the power to receive and transfer ahead the data in form of packets to its neighbours or close to destination node. Movability is the feature of ad hoc networks [3]. So, the network communication topology modifies as per time. Mobile wireless network has a lot of strong applications such as emergency saving functions, assembly events, conferences, and combat zone communication while moving vehicles and/or troopers and make the mobile networks for much larger scope.

2   Idea About Work

The idea of this paper is to introduce differences between two different ad hoc protocols using different performance metrics. For this one, authors have been taken various parameters on the simulators. ZRP and AODV, both are the two protocols of ad hoc routing. First one is the *hybrid protocol*, while other one is the *reactive protocol*.

A hybrid protocol suggests that the integration of reactive and proactive protocols and takes edges of those two necessary protocols. Therefore, routes are created fastly within the routing zones. ZRP is associate example of this one. In reactive sort protocol, associate data formatting of a route-discovery approach by the origin node to search out the route is formed to the destination node, once the origin node has knowledge packets to send. once a route is made, the route-preservation is initiated to keep up this route till it's now not needed or the destination isn't approachable. The most important benefits of such quite protocols is that overhead messaging is minimize [4–6].

There is novel approach to maintain routing scheme in an AODV protocol. It supports the routing boards as single entry per destination. AODV has come from DSR protocol, in which many routing cache entries for every destination can maintain easily. Path addressing is no required at the time of propagate RREP in relies on routing tables for every entry for back to the source and vice versa. AODV deals with sequence numbers at every time to maintain the updation of scattering information as routing and avoid routing iterations again and again. All packets use sequence number to manage routing scheme in the required routing boards [7, 8].

Most important feature of AODV to maintain time-based records in each entry containing by proper utilization of routing table entries. It can be expired after no such longer is used. Collection of antecedent node is maintained by data packet, neighbour node details and routing table entry etc. An error may come and RERR help to find the errors when hop link broke down. Every antecedent node carries ahead the RERR to its self collection of predecessors to make effectively removing all broken down links an its routes [9, 5].

3 Simulation Setup

This paper has two types of simulations. One is based on the OPNET Simulator for AODV protocol and other simulation is based on EXATA (a variant of QualNet) emulator for ZRP protocol.

OPNET is a modeler tool and it speeds up the research and development steps for analyzing and designing different communication networks, devices, protocols, and applications as demand. Designed networks can easily analyze and make high and huge impact over the technologies supported by the tool. It also includes a development environment to access all modelling methodology such as network types and network technologies supported by defined protocols [10].

EXata is an evaluation tool that creates at most same digital copy of physical networks in virtual environment that are indiscernible to various applications, devices, or users. The tool is specially designed to interoperate with all physical components including applications, devices, management tools and users related to the specified networks [11].

In this paper, OPNET simulation set up of a network with 40 wireless nodes moving at random, each with various speeds between 1 and 10 m/s. Each of the objects can move in a random direction, stop for some time (per the *pause time*), and then change its direction at random and move again. The *traffic pattern* models the voice based data transferred from one node to the other [8, 12, 13]. The data are sent at a rate of 2 kbps to represent compressed voice data. The number of data source nodes is chosen based on the supposition that a half of the nodes send the data and a rest of the nodes receive the data. The data destination is determined at random to represent the real situations. The simulation scenario is summarized below and as shown in the Figs. 2 and 3 on the basis of defined parameters, which are given in the Table 1:

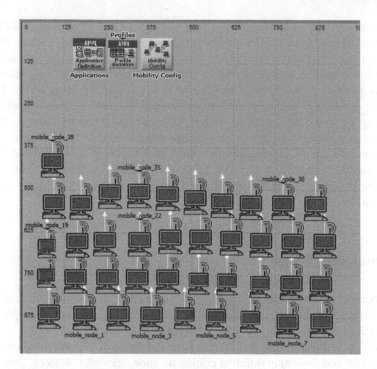

Fig. 2 AODV simulation under OPNET with 40 nodes

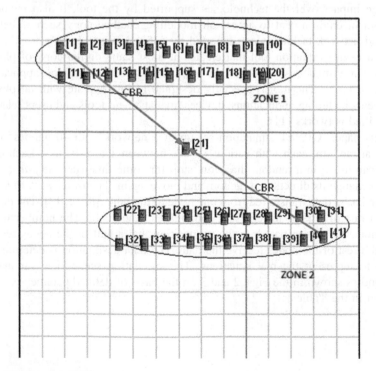

Fig. 3 ZRP simulation under EXata with 40 nodes

Table 1 Evaluation setup environment for both protocols

No.	Details	Value
1	Environment area	1000 × 1000 m
2	Mobility model	Random waypoint
3	Running time	10 min
4	Number of nodes	40
5	Node speed	1–10 m/s
6	Traffic type	CBR (voice)
7	Size of packet	512 bytes
8	Sending frequency	4 packets/s
9	Traffic destination	Random

4 Result and Analysis

The simulation is done using AODV and ZRP protocol into OPNET Simulator and EXata emulator, respectively. Practically the original values of performance metrics in the designed environment may affected due to many factors such as speed of node, node's moving direction, traffic of the node, flow of the packet, congestion at a definitive node etc. Therefore, it is hard to calculate the interpretation of a protocol by measuring the gained values from the each and every scenario. Finally, we got the different-2 values of various running simulations [14, 8]. In the Fig. 4, Number of Hops per Route have been shown.

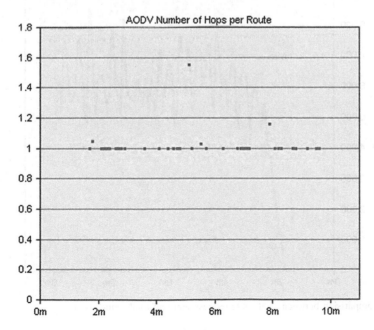

Fig. 4 Number of hops per route (OPNET)

The above results are from OPNET Simulator and the results would be compared with ZRP and see how to make our network as a better network for choosing the ad hoc protocol. With the help of Fig. 5, we can see the total route request sent and total replies sent using AODV under OPNET. From the Figs. 6 and 7, it is easy to see about comparision between routing traffic received and sent and total dropping packets in AODV protocol, respectively.

Here we are approaching our work with the EXata emulator. From Figs. 8 and 9, we can see first and last packet received.

In the above Figs. 10 and 11, we have found out the mean one way delay and mean jitter of ZRP using EXata emulator. From Fig. 12, we got the throughput of ZRP after working on 40 nodes.

4.1 Summary and Enhancement of the Work

This article has given just idea about difference between ZRP and AODV protocol based on two different model, one is simulator and other one is emulator. The work is origin from analysis and study of ad hoc network. In this work, the simulation is

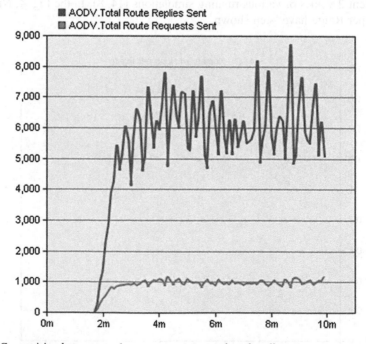

Fig. 5 Comparision between total route request sent and total replies sent

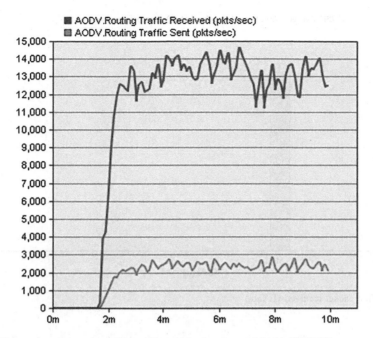

Fig. 6 Comparison between routing traffic received and sent (pkts/s) (OPNET)

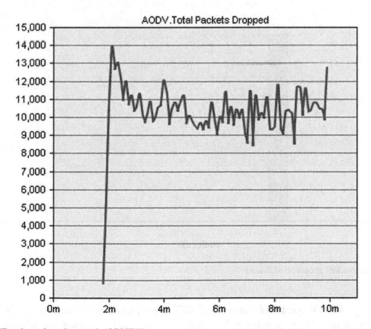

Fig. 7 Total packet dropped (OPNET)

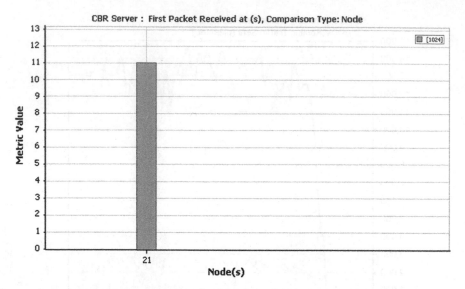

Fig. 8 First packet received (EXata)

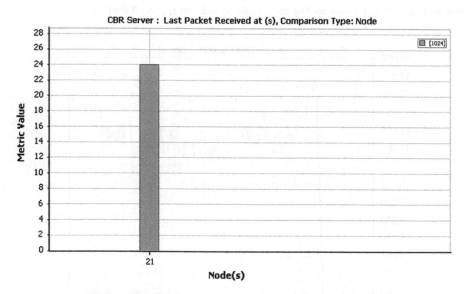

Fig. 9 Last packet received (EXata)

done using AODV protocol under OPNET simulation environment while ZRP protocol uses the EXata emulator. The future work can be carried out to secure the network using network security protocols. The next goal of the authors would be to prevent dropped packets in the AODV and maximizes the throughput for the ZRP.

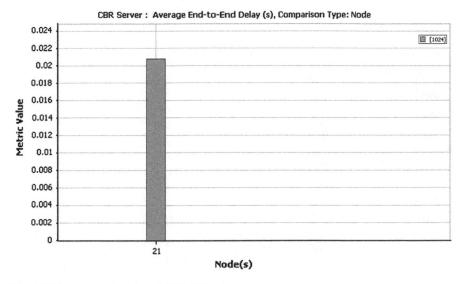

Fig. 10 Mean one way delay of ZRP (EXata)

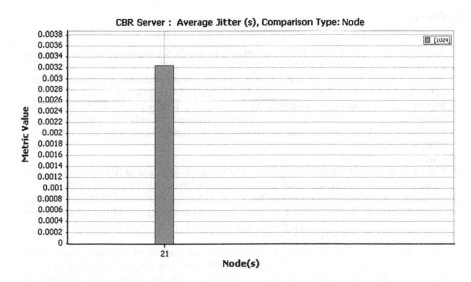

Fig. 11 Mean jitter of ZRP (EXata)

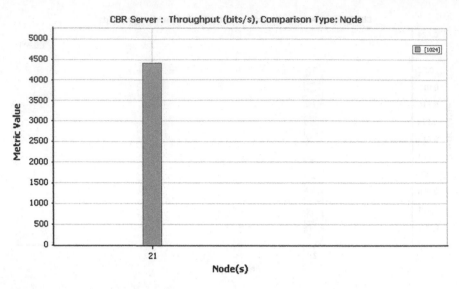

Fig. 12 Throughput of ZRP (EXata)

References

1. Apte, S., Katkar, V.: Review of ad hoc routing protocols for cognitive radio network. Int. J. Eng. Res. Appl. (IJERA). ISSN:2248-9622
2. Hu, Y.-C., Perrig, A., Johnson, D.B.: Ariadne: a secure on-demand routing protocol for ad hoc networks. In: MobiCom 2002, September 23–28, 2002, Atlanta, Georgia, USA
3. Razouqi, Q., Boushehri, A., Gaballah, M., Alsaleh, L.: Extensive simulation performance analysis for DSDV, DSR and AODV MANET routing protocols. In: 27th International Conference on Advanced Information Networking and Applications Workshops (06550420) (2013)
4. http://www.opnet.com/
5. Soni, S.J., Nayak, S.D.: Enhancing security features & performance of AODV protocol under attack for MANET. IEEE Int. Conf. Intell. Syst. Signal Proc. (ISSP) 325–328 (2013)
6. Giannoulis, S., Antonopoulos, C., Topalis, E., Koubias, S.: ZRP versus DSR and TORA: a comprehensive survey on ZRP performance. Emerging technologies and factory automation, 2005. ETFA 2005. In: 10th IEEE Conference on, vol. 1, p. 8, pp. 1024, 19–22 September 2005. doi:10.1109/ETFA.2005.1612635
7. Das, S.R., Perkins, C.E., Royer, E.M.: Performance comparison of two on-demand routing protocols for ad hoc networks. In: Proceedings IEEE Infocom, pp. 3–12 (March 2000)
8. Memon, S., Kazi, H., Ibrahim Channa, M., Arshad Shaikh, M.: Performance evaluation of MANET's reactive & proactive routing protocols in high speed VANETs. Int. J. Mod. Commun. Technol. Res. (IJMCTR) 2(12), (December 2014). ISSN:2321-0850
9. http://www.superinst.com/products/network_evaluation_software/exata.html
10. Lo, N.-W., Liu, F.-L.: A Secure Routing Protocol to Prevent Cooperative Black Hole Attack in MANET, pp. 59–64. Springer (2013)
11. Lee, G., Kim, W., Kim, K., Oh, S., Kim, D.: An Approach to Mitigate DoS Attack Based on Routing Misbehavior in Wireless Ad Hoc Networks. Springer (May 2013)

12. Khan, L.U., Khan, F., Khan, N., Naeem Khan, M., Pirzada, B.: Effect of network density on the performance of MANET routing protocols. In: International Conference on Circuits, Power and Computing Technologies [ICCPCT-2013] (2013)
13. Gouda, B.S.: A comparative analysis of energy preservation performance metric for ERAODV, RAODV, AODV and DSDV routing protocols in MANET. Int. J. Comput. Sci. Eng. Technol. (IJCSET) 3, 516–524
14. Lin, X., HamedMohsenian Rad, A., Wong, V.W.S., Song, J.-H.: Experimental comparisons between SAODV and AODV routing protocols. In: Proceedings of the 1st ACM Workshop on Wireless Multimedia (2005)

Optimal Probabilistic Cluster Head Selection for Energy Efficiency in WSN

Madhukar Deshmukh and Dnyaneshwar Gawali

Abstract Conventional Low Energy Adaptive Clustering Hierarchy (LEACH) is a cluster based routing protocol for Wireless Sensor Networks (WSN), which is effective in enhancing lifetime of the nodes thereby increasing the entire network life. The protocol is based on functionalities such as spatially distributed cluster formation, random selection of cluster heads, processing of data locally in the clusters and transmission of aggregated data to the base station (BS). Further, the cluster-head (CH) is selected from the member nodes (MN) from each of the cluster based on remaining energy at the node. In literature, various versions of LEACH with enhanced network life are presented. In this paper, an Efficient LEACH protocol is proposed which includes selection of CH for every round of CH selection based on results on Voronoi tessellations from stochastic geometry and remainant energy in the member node devices. In proposed protocol, a novel method is used to choose the CHs wherein the CHs and member nodes (MNs) of clusters are distributed as two independent homogeneous spatial Poisson Point Processes (PPPs). Probability of selecting the CHs and threshold is derived using results from spatial statistics. The Proposed algorithm selects optimum number of CHs leading to reduction in total energy spent in the network compared to conventional LEACH and other such algorithms. The network life is measured by number of rounds. Monte-Carlo simulations are carried out for performance analysis of LEACH, TEEN and other PPP based protocols. Furthermore, total energy dissipated in the network for each round is fairly constant throughout the network life i.e. distribution of total energy consumption by the network is fairly uniform over the rounds.

Keywords Cluster head · Energy efficiency · LEACH · Voronoi cluster

---

M. Deshmukh (✉) · D. Gawali
Trinity College of Engineering and Research (Savitribai Phule Pune University), Pune, India
e-mail: mmd@es.aau.dk

D. Gawali
e-mail: dsgawali@gmail.com

M. Deshmukh
CTIF, Department of Electronic Systems, Aalborg University, Aalborg, Denmark

© Springer India 2016
A. Nagar et al. (eds.), *Proceedings of 3rd International Conference on Advanced Computing, Networking and Informatics*, Smart Innovation, Systems and Technologies 44, DOI 10.1007/978-81-322-2529-4_20

197

1 Introduction

In WSN, sensors are randomly distributed over the two dimensional space. The sensor nodes are small in size, with low battery power, low memory power and small transmission range. To minimize the energy consumption of individual sensor node a hierarchical clustering approach is used. The clustering approach allows good scalability for relatively large sized networks and extends the network life-time. In this clustering approach, other nodes in the cluster are called as member nodes (MNs) which communicate with only the CH. It is important to note that the CH spends larger energy than MNs in transmitting the aggregated information to the BS thereby saving energy consumption at the MN. Furthermore, in the next round all the clusters are reformed based on their energy status and CH in the previous round may not be CH in the next round. Many clustering based protocols have been proposed such as Low Energy Adaptive Clustering Hierarchical (LEACH) [1–4].

The LEACH is an important algorithm for hierarchical sensor networks. In all types of clustering protocols, the entire load of data aggregation and data trans-mission is done by CHs. LEACH [1] directly transmits the collected aggregated data to the BS. Due to extensive usage of node as a CH, the CH node may be dead earlier than other sensor nodes in the network. In [1] there is no discussion about, how to select how many clusters or CHs are selected for a round. In [5–7] nodes are assumed to be distributed as Poisson Point Process (PPP) to devise energy efficient protocol because it can give optimal number of CHs per round. In literature, it is found that the results in spatial statistics helps deciding the optimal number of clusters per round which reduces the probability of selecting the same node as a CH node for the round.

In this proposed work, distribution of sensor nodes is assumed to be homoge-neous spatial PPP. The CHs and MNs of clusters are distributed as two independent homogeneous spatial Poisson Point Processes Ψ_1 and Ψ_0. Their respective intensities are assumed to be λ_1 and λ_0 respectively. In [2], first order radio model is used only with free space path loss (d^2), for communicating CH and MN as well as CH and BS. In this paper, same radio model is considered, but with free space path loss (d^2) between CH and MN and multi-path loss (d^4) between CH and BS, as these distances are different; also the radio channel is considered to be symmetric. Main objectives of proposed algorithm are as follows: computing the optimal number of CH and MN nodes using PPP [5, 7] and present a modified mathematical model which optimally selects the CHs in order to minimize energy consumption of network thereby increasing network lifespan [5, 6]. In proposed algorithm, the average distances between MN and CH and CH and BS computed using the results given in [9, 10]. Finally, the experimental results due to proposed method are compared that due to other protocols [2, 5, 7, 8]. Simulations are carried out using modified mathematical model. The experimental result shows, improvement in the duration of alive node for maximum time. This paper is organized as follows. Section 2 presents state of the art on clustered WSN. Section 3 describes methodology and experimental setup in detail. Section 4 presents the proposed algorithm along with the derivations. In Sect. 5, simulation experiments and results are discussed. The paper is concluded in Sect. 6.

2 Background on Clustered WSN

Present work describes the investigation on clustered WSNs. Notable contribution in this context is done by Heizelman [2] who has proposed first clustering algorithm called LEACH which employees randomization for even distribution of the energy load among the nodes in the network. LEACH operation is split into rounds. Every round commences with a set-up phase [2], where the clusters and CHs are formed. The messages are broadcasted by CH nodes to MN nodes to inform that they are CH nodes. This is accomplished in set up phase. In the steady state phase, MN nodes transmitts their data to CH nodes for aggregation. A random number m in the range of 0–1 for node n is generated for CH node those were not cluster heads in past $1/p$ rounds. Further in case the generated random number is less than threshold $\Gamma(n)$, nodes are selected as CH.

According to LEACH protocol, the threshold value is set as below
If $m < \Gamma(n)$ for node n, the node is selected as a CH where [2, 3, 7].

$$\Gamma(n) = \begin{cases} \frac{p}{1-pc\left|\frac{1}{p}\right|} & \text{for } n \in \chi \\ 0 & \text{for otherwise} \end{cases} \quad (1)$$

where, c = present round number and p is CHs in terms of desired percentage. It may be noted that the node become CH in present round and will be selected as CH in next $1/p$ rounds, and χ is set of non-CH member nodes in the $1/p$ rounds. In TEEN [8] hierarchical scheme along with a data dependant mechanism is employed. TEEN uses similar radio model as LEACH. TEEN describes soft and hard thresholds [8] which minimises the transmissions from MNs. Initially, parameter to hard threshold value in TEEN. Thereafter node transmits the data coming from sensor. The latest data is forwarded when the data attains the soft threshold value. There are many limitations in LEACH and TEEN protocols. It is quite obvious that optimal number of clusters and CHs for the given round may not be defined by Heinzelman [2] at all times. Furthermore, the residual energy in sensor nodes of clusters is not considered while choosing the CH [7]. Radio model takes into account the contention- and error-free environment which is not always true [6]. LEACH was confined to small sensing field and in TEEN protocol in case the thresholds mentioned earlier are not attained, the nodes will never be able to transmit data to BS. In [5–7], PPP approaches have been employed to execute energy efficient protocol because of its ability to provide optimal CHs for each round. Hence in this paper, sensor nodes are assumed to be distributed accordingly as homogeneous Spatial Poisson Point Processes and results from stochastic geometry are used to find the optimal number of CH per round. This scenario can be further modeled under the assumption that nodes are distributed as spatial cluster process.

3 System Model for Proposed Algorithm

System Model is proposed along with assumptions and radio model in the following subsections.

3.1 Assumptions for Proposed Algorithm

- All nodes in the sensing network in $\mathbb{R}^2$ space of sides $2a$ ($Area, A = 4a^2$) are stochastically distributed as a homogeneous Spatial Poisson Point Process Ψ having intensity λ.
- n is a Poisson random variable with mean λA describing number of sensors in $\mathbb{R}^2$ field.
- Furthermore, CHs and MNs are distributed as two independent homogeneous spatial Poisson [2] Ψ_0 and Ψ_1 processes with intensities λ_0 and λ_1 respectively such that $\lambda = \lambda_0 + \lambda_1$.
- The CH aggregate the data received from all member nodes within the cluster, and forward it to the BS.
- Radio range r is assumed to same for all sensor nodes.
- Transmit energy and receive energy required for each sensor node is considered to be 1 J/unit of data.
- A distance d between sensor node and respective CH is equivalent to d/r hops.
- First Order Radio model along with free space path loss d^2 between CH and MN and multi-path loss d^4 between CH and BS. ...

3.2 Radio Model

The radio characteristics for first order radio communication model as follows: the radio hardware energy consumes $E_{elec} = 50$ nJ/bit [2] to run the transmitter or receiver electronics and the receiver dissipates energy to run the radio electronics. Two radio models are assumed for communication between MN to CH and CH to BS; due to the fact that the average distance between MN to CH is smaller than that of CH to BS. Hence, d^2 a free-space path loss; but the distance between CHs and BS is larger, hence, d^4 a Multi-Path is used. Therefore the transmit amplifier consumes $\epsilon_{fs} = 10$ pJ/bit for path loss model d^2 and $\epsilon_{mp} = 0.0013$ pJ/bit for path loss model d^4. Thus for transmitting and receiving l-bit data at a distance d from the receiver, sensor radio device expends:

$$E_{txr}(l, d) = E_{txr-elect}(l) + E_{tx-amp}(l, d) = \begin{cases} E_{elec} + ld^2\epsilon_{fs} & \text{for } d < d_0 \\ lE_{elec} + ld^4\epsilon_{mp} & \text{for } d \geq d_0 \end{cases} \quad (2)$$

$E_{rx} = lE_{rx-elec} = lE_{elec}$ where $d_0 = \sqrt{\frac{\epsilon_{fs}}{\epsilon_{mp}}}$ In LEACH, there is no provision for deciding the number of CHs. Earlier radio model (LEACH) depends on the intensity of node distribution in a network. In LEACH, energy spent is directly proportional to the distance between the CH and MNs in a cluster and that of between CHs and BS. These distances depend on number of sensor nodes, number of clusters, size of sensing region and location of BS. Thus in this paper, the focus is on defining the optimal number of CH per round and average distance between MN to CH and that between CH to BS for minimizing the energy dissipation in the network. The setup and steady state phases of proposed protocol are assumed to be same as that of LEACH protocol.

4 Proposed Algorithm

4.1 Probability Optimization for Selection of Cluster Head

As described in the above section, in this work, it is considered that, n sensor nodes are distributed according to a homogeneous spatial Poisson Point Process Ψ in $\mathbb{R}^2$ field with mean λA [1]. For the given area the expected number of clusters K and CHs can be computed using Voronoi tessellation. There are on an average np sensor nodes which become CH with the optimal probability p_{opt}, if Voronoi clusters are assumed to k. Therefore for np_{opt} clusters, each cluster consists of on an average n/k . A Voronoi tessellation is formed over R^2 by member nodes that are close to CH of particular Voronoi cluster Each cell corresponds to a PPP Ψ_1 with intensity λ_1 and statistical center of the cell is called as nucleus of Voronoi cells as shown in Fig. 1. Numbers of clusters are equal to number of CHs per round. Let $N_d = \mathcal{N}(\Psi_0)$, the number of process points in a particular Voronoi cluster be a random variable and let the total number of sensor nodes in this region be N, then

$$\mathbb{E}(N_d|N = n) \approx \mathbb{E}(N_d) = \frac{\lambda_0}{\lambda_1} \quad (3)$$

Let L_{chi} be the average sum-length of all segments connecting the points in Ψ_0 (MNs to the CH) and the nuclei of a Voronoi clusters, then according to the results in [9]; the average of the sum of the distances between cluster member node (MN) and the CH is (i.e. distance between MN to CH)—d_{toch}:

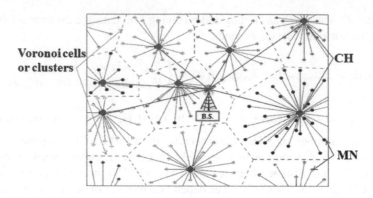

Fig. 1 Architecture for proposed protocol using Voronoi tessellation

$$\mathbb{E}(L_{chi}|N=n) \approx \mathbb{E}(L_{chi}) = \frac{\lambda_0}{2\lambda_1^{3/2}} \qquad (4)$$

Let L_{cbi} be a random variable representing the lengths of all the segments from the node at (x_i, y_i), $i = 1, \ldots n$ to the BS, i.e. Lcbi is the total distance of all the CH of PPP Ψ_1 process with intensity λ_1. Here BS is considered to be located at the center $\mathcal{O}(0,0)$. Then expected distance of a CH node in the Voronoi cell from the BS is (i.e. distance between CH to BS $-d_{toBS}$) [10]:

$$\mathbb{E}(L_{cbi}|N=n) \approx \mathbb{E}(L_{cbi}) = \sqrt{\frac{k}{\pi\lambda_1}} \qquad (5)$$

As there are $k = np_{opt}$ clusters per round therefore, the average total length of segments between all these CHs and the BS is $np_{opt}L_{cbi}$. According to considered radio model, during transmission of each l bit data, energy dissipated at each Voronoi cluster consists of two parts: energy spent by one CH and its members, MN nodes. Energy spent by MN for sensing, processing and transmitting the data to CH node, as the distance between MN and CH is small, the free-space path $(d^2$ power loss) model is used. On the other hand, as the distance between CH and BS is long, energy spent by CH node for receiving the data, aggregation of data and transmission of this data to the BS, multi-path (d^4) model is used. Expected energy consumed by the CH node for k clusters during one frame duration is:

$$\mathbb{E}(c_{ch}|N=n) \approx \mathbb{E}(c_{ch}) = kl(E_{elec} + E_{DA}) + l\epsilon_{mp}L_{cbi}^4 \qquad (6)$$

Where l represents number of bits in a message. Let c_v be the energy spent by the members (MNs and one CH) belonging to a Voronoi cluster to send one unit of data to the CH. Then expected energy spent by Voronoi cluster is:

$$\mathbb{E}_v(c_v|N = n) = l E_{elec}\mathbb{E}(N_d|N = n) + l E_{DA}\mathbb{E}(N_d|N = n) + l \epsilon_{fs}(\mathbb{E}(L_{chi}|N = n)^2 \tag{7}$$

Therefore from Eqs. (3) and (4) the expected total energy spent by Voronoi cluster is:

$$\mathbb{E}_v(c_v|N = n) = l(E_{elec} + E_{DA})\left(\frac{\lambda_0}{\lambda_1}\right) + l\epsilon_{fs}\left(\frac{\lambda_0^2}{4\lambda_1^3}\right) \tag{8}$$

For the given $\mathbb{R}^k$ field, there are k clusters, therefore the expected energy spent by the members of Voronoi cell is:

$$\mathbb{E}(c_{mn}|N = n) = kc_v \tag{9}$$

Then c_{vt} is the expected total energy spent by the MN and CH for a Voronoi cell per round is given using Eqs. (6) and (9) is: $\mathbb{E}(c_{vt}|N = n) = \frac{\mathbb{E}(c_{mn}) + \mathbb{E}(c_{ch})}{c}$

$$\mathbb{E}(c_{vt}|N = n) = \frac{l}{c}(E_{elec} + E_{DA})\left(\frac{\lambda_0}{\lambda_1} + k\right) + \frac{l\epsilon_{fs}}{c}\left(\frac{\lambda_0^2}{4\lambda_1^3}\right) + \frac{l\epsilon_{mp}}{c}\left(\frac{k}{\pi\lambda_1}\right)^2 \tag{10}$$

As described in earlier sections, MNs and CHs are assumed to be distributed as two independent homogeneous Spatial Poisson Point processes Ψ_0 and Ψ_1 with intensities $\lambda_0 = (1 - p_{opt})\lambda$ and $\lambda_1 = p_{opt}\lambda$ where $\lambda = \frac{n}{A}$. A is area of the field and p_{opt} be optimal probability of selection as a CH node. Then the mean energy spent by Voronoi cluster is Eq. (10) becomes:

$$c_{vt} = \frac{l}{c}(E_{elec} + E_{DA})\left(\frac{1}{p} - 1 + np\right) + \frac{l\epsilon_{fs}}{c}\left(\frac{(1-p)^2}{4p\lambda}\right) + \frac{l\epsilon_{mp}}{c}\left(\frac{n}{pi\lambda}\right)^2 \tag{11}$$

Differentiating above Eq. (11) w. r. t. p and equating it to 0 such that $\frac{dc_{vt}}{dp} = 0$. The value of p at which there is minima is optimal solution of p_{opt}

$$m_4 p_{opt}^4 - m_3 p_{opt}^3 + m_2 p_{opt}^2 + m_1 p_{opt} - m_0 = 0 \tag{12}$$

where, $m_4 = \frac{l\epsilon_{fs}}{16c\lambda}$, $m_3 = \frac{l\epsilon_{fs}}{6c\lambda}$, $m_2 = \left(\frac{l(E_{elec} + E_{DA})}{2c} - \frac{l\epsilon_{fs}}{8c\lambda}\right)$, $m_0 = \frac{nl(E_{fs} + E_{DA})}{c}$ and $p_{opt} = \frac{1}{p}$
The above Eq. (12) has one real root and two imaginary roots. Then the optimal probability of selecting the CH for the Voronoi cluster is decided by solution of Eq. (12) for p_{opt} in terms of real roots whereas imaginary roots are rejected. Optimal number of clusters and CHs and optimal probability of selecting a Cluster Head is taken from the solution of Eq. (12) for different number of nodes as given in Table 1. Here a = 50, and there can be 100, 200, 300, 500 nodes if $\lambda = 0.01, 0.02, 0.03$ and 0.05 respectively. For the proposed experimental results, an optimal

Table 1 Probability p and corresponding energy dissipation based on the proposed protocol

Sr. no.	No. of nodes (n)	PPP intensity (λ)	Prob. of selecting CH node (p)	No. of clusters (k) or CHs/round
1	100	0.01	0.0271	2.713
2	200	0.02	0.0571	11.4111
3	300	0.03	0.0814	24.4224
4	500	0.05	0.1357	67.8468

probability is used for deciding optimal CHs. The threshold equation computed by Eq. (1) is used for choosing the CHs stochastically per c round. The algorithm is designed so that each node becomes a CH at least once. The total energy dissipated in the network is in accordance with parameters p_{opt} and k_{opt}.

4.2 The Proposed Algorithm

Stages of clusters establishment for the proposed algorithm are presented in the following.

- Step 1: Deploy the sensor nodes according to a homogeneous spatial PPP of λ intensity in 2D field with mean λA.
- Step 2: Consider the optimal probability p_{opt} as derived in Eq. (12).
- Step 3: Calculate the total number of Voronoi k ($k = k_{opt}$) [11] clusters and CHs per round, as $k_{opt} = (np_{opt})$.
- Step 4: Select the CH node according to the threshold, defined by Eq. (1) replacing $p = p_{opt}$ in this equation.
- Step 5: The selected CHs broadcast a short message in its cluster to acknowledge other members (MN nodes) of cluster.
- Step 6: When MN nodes receive the broadcast from the CH node, it transmits its own information to CH, such as current relative position and ID number.
- Step 7: The MN nodes sends a message consisting cluster and node IDs to CH. At any instant of time a single node can be a member of only one cluster.
- Step 8: Repeat steps 4–7 till the number of clusters attains the value k_{opt}.
- Step 9: If the number of selected cluster reaches to k_{opt}, MNs corresponding to each cluster selects their Cluster Head. ...

The node which are selected as CHs, broadcast their status as a CH for the current round c and MNs receive the message from their respective CH and joins as a member in the particular cell. A CH broadcasts a Time Division Multiple Access (TDMA) time-schedule to the sensor nodes in the cluster. After set time, data aggregation is performed by CH and transmits aggregated data to the BS. The whole network enters into Cluster formation phase and again starts selection process of CH.

5 Experimental Setup for Simulations

Monte-Carlo simulations for the proposed algorithm described in Sect. 4 along with other important existing algorithms have been carried out in Matlab. The optimal probability is decided by Eq. (12). The cost of transmission and reception was calculated by the equations mentioned in Sect. 3.2. The simulation parameters are directly taken from the work of Heizelman [2] (refer Table 2) in view to have comparative understanding of the performance of proposed algorithm. The results of proposed algorithm are compared with the results of existing algorithms reported in the literature [2, 5, 7, 8].

5.1 Experimental Setup

Parameters related to energy in Table 2 are taken from Eq. (2). For the proposed experiment; 300 nodes are used in the sensor network over a two dimensional field of sides $2a$ between $(x = 0, y = 0)(x = 100, y = 100)$ with the BS located at center of the network (i.e. at $x = 50, y = 50$).

5.2 Results and Discussion

Figure 2a shows numbers of alive nodes per round. Alive nodes for protocols are represented as shown in red curve [2], green curve [8], black curve [5], purple curve [7]. The proposed protocol is indicated by blue curve in Fig. 2. It may be noted that Fig. 2b–e have been shown with the curve of same color (refer Fig. 2a). From Fig. 2a, more number of nodes remain alive due to the proposed protocol than protocols presented in [2, 5, 8]. Figure 2a show number of alive nodes due to [2] starts decreasing at around 1000 rounds, due to [8] at 1400 rounds, for [5] at 2442

Table 2 Simulation Parameters Used [2]

Sr. no.	Parameter	Value
1	Environment size (2a × 2a sq. m field)	100 m × 100 m
2	Number of nodes	300
3	Packet size (l)	4000 bits
4	Number of rounds (c_{max})	5000 rounds
5	Initial energy per node (E_0)	0.5 J
6	E_{elec} = Ebit = E_{tx} = E_{rx}	50 nJ/bit
7	E_{fs}	10 pJ/bit/m^2
8	E_{mp}	0.0013 pJ/bit/m^4
9	E_{DA}	5 nJ/bit/signal

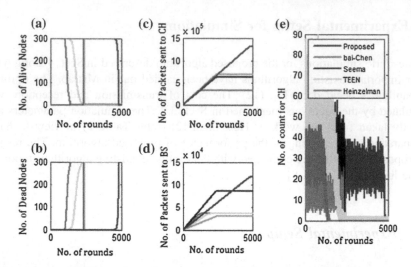

Fig. 2 Simulation results

rounds and due to [7] at 2470 rounds where as in the proposed protocol it starts reducing at 4674 rounds. It can be observed from the graphs that performance of proposed algorithm is comparatively better in terms of the network life. Number of alive nodes are limited by residual energies of individual nodes. Figure 2b plots the graph of dead sensor nodes per round. The graph of sensor nodes dead during each round in Fig. 2b is a complement of the fig. shown by Fig. 2a. Here also the performance of the proposed protocol has good result. Figure 2c plots the data packets send to the CH node during each round and Fig. 2d plots the data packets send to the BS node during each round. From both the figures, packets send to CH are more and packets send to BS are less; due to the CH aggregated the data packets and sent these aggregated data only. Due to clustering of sensor nodes, numbers of transmissions are minimized. Figure 2e shows the simulation result of number of counts of CH per round. It also shows that in the proposed algorithm, the proportions of formation of clusters as well as CHs based on derived optimal probability are constant till the last node. The better performance is due to the novel method of CH selection in the proposed algorithm.

6 Conclusion

In this work, the close form expression for optimal CH selection based on results in Stochastic Geometry is presented. In the proposed work sensor nodes are deployed as per homogeneous PPP Ψ of intensity λ in $\mathbb{R}^2$ space and optimal probability was computed. Based on this probability the optimal numbers of clusters as well as CHs per round were calculated. The main focus of this work is to examine the performance of proposed algorithm and protocol in context to energy efficiency and throughput. The experimental results show that the proposed efficient LEACH like protocol maximizes the network lifetime as compared to classical LEACH routing protocol and consequentially the network throughput also.

References

1. Akyildiz, I., Su, W.: A survey on sensor networks. IEEE Commun. Mag. **40**(8), 102–114 (August 2002)
2. Heizelman, W., Chandrakasan, A., Balakrishman, H.: Energy efficient routing protocols for wireless micro sensor networks. In: Proc 33rd HICSS-2000 (2000)
3. Heinzelman, W., Chandrakasan, A., Balakrishnan, H.: An application-specific LEACH protocol architecture for wireless microsensor networks. IEEE Trans. Wireless Commun. **1**(4), (October 2002)
4. Vinodh Kumar, S., Ajit Pal.: Assisted-Leach (A-Leach) energy efficient routing protocol for wireless sensor networks. Int. J. Comput. Commun. Eng. **2**(4), (July 2013)
5. Bandyopadhaya, S., Coyle, E.J.: An energy efficient hierarchical clustering algorithm for wireless sensor networks. In: Proceeding of IEEE INFOCOM, Piscataway, USA. IEEE (2003)
6. Bandyopadhaya, S., Coyle, E.J.: Minimizing communication cost in hierarchically clustered algorithm for wireless sensor networks. Comput. Netw. **44**, (2004)
7. Chen, B., Zhang, Y., Li, Y., Hao, X., Fang, Y.: A clustering algorithm of Cluster-head Optimization for wireless sensor networks based on energy. J. Inf. Comput. Sci. **8**(11), 2129–2136 (2011)
8. Manjeshwar, A., Agrawal, D.P.: TEEN: A routing protocol for enhanced efficiency in wireless sensor networks. Int. Parallel Distrib. Proc. Symp. **3**, 30189a (2001)
9. Foss, S.G., Zuyev, S.A.: On a Voronoi aggregative process related to a bivariate poisson process. Adv. Appl. Probab. **28**(4), 965–981 (1996)
10. Baccelli, F., Baszczyszyn, B.: Stochastic geometry and wireless networks. Volume I Theory, vol. 3, No 3–4 of Foundations and Trends in Networking. NoW Publishers (2009)
11. Kim, K.T., Youn, H.Y.: A stochastic and optimized energy efficient clustering protocol for wireless sensor networks. In: A Research Article in International Journal of Distributed Sensor Networks. (March 2014)

Firefly Algorithm Approach for Localization in Wireless Sensor Networks

R. Harikrishnan, V. Jawahar Senthil Kumar and P. Sridevi Ponmalar

Abstract Development of sensor technology has led to low power, low cost and small sized distributed wireless sensor networks (WSN). The self organizable and distributed characteristics of wireless sensor networks had made it for monitoring and control applications at home and other environment. In most of these applications location information plays a crucial role in increasing the performance and reliability of the network. This paper makes an attempt in analyzing and implementation of a novel nature based algorithm known as firefly localization algorithm. This is a distributed algorithm which uses range based trilateration method for distance measurement required for estimating the location of the sensor node. The algorithm is simple to implement and has better convergence and accuracy.

Keywords Firefly localization algorithm · WSN · Biological inspired algorithm · Landmark node · Sensor node · Firefly · Attractiveness · Brightness

1 Introduction

Intelligent systems require sensors for monitoring the various parameters of the environment. These sensors with particular topological and infrastructure form wireless sensor network and collaborate with each other, to be suitable for a particular application. The wireless sensor networks find applications like health care

R. Harikrishnan (✉)
Sathyabama University, Chennai, India
e-mail: rhareish@gmail.com

V. Jawahar Senthil Kumar · P. Sridevi Ponmalar
CEG, Anna University, Chennai, India
e-mail: veerajawahar@gmail.com

P. Sridevi Ponmalar
e-mail: sridevi_ponmalar@yahoo.co.in

© Springer India 2016
A. Nagar et al. (eds.), *Proceedings of 3rd International Conference on Advanced Computing, Networking and Informatics*, Smart Innovation, Systems and Technologies 44, DOI 10.1007/978-81-322-2529-4_21

monitoring, military surveillance, home automation, livestock monitoring, traffic control, precision agriculture, environment monitoring, pollution control, and forest fire detection. Most of these applications demand localization of sensor nodes. The information gathered by the network would be meaningful only if it has the location information of the sensor nodes. The localization is considered as a minimizing optimization problem, where the aim is to minimize the localization error by the firefly localization algorithm and estimate the location of the sensor node by itself.

An analytical optimization method requires more computation and the complexity increases exponentially as the network infrastructure expands. This motivated to consider bio inspired and nature inspired algorithms for resource constraint wireless sensor networks [1]. The nature adapt itself to resolve complex relationship and find the optimal solution for balancing ecosystem and maintain the diversity. The principle strategy used is simple which yields better results. Bio inspired optimization uses heuristic approach, that imitate nature and many processes in the nature can be considered as an optimization problem [2].

Biologically inspired algorithm is based on certain phenomenon of the nature that can optimize itself for their needs and survival. Biological inspired algorithm are considered powerful and are analyzed for solving complex numerical optimization problem. Firefly algorithm is one such nature inspired algorithm and has shown better accuracy for optimization problems [3]. Fireflies flash lights, to search for food by means of attracting their prey, to find their mates, and safeguard themselves from their predators for survival. Firefly algorithm depends on the intensity of light emitted by fireflies. Flashing light of firefly can be easily formulated as objective or fitness function of the sensor node localization problem. The firefly algorithm is simple in concept and easy to implement [4].

This paper proposes a nature inspired swarm intelligence algorithm which gets idea from firefly insect behavior. It is based on bio luminescent communication used by fireflies for their social behavior. Photogenic organs present in the body of fireflies are responsible for this flashing light [5].

The proposed algorithm is known as firefly localization algorithm which aims for simple implementation with less computation time. In this algorithm the sensor nodes with unknown location information estimates its location from the landmark nodes which knows its location. It uses trilateration range based method for distance calculation between landmark node and sensor node.

2   Sensor Node Localization Problem Formulation

Localization scenario of 40 sensor nodes is considered. In this 8 landmarks with known location information and 32 sensor nodes with unknown location is considered. The aim is to find the location of these 32 unknown sensor nodes with the location knowledge of landmarks. Range based method is used which implies trilateration method. Three known landmarks which are nearer to the selected unknown node are chosen. With the help of these landmarks and its location information, the distance of these landmarks with the chosen unknown node and the location of unknown node are found.

First the three landmarks positions are taken. The distance between the three landmarks and unknown nodes are calculated by using the Eq. (1).

$$d_{ist} = \sqrt{\left(\left(x_{est} - x_{landmark} \right)^2 + \left(y_{est} - y_{landmark} \right)^2 \right)} \tag{1}$$

The localization function is calculated by using Eq. (2)

$$f(X_u, Y_u) = \left[\sqrt{(X_{est} - X_{anc})^2 + (Y_{est} - Y_{anc})^2} - d_{ist} \right]^2 \tag{2}$$

3 Firefly Localization Algorithm

This algorithm is based on firefly algorithm developed by Dr. Xin-She Yang in 2007 which uses flash light nature of fireflies [6].

Firefly localization algorithm is based on the following ideal social behavior rules of fireflies.

1. All fireflies are unisex. One firefly move towards another brighter firefly regardless of sex.
2. Attractiveness is proportional to brightness and inversely proportional to distance. If no brighter firefly is found firefly movement is random in nature.
3. The brightness or light intensity of the firefly is found from the sensor node localization objective function given by the Eq. (2).

Based on the above rules the firefly localization algorithm is designed [7]. The basic operations are converting firefly algorithm into sensor node localization optimization problem, attractiveness calculation, distance calculation and position movement calculation of the firefly.

The design process of firefly localization algorithm is as follows.

Localization estimation variables are selected as fireflies and population size of 20 is taken. Initialize this vector population within the solution space. Minimization of localization error is taken as fitness function for firefly localization algorithm represented by Eq. (2). This objective function is used to find the brightness of each firefly. Firefly localization algorithm evaluates fitness value for each vector in the population. Next is the calculation of attractiveness process.

Attractiveness is the ith firefly move towards more attractive brighter jth firefly. This attractiveness function is given by the Eq. (3)

$$A(d) = A_0 * \exp(-\gamma \, d^m) \text{ With m} \gg 1 \tag{3}$$

d   Distance between two fireflies.

A_0 Initial attractiveness at d = 0.

γ Absorption coefficient that controls the light intensity.

The attractiveness decreases with the increase in distance because of the absorption factor in nature. Next is the distance calculation process between the ith firefly and the jth firefly. The distance of attraction of brightest firefly from the current firefly is calculated by the Eq. (4)

$$d_{ij} = \sqrt{\left((a_i - a_j)^2 + (b_i - b_j)^2 \right)} \tag{4}$$

Next is the movement process. The movement of ith firefly towards brighter jth firefly is based on attractiveness and distance between them and is given by the Eq. (5)

$$m_i^{k+1} = m_i^k + A_o * \exp\left(-\gamma \, d^2\right) * \left(a_j^k - a_i^k\right) + \propto *rand_i^k \tag{5}$$

m_i^k Initial position of ith firefly

$A_o * \exp\left(-\gamma \, d^2\right) * \left(a_j^k - a_i^k\right)$ Attractiveness towards jth firefly by ith firefly

$\propto *rand_i^k$ Random movement of ith firefly

If the firefly does not find any brighter firefly, it will move randomly. The above processes repeat up to the stopping criterion is achieved. That is when 100 iterations are over or convergence is achieved. The brightest firefly position is the estimated location of the sensor node. The design parameters of firefly localization algorithm is shown in the Table 1. The above procedure is summarized as the following firefly localization algorithm steps.

Table 1 Design parameters of firefly localization algorithm	Parameters	Values
	Maximum iteration	100
	Space size	99
	Landmarks	8
	Unknown nodes	32
	Total nodes	40
	Number of fireflies	20
	Randomizing coefficient, α	0.2
	Initial attractiveness at d = 0, A_o	1.0
	Absorption coefficient, γ	0.96

Step 1: Firefly is a set of control variables of sensor node localization function
Step 2: Initialize fireflies' population size of 20 in the solution space
Step 3: Objective function of sensor node localization is calculated to find the brightness of the fireflies using Eq. (2)
Step 4: Attractiveness of firefly towards other fireflies is calculated using Eq. (3)
Step 5: Distance between the fireflies is calculated using Eq. (4)
Step 6: Firefly i move towards brighter j firefly using Eq. (5)
Step 7: Rank the fireflies and find the current global best
Step 8: Repeat the step 4, step 5, step 6, step 7 till the stopping criterion is satisfied
Step 9: Estimate the optimal location information of the sensor node.

The output of the firefly localization algorithm is shown in the Fig. 1.

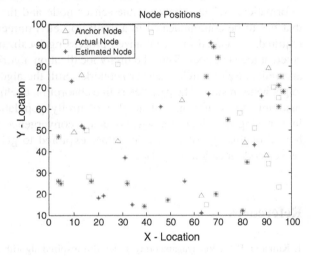

Fig. 1 Output of the firefly localization algorithm

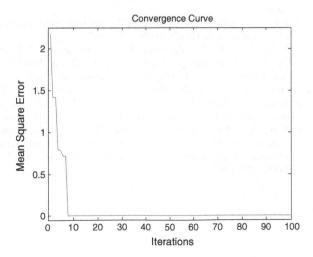

Fig. 2 Convergence curve of the firefly localization algorithm

The convergence curve of the firefly localization algorithm is shown in the Fig. 2.

The convergence curve shows that the firefly localization algorithm converges at the 8th iteration.

4 Conclusion

The firefly localization algorithm discussed in this paper provides better convergence and accuracy for the estimation of sensor node location information. The algorithm is simple, takes less computation time and it is a distributed localization algorithm. In communication point of view this algorithm requires less location information exchange between the sensor node and the sink node or base station that has to be communicated in a multi hop environment. Thus it reduces energy required, compared to a centralized processing localization algorithm that is executed at the sink node. Thus the firefly localization algorithm increases the life time and reliability of wireless sensor networks. Still the algorithm can be further tuned to get better results. The parameters like absorption coefficient, initial attractiveness, randomizing coefficient and number of firefly population size can be varied for improvising the algorithm towards better convergence and accuracy. And also a hybrid of this algorithm can be further explored to get satisfactory estimation of sensor node location information.

References

1. Kulkarni, R.V., Venayagamoorthy, G.K.: Bio-inspired algorithms for autonomous deployment and localization of sensor nodes. IEEE Trans. Syst. Man Cybern. Part C: Appl. Rev. **40**(6), (2010)
2. Binitha, S, Sathya, S.S.: A survey of bio inspired optimization algorithms. Int. J. Soft Comput. Eng. (IJSCE) **2**(2), (2012). ISSN:2231-2307
3. Yang, X.S.: Firefly algorithm, levy flights and global optimization. In: Research and Development in Intelligent Systems, vol. XXVI, pp. 209–218. Springer, London (2010)
4. Apostolopoulos, T., Vlachos, A.: Application of the firefly algorithm for solving the economic emissions load dispatch problem. Int. J. Comb. **2011**, (2011). Hindawi Publishing Corporation
5. Basu, B., Mahanti, G.K.: Fire fly and artificial bees colony algorithm for synthesis of scanned and broad-side linear array antenna. Prog. Electromagnet. Res. B **32**, 169–190 (2011)
6. Yang, X.S.: Nature-Inspired Meta-Heuristic Algorithms. Luniver Press, Beckington (2008)
7. Yanga, X.-S., Hosseinib, S.S.S., Gandomic, A.H.: Firefly algorithm for solving non-convex economic dispatch problems with valve loading effect. Elsevier Appl. Soft Comput. **12**, 1180–1186 (2012)

Performance Analysis of Location and Distance Based Routing Protocols in VANET with IEEE802.11p

Akhtar Husain and S.C. Sharma

Abstract A new sub-category of MANET is vehicular ad hoc network where vehicles communicate with each other. Because of constraint roads and very high speed of vehicles, routing is an issue in VANET. Most of the papers analyzed the performance of topology based routing protocols. This paper analyzed the performance of distance-effect routing algorithm for mobility (DREAM) and location aided routing (LAR) protocols for city and highway environment. Packet delivery ratio, throughput and delay metrics are considered for analysis of DREAM and LAR routing protocols using intelligent driver model (IDM) based VanetMobiSim and ns2 with IEEE802.11p.

Keywords MANET · VANET · DREAM · LAR · IDM · Vanetmobisim · ns2

1 Introduction

In vehicular network systems, VANET plays an important role, which is formed by vehicles like car, bus, truck, cab etc. to convey some information to neighbor vehicles and road-side units. It is assumed that these vehicles have devices for computation, event recording, transceiver and GPS system. On the road, VANET makes possible for vehicles to use applications like safety application [1], collision control warning, Internet surfing and advertisement while moving. Because vehicles move on predefined roads, they have a restricted mobility path [2]. In VANET environment mostly topology-based protocols AODV [3], DSR [4] are compared with location-based protocol LAR [5] and distance effect protocol DREAM [6].

A. Husain (✉) · S.C. Sharma
Indian Institute of Technology Roorkee, Roorkee, Uttarakhand 247667, India
e-mail: ahr13dpt@iitr.ac.in

S.C. Sharma
e-mail: scs60fpt@iitr.ac.in

© Springer India 2016
A. Nagar et al. (eds.), *Proceedings of 3rd International Conference on Advanced Computing, Networking and Informatics*, Smart Innovation, Systems and Technologies 44, DOI 10.1007/978-81-322-2529-4_22

215

2 Related Research Work

Routing protocols in ad hoc network for VANET applications based on IDM [7] are categorized in topology and location based [8]. The routing performance of GSR, AODV and DSR protocols studied with a city vehicular network by Lochert et al. in [9]. With delivery ratio and delay metrics GSR outperforms DSR and AODV. Authors in [10] did analysis of TORA, FSR, DSR and AODV in city scenario with IDM mobility model. Husain et al. [11] analyzed the performance of topology as well as location based protocols in VANET for city and highway network. Bakhouya et al. [12] did simulation of DREAM protocol using MOVE [13, 14] with traffic simulator TraNS [15]. Ko et al. [16] perform the simulated study of LAR routing protocol with IDM and it shows that LAR is reasonably good candidate for VANET. Lobiyal [17] consider the JNU real map and divide into smaller routes to study the mobility impact in VANET. Tee and Lee [18] analyzed the routing performance of GPSR, AODV and DSR using NS-2 and IDM. Multi-path doppler routing protocol (MUDOR), DSR and DSDV has been simulated by Xi et al. [19]. Simulation of position and topology based protocols done by Azarmi et al. [20] and position based routing protocols found better.

3 Location Based Routing Protocols for VANET

3.1 Distance-Effect Routing Algorithm for Mobility (DREAM)

DREAM [12] routed the data packets in the network by the use of distance and geographical location of the nodes. This geographical location is used to discover the route and bound the flooding in a small limited region. A proactive scheme is used in the routing process of DREAM. Each node of the network stores the location of all the nodes in a location table.

3.2 Location Aided Routing (LAR)

The main objective of LAR [16] is to lower the overhead caused by routing process, for which this protocol uses information about location of the nodes with the help of GPS or some other location service.

4 Simulation and Evaluation Setup

Mostly routing protocols are simulated using network simulators like ns-2 [21], OMNeT++ [22], J-SIM [23], and JiST/SWANS [24]. Movements of vehicle patterns are generated using traffic simulators like CORSIM, VISSIM, SUMO, and VanetMobiSim. Analysis of DREAM and LAR in this paper is done using the network simulator ns-2.33 with IDM_IM based VanetMobiSim [25] traffic simulator.

4.1 System Model

1100 m × 1100 m rectangle area is considered for simulation. The parameters for vehicles mobility of VanetMobiSim simulator are shown in Table 1.

The parameters used in vehicular ad hoc network simulation in ns-2 are shown in Table 2.

5 Result and Discussion

The following performance metrics are considered to evaluate and analyze the performance of DREAM and LAR routing protocols:

5.1 Packet Delivery Ratio (PDR)

PDR is the ratio of the total number of data packets successfully delivered divided to the total data packets transmitted by all the nodes in a network. Mathematically, packet delivery ratio (PDR) = Sa/Sb

Sa = Sum of the data packets received by each destination
Sb = Sum of the data packets generated by each CBR source

Table 1 Mobility model parameters

Parameter	Value
Threshold acceleration	0.2 m/s^2
Politeness factor of driver	0.5
Safe deceleration	4 m/s^2
Safe headway time	1.5 s
Jam distance	2 m
Comfortable deceleration of movement	0.9 m/s^2
Maximum acceleration of movement	0.6 m/s^2
Vehicle length	5 m

Table 2 Network simulation parameters

Parameter	Value
Simulation tool	NS-2.33
MAC protocol	IEEE802.11p
Mobility model	IDM
Transmission range	250 m
Simulation area	1100 m × 1100 m
Channel	Wireless
Simulation time	1000 s
Packet length	512 Bytes
Data rate	8 packets/s
Bandwidth	2 Mbps
Type of traffic	CBR
Vehicle speed	25 km/hr (city),120 km/hr (highway)
Type of interface queue	Drop tail/CMU priority queue
Size of interface queue	50 packets
Number of vehicles	5–40
Routing protocols	LAR and DREAM
Maximum connections	65 %

Figure 1 represents the ratio of packets that are transmitted during simulation by ns-2. It has been shown that in starting PDR increases with the increase in number of nodes. It is because when number of nodes is very low, it may not be possible to establish a communication path from the source to destination.

5.2 Throughput

Throughput is the sum of bits received successfully by all the vehicles. It is represented in kilo bits per second (kbps). Mathematically,

Fig. 1 Packet delivery ratio in city and highway scenario

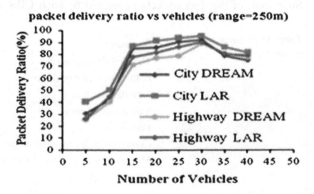

packet delivery ratio vs vehicles (range=250m)

Fig. 2 Throughput in city
and highway scenario

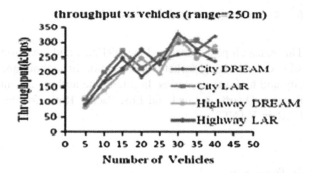

Throughput = Sum of the packets transmitted successfully/Last packet transmit time

From Fig. 2, it is observed that when node density is up to 15 the throughputs of LAR in city scenario is higher than DREAM in both scenario and LAR in highway scenario. However when density is increased from 15 to 30 nodes, LAR in highway scenario is higher than DREAM in both scenario and LAR in city scenario.

5.3 End to End Delay (EED)

EED is the sum of buffering time, queuing time, MAC layer retransmission time of packets and delay in propagation.

Mathematically, average end-to-end delay = S/N
S = Sum of time spent to deliver packets for each destination, and
N = Number of packets received by all the destination nodes.

Figure 3 summarizes the variation of end to end delay by varying node density. It is also evident that average delay increases with increasing the number of nodes. At low node density, DREAM in highway scenario consistently presents the highest delay.

Fig. 3 End2End delay in city
and highway scenario

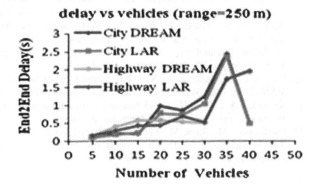

6   Conclusions and Future Work

This research paper, analyzed the DREAM and LAR routing protocols in vehicular ad hoc considering packet delivery ratio, throughput and end2end delay metrics in city and highway scenarios. In future, the authors are intended to add two metrics: Routing Overhead Load and Lost Packet Ratio for the analysis of DREAM and LAR in VANET.

References

1. Yang, X., Liu, J., Vaidya, N., Zhao, F.: A vehicle-to-vehicle communication protocol for cooperative collision warning. In: 1st IEEE International Conference on Mobile and Ubiquitous Systems: Networking and Services, Boston, pp. 114–123 (2004)
2. Bernsen, J., Manivannan, D.: Unicast routing protocols for vehicular ad hoc networks: a critical comparison and classification. Pervasive Mob. Comput. 5, 1–18 (2009)
3. Perkins, C.E., Royer, E.M.: Ad-hoc on-demand distance vector routing. In: 2nd IEEE Workshop on Mobile Computing Systems and Applications, Orleans, pp. 90–100 (1999)
4. Johnson, D.B., Maltz, D.A.: Dynamic source routing in ad hoc wireless networks. In: Mobile Computing, ed: Springer, pp. 153–181 (1996)
5. Camp, T., Boleng, J., Williams, B., Wilcox, L., Navidi, W.: Performance comparison of two locations based routing protocols for ad hoc networks. In: 21st IEEE Joint Conference of Computer and Communications Societies, vol. 3, pp. 1678–1687. New York (2002)
6. Basagni, S., Chlamtac, I., Syrotiuk, V.R., Woodward, B.A.: A distance routing effect algorithm for mobility (DREAM). In: 4th ACM/IEEE International Conference on Mobile Computing and Networking, pp. 76–84. Dallas (1998)
7. Haerri, J., Filali, F., Bonnet, C.: Performance comparison of AODV and OLSR in VANETs urban environments under realistic mobility patterns. In: 5th IFIP Mediterranean Ad-Hoc Networking Workshop (Med-Hoc-Net-2006). Lipari, Italy (2006)
8. Mauve, M., Widmer, J., Hartenstein, H.: A survey on position-based routing in mobile ad hoc networks. IEEE Netw. Mag. 15, 30–39 (2001)
9. Lochert, C., Hartenstein, H., Tian, J.: A routing strategy for vehicular ad hoc networks in city environments. In: IEEE Intelligent Vehicles Symposium, pp. 156–161. Columbus (2003)
10. Jaap, S., Bechler, M., Wolf, L.: Evaluation of routing protocols for vehicular ad hoc networks in typical road traffic scenarios. In: 11th EUNICE Open European Summer School on Networked Applications, pp. 584–602. Colmenarejo (2005)
11. Husain, A., Kumar, B., Doegar, A.: Performance evaluation of routing protocols in vehicular ad hoc networks. Intl. J. Internet Protoc. Technol. 6, 38–45 (2011)
12. Bakhouya, M., Gaber, J., Wack, M.: Performance evaluation of DREAM protocol for inter-vehicle communication. In: 1st International Conference on Wireless Communication, Vehicular Technology, Information Theory and Aerospace & Electronics Systems Technology, pp. 289–293. Denmark (2009)
13. Lan, K.C., Karnadi, F.K., Mo, Z.H.: Rapid generation of realistic mobility models for VANET. In: IEEE Wireless Communication and Networking Conference. Hong Kong (2007)
14. Karnadi, F.K., Mo, Z.H.: Rapid generation of realistic mobility models for VANET. In: IEEE Wireless Communications and Networking Conference, pp. 2506–2511. Hong Kong (2007)
15. Piorkowski, M., Raya, M., Lugo, A.L., Papadimitratos, P., Grossglauser, M., Hubaux, J.P.: TraNS: realistic joint traffic and network simulator for VANETs. ACM SIGMOBILE Mobile Computing and Communications Review, vol. 12, pp. 31-33 (2008)

16. Ko, Y.B., Vaidya, N.H.: Location aided routing (LAR) in mobile ad hoc networks. Wireless Netw. **6**, 307–321 (2000)
17. Lobiyal, D.: Performance evaluation of realistic Vanet using traffic light scenario. arXiv preprint arXiv:1203.2195 (2012)
18. Tee, C., Lee, A.C.: Survey of position based routing for inter vehicle communication system. In: 1st International Conference on Distributed Framework and Applications, pp. 174–182 (2008)
19. Xi, S., Li, X.M.: Study of the Feasibility of VANET and its Routing Protocols. In: 4th International Conference on Wireless Communication, Networking and Mobile Computing, pp. 1–4 (2008)
20. Azarmi, M., Sabaei, M., Pedram, H.: Adaptive routing protocols for vehicular ad hoc networks. In: International Symposium on Telecommunication, pp. 825–830 (2008)
21. Fall, K., Varadhan, K.: The network simulator. http://www.isi.edu/nsnam/ns (2007)
22. Varga, A.: The OMNeT++ discrete event simulation system (2005). http://www.omnetpp.org/download/docs/papers/esm2001-meth48.pdf
23. Sobeih, A., Hou, J.C., Kung, L.C., Li, N.: J-Sim: a simulation and emulation environment for wireless sensor networks. IEEE Wireless Commun. **13**, 104–119 (2006)
24. Barr, R., Haas, Z.J., Van, R.: Jist: embedding simulation time into a virtual machine. In: EuroSim congress on modelling and simulation (2004)
25. Harri, J., Fiore, M.: VanetMobiSim–Vehicular Ad hoc Network mobility extension to the CanuMobiSim framework. Institut Eurécom Department of Mobile Communication, vol. 6904 (2006)

Fuzzy Based Analysis of Energy Efficient Protocols in Heterogeneous Wireless Sensor Networks

Rajeev Arya and S.C. Sharma

Abstract Recent technological advances in wireless sensor network (WSN) lead to many new applications where energy awareness consideration is essential. There are some constrained to use wireless sensor network like energy, environmental effect like temperature, pressure, sound. To overcome these issues many new protocols are developed where energy awareness consideration is important. Most of the work reputed deals with modification, design, development of new routing protocols; which works for different applications of wireless network architectures. This paper analyzes the energy efficient operation in wireless sensor nodes. For this purpose, Low Energy Adaptive Clustering Hierarchy (LEACH), Hierarchical Cluster based Routing (HCR), Stable Election Protocol (SEP) and Gateway Cluster Head Election–Fuzzy Logic (GCHE-FL) protocols were analyze and compare. Clustering algorithm has been used because it reduces the energy of the sensor network. In the present work, Fuzzy logic with three input parameters and single output parameter; is used in cluster head election and protocol gateway heterogeneous wireless sensor network. It has been observed that in GCHE-FL the sensor node being alive for much more time as compare to LEACH, SEP, HCR.

Keywords Fuzzy logic · Gateway · Cluster head · Energy · Wireless sensor network area

R. Arya (✉) · S.C. Sharma
Electronics and Communication Discipline, DPT,
Indian Institute of Technology, Roorkee, India
e-mail: rajeev.arya.iit@gmail.com

S.C. Sharma
e-mail: scs60fpt@iitr.ernet.in

© Springer India 2016
A. Nagar et al. (eds.), *Proceedings of 3rd International Conference on Advanced Computing, Networking and Informatics*, Smart Innovation, Systems and Technologies 44, DOI 10.1007/978-81-322-2529-4_23

223

1 Introduction

The group of sensors builds a novel type of technology for the recent years called wireless sensor network and the modern advances in (MEMS) micro electro-mechanical system technology results low energy digital circuitry. Wireless sensor network area is one of the probable computing technologies, which offered extraordinary opening in several area ranging from military applications of sensor node include battlefield surveillance and monitoring, sensor is also used in environmental application, industrial control health monitoring, temperature, sound, pressure and home network [1–4]. Each sensor has competence of wireless communication and a certain level of intelligence for signal processing and networking of the data and communicates using radio link.

2 Research Review

Heinzelman et al. Suggested Low energy adaptive clustering (LEACH) protocol [5–7] is separated into different rounds. The each round consists two phases, such as study state phase and setup phase. Gupta et al. [8], Fuzzy logic protocol, select Cluster Head by fuzzy logic approach. Fuzzy logic interference scheme designer considered three descriptors Centrality, concentration and energy level and each descriptor is three parts, and chance output [9]. Cluster heterogeneous WSN by using a stable election protocol [12–14]. Hierarchical Cluster-based Routing (HCR) method is an expansion of the LEACH protocol that is a self structured cluster-based approach for uninterrupted monitoring [10, 11]. Tashtoush et al. suggested the Clustering distributed WSN based on fuzzy inference scheme. The first election is Gateway Election and the capable node is selected based on their energy, centrality and proximity to the base station. The second election is cluster head election used three parameters. These parameters are efficiency, cluster distance and concentration. In the present paper, we analyzed the different protocol (LEACH, SEP, HCR, GCHE-FL) using fuzzy logic scheme and Matlab simulation. The goal of the present work is to analysis presentation of different energy aware routing algorithms to study under specified constraints using fuzzy logic scheme.

3 LEACH, HCR, Stable Election Protocol, Gateway Cluster Head Election-Fuzzy Logic (GCHE-FL) Protocol

Heinzealman et al. suggested LEACH protocol [5–7] which is using TDMA as MAC protocol. It consists of two phases, such as study state phase and setup phase. The steady state phase consists of transmit and sending of data to the cluster head (CH) and base stations (BS).

The proposed HCR method is expansion of the LEACH [12–14] protocol that is auto structured clustering procedure for uninterrupted scheme. In HCR, every cluster is controlled by a group of acquaintances energy proficient clusters are maintained for longer time duration. The energy-proficient clusters are located using heuristics method.

SEP a heterogeneous-aware protocol which extend the time period earlier than the extinction of the primary sensor node, which is critical for lots of use wherever the advice as of the network have to be consistent [9]. SEP is supported on probabilities of every node to turn into cluster head according to the left over energy in all nodes.

Gateway cluster head election-Fuzzy logic uses if then rule to maximize the life time of wireless sensor network. It used two elections fuzzy logic to estimate the merit of sensor to befall gateway and cluster head.

In the gateway election, the node's possibility for being a gateway is calculated according to its physical properties such as its interior energy, its proximity to the base station. Such as traffic load is intense at most on cluster heads [8].

$$T(n) = \begin{cases} \frac{q}{1-q\times(r\,\mathrm{mod}^{1}/q)_s} & \times chance\ if\ n \in G \\ 0 & otherwise \end{cases} \qquad (1)$$

Gateway Cluster Head Election-FL configures clusters in all rounds [8]. In all round, every non-gateway wireless sensor node creates a random number between 0 and 1, if these random numbers are lesser than q_{opt}, the wireless sensor node analyzes the chance with fuzzy if then rule and presented an applicant Message through the chance. K_{opt} = optimal number of the cluster head, δ = constant Value, b = fraction of the all node

$$q_{opt} = \delta(k_{opt}/n \times (1-b)) \qquad (2)$$

$$k_{opt} = (\sqrt{n \times (1-b)} \times \sqrt{(\varepsilon fs/\varepsilon r)} \times \sqrt{\frac{area}{0.75 \times 0.5 \times \sqrt{area}}} \qquad (3)$$

4   Fuzzy Logic Approach

The system of fuzzy logic comprised of a fuzzifier, fuzzy system, fuzzy inference, and a defuzzifier. In the present analyses we have applied the usually fuzzy inference method called Mamdani due to its ease. The fuzzy logic (FL) approach consists of four step that is fuzzification of the input variables, Rule assessment (inference), Aggregation of the rule outputs (composition) and Defuzzification [10, 11]. In this part, we initiate GCHE-FL that uses fuzzy if-then rule to exploit the lifetime of WSN. GCHE-FL uses two-election fuzzy logic to judge the qualification of sensors to become the cluster head and a gateway.

5 System Architecture

Sensor node is at check extending in the surroundings node. BS (Base station) is positioned in the focal point of the surroundings WSN. Once arranged, the node does not travel. All wireless nodes have the similar ability and energy. Base station arranged the id and location for all nodes. Every node is dispersed homogeneously over sensor area (Fig. 1).

Energy spending model has been used in this paper similar to energy model in the LEACH protocol [5–7]. In which Every node to throw 1 byte data to space of it consumes a lot E_s energy, which is gated from this equation. Also the full quantity of energy that is used in the receiver for getting k bit node is gated from equation.

$$E_s = \begin{cases} l \times E_{elect} + l \times \varepsilon_{fs} \times d^2 & d > d_{co} \\ l \times E_{elect} + l \times \varepsilon_{mp} \times d^2 & d \geq d_{co} \end{cases} \tag{4}$$

Also the full quantity of energy that is used in the receiver for getting k bit node is gated from equation.

$$E_r = l \times E_{elect} \tag{5}$$

5.1 Fuzzy Analysis of Gateway Election and Cluster Head Election

Therefore, in the gateway election, these three parameters, energy, centrality and proximity to BS are fuzzy logic input and sensor chance output. These are input parameter first on energy (poor, good excellent) second on proximity to base (poor, good excellent) and third on the centrality (close, adequate, far). Then $3^3 = 27$ rules

Fig. 1 Snap shot of wireless sensor network area

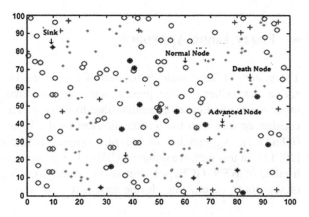

are used for fuzzification to sensor chance output. The centroid based defuzzification is used. The if then rule for gate way election.

To calculate a chance, we have used three fuzzy logics variables concentration Cluster Distance and efficiency. These are input parameter first on efficiency (Low, medium, high) second in the Cluster Distance (low, medium, high) and third on concentration (Low medium, high). Then $3^3 = 27$ rules are used for fuzzification to sensor chance output. The centroid based defuzzification is used. The if then rule for Cluster head election.

6 Simulation Parameter

They have been considered for LEACH, SEP, Hierarchical Cluster based Routing and Gateway Cluster Head Election-FL used fuzzy tool (MATLAB R2008b). The area of wireless sensor network is chosen as of 100 m × 100 m and sink located on center of sensing field (Table 1).

Table 1 Parameter

Parameter	Value
Total energy ($E_{elect.}$)	50 nJ/bit
Data aggregation energy (E_{DA})	5 nJ/bit/message
Initial energy (E_o)	0.5 J
Maximum number of rounds (R)	4999
Number of nodes in the field (n)	200
Transmit energy (E_{TX})	50 nJ/bit
Receive energy (E_{RX})	50 nJ/bit
Transmit amplifier types (ε_{fs})	10 pJ/bit/(m2)
Transmit amplifier types (ε_{mp})	0.0013 pJ/bit/(m4)

Fig. 2 Number of death node per round for different protocols

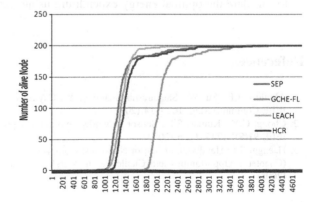

Table 2 Comparative analysis of different protocol (with m = 0.1, a = 1)

	LEACH	SEP	HCR	GCHE-FL
First node dies	951	998	1122	1775
50 % node alive	1264	1298	1365	2148
10 % node alive	1515	1688	1789	2721

7 Result and Analysis

In this paper, we compare performance of four protocols based on fuzzy logic using Matlab. Wireless sensor network area (100 m × 100 m), 200 nodes and 4999 numbers of rounds. The Fig. 2 based on attribute of nodes show that GCHE-FL far better than SEP, HCR and LEACH due to change in fuzzy rules because of extra fuzzy input (Table 2).

It is observed from the fuzzy analysis that first node die after 951, 998, 1122 and 1775 numbers of rounds in LEACH, SEP, HCR and GCHE-FL respectively. Half of the node being alive in LEACH, SEP, HCR, GCHE-FL after 1264, 1298, 1365, 2148 number of rounds respectively, in wireless sensor network areas only 10 % nodes being alive after 1515,1688, 1789 and 2721 numbers of rounds in LEACH, SEP, HCR and GCHE-FL respectively. On the base of simulation result we found that GCHE-FL is better than LEACH, SEP and HCR.

8 Conclusions

In this paper we have analyzed four different protocols (LEACH, SEP, HCR, GCEH-FL) for Energy of sensor node. It is observed that in GCEH-FL, the sensor node being alive for much more time as compared to LEACH, SEP, HCR which indicate that energy expenditure of node is less as compared with LEACH, SEP and HCR. Which implies that GCHE-FL of performed well in wireless sensor network. The simulation analysis of protocol consumes less energy and prolong lifetime of the network compared with others protocol. The future direction of this work will be to calculate the optimal energy expenditure using fuzzy logic system.

References

1. Akyildiz, I.F., Su, W., Sankarasubramaniam, Y., Cayirci E.: A survey on sensor networks. IEEE Commun. Mag. 102–114 (2002)
2. Chong, C.Y., Kumar, S.P.: Sensor networks: evolution opportunities, and challenges. Proc. IEEE **91**(8), 1247–1256 (2003)
3. Haenggi, M.: Handbook of Sensor Networks: Compact Wireless and Wired Sensing Systems, (Chapter 1 Opportunities and Challenges in Wireless Sensor Networks, 1–11). CRC Press (2005)

4. Estrin, D., Girod, L., Pottie, G., Srivastava, M.: Instrumenting the world with wire-less sensor networks. In: International Conference on Acoustics, Speech, and Signal Processing (ICASSP 2001), pp. 2033–2036 (2001)
5. Heinzelman, W.R., Chandrakasan, A.P., Balakrishnan, H.: An application-specific protocol architecture for wireless microsensor networks. IEEE Trans. Wireless Commun. 1(4), 660–670 (2002)
6. Ben Alla, S., Ezzati, A., Beni Hssane, A., Hasnaoui, M.L.: Improved and Balanced LEACH for heterogeneous wireless sensor networks. Int. J. Comput. Sci. Eng. 2(8), 2633–2640 (2010)
7. Qing, L., Zhu, Q., Wang, M.: Design of a distributed energy efficient clustering algorithm for heterogeneous wireless sensor networks. Elsevier, Comput. Commun. 29, 2230–2237 (2006)
8. Gupta, I., Riordanand, D., Sampalli, S.: Cluster-head election uses fuzzy logic for wireless sensor networks. In: Proceedings of the 3rd Annual Communication Networks and Services Research Conference, pp. 255–260, May 2005
9. Smaragdakis, G., Malta, I., Bestavros, A.: A stable election protocol for clustered heterogenous wireless sensor network. In: Second International Workshop on Sensor and Actor Network Protocols and Applications (SANPA 2004) (2004)
10. Tashtoush, Y.M., Okour, M.A.: Fuzzy self-clustering for wireless sensor network. In: IEEE/IFIP International Conference on Embedded and Ubiquitous Computing (2008)
11. Daliri, Z., et al.: Railway security through the use of wireless sensor networks based on fuzzy logic. Int. J. Phys. Sci. 6(3), 448–458 (2011)
12. Heinzelman, W.R., Chandrakasan, A., Balakrishnan, H.: Energy efficient communication protocol for wireless microsensor networks. In: Proceedings of the Hawaii International Conference on System Sciences, January 2000
13. Attea B.A., Khalil E.A.: A new evolutionary based routing protocol for clustered heterogeneous wireless sensor network. Elsevier, App. Soft Comput. 12, 1950–1957 (2012)
14. Hussain, S., Matin, A.W.: Hierarchical cluster-based routing in wireless sensor networks. IEEE/ACM International Conference on Information Processing in Sensor Networks (IPSN) Work-in-progress track, IEEE Computer Society, Nashville, TN, 19–21 April 2006

Node Application and Time Based Fairness in Ad-Hoc Networks: An Integrated Approach

Tapas Kumar Mishra and Sachin Tripathi

Abstract In ad hoc networks, balancing overall delay among multiple flows with different priorities is a major challenge. Normally, packets having higher priority are scheduled to dispatch first than the packets having lower priority. However, priority is calculated by considering the weights of either application or region. As a result, flows with lower priority reach to the destination after a long delay or dropped because of starvation. So an efficient mechanism must be framed for achieving delay bounded service in heterogeneous applications of ad hoc network. In this paper, we propose an integrated priority measurement technique for minimize overall delay among multiple flows. The proposed model uses the elapsed time as a measure factor to calculate priority. The packets suffered a larger as well as less delay are assigned a lower priority as they are either less important at sink node or can be delayed to reach at destination within time bound. Performance evaluation of this protocol moderates the end to end latency to the flows having lower priority.

Keywords Ad hoc networks · Fairness · TCP · WSNs · Reliability

1 Introduction

In the recent years, ad hoc networks are used in broad areas of communication and forecasting like emergency search, agriculture, rescue operations, military control and command operations in a battlefield, etc. [1–5]. In order to achieve a smooth communication with wireless network, the most important point to consider is the reliability and timeliness. Based on the literature [1, 2, 4] it is clear that even though

T.K. Mishra (✉) · S. Tripathi
Indian School of Mines, Dhanbad, Jharkhand 826004, India.
e-mail: kmtapas@gmail.com

S. Tripathi
e-mail: var_1285@yahoo.com

© Springer India 2016   231
A. Nagar et al. (eds.), *Proceedings of 3rd International Conference on Advanced Computing, Networking and Informatics*, Smart Innovation, Systems and Technologies 44, DOI 10.1007/978-81-322-2529-4_24

there are a huge number of protocols proposing fairness [3, 5–17] have been proposed. However, none of them have considered to reduce the end-to-end delay for the flows bearing lower priority. Apart from this, several communication challenges such as: reliability, fairness among applications, congestion, etc. exists in real time services.

Among the pool of challenges, the most important issue is fairness for all applications. As different applications have different requirements, they should not be transmitted in same rate, which introduces the priority among the different applications. Now, there are many nodes and at the same time, each node executes with different applications having a different priority level. As priority is set to applications and the highest priority application throws more packets to the network, so the packets having lower priority gets dropped in the middle of the transmission. When priority is set according to the nodes of different area, the important data from lower priority occupied nodes cannot be received at the receiver which may cause more damage in the analysis. Sometimes, packets may go for starvations due to high injection of higher priority packets and cannot give an effective result for lower priority applications in the analysis at the destination.

In this paper, we propose a priority calculation model that assigns priority to packets dynamically. Here, the packets, that have suffered a long delay are assigned to lower priority as their data might be less informative at the destination. Likewise, the packets have been created recently also assigned lower priority as they can be delayed for a very short time and then be transmitted so that it will reach at destination within specified time bound.

The rest of this paper is organized as follows. The literature related to priority are discussed in Sect. 2 and the proposed model is presented in Sect. 3. In Sect. 4, performance of the model is presented with simulation results. Finally, conclusion and future work are sketched in Sect. 5.

2 Related Work

In this section, we present the mechanisms, attempted to provide fairness among different flows. In PCCP [13], the scheduler differentiates packets according to their application type. It maintains two queues, one for current node and another for packets from transit traffic. It claims to reduce packet loss and maintain fairness in terms of nodes with the support of multi path routing. Three techniques as the Intelligent Congestion Detection (ICD), Implicit Congestion Notification (ICN) and Priority-based Rate Adjustment (PRA) are used in this protocol to achieve the fairness.

In PHTCCP [18], the classifier differentiates packets like [13]. However, it creates a of queues for each flow. It uses Weighted Fair Queuing (WFQ) for scheduling packets. Also, it follows multi path to transmit packets and maintains link utilization when some nodes in a particular route are inactive or in sleep mode. Some cases the packets from a given source of same application type reaches at a transit node in different times and can not get higher priority even if the second packet has faced some transit delay. This particular situation can not serve for a better reliability.

STCP [15], modifies the packet format. It was implemented for multiple flows. The packet header contains Flow Id as one of the fields to differentiate packets from different flows. In this protocol, only one queue was maintained, but the priority of the packet was set according to their Flow Id. The Flow id guides the scheduler to select a packet for dispatch. First in first out (FIFO) scheduling policy is used to dispatch Packets having same Flow Id.

In CCF [16], two types of queues are maintained in every node. The first one is Per-child queue, one queue for each child node and one more queue for its own. The node maintains a per child tree size, which helps the node to select the packet that to be transmitted further. Two sophisticated mechanisms Probabilistic Selection (PS) and Epoch-based Proportional Selection (EPS) are used to calculate the priority of the packet which determines the probability of the packet can be dispatched.

In RT2 [11], Time-Critical Event First (TCEF) scheduling policy is used to determine priority of the packet. Like [15], one more field is associated with each packet named Time Elapsed. This field signifies the remaining time of deadline. The priority of the packet is measured using this elapsed time field. Total communication is differentiated into two types in this paper. They are Sensor Actor communication to gather information from the sensors and Actor–Actor communication to manage or reconfigure the nodes for an optimal use of sensors. Hence actor–actor communication packets are nominated with higher priority.

3 Proposed Model

In this proposed model, we present an integrated scheduling policy with a brief description. This policy claims that service rate of packets of each application type will be equivalent to their priority and average latency of the packets having lower priority will be reduced. As, the priority is being complicated to define among various nodes and applications, we use normalization (time is mapped to [0–1]) to determine the priority in dynamic to consider transmission and queuing delay of the packets.

3.1 Problem Motivation

As described in the Fig. 1, the nodes of different geographical locations carry different priorities. At the same time, each node deals with multiple applications where each application behaves different priorities also. Hence, the priority of a packet is being complicated to calculate. Again, as fairness is required to maintain

Fig. 1 Network design
showing the tree structure
node configuration with
region wise priority of nodes

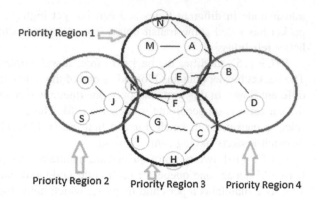

reliability, we must have to consider the time taken by a packet to complete its path between source and destination.

As an example, consider the network given in Fig. 1, suppose two packets from a given node say S are sent to a given destination D. In the middle of the path, it touches several nodes as J, G, C. All the intermediate nodes might be connected to many descendant nodes. At a given instance, it may happen that a particular packet may be always stored in the last in the queue. In this situation, the packet stores always at the last of the queue, may suffer a node delay at each intermediate node. Integrating all the delays at the nodes, the packet will suffer a lot till the end. Sometimes the time difference between two packets of different priority from a single node becomes high; this forms a real challenge for analysis at the destination end.

3.2 Solution Model

In this section, the priority calculation policy of a packet with its scheduling are presented. According to this model, the packet can recover its end-to-end communication delay, suffered at the intermediate nodes. The model is a threefold approach of (a) calculation of priority, (b) injection into the queue, and, (c) dispatch from queue. The priority of the packet is calculated by integrating other factors. The packets created just before a few times, can be placed in the less priority queues to be scheduled at a later time and the packets created before a long time also be placed in less priority queues as the contents of the packet may carry less importance for the destination. Only the packets of middle aged can accumulate a higher fractional ratio and claims a higher priority. The various symbols used during calculation is described in Table 1.

Priority Calculation During calculation of integrated priority of a packet, we consider all other related priority fields, (i.e., priority of node, application, and time elapsed which are presented in Table 1). We calculate percentage of times elapsed

Table 1 Notations and descriptions

Symbols	Notations
ϵ	Elapsed time
σ	Total estimated time to reach
K	Number of queues in a node
P	Integrated priority
α	Priority of node
β	Priority of application
η	Weighted time
TTL	Time to live field of data packet

by the packet (ϵ), to get the value within the range [0–1], to reach at the current node. Total estimated time to reach at destination, σ, is always greater than TTL. Hence, ϵ will a result as a fraction.

$$\epsilon = \frac{TTL}{\sigma}. \tag{1}$$

Then, we take the minimum value of time elapsed and remaining time of the packet to calculate the weight of the packet according to time. That is,

$$\eta = \min(\epsilon, 1 - \epsilon). \tag{2}$$

Now the integrated priority of the packet based on node (α), application based priority (β), and time (η) is,

$$P = (\alpha + \beta)^{\eta}. \tag{3}$$

Injection into Queue Using Hashing Suppose, there exist K number of queues in a given node. As the total system is a heterogeneous collection of nodes, the number of queues varies from node to node. Then, we map integrated priority with K and calculate the indexing priority for injection into queue. The maximum possibility of the priority ($Priority_{\max}$) becomes $\sqrt[2]{\max(\alpha) + \max(\beta)}$ as the maximum value of η by Eq^n 2 ranges [0–0.5], So, $Priority_{\max}$ divides K to get slot size. That is,

$$slot\text{-}size = \frac{Priority_{\max}}{K}. \tag{4}$$

$$Indexing\text{-}priority = \frac{P}{slot\text{-}size}. \tag{5}$$

Now the indexing priority will range from 0 to $K-1$ and we insert the packet into the queue according to its indexing priority. However, the packet would not be dropped there in case of full of any queue, rather, it would be placed in the next prior queue.

Queue Dispatch Using the Priority Queue Packet scheduler selects packets from queues of higher priorities in a weighted round robin fashion [19]. In case of any vacancy slot of packet in the buffer, the scheduler simply moves its pointer to the next slot. After completion of one round the scheduler starts again from the beginning of the highest priority queue.

4 Performance Analysis Through Simulation

A number of extensive simulations are carried on to evaluate the performance of our protocol through NS-2 [20]. During simulation, feasible parameters suggested by [1] are considered for benchmarking. The proposed model reduces average latency of flows having less priority that can make the data more effective during processing at end user which is clearly visible in Fig. 2. It is seen that flows having less priority suffer more transmission delay due to starvation at the intermediate nodes. However, the flows having higher priority can transmit easily as compared with other flows. This makes a significant difference in latency even if they travel the same distance. But, the proposed model minimizes the difference between end-to-end delay among different flows by reducing the queuing delay which is shown in Fig. 3.

Fig. 2 Measurement of average end-to-end delays of two flows

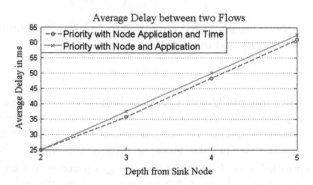

Fig. 3 Standard deviation measures of end-to-end delays of two flows

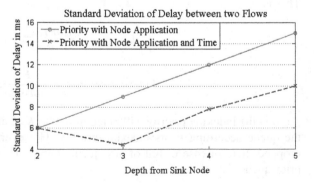

5 Conclusion and Future Work

In this paper, we have presented integrated approach for calculating priority of a packet based on its position of node, application type, and time. This approach attempts to collect all packets at destination by a little margin of differences of end-to-end delay. We have analyzed through simulation that the model achieves; (1) desired throughput according to the priority set and monitored by the base station, (2) fairness among the applications using an integrated approach of priority calculation, (3) minimum average of end-to-end delay among different sources. As our future work, we would like to work on energy efficiency and large scale integration of this model in time bound data communications in real time scenario.

References

1. Rathnayaka, A.J.D., Potdar, V.M.: Wireless sensor network transport protocol: a critical review. J. Netw. Comput. Appl. (2013)
2. Ganesh, P., Chockalingam, S.: A collection of open research problems in rate based transport protocols for multihop ad hoc wireless network. In: IEEE-International Conference on Advances in Engineering, Science and Management (2012)
3. Francisco, J., Chai, K., Cano, J., et al.: A survey and comparative study of simulators for vehicular ad hoc networks (VANETs). Wireless Commun. Mob. Comput. **11**(7), 813–828 (2011)
4. Akyildiz, I.F., Su, W., Sankarasubramaniam, Y., Cayirci, E.: Wireless sensor networks: a survey. IEEE Commun. Mag. (2002)
5. Liaojun, P., Huixian, L., Qingqi, P., Nengbin, L., Yumin, W.: Fair data collection scheme in wireless sensor networks. Commun. Soft. (2013)
6. Chen, Z., Yang, G., Chen, L., et al.: Data aggregation scheduling with guaranteed lifetime and efficient latency in wireless sensor networks. China Commun. **9**(9), 11–21 (2012)
7. Hamid, M.A., Alam, M.M., Isiam, M., et al.: Enforcing fairness for data collection in wireless sensor networks. In: Proceedings of the 8th Annual Communication Networks and Services Research Conference, 11–14 May 2010, pp. 192-198. Montreal, QC, Canada, IEEE press (2010)
8. Hamid, M.A., Alam, M.M., Isiam, M., et al.: Fair data collection in wireless sensor networks. Anal. Protoc. Ann. Telecommun. **65**(7), 433–446 (2010)
9. Wakuda, K., Kasahara, S., Takahashi, Y., Kure, Y., Itakura, E.: A packet scheduling algorithm for max-min fairness in multihop wireless LANs. Comput. Commun. (2009)
10. Sridharan, A., Krishnamachari, B.: Maximizing network utilization with max-min fairness in wireless sensor networks. Wireless Netw. **15**(5), 585–600 (2009)
11. Gungor, V.C., Akan, O.B., Akyildiz, I.F.: A real-time and reliable transport (RT2) protocol for wireless sensor and actor networks. IEEE/ACM Trans. Netw. (2008)
12. Rangwala, S., Gummadi, R., Govindan, R., et al: Interference-aware fair rate control in wireless sensor networks. In: Proceedings of the ACM Conference on Applications, Technologies, Architectures, and Protocols for Computer Communications, pp. 63–74 (2006)
13. Wang, C., Sohraby, K., Lawrence, V., Li, B., Hu, Y.: Priority-based congestion control in wireless sensor networks. In: Proceedings of the IEEE Conference on Sensor Networks, Ubiquitous, and Trustworthy Computing (2006)
14. Yuan, X., Duan, Z.: Fair round robin: a low complexity packet scheduler with proportional and worst-case fairness (2006)

15. Iyer, Y.G., Gandham, S., Venkatesan, S.: STCP: A generic transport layer protocol for wireless sensor networks. In: Proceedings of the 14th IEEE International Conference on Computer Communications and Networks (2005)
16. Ee, C.T., Bajcsy, R.: Congestion control and fairness for many-to-one routing in sensor networks. In: Proceedings of the 2nd International Conference on Embedded Networked Association for Computing Machinery Sensor Systems (2004)
17. Ramabhadran, S., Pasquale, J.: Stratified round robin: a low complexity packet scheduler with bandwidth fairness and bounded delay (2003)
18. Monowar, M., Rahman, O., Pathan, A.K., Hong, C.S.: Prioritized heterogeneous traffic-oriented congestion control protocol for WSNs. Int. Arab J. Inf. Technol. (2012)
19. Mishra, T.K., Tripathi, S.: Fair and reliable transmission control protocol for ad hoc networks. In: International Conference on Advances in Computing and Communications (2014)
20. NS-2. http://www.isi.edu/nsnam/ns, NS-2.35. Accessed 4 Nov 2011

Bee Colony Optimization for Data Aggregation in Wireless Sensor Networks

Sujit Kumar and Sushil Kumar

Abstract Energy constraint nature of wireless sensor networks has led to the need of data aggregation. Problem of optimal data aggregation scheme is a NP-hard problem. Bee colony System, a metaheuristic algorithm, imparts inherent and natural means of optimization for optimal data aggregation. In this paper, Bee Colony Optimization (BCO) using Bee Fuzzy System is used for data aggregation in wireless sensor networks (WSNs). Simulation is done using MATLAB. The performance shows the considerable improvement in energy optimization of wireless sensor networks.

Keywords Wireless sensor networks · Data aggregation · Bee colony optimization · Energy-efficiency

1 Introduction

Recent trends in WSNs have shown tremendous use of energy efficient algorithms for maximization of network lifetime. A WSN works in an unreachable and unhandled environment with limited battery power. So, energy optimization is a basic and important requirement of sensor network [1]. Designing low power consuming protocols for WSN is a challenging task. In heterogeneous WSN multihop routing algorithm consumes more energy resources. So, the lifetime of sensor network is reduced. Solution to this problem is use of multihop routing with data

S. Kumar (✉) · S. Kumar
School of Computer and Systems Sciences, Jawaharlal Nehru University,
New Delhi, India
e-mail: sujit92_scs@jnu.ac.in

S. Kumar
e-mail: skdohare@mail.jnu.ac.in

© Springer India 2016
A. Nagar et al. (eds.), *Proceedings of 3rd International Conference on Advanced Computing, Networking and Informatics*, Smart Innovation, Systems and Technologies 44, DOI 10.1007/978-81-322-2529-4_25

239

aggregation techniques to reduce redundant transmission along with saving energy. In these techniques, sensor nodes after gathering data send it to the aggregator nodes which aggregate them. Finally the aggregated data is sent to the target node or base station. Thus data aggregation approaches help in minimizing the amount of information dissemination [1–3].

The remaining part of the paper is arranged as follows: Relevance of this work is presented in next section. Section 3 describes the related work in the field briefly. In Sect. 4 problem definition and proposed work has been discussed. In the rest of sections Bee Colony Optimization for data aggregation, its implementation, results and conclusion are provided.

2 Data Aggregation: Problem Modelling

We assumed single target and multiple source nodes for the application model development. In WSN nodes communicates with its neighbours through multihop communication so as to transfer the gathered information to the final target node. It is also considered that high density of network provides a connected graph of nodes. In-network data aggregation is applied to communicate throughout the network. The aggregation can be lossy or lossless which can be measured depending upon the number of messages lost after aggregation of received data [4, 5]. Aggregation algorithm compares in terms of processing delay and energy gain. In case of correlated data sensing, number of source nodes determines the energy efficiency of the network [1, 3].

3 Related Work

In heterogeneous WSN energy constraints provides the way for correlated data sensing such that data aggregation occur at intermediate nodes. Data Aggregation in terms reduces the number of redundant transmissions which consumes lots of network energy. It leads to lifetime maximization of WSN [2]. Optimal data aggregation tree construction is a NP-hard problem [6]. There are various approaches applied for finding the minimal or optimal aggregation tree like Greedy approach, Dynamic programming approaches, etc. Research is going on in the field of sensor networks to find energy efficient algorithm for this. Combinatorial optimization problem is used to model optimal aggregation problem and Bee Colony optimization is used for solving it in our work. BCO has been used previously for routing purposes in sensor networks [6–8]. We are applying the Bee Colony Optimization with Bee Fuzzy System [6] for finding the solution to optimal aggregation problem in wireless sensor networks.

4 Problem and Proposed Approach

4.1 Problem Definition

Problem is to find the optimal aggregator nodes in a wireless sensor network. As combinatorial searching is applied for search space hence optimal aggregation tree is NP-hard problem in WSN. Thus Swarm Intelligence based approach is applied in finding the solution which is a collective intelligence approach.

4.2 Proposal for Solution

1. Bee colony System optimization with Bee fuzzy System is used for the problem solution.
2. Bee-Aggregation algorithm, builds aggregation tree iteratively which provides the minimal cost solution. MATLAB is used for simulation.
3. Minimal aggregation and other techniques are compared for performance evaluation [9, 10].

5 Bee System and Optimization

In Bee Colony Optimization (BCO) [6], artificial bees coordinate so that combinatorial optimizations problem can be solved. Initially it is assumed that all artificial bees are in hive at the start of the search process for nectar. BCO mechanism deals with the foraging behavior of bees. This behavior helps bees in finding the minimal path from the nectar to hive. Initially, the foraging bee start the search for food randomly which is nearby their hive. After finding nectar, it calculates quality and quantity of nectar or food source and carries some food back to the hive. Within hive it transfers the food for storage. There is an area in bee hive called dance floor where the bees which have found nectar source dance, so as to convince other bees to follow them to the food source. After return, the foraging bee can now:

1. Become an uncommitted follower by abandoning the old food resource.
2. Continue the same work of foraging of nectar without recruiting the new hive mates.
3. Dance on the dance floor to recruit new hive mates to follow for the nectar source.

So, the direct communication between the bees is through dance which allows them to find shortest distance between food source and the hive. This nature of working of real bee colonies is used in artificial bee colony for solving optimization problems. In BCO algorithm the opting of choice are simulated using parameterized combinatorial model along with approximate reasoning. It is also called Fuzzy Bee System (FBS) [6]. Bees in their communication and actions within the FBS use rules of fuzzy logic and approximate reasoning.

Thus in general BCO tries to give solution to optimization problem in two steps:

1. Partial solutions are constructed using the combinatorial probabilistic model over the search space.
2. Earlier iterations for partial solutions are used to modify the results and thus lead to high quality solutions.

6 BCO for Data Aggregation

6.1 BCO for Optimal Aggregation Tree

Our algorithm consists of two passes. In this while flying our bees performs two passes forward pass and backward pass. In forward pass, from the collective past experiences and individual new exploration, partial solutions are created by bees. While in backward pass, after returning to hive they all take part in decision making. Here we assume that all bees have information about the partial solutions of other bees. Thus comparison of solutions is done accordingly. Hence depending upon the quality of partial solution, bees have to decide about the three choices available to them.

6.2 Reasoning Approach for Optimization of Aggregation Tree

Finding an optimal aggregation tree is a NP-hard problem for a dense WSN. So this problem is reduced to weighted set cover problem. In our Bee-aggregation algorithm, give an option of various source nodes we construct an aggregation tree which gives the local best solution. Algorithm tries to find the global best solution from the large combinatorial search space for minimal aggregation tree. Earlier iterations by bee are also need to be remembered.

Bee Colony Algorithm
Input: Is a weighted graph of all nodes, neighborhood information
2.**While** End condition is not met do
a. Calculate initial partial solution and the node distance potential
b. Arrange the activities accordingly
c. Bee System based solution designing
d. Partial Solution update
e. Node distance potential update
3. **End** activities
4. Best Solution ← Best solution found in all the cases
Output: Best candidate for the optimal solution

6.3 Bee Aggregation Algorithm

We considered a WSN model where it is modeled a weighted group G (V, E) where the source nodes s ϵ V and a target node t ϵ V. Edge cost is estimated by Euclidean distance within direct communication range. All source nodes s ϵ V is input for the algorithm. Bees are assigned to the source nodes. Bees search the path and communicate with dance on the dance floor. Bees try either to find the shortest path to target or follow the other bee's path to shortest aggregation point. Node potential gives the value of distance to target node. Each bee iterates for constructing aggregation tree using the internal nodes as aggregator points, as this gives the local minimal aggregation tree. Thus to find the global minimal aggregation tree algorithm iterates for different permutation values. Our algorithm runs in two passes: Bees from the hive moves towards the target searching a path to target to hive and updates through decision making.

The control equations for the algorithm are as follows: Artificial bees are placed at source nodes randomly and thus during iteration calculates the solution component attractiveness.

Let us consider x_i as attractiveness value for ith bee. The probability of choosing ith bee solution for the partial solution addition is, p_i:

$$p_i = x_i / \sum (x_k) \tag{1}$$

where, p_i is the ratio of particular option attractiveness to all option attractiveness that is for all value of k.

Bee's partial solution can be compared to give the badness of partial solution. This partial solution badness can be defined as:

$$P_k = (P^{(k)} - P_{min})/(P_{max} - P_{min}) \tag{2}$$

where,

P_k kth bee partial solution badness,
$P^{(k)}$ kth bee discovered partial solution function,
P_{min} best discovered partial solution function,
P_{max} worst discovered partial solution function

In this paper we assumed that probability of bee flying to same path (p*) without selecting the new nodes is very less (p* $\ll$ 1). Thus, bee dances on the dance floor with probability $(1 - p*)$.

Thus, Node potential (t_{ij}) can be found by a weighted function for either following other to reach aggregator or minimal route to target as given below:

$$t_{ij} = y * X + z * X_1 + u * X_2 \qquad (3)$$

where,

X is cost of following nearest aggregator;
X_1 is cost of shortest route to the target;
X_2 is cost of correlated data
y is the weight for the nearest aggregator
z is weight for the shortest route to aggregator
u is weight for data correlation

7 Implementation and Results

The Bee-Aggregation algorithm is simulated in MATLAB by setting up a 100 nodes sensor network. Random topology is applied for the neighborhood. Various sets of source nodes and a single target node is used for constructing the optimal aggregation tree using Bee-Aggregation. By varying the various parameters of BCO and weights of aggregation, minimal distance and data correlation it is simulated.

BCO parameters	Simulation values
y	0.6
z	0.5
u	0.2

Energy efficiency can be measured as cost comparison to the aggregation cost for different data aggregation strategies. Optimal aggregation performs better as it saves 35 % (for the average number of source nodes) to 20 % (in dense number of source nodes) in case of greedy techniques. Also in case of opportunistic aggregation, optimal aggregation saves 50 % (on the average number of source nodes) to 25 % (dense number of source nodes) of energy. Figure 1 shows the above stated result.

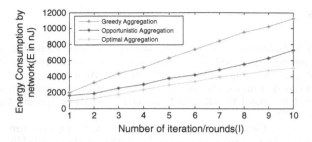

Fig. 1 Energy consumption of wireless sensor network per rounds

Fig. 2 Total aggregation cost comparison for different strategies

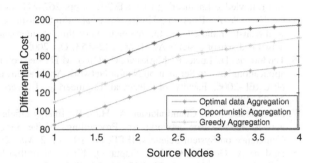

In Fig. 2 differential cost is plotted against source to show the performance of optimal aggregation in comparison to other strategies for aggregation tree. Thus with average number of nodes Bee optimal aggregation performs better in comparison to other approximation algorithms.

8 Conclusion

Optimal aggregation problem is implemented and explored by Bee Colony Optimization. Simulation has been done to find the appropriateness of approach. It has been observed that the energy efficiency depends on the number of source nodes. Our result shows that the optimal aggregation improves and thus saves energy up to 50 % for average node density.

Energy optimization through various optimization techniques is beneficial in increasing the lifetime of wireless sensor network. Various techniques like Monkey search algorithm, Particle Swarm intelligence can be applied to modify the cluster level energy consumption.

References

1. Al-Karaki, J.N., Ul-Mustafa, R., Kamal, A.E.: Data aggregation in wireless sensor networks—exact and approximate algorithms. In: The proceedings the International Workshop on High-Performance Switching and Routing, Phoenix, AZ, April 2004
2. Krishanamachari, B., Estrin, D., Wicker, S.: The impact of data aggregation in wireless sensor networks. In: International Workshop of Distributed Event Based Systems (DEBS), Vienna, Austria, July 2002
3. Intanagonwiwat, C., Estrin, D., Govindan, R., Heidemann, J.: Impact of network density on data aggregation in wireless sensor networks. Technical Report 01–750, University of Southern California, Nov 2001
4. Bidar, M., Kanan, H.R.: Jumper firefly algorithm. In: International Conference on Computer and Knowledge Engineering (ICCKE-2013), pp. 267–271, Oct–Nov 2013
5. Geoffrey, W.-A., Patel, G., Tiwari, A., et al.: Firefly-inspired sensor network synchronicity with realistic radio effects. In: Proceedings of the 3rd international conference on Embedded networked sensor systems.ACM, pp. 142–153, Oct 2005
6. Teodorovic, D., Lucic, P., Markovic, G., Orco, M.D.: Bee colony optimization: principles and applications. In: 8th Seminar on Neural Network Applications in Electrical Engineering, IEEE NEUREL-2006, Faculty of Electrical Engineering, University of Belgrade, Serbia, Sept 25–27, 2006
7. Heinzelman, W.R., Chandrakasan, A., Hari, B.: Energy efficient communication protocol for wireless microsensor networks. In: Proceedings of the 33rd Annual Hawaii International Conference on System Sciences. IEEE, pp. 10, vol. 2, May 2000
8. Hashmi, A., Goel, N., Goel, S., Gupta, D.: Firefly algorithm for unconstrained optimization. IOSR J. Comput. Eng. (IOSR-JCE) **11**, 75–78 (2013)
9. Wang, P., He, Y., Huang, L.: Near optimal scheduling of data aggregation in wireless sensor networks. In: Ad Hoc Networks, Elsevier, vol. 11, pp. 1287–1296, Nov 2013
10. Huang, S.C.-H., Wan, P.J., Vu, C.T., Li, Y., Yao, F.F.: Nearly constant approximation for data aggregation scheduling in wireless sensor networks. In: Proceedings of IEEE INFOCOM, pp. 105–145, Oct 2013

Part IV
Cryptography and Security Analysis

Extended Visual Cryptography Scheme for Multi-secret Sharing

L. Siva Reddy and Munaga V.N.K. Prasad

Abstract This paper proposes a novel user-friendly visual cryptography scheme for multiple secret sharing. We have generated meaningful shares for multiple secret images using cover images. These meaningful shares are shared among participants. All the shares are required to recover secret images that are shared. The proposed scheme uses Boolean-based operations for generating meaningful shares and recovering all secret images that are used. The proposed scheme achieved lossless recovery of multiple secrets and overcomes the problem of management of meaningless shares.

Keywords Visual cryptography · User-friendly visual cryptography scheme · Multiple secret sharing · Extended visual cryptography scheme · Meaningful shares

1 Introduction

Secret sharing scheme was initially proposed by Blakely and Shamir in 1979 [1, 2, 3]. In general this secret sharing scheme is also called (k,n) secret sharing. In this scheme secret information is divided into n shares and distributed among n participants and it is an encryption operation. At least k participants are needed to recover original secret information and it is a decryption operation. Participants less than k cannot recover secret information. Naor and Shamir proposed (k,n) visual cryp-

L. Siva Reddy (✉) · M.V.N.K. Prasad
Institute for Development and Research in Banking Technology, Hyderabad, India
e-mail: sivareddy09503@gmail.com

M.V.N.K. Prasad
e-mail: mvnkprasad@idrbt.ac.in

L. Siva Reddy
School of Computer and Information Sciences, University of Hyderabad,
Hyderabad, India

© Springer India 2016
A. Nagar et al. (eds.), *Proceedings of 3rd International Conference on Advanced Computing, Networking and Informatics*, Smart Innovation, Systems and Technologies 44, DOI 10.1007/978-81-322-2529-4_26

tography scheme (VCS) to encode secret which is in the image form [4]. In this scheme computational devices are used to encrypt secret image into n shares. Decryption can be done by human visual system after stacking k shares. This scheme is useful if there are no computational devices for decryption. Random-grid based VCS is proposed by Kafri et al. and Shyu [5, 6] to overcome the drawback of pixel expansion of conventional scheme. In this scheme each share acts as random grid.

Multiple Visual Cryptography Secret (MVCS) sharing scheme was proposed to encrypt multiple secrets at a time. Wu and Chen [7] proposed (2,2) MVCS to encrypt two secrets into two square shares SH1 and SH2. The first secret S1 is recovered by stacking two shares SH1 and SH2. The second secret image S2 is recovered by stacking share SH1 with 90° rotated share SH2 [7]. The rotation angle can be of q × 90 where $1 \le q \le 3$. Wu and Chang [8] proposed multi-secret sharing scheme by encrypting secrets in the form of circular shares. Generation of secret S2 can be done by stacking share SH1 with 360° multiple rotated share SH2. In this scheme the limitation of 90, 180, 270° rotation is removed. Shyu et al. [9] proposed MVCS for more than two secrets using two circular shares with different rotation on one of shares. Feng et al. [10] proposed another MVCS scheme for sharing multiple secrets. In this scheme two circular ring shares R1 and R2 are used to encrypt multiple secrets. Some of the other authors who work on multiple secret sharing are Shyu et al. [11], Chang et al. [12], Wu et al. [13].

All the above mentioned schemes for MVCS have many drawbacks. One of the main drawbacks is lack of visual quality of the recovered secret image. The second drawback is distortion of the recovered image from the original image (pixel expansion). These schemes didn't prepare meaningful shares for users. These meaningful shares are mainly used for identification and management by users and to avoid suspicion of attackers who may focus on meaningless shares.

The disadvantages of meaningless shares can be minimized by using extended visual cryptography scheme (EVCS). It is also called as user-friendly visual cryptography scheme. The EVCS generates meaningful shares by stacking cover images onto meaningless shares. By using EVCS dealers can identify each share easily. Ateniese et al. proposed (k,k)-threshold EVCS which generates meaningful shares using binary images [14]. In this sheme k meaningful shares are generated and all are needed to recover secret image. Fang [15] proposed EVCS for conventional VC with a progressive decryption effect. Here progressive decryption effect means quality of secret image gets improved when number of shares is increased in overlapping. Chen et al. [16] proposed EVCS for random-grid-based techniques. Random-grid based techniques are pixel expansion less schemes. Wang et al. developed an EVCS using matrix extension algorithm which generates random matrix shares instead of meaningless shares [17]. Here shares are matrices which have pixel values of secret image.

Chen and Wu proposed an (n,n) secure Boolean-based multi-secret sharing scheme [18]. In this scheme n shares are generated using n secret images. This scheme mainly uses Boolean-based operations like XOR and bitwise circular shift operation. N shares are required to recover n secrets. Chen and Wu scheme [18] uses random image R in order to randomize the original secret images. It is generated by using two functions XOR operation and bitwise shift operation. This scheme for MVCS faces some drawbacks. The first drawback is generation of meaningless shares in which dealers feel difficulty to manage those shares. The second drawback is to use minimum number of secrets (4 or more) for sharing in order to ensure relative security. If we use two secrets for share generation then shares may have secret information as they are generated using XOR calculation.

This paper proposes extended visual cryptography scheme for multiple secret sharing that develops meaningful shares from multiple secrets and cover images. The proposed scheme uses Boolean-based operations for generating meaningful shares using cover images. Lossless recovery of multiple secrets is achieved using proposed scheme. The proposed scheme is free from pixel expansion. The experimental results of the proposed work shows that MVCS sharing scheme gets improved with meaningful shares.

The organization of this paper is as follows. Proposed scheme is discussed in Sect. 2. Sections 2.1 and 2.2 explains proposed scheme share generation and its recovery, respectively. Section 2.3 discusses on proposed scheme. Section 3 shows experimental results and comparison with related work. Section 4 completes the proposed scheme with conclusion.

2 Proposed Scheme

This section explains proposed work of extended visual cryptography scheme for multi-secret sharing. This scheme generates meaningful shares using multiple secrets and cover images. Boolean based operations are mainly used in this scheme. Lossless recovery of multiple secrets is achieved using the proposed work. Sections 2.1 and 2.2 explains about share generation and secret recovery of the proposed scheme. Section 2.3 discusses on proposed scheme.

2.1 Share Generation

This section explains how to generate meaningful shares using multiple secret images and cover images. The proposed scheme for share generation uses random image generation function which was proposed by Chen and Wu [18]. This random image generating function mainly involves two functions F_1 and F_2. Function F_1 uses XOR operation and function F_2 uses bitwise circular shift operation. The algorithm for share generation is as follows:

1. Let $I_1, I_2, \ldots, I_n$ are n multiple secret images that are used as input for share generation.

2. Now random image R_1 is generated using secret images $I_1, I_2, \ldots, I_n$ as follows:

$$R_1 = F_1(F_2(I_1, I_2, \ldots, I_n));$$
$$\text{where } F_2 = I_1 \oplus I_2 \oplus \ldots \oplus I_n;$$
$$F_1 = \text{bitwise circular shift of } F_2.$$

3. Meaningless images $M_1, M_2, \ldots, M_n$, are generated using XOR operation of given multiple secret images with random image R_1 respectively.

4. Now meaningful shares $S_1, S_2, \ldots, S_k$ are generated from meaningless images and n cover images $C_1, C_2, \ldots, C_n$ using Boolean OR and Boolean AND operations as follows:

$$S_1 = M_1 \otimes C_1$$
$$S_2 = M_1 \odot C_1$$
$$S_3 = C_1$$
$$S_4 = M_2 \otimes C_2$$
$$S_5 = M_2 \odot C_2$$
$$S_6 = C_2$$
$$\vdots$$
$$S_{k-2} = M_n \otimes C_n$$
$$S_{k-1} = M_n \odot C_n$$
$$S_n = C_n$$

where $1 \leq k \leq 3 \times n$

5. Finally k meaningful shares are generated for the given multiple secret images.

Figure 1 shows share generation phase of the proposed scheme. In this phase n multiple secret images $I_1, I_2, \ldots, I_n$ are used. A random image is generated from these secret images using random image generation function. Then N meaningless images $M_1, M_2, \ldots, M_n$ are generated by applying XOR operation on the secret images and random image. Finally, K meaningful shares $S_1, S_2, \ldots, S_k$ are developed by performing Boolean operations AND, OR between meaningless images and cover images.

2.2 Secret Images Recovery Procedure

The recovery procedure uses k meaningful share images from share generation as an input for recovering original secret images. Boolean XOR operation is used on these shares to generate n meaningless shares. Random image R_1 is generated from these meaningless share images using random image generating function. Finally n secret images are recovered by applying Boolean XOR operation between random image and meaningless shares. The algorithm for secret images recovery is as follows:

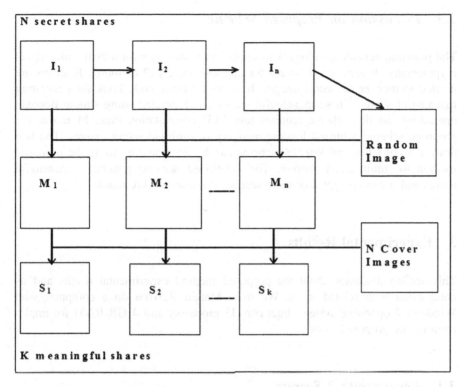

Fig. 1 N meaningful share images generation from n secret images

1. Recover meaningless shares $M_1, M_2, \cdots, M_n$ from k meaningful shares as follows:

$$M_1 = S_1 \oplus S_2 \oplus S_3;$$
$$M_2 = S_4 \oplus S_5 \oplus S_6;$$
$$\vdots$$
$$M_n = S_{k-2} \oplus S_{k-1} \oplus S_k;$$

where $1 \leq k \leq 3 \times n$

2. Random image R_1 is generated using meaningful shares.

3. Finally, n secret images $I_1, I_2, \cdots, I_n$ are generated as follows:

$$I_1 = M_1 \oplus R_1;$$
$$I_2 = M_2 \oplus R_1;$$
$$\vdots$$
$$I_n = M_n \oplus R_1;$$

2.3 Discussion on Proposed Scheme

The proposed scheme develops (k,k) multi-secret sharing with user-friendly visual cryptography. It generates k shares for multiple secrets (2 or more). K shares are needed to recover all secret images. It doesn't require code book for generating meaningful shares. These meaningful shares are generated using simple Boolean operations. So this scheme requires less CPU computation time. Moreover the proposed scheme achieved lossless recovery of multiple secret images. The two drawbacks of Chen and Wu [18] scheme can be overcome by using the proposed method for multi-secret sharing. The developed scheme generates meaningful shares and it can encrypt 2 or more secrets in a more secure manner.

3 Experimental Results

This section discusses about the proposed method experimental results and its comparison with related work. We used Matlab R2010a on a computer with Windows 7 operating system, Intel core I3 processor and 4 GB RAM for implementing the proposed work.

3.1 Results with 2 Secrets

Figure 2 shows experimental results of the proposed work. Here we took two secret images: Pepper, Barbara and two cover images: Lena, Cameraman. All these images are of 512 × 512 size. Figure 2e–j shows the meaningful shares generated by applying proposed work share generation operation. Figure 2k, l shows recovered secret images generated by applying secret recovery procedure. Lossless recovery of the secret images is achieved with meaningful shares.

Peak Signal-to-Noise Ratio (PSNR) is the accuracy measure for the image quality of generated shares. Table 1 shows PSNR values between share images and original images. In general, PSNR value of any image to be visible is more than 30 dB. The proposed work achieved PSNR value of the recovered images approaches to infinite which means that lossless recovery of secret images. Meaningful shares generated by the proposed scheme have PSNR values more than 50 dB. So the generated shares can be easily managed by the dealer as they are visible to human visual system. In this way proposed method works as extended visual cryptography scheme for multiple secret sharing.

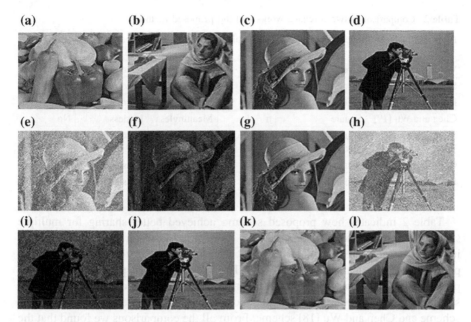

Fig. 2 Two secret images (**a, b**), two cover images (**c, d**), six meaningful shares (**e–j**), recovered images (**k, l**)

Table 1 PSNR values between share images and original images

Image	PSNR value
Share 1 versus cover image 1	57.6077
Share 2 versus cover image 1	57.8255
Share 3 versus cover image 1	Infinite
Share 4 versus cover image 2	56.9659
Share 5 versus cover image 2	57.8235
Share 6 versus cover image 2	Infinite
Recovered image 1 versus secret image 1	Infinite
Recovered image 2 versus secret image 2	Infinite

3.2 Comparison of the Proposed Method with Related Schemes

This section explains the comparison between the proposed work and the related schemes. Comparison is made in different areas like: share shape, shareable secrets, quality of share, recovery type, and pixel expansion. Share shape indicates type of generated share like square, rectangle, and circle. Shareable secrets mean number of secrets that can be shareable by using the scheme. Quality of share explains about generated share visual quality. Recovery type indicates that how secret images are recovered for the scheme. Pixel expansion explains about distortion of recovered image from the original image.

Table 2 Comparison between related works and the proposed method

Proposed method	Share shape	Shareable secrets	Share quality	Type of recovery	Pixel expansion
Wu and Chang [8]	Circle	2	Meaningless	Recognizable	Yes
Shyu et al. [9]	Circle	n	Meaningless	Recognizable	Yes
Feng et al. [10]	Square	n	Meaningless	Recognizable	Yes
Chen and Wu [19]	Square	n − 1	Meaningless	lossless	No
Shyu and Chen [20]	Square or rectangle	2 or 4 or 8	Meaningless	Recognizable	Yes
Chen and Wu [18]	Square	n	Meaningless	Lossless	No
Proposed scheme	Square	n	Meaningful	Lossless	No

Table 2 indicates how proposed scheme achieved better sharing for multiple secret images. It generates square shape share images which are mainly used in image processing. It can share multiple secrets without distortion which is not possible for schemes of [8–10, 19, 20]. The most important property of the proposed method is meaningful share generation which is impossible for all other schemes. Lossless recovery of all the secrets is only possible with the proposed scheme and Chen and Wu [18] scheme. From all the comparisons we found that the proposed work is better useful scheme for multiple secret sharing.

4 Conclusion

In this paper we proposed a novel algorithm for meaningful share generation of multiple secret sharing scheme. The experimental results show that proposed scheme is a better suitable scheme for multi-secret sharing than previous approaches. It also indicates that lossless recovery of multiple secrets is achieved with our scheme. As the scheme requires only Boolean based operations it is computationally costless scheme. There are many advantages applicable for the proposed work. The first one is generating meaningful shares which enable dealers to manage the shares effectively. The second one is its distortion free recovery of multiple secrets. The third one is its ability to share many secrets. The main contribution of our proposed work is applying extended visual cryptography scheme or user-friendly cryptography for the multi-secret sharing schemes with lossless recovery of secrets.

References

1. Blakely, G.R.: Safeguarding cryptography keys. Proc. Nat. Comput. Conf. **48**, 313–317 (1979)
2. Shamir, A.: How to share a secret. Commun. ACM **22**, 612–613 (1979)
3. Cimato, S., Yang, C.-N.: Visual cryptography and secret image sharing. In: Digital Imaging and Computer Vision Series (2012)

4. Naor, M., Shamir, A.: Visual cryptography, In: Proceedings of Advances in Cryptology, pp. 1–12 (1995)
5. Kafri, O., Keren, E.: Encryption of pictures and shapes by random grids. Opt. Lett. **12**, 377–379 (1987)
6. Shyu, S.: Image encryption by random grids. Pattern Recogn. **40**, 1014–1031 (2007)
7. Wu, C.C., Chen, L.H.: A study on visual cryptography. Master Thesis, Institute of Computer and Information Sciences, National Chaio Tung University, Taiwan (1998)
8. Wu, H.C., Chang, C.C.: Sharing visual multi-secret using circle shares. Comput. Stan. Interfaces **28**, 123–135 (2005)
9. Shyu, S.J., Huang, S.Y., Lee, Y.K., Wang, R.Z.: Sharing multiple secrets in visual cryptography. Pattern Recogn. **40**(12), 3633–3651 (2007)
10. Feng, J.B., Wu, H.C., Tsai, C.S., Chang, Y.F., Chu, Y.P.: Visual secret sharing for multiple secrets. Pattern Recogn. **41**, 3572–3581 (2008)
11. Shyu, S.J., Huang, S.Y., Lee, Y.K., Wang, R.Z.Chen: Sharing multiple secrets in visual cryptography. Pattern Recogn. **40**, 3633–3651 (2007)
12. Chang, C.-C., Tu, N.T., Le, H.D.: Lossless and unlimited multi-image sharing based on Chinese remainder theorem and Lagrange interpolation. Sig. Process. **99**, 159–170 (2014)
13. Wu, H.C., Chang, C.C.: Sharing visual multi-secrets using circle shares. Comput. Stand. Interfaces **28**, 123–135 (2005)
14. Ateniese, G., Blundo, C., Santis, A.D., Stinson, D.R.: Extended capabilities for visual cryptography. Theoret. Comput. Sci. **250**, 143–161 (2001)
15. Fang, W.P.: Friendly progressive visual secret sharing. Pattern Recogn. **41**, 1410–1414 (2008)
16. Chen, T.H., Lee, Y.S.: Yet another friendly progressive visual secret sharing scheme. In: Proceedings of 5th International Conference on Intelligent Information Hiding and Multimedia Signal Processing, pp. 353–356 (2009)
17. Wang, D., Yi, F., Li, X.: On general construction for extended visual cryptography schemes. Pattern Recogn. **42**, 3071–3082 (2009)
18. Chen, T.H., Wu, C.S.: Efficient multi secret image sharing based on Boolean operations. J. Syst. Softw. **92**, 107–114 (2014)
19. Chen, T.H., Wu, C.S.: Efficient multi-secret image sharing based on Boolean operations. Sig. Process. **91**, 90–97 (2011)
20. Shyu, S.J., Chen, K.: Visual multiple secret sharing based upon turning and flipping. Inf. Sci. **181**, 3246–3266 (2011)

Identifying HTTP DDoS Attacks Using Self Organizing Map and Fuzzy Logic in Internet Based Environments

T. Raja Sree and S. Mary Saira Bhanu

Abstract The increasing usage of internet resources may lead to more cyber crimes in the network domain. Among the various kinds of attacks, HTTP flooding is one of the major threats to uninterrupted and efficient internet services that depletes the application layer. It is hard to find out the traces of this attack because the attacker deletes all possible traces in the network. Thus, the only possible way to find the attack is from the trace log file located in the server. This paper proposes a method using Self Organizing Map (SOM) and fuzzy association rule mining to identify the attack. SOM is used to isolate the unknown patterns and to identify the suspicious source. The attacks are identified using fuzzy association rule mining. The statistical test has been carried out to measure the significance of features to identify the legitimate or intrusive behavior.

Keywords Self organizing map · Fuzzy association rule mining · HTTP flood

1 Introduction

Today, the world is extremely dependent on the accessibility of the cyberspace. The cyber criminals flood large amount of packets targeting the computer networks or servers. This may result in various kinds of attacks such as HTTP flood, TCP flood, SYN flood etc. HTTP flood is one of the major threats to internet service because it depletes the resources such as network bandwidth, CPU cycles and application or services with the overflow of messages. For example, Sony play—station network and Sony entertainment network were hijacked by attackers [1] and it leads to

T. Raja Sree (✉) · S. Mary Saira Bhanu
Department of Computer Science and Engineering, National Institute
of Technology, 620015 Tiruchirappalli, Tamil Nadu, India
e-mail: 406112001@nitt.edu

S. Mary Saira Bhanu
e-mail: msb@nitt.edu

© Springer India 2016

A. Nagar et al. (eds.), *Proceedings of 3rd International Conference on Advanced Computing, Networking and Informatics*, Smart Innovation, Systems and Technologies 44, DOI 10.1007/978-81-322-2529-4_27

259

financial losses to the targeted companies. Moreover, the attackers delete all the traces from where the attack has originated. Thus, it is necessary to discover the traces of an attack from the log files located in the server. Machine learning techniques have been used to identify the attacks from the trace log file located in the server.

The anomalous behaviors are difficult to determine using supervised learning techniques. The unknown attacks cannot be observed because of the secular changes in the network patterns and their characteristics often renders the classification techniques to become ineffective [2]. Therefore, the unsupervised learning techniques such as k-means, SOM, Art2 etc., are used for detecting the application layer attacks. Unsupervised learning techniques for attack detection are focused in [3, 4]. Even though, little research has been borne out in unsupervised learning to investigate the new attacks at the network layer, but it has not concentrated on the application layer attacks. Hence, the proposed method uses unsupervised learning technique (SOM) and fuzzy association rule mining for attack detection.

The SOM is used to isolate the unseen patterns from the neighborhood map and also it reduces the high dimensional input data into two dimensional representation spaces. The fuzzy association rule mining techniques form the set of rules using the features (IP address, Time stamp, requested URL, etc.,) and the support and confidence are measured for the usefulness of each rule in attack detection. When an attack occurs, the confidence of the new rules is compared to the minimum confidence of the rules in the rule base to find the pattern. If the new pattern is not found in fuzzy base, then this rule is updated into the fuzzy base. The statistical test has been used to identify the significance of the features to determine whether the data is normal or anomalous.

The remainder of the paper is formed as follows: The related work is discussed in Sect. 2. Section 3 illustrates the proposed work. Section 4 explains the experimental work carried out in this paper and depicts the observational outcomes. Section 5 provides the summary and future work.

2 Related Work

2.1 DDoS Attack

Distributed Denial of Service (DDoS) attacks have the set of compromised hosts to flood the traffic to the targeted resources and make the target's service unavailable [5]. DDoS attacks can be launched in two forms: Direct attacks and Reflector attacks [6, 7]. In direct DDoS attacks, the attacker directly sends a massive amount of packets to the victim host or server through multiple compromised hosts or machines. In reflector-based DDoS attacks, the attacker sends requests to a reflector host to forward a massive amount of attack traffic by spoofing IPs of victim host(s).

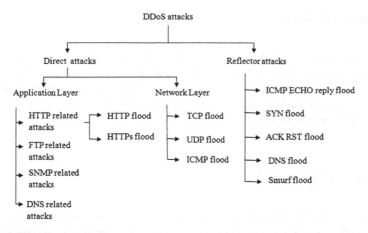

Fig. 1 Taxonomy of DDoS attacks

As a consequence, the reflector hosts send their responses to the victim host, flooding the traffic. The taxonomy of DDoS attacks is depicted in Fig. 1.

The most important application layer attack is HTTP flooding, which the attacker establishes the connection by sending valid or invalid HTTP requests to exhaust the server resources such as CPU, memory, disk or database bandwidth or I/O bandwidth etc. [8]. Rate based engines are not successful in detecting HTTP flood attacks because the volume of HTTP traffic may be under the detection threshold [9]. An HTTP GET request attacks cannot be detected using SYN rate limit method. Consequently, various parameters such as rate based and rate invariant attacks are employed for the detection of attacks and it passes to the high number of false positives. Yatgai et al. proposed an approach based on the browsing order of the page and the correlation analysis with a browsing order to page information size. The outcomes indicate that this method has not addressed the usage of large access log files to discover attacks and it may result in false positives [10].

Hayoung et al. proposed the real time intrusion detection scheme based on clustering by SOM from the traffic of the historical data and the map units are labeled using the correlation of features to detect the DDoS attack [3]. But the labeling of map units may reduce the detection accuracy in predicting the attack as normal. Ajanta et al. combined the concept of [3] with fuzzy logic to obtain the high detection rate with low false positive [4]. They used unsupervised learning (SOM) for predicting the suspicious nature of unseen patterns and modeling the fuzzy rule from each neighboring map unit. If a new attack occurs, only the new rules corresponding to the map units will be updated rather than the entire model in the fuzzy rule base. The methods employed in the literature have not addressed the detection of HTTP flood attacks from the trace log file located in the server. Therefore, the proposed approach is used for the detection of abnormal behavior based on the unsupervised learning (SOM) and fuzzy association rule mining.

3 Proposed Work

The architecture of the proposed work is shown in Fig. 2. It consists of two stages, namely clustering and rule generation. The trace of log information is preprocessed and is fed to the clustering phase for the identification of patterns from the network traffic. The retrieved data is then modeled using the fuzzy association rule to predict the suspicious nature of the source.

3.1 Data Preprocessing

The legitimate and illegitimate data found in web server access log files are passed as input to the data preprocessing to get rid of the un-cleaned data and for selecting the relevant features by examining the log files. The functions of the preprocessor [11] are as follows:

- Remove the irrelevant records such as images, multimedia files, administrative action files, and page style files (*.JPG, *.GIF, etc.).
- Identification of individual users by IP address, domain, HTTP login, referred page, client type and cookie information.
- Session identification and creative by time gap, duration on page, backtracking is observed. The cleaned, modified dataset is obtained.

After receiving the modified dataset, parsing and analyzing have been performed in the log files to identify the relevant features (IP destination, time stamp of the request, requested URI of the page and the referral page of the user) for the detection of DDoS attack. These relevant features are transformed into numerical data for further processing. Hence it is necessary to change the relevant features to numeric values because SOM resolves only the numerical data to perform clustering on the web log data.

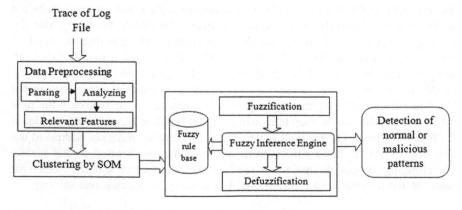

Fig. 2 Architecture of proposed HTTP flood attack detection

3.2 Clustering by SOM

Kohonen introduced a feed forward neural network structure consisting of neurons that are arranged in rows and columns [3, 12–14]. The 'n' input patterns are connected to each of the 'm' output cluster units in 2-dimensional space and the weight vector serves as the exemplar of the input patterns. It groups similar input patterns and this pattern is compared to the weight vector of each neuron and the closest neuron is declared as the winner. The nearest topological neurons and the winner's weight neuron are adjusted to find the best matching unit (BMU).

3.2.1 Steps of SOM Algorithm

- *Initialize the network* For each node j, choose the random values to the initial weight vector w_j.
- *Activation of input* Activate the input vector X to all the nodes in the network from the input space.
- *Calculate the winning node* The smallest distance between the weight vector and the associated input vector, i.e., $\min\left\{d_j(x)\right\} = \sum_{i=1}^{D}(x_i - w_{ji})^2$ are selected as the winning neuron Z(x). The node which has the smallest distance is the BMU of the node.
- *Weight updation* The weights are updated by using the equation $\Delta w_{kl} = \gamma(t)T_{k,Z(x)}(t)(x_k - w_{lk})$. Where $T_{l,k}(t)$ is the Gaussian neighborhood, $\gamma(t)$ is the learning rate.
- Iterate the steps until the minimum distance criterion is satisfied or further no changes in the feature map.

3.3 Fuzzification

The process of converting crisp values into the linguistic values of fuzzy sets in the rule base is called Fuzzification. The degree of membership values in a fuzzy set is assigned between 0 and 1 [15]. Table 1, shows the linguistic variables and the fuzzy numbers are assigned for each parameter used in the fuzzy rule base.

The commonly used membership function associated with the fuzzy sets is triangular membership function. It is defined by using three parameters y_1, y_2, y_3 [16] are expressed in Eq. (1):

$$\mu_{A(z)} = triangular(z; y_1, y_2, y_3) = \begin{cases} 0 & z \leq y_1 \\ \frac{z-y_1}{y_2-y_1} & y_1 \leq z \leq y_2 \\ \frac{y_3-z}{y_3-y_2} & y_2 \leq z \leq y_3 \\ 0 & y_3 \leq z \end{cases} \tag{1}$$

Table 1 Linguistic variable used in fuzzy rule base

Linguistic variable	Fuzzy numbers				
	IP	Request time	Requested URL	Requested file size	Referral URL
Very low (VL)	(−20 0 20)	(−4 0 2)	(−20 0 10)	(−500 0 500)	(−4 0 2)
Low (L)	(10 30 50)	(1 3 5)	(5 15 25)	(250 750 1250)	(1 3 5)
Medium (M)	(40 60 80)	(4 6 8)	(20 30 40)	(1000 1500 2000)	(4 6 8)
High (H)	(70 90 110)	(7 10 14)	(35 45 65)	(1750 2250 2750)	(7 9 11)

3.3.1 Fuzzy Inference Engine and Fuzzy Rule Base

Fuzzy rule base is mainly used for storing rules in order to obtain a new fact. It uses fuzzy if-then rules of the form for pattern classification problems to detect attacks. The fuzzy association rules for the patterns are expressed in Eq. (2).

$$Rule: If\ y_1\ is\ a_{1p}\ and\ y_2\ is\ a_{2p}\ldots and\ y_n\ is\ a_{np}\ then\ class\ c_n\ is\ z_p \qquad (2)$$

where $y_1, y_2, \ldots, y_n$ represent the antecedent attributes in the pth rule; a_{jp} (j = 1, 2, ..., n, p = 1, 2, ..., m) is the value of the jth antecedent part in the pth rule; c_n represents the pth output consequent part, and the value is z_p. Association rule mining is an efficient method for discovering the interesting patterns that can be found in large database. The confidence $C_{rule(y=>z)}$ and the support $S_{rule(y=>z)}$ of the fuzzy association rule [17] are expressed in Eqs. (3) and (4).

$$Confidence\ C_{rule(x=>y)} = \frac{\sum_{p=1}^{m} \mu_x(x_p) * \mu_y(y_p)}{\sum_{p=1}^{m} \mu_x(x_p)} = \frac{\sum_{p\epsilon class}(y_p) * \mu_x(x_p)}{\sum_{p=1}^{m} \mu_x(x_p)} \qquad (3)$$

$$Support\ S_{rule(x=>y)} = \frac{\sum_{p=1}^{m} \mu_x(x_p) * \mu_y(y_p)}{m} = \frac{\sum_{p\epsilon class}(y_p) * \mu_x(x_p)}{m} \qquad (4)$$

where $\mu_x(x_p)$ denotes the antecedent part of X, $\mu_y(y_p)$ denotes the consequent part of Y, m denotes the number of datasets.

Initially, the access log data is collected from the web server. Then, the relevant features are identified from the web server access log file and the support and confidence values of these features are calculated using fuzzy association rule mining for the detection of an attack. When an attack occurs, the new rule confidence is compared to the minimum confidence in the fuzzy rule base, and the resulting rule is added and updated in the fuzzy rule base for the identification of attacks. With this type of classification rule, it is possible to monitor the attack sequences in real time to discover the anomalous attacks.

4 Experimental Results

4.1 Attack Generation

The DDoS attack log files are created using different machines of a LAN. Apache server is installed in the server machine. HULK [18], HTTP DOS [19], HOIC [20] etc. are some of the attack tools and attack scripts that are used to launch the HTTP DDoS attacks. These attack tools are installed on several machines by sending several malicious packets to the targeted server. The normal traffic has been induced by the normal browsing activities carried out by the different machines. The attack traffic has been generated by using attack tools and the attack is reflected in the log file. Unix Logs are found in /usr/local/. Linux logs are found in /var/log and / usr/adm. However, DDoS attack has been reflected in /var/log/apache2/access.log and /var/log/apache2/error.log. These attacks have been found by taking the correlation of access log and error log.

Figure 3 displays the SOM sample hits obtained with 10-by-10 neuron SOM. The region inside the hexagonal map shows the actual distribution of data in SOM. The crowded data in the map shows attack whereas in other places data are unevenly spread demonstrates no attack. SOM is used to produce natural clustering for the given input data set. The topological preserving property of the input neuron is maintained based on the grouping of nearby neurons present in the dataset. The clustered results are applied to distinguish the anomalous and normal behavior of the user.

The ranges of fuzzy rules are defined based on the minimum and maximum value of each data source. The degree of the membership value of an object in the fuzzy set lies between 0 and 1. The membership function of the linguistic variable

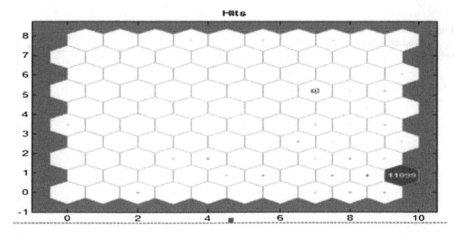

Fig. 3 SOM Clustering

in the referral_URL data source is depicted in Fig. 4. The classification of normal and attack instances of the sample is shown in Fig. 5. The rules are framed from each data source with the help of fuzzy association rule mining is described in Table 2. The classification of normal and anomalous behavior is predicted based on the association rule mining. The patterns are organized based on the association of support and confidence of each particular set. The possible rules are generated from the different data items with support and confidence. The novel and anomalous attack are detected by extracting the rules from the fuzzy rule base. When a new rule is found, then compare the new rule obtained has minimum confidence C_{min} with the existing rules in the rule base and if it is not available in the rule base then the new rules are added by updating the contents into the rule base. The rule generated has different patterns to predict whether the data is normal or attack.

Fig. 4 Membership function of Referral_URL

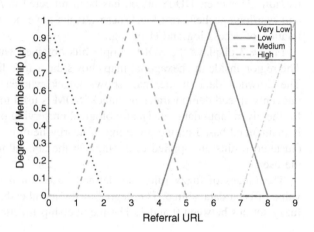

Fig. 5 Classification of normal and abnormal behavior

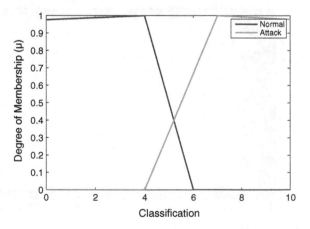

Table 2 Sample content in fuzzy rules

Sl. no.	Rule content	Support	Confidence
1	If(sourceIPaddress = 10.1.49.24) ≥ high & ((Destination IPaddress = 10.1.49.48) ≥ high& (protocol = TCP) ≥ high & ((port = (60560, 80)) ≥ medium then unknown (suspicious) attack	0.74	0.93
2	If(sourceIPaddress = 10.1.49.45) ≥ medium & (Destination IPaddress = 10.1.49.48) ≥ low& (protocol = TCP) ≥ low & (port = 60650) ≥ low then normal	0.7	0.87
3	If(sourceIPaddress = 10.1.49.25) ≥ high & (Destination IPaddress = 10.1.49.48) ≥ high&(protocol = TCP) & (requested − url = high) & (requested − file − size = high) then attack	0.65	0.85
4	If(sourceIPaddress = 10.1.49.29) ≥ low & (Destination IPaddress) ≥ high & (time = medium) & (requested − url = medium) & (requested − file − size = low) then normal	0.7	0.82

4.2 T-Test

The t-test is to evaluate the significance of individual feature values in the web log data and it is computed with the Eq. (5), let x_i, (i = 1, 2, ..., n) is a random sample of size n with mean μ and variance s^2. The t-test is defined by the statistic [21]

$$t = \frac{\bar{x} - \mu}{\frac{s}{\sqrt{n}}}, \text{ where } \bar{x} = \frac{1}{n}\sum_{i=1}^{n} x_i \text{ and } s^2 = \frac{1}{n-1}(x_i - x)^2 \tag{5}$$

where $\bar{x}$ and s^2 are the mean and variance of the feature values, n be the number of elements in the features with the degrees of freedom (n − 1).

The hypothesis H_0 to identify the relevant features used for attack detection is as follows:

$$H_0 : \begin{cases} \text{All the features (Remotehost, Req.time, Req.url, Req.file.size, Referrer} \\ -URL) \text{ in sample space are significant} \\ \text{All the features (Remotehost, Req.time, Req.url, Req.file.size, Referrer} \\ -URL) \text{ in sample space are significant} \end{cases}$$

From the Eq. (5), the values of each feature considered significantly important with 95 % confidence. The significance of test samples is depicted in Table 3.

Table 3 Significant of t-test

	T	Norma 1 (sig.2–tailed)	Attack (sig.2–tailed)
Remote host	66.895	0.000	0.010
Request time	2.01053	0.000	0.039
Request URL	4.598	0.000	0.051
Request-file-size	231.985	0.000	0.023
Referrer	2.384	0.017	0.034

5 Conclusion

The HTTP DDoS attack has been identified by using SOM and fuzzy association rule mining. The attacks are generated by using various attacking tools. When an attack occurs, unknown patterns are found by comparing the confidence of new rule with the minimum confidence of the entire rules stored in the fuzzy rule base. If the pattern is not found, and so the resulting patterns are appended and updated in the rule base to see the pattern as either attack or normal. The result shows that the new attacks can be found easily from web server log files using association rule mining. The statistical test has been carried out to employ the significance of the parameters used in determining the normal or anomalous behavior. In future, the various pattern recognition methods will be used for efficient identification of the DDoS attack.

References

1. DDoS attack. http://www.businessweek.com/articles/2014-08-26/ddos-attacks-are-soaring
2. Sebyala, A.A., Olukemi, T., Sacks, L., Sacks, D.L.: Active platform security through intrusion detection using naive bayesian network for anomaly detection. In: International Symposium on Communications, pp. 1–5. London (2002)
3. Oh, H., Chae, K.: Real-time intrusion detection system based on self-organized maps and feature correlations. In: 3rd International Conference on Convergence and Hybrid Information Technology ICCIT'08, vol. 2, pp. 1154–1158. IEEE Press, (2008)
4. Konar, A., Joshi, R.C.: An efficient intrusion detection system using clustering combined with fuzzy logic. In: Ranka, S., Banerjee, A., Biswas, K., Dua, S., Mishra, P., Moona, R., Poon, S. H., Wang, C.-L. (eds.) Contemporary Computing 2010, LNCS, vol. 94, pp. 218–228. Springer, Heidelberg (2010)
5. Byers, S., Rubin, A.D., Kormann, D.: Defending against an Internet-based attack on the physical world. ACM Trans. Internet Technol. (TOIT) 4(3), 239–254 (2004)
6. Beitollahi, H., Deconinck, G.: Analyzing well-known countermeasures against distributed denial of service attacks. J. Comput. Commun. 35, 1312–1332 (2012)
7. Bhuyan, M.H., Kashyap, H.J., Bhattacharyya, D.K., Kalita, J.K.: Detecting distributed denial of service attacks: methods, tools and future directions. J. Comput. 57, 537–556 (2014)
8. Siaterlis, C., Maglaris, V.: Detecting incoming and outgoing DDoS attacks at the edge using a single set of network characteristics. In: 10th IEEE Symposium on Computers and Communications ISCC, pp. 469–475. IEEE Press (2005)

9. HTTP flood Attacks Danger and Security, http://security.radware.com/knowledge-center/DDoSPedia/http-flood/
10. Yatagai, T., Isohara, T., Sasase, I.: Detection of HTTP-GET flood attack based on analysis of page access behavior. In: IEEE Pacific Rim Conference on Communications, Computers and Signal Processing, pp. 232–235. IEEE Press (2007)
11. Pabarskaite, Z.: Enhancements of preprocessing, analysis and preparation techniques in web log mining. Vilnius Technikes, Vilnius (2009)
12. Kohonen, T.: Self-organized formation of topologically correct feature maps. J. Bio. cybern **43** (1), 59–69 (1982)
13. SOM Toolbox for Matlab. http://www.cis.hut.fi/projects/somtoolbox/
14. Dusan, S., Natalija, V., An, A.: Detection of malicious and non-malicious website visitors using unsupervised neural network learning. J. Appl. Soft Comput. **13**, 698–708 (2013)
15. Liao, N.: Network forensics based on fuzzy logic and expert system. J. Comput. Commun. **32**, 1881–1892 (2009)
16. Fuzzy logic. www.dma.fi.upm.es/java/fuzzy/fuzzyin/funpert_en.html
17. Ishibuchi, H., Yamamoto, T., Nakashima, T.: Determination of rule weights of fuzzy association rules. In: The 10th IEEE International Conference on Fuzzy Systems, vol. 3, pp. 1555–1558. IEEE Press (2001)
18. HULK attack. http://github.com/grafov/hulk
19. OWASP HTTP GET DDoS attack. www.exploiterz.blogspot.in/2013/07/owasp-http-getpost-ddos-attacker-tool.html
20. HOIC attack tool. www.thehackersnews.com/2012/03/another-ddos-tool-from-anonymous-hoic.html
21. Gupta, S.C., Kapoor, V.K.: Elements of Mathematical Statistics, 3rd edn. (2003)

Improving the Visual Quality of (2, *n*) Random Grid Based Visual Secret Sharing Schemes

Lingaraj Meher and Munaga V.N.K. Prasad

Abstract Visual secret sharing scheme (VSS) based on the concept of random grid (RG) that suffers from low visual quality of the decrypted secret. In this work a (2, *n*) RG based VSS has been proposed to improve the contrast. The proposed scheme supports the traditional stacking decryption method. To improve the visual quality further a decryption algorithm is proposed. The experimental result and comparison of visual quality with related methods has shown that the proposed schemes perform well.

Keywords Secret · Random grid · Visual secret sharing · Encryption · Decryption · Contrast · Visual quality

1 Introduction

There is an exponential increasing with the popularity of the information technology in last decades. So the security of the information has become a critical issue now. Nour and Shamir proposed a secure cryptography method, called the Visual Cryptography (VC) or Visual Secret Sharing scheme (VSS), to hide the secret in the form of images [1]. In the encoding process, devices were needed to encrypt the

L. Meher (✉)
Institute for Development and Research in Banking Technology,
Hyderabad 500057, India
e-mail: sanjib29meher@gmail.com

M.V.N.K. Prasad
School of Computer and Information Sciences, University of Hyderabad,
Hyderabad 500046, India
e-mail: mvnkprasad@idrbt.ac.in

© Springer India 2016   271
A. Nagar et al. (eds.), *Proceedings of 3rd International Conference
on Advanced Computing, Networking and Informatics*, Smart Innovation,
Systems and Technologies 44, DOI 10.1007/978-81-322-2529-4_28

secret image into *n* shared image for n participants. The encrypted secret image can be decrypted by directly stacking n secret images and can be recognised by the human visual system.

Extensive investigation on VC and its related properties is done based on [1]. Study to enhance the visual quality of the decrypted secret image is done in [2, 3]. The gray level/color encoding method for images is introduced in [4]. Studies on achieving optimal contrast can be found in [5]. A strategy to deal with the misalignment problem of VSS is discussed in [6]. However, these methods still unable to solve the problem of pixel expansion, optimal contrast and hence the less visual quality of the decrypted secret image is obtained. Codebook is needed for the encryption process of VSS. Sometimes, designing such a code book is difficult to realize. Random grids (RG), introduced by Kafri and Karen in 1987 [7] is the alternate way to solve the pixel expansion problem. RG based VSS method has advantages that it does not require pixel expansion and need no code book design. Unfortunately these methods are only for VSS based on RG. Inspired by Kafri et al. [8] and Chen and Tsao [9] independently proposed RG-based VSS in 2009. In 2011 Chen and Tsao [10] proposed the construction of RG-based VSS. Recently in 2013 T Gou et al. proposed methods to improve the contrast of [11].

2 Related Work

The basic visual secret sharing scheme proposed by Nour and Shamir [1] assumes that the message consists of a collection of black and white pixels. Each pixel is handled separately. Based on general access structure Ateniese et al. designed a more general model for VSS schemes [12]. The participants of the qualified subset are eligible to recover the secret image while the participants in the forbidden subset are not.

A lot of research has been done to remove problems in the traditional methods such as probabilistic methods proposed by Ito et al. [13]. The problem exists in this solution is that it still requires the tailor-made code book design. Similarly Yang et al. proposed a model to solve this problem by using the column matrix to encode the original pixel [14], further Cimato et al. extended the method proposed by Yang [14] to make it the generalised probabilistic approach [15]. Random grid (RG) serves as an alternative approach to solve the problems with some more advantage over the size invariant VSS methods. Random grid based VSS is the emerging field of research in the area of visual cryptography.

The three algorithms proposed by Kafri and Karen [7] support only for binary images. Shyu in 2007 [8] proposed method for encryption of grey image by using the three algorithms. The idea of their method is that a grey-level image is converted into a half-tone image and to exploit the encryption method for binary image. The half-tone technology [16, 17] utilises the density of binary dots to simulate the grey level image. Error Diffusion algorithm [17] is used to implement the half-tone technique.

Chen [9] proposed a method for (2, *n*) VSS by using the random grid. Similar to their (*n*, *n*) scheme this scheme also free from the problems in the existing traditional VSS. It satisfies the contrast condition and security conditions. Recently in 2013 Chen and Lee et al. [18] proposed a quality adoptive (2, *n*) Visual Secret Sharing Scheme based on Random Grid. Their method based on the idea that better visual quality of the reconstructed image can be obtained if stacked image area pertaining to the white pixels of the secret image contains fewer black pixel as compared to white pixels. This method improves the visual quality of the constructed secret image without affecting the security of the secret message in the original image.

The described schemes satisfy all the requirements of a threshold VSS, but the contrast of the stacked image obtained by stacking any k random grids is very low. Hence, to improve the contrast of the threshold VSS scheme Guo and Leu [11] extend this method a proposed threshold VSS scheme for random grid with improved contrast.

3 Proposed Method

To increase the visual quality proposed method is designed so that it does not darken the stacked image area pertaining to white pixels in the secret image.

3.1 Encryption Method

A binary secret image S with the size $w \times h$ is given as input. The proposed method outputs n random grids RG_k, $k = 1, 2, \ldots, n$ with secret image of equal size. Pixel value 0 is used to represent the transparent (white) pixel and 1 is used to represent the opaque (black) pixel. For a white secret pixel $S(i, j)$ a random number C is generated by a random function which can generate 0 and 1 with equal probability. Generated random value C is assigned to the $RG_k(i, j)$, $k = 1, 2, \ldots, n$. For a black secret pixel involves the generation of n bits say $[t_1, t_2, t_3, \ldots, t_n]$. The first bit is generated by assigning the value 0 or 1 by a random function with equal probability. The second bit is obtained by complementing the first bit. This process continues until we get the $\lfloor \frac{n}{2} \rfloor$th bit. The remaining $n - n \times \lfloor \frac{n}{2} \rfloor$ bits are generated according to the uniform distribution. The generated n bits are randomly arranged in $RG_k(i, j)$, $k = 1, 2, \ldots, n$. The proposed method is given as algorithm 1.

Algorithm 1: Encryption Process

Input: A secret binary image S .
Output: n Random Grids $RG_1, RG_2, ..., RG_n$
 1: **for** each pixel of $S(i,j)$ **do**
 2:     **if** S(i,j) = black **then**
 3:         Generate n random bit $[t_1, t_2, t_3, ..., t_n]$
 4:         generate $t_1 = CoinFlipping()$
 5:         generate $t_2 = \overline{t_1}$
 6:         In the same way generate t_3, t_4
 7:         Repeat the process until $[t_1, t_2, t_3, ..., t_{\lfloor \frac{n}{2} \rfloor}]$ are generated
 8:         remaining $n - n \times \lfloor \frac{n}{2} \rfloor$ bits are generated according to the uniform distribution
 9:     **else**
10:         Generate a random bit 0 or 1 with equal probability
11:     **end if**
12:     The above n bits is randomly arranged in $RG_k(i,j), k = 1, 2, ..., n$
13: **end for**
14: Output $RG_1, RG_2, ..., RG_n$

3.2 Decryption Method

For decryption of the secret image we proposed two methods.

Method 1 In this method the decryption is done by stacking. Stacking can disclose the secret image from any two share image. When all the random grids are stacked together then the black pixels of the secret image are perfectly constructed in the stacked image.

Method 2 The proposed method 2 is given as algorithm 2. In the encryption phase the white pixel in the secret image is encrypted by assigning the same value 0 or 1 to all corresponding pixel in each shares, but the value is chosen randomly. The step 4 of the algorithm 2 will be true for the pixel p in the original image if all the pixel value in each share corresponding to the pixel p are same, i.e. either all are black or all are white, so we interpret it as a white pixel for the reconstruction. Decryption by using the method 2 will give the lossless recovery of the secret image when all the shares given as input. When any two share image given input it will reveal the secret image with a higher contrast than the stacking method.

Algorithm 2: Decryption Process

Input: k number of Random Grids Where $2 \leq k \leq n$

Output: Recovered secret image.

1: step 1. Choose k no of input shares Where $2 \leq k \leq n$

2: **for** 1 to n **do**

3:     Check each pixel value at position (i, j) of $RG_1, RG_2, ..., RG_n$

4:     **if** pixel value at position (i, j) of $RG_1, RG_2, ..., RG_n$ is same **then**

5:         S(i,j) = WHITE

6:     **else**

7:         S(i,j) = BLACK

8:     **end if**

9: **end for**

10: Output Recovered secret image S'

---

4 Experimental Result

To evaluate the visual quality of the reconstructed image B for the original secret image S, the relative difference of the light transmissions between the transparent and opaque pixels in the reconstructed image B is calculated which is called contrast of the reconstructed image B. A standard image Lena of size 512 × 512 shown in Fig. 1a. The four random grids with the same size of secret images are generated from encoding process which is shown in Fig. 1b–e. In the decoding process by method 1, by superimposing any pair of random grids, the secret information can be disclosed with the contrast value of 0.2858 which is shown in Fig. 1f–k. Figure 1l–o shows the reveled secret image by three random grids. The Fig. 1p shows the reconstructed image obtained from all the four random grids. Similarly the output obtained by applying method 2 is shown in Fig. 2.

4.1 Comparison

To evaluate and compare the visual quality comparison of contrast for different cases between the proposed methods and related RG based VSS algorithm are demonstrated. Table 1 shows the comparison of the contrast for (2, 4) cases. From the value in the Table 1 it can be observed that the improved contrast is achieved by the proposed methods. For two RGs the contrast value obtained by applying decryption by the proposed method 1 is 0.28 and by method 2 is 0.5. Method 2 gives contrast value 1 which indicates the lossless recovery of the secret image.

Table 2 shows a comparison of contrast with the other related (2, *n*) VSS schemes by increasing the value of n. Here n indicates the total number of shares.

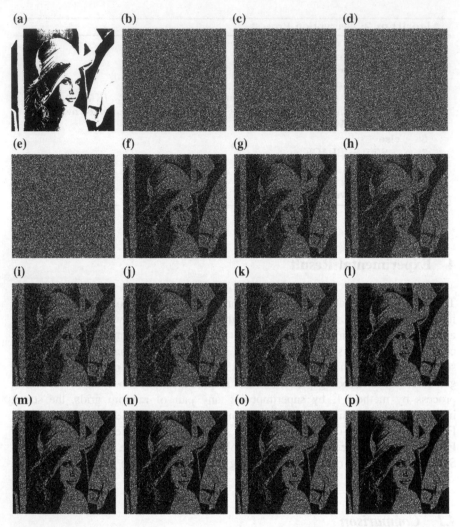

Fig. 1 The experimental result of the output of method-1 for Lena binary image of size 512×512 **a** Secret**b** RG_1**c** RG_2**d** RG_3**e** R G_4**f** $RG_1 \otimes G_2$**g** $RG_1 \otimes RG_3$**h** $RG_1 \otimes RG_4$**i** $RG_2 \otimes R$ G_3**j** $RG_2 \otimes RG_4$**k** $RG_3 \otimes RG_4$**l** $RG_1 \otimes RG_2 \otimes RG_3$**m** $RG_1 \otimes RG_2 \otimes RG_4$**n** $RG_2 \otimes RG_3 \otimes RG_4$**o** $RG_1 \otimes RG_3 \otimes RG_4$**p** $RG_1 \otimes RG_2 \otimes RG_3 \otimes RG_4$

The value of n has been taken as n = 3, n = 4, n = 5 and n = 6. Contrast has been calculated for different values of n for the proposed methods and existing methods [9, 18, 19]. The contrast value obtained by other existing methods and the proposed methods, goes on decreasing with the increase of value of n. The advantage of the proposed method is that, the rate of decrease in the contrast value with the increased value of n for the proposed methods is small.

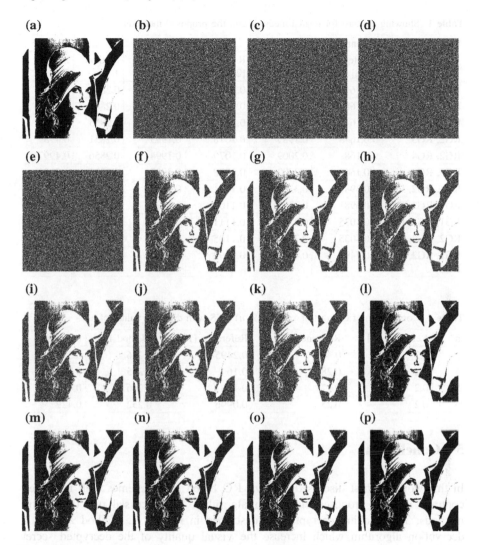

Fig. 2 The experimental result of the output of method-2 for Lena binary image of size 512×512 **a** Secret **b** RG_1 **c** RG_2 **d** RG_3 **e** R G_4 **f** $RG_1 \oplus RG_2$ **g** $RG_1 \oplus RG_3$ **h** $RG_1 \oplus RG_4$ **i** $RG_2 \oplus R$ G_3 **j** $RG_2 \oplus RG_4$ **k** $RG_3 \oplus RG_4$ **l** $RG_1 \oplus RG_2 \oplus RG_3$ **m** $RG_1 \oplus RG_2 \oplus RG_4$ **n** $RG_2 \oplus RG_3 \oplus RG_4$ **o** $RG_1 \oplus RG_3 \oplus RG_4$ **p** $RG_1 \oplus RG_2 \oplus RG_3 \oplus RG_4$

Table 1 Showing contrast for related methods and the proposed methods

Recovery	Contrast					
	Reference [18]	Reference [20]	Reference [21]	Reference [9]	Method 1	Method 2
RG1, RG2	0.1642	0.1994	0.1079	0.2002	0.2853	0.4987
RG1, RG3	0.1682	0.1976	0.1088	0.1994	0.2866	0.5001
RG1, RG4	0.1648	0.2011	0.1081	0.1992	0.2855	0.5002
RG2, RG3	0.1648	0.1985	0.1078	0.1993	0.2851	0.4989
RG2, RG4	0.168	0.2009	0.1079	0.1994	0.2856	0.4997
RG3, RG4	0.1646	0.2009	0.1078	0.2002	0.2852	0.5001
RG1, RG2, RG3	0.3987	0.3331	0.1111	0.3335	0.4271	0.7779
RG1, RG2, RG4	0.3981	0.3332	0.1107	0.3329	0.4169	0.7678
RG1, RG3, RG4	0.4035	0.3329	0.1111	0.3328	0.4209	0.7788
RG2, RG3, RG4	0.4035	0.3332	0.1105	0.3337	0.4209	0.7767
RG1, RG2, RG3, RG4	0.7496	0.4578	0.3792	0.4128	0.5909	1

Table 2 Comparison of contrast of various $(2, n)$ VSS with different values of n

n	Reference [9]	Reference [19]	Reference [18]	Method 1	Method 2
3	0.2	0.4	0.2495	0.286	0.5
4	0.2	0.286	0.164	0.285	0.5001
5	0.2	0.273	0.128	0.246	0.427
6	0.2	0.25	0.0996	0.252	0.429

5   Conclusion

In this work we first discussed a $(2, n)$ RG based VSS schemes with improved contrast. For the proposed encryption algorithm, we discussed two decryption methods. The stacking decryption is described in method 1. Method 2 gives a decryption algorithm which increase the visual quality of the decrypted secret image. The first method need no computation because it involves only stacking. The second method requires little computation but by this method lossless recovery of secret image is obtained.

References

1. Naor, M., Shamir, A.: Visual cryptography, in: Advances in Cryptology EURO—CRYPT'94, Springer, pp. 1–12 (1995)
2. Blundo, C., De Santis, A.: Visual cryptography schemes with perfect reconstruction of black pixels. Comput. Graphics **22**(4), 449–455 (1998)

3. Koga, H., Ueda, E.: Basic properties of the (t, n)-threshold visual secret sharing scheme with perfect reconstruction of black pixels. Des. Codes Crypt. **40**(1), 81–102 (2006)
4. Lin, C.-C., Tsai, W.-H.: Visual cryptography for gray-level images by dithering techniques. Pattern Recogn. Lett. **24**(1), 349–358 (2003)
5. Krause, M., Simon, H.U.: Determining the optimal contrast for secret sharing schemes in visual cryptography. Comb. Probab. Comput. **12**(03), 285–299 (2003)
6. Yang, C.-N., Peng, A.-G., Chen, T.-S.: Mtvss:(m) isalignment (t) olerant (v) isual (s) ecret (s) haring on resolving alignment difficulty. Sig. Process. **89**(8), 1602–1624 (2009)
7. Kafri, O., Keren, E.: Encryption of pictures and shapes by random grids. Opt. Lett. **12**(6), 377–379 (1987)
8. Shyu, S.J.: Image encryption by random grids. Pattern Recogn. **40**(3), 1014–1031 (2007)
9. Chen, T.-H., Tsao, K.-H.: Visual secret sharing by random grids revisited. Pattern Recogn. **42** (9), 2203–2217 (2009)
10. Chen, T.-H., Tsao, K.-H.: Threshold visual secret sharing by random grids. J. Syst. Softw. **84** (7), 1197–1208 (2011)
11. Guo, T., Liu, F., Wu, C.: Threshold visual secret sharing by random grids with improved contrast. J. Syst. Softw. **86**(8), 2094–2109 (1996)
12. Ateniese, G., Blundo, C., De Santis, A., Stinson, D.R.: Visual cryptography for general access structures. Inf. Comput. **129**(2), 86–106 (1999)
13. Kuwakado, H., Tanaka, H.: Image size invariant visual cryptography. IEICE Trans. Fundam. Electron. Commun. Comput. Sci. **82**(10), 2172–2177 (1999)
14. Yang, C.-N., Laih, C.-S.: New colored visual secret sharing schemes. Des. Codes Crypt. **20**(3), 325–336 (2000)
15. Cimato, S., De Prisco, R., De Santis, A.: Optimal colored threshold visual cryptography schemes. Des. Codes Crypt. **35**(3), 311–335 (2005)
16. Sullivan, J., Miller, R., Pios, G.: Image halftoning using a visual model in error di_usion. JOSA A **10**(8), 1714–1723 (1993)
17. Ulichney, R.A.: Review of halftoning techniques. In: Electronic Imaging, International Society for Optics and Photonics, pp. 378–391 (1999)
18. Chen, T.-H., Lee, Y.-S., Huang, W.-L., Juan, J.S.-T., Chen, Y.-Y., Li, M.-J.: Quality adaptive visual secret sharing by random grids. J. Syst. Softw. **86**(5), 1267–1274 (2013)
19. Chen, S.-K., Lin, S.-J.: Optimal (2, n) and (2, infinity) visual secret sharing by generalized random grids. J. Vis. Commun. Image Represent. **23**(4), 677–684 (2012)
20. Wu, X., Sun, W.: Improving the visual quality of random grid-based visual secret sharing. Sig. Process. **93**(5), 977–995 (2013)
21. Wu, X., Sun, W.: Random grid-based visual secret sharing with abilities of or and xor decryptions. J. Vis. Commun. Image Represent. **24**(1), 48–62 (2013)

An Efficient Lossless Modulus Function Based Data Hiding Method

Nadeem Akhtar

Abstract Data hiding methods are useful for secure communication over the internet. Important data is hidden in cover data that is used as decoy. A lossless data hiding method is judged by the maximum size of secret data that can be hidden without distortion and similarity between the original cover image and stego image. In this work, module based least significant bit substitution methods are studied and a new enhancement is proposed to improve the quality of stego image. The new method divides the cover and secret images in several blocks and compressed secret block is hidden into corresponding cover block using modulo operation. The extra data is stored in the free space that is generated after compression. The quality of stego image in proposed method is better than that of existing methods that hides the secret data using modulo operation.

Keywords Data hiding · Lossless · Modulus function · Block processing · Compression

1 Introduction

The goal of a good lossless data hiding method is (1) to hide the existence of secret data (2) to hide as much data as possible without distortion (3) to keep the distortion in cover image minimal and (4) to provide robustness i.e. to be able to recover the secret data after cover image has been modified [1, 2]. A good data hiding method should provide high values for these four parameters. But often, there are trade-offs between these parameters. For example, when the size of secret data is increased, distortion in cover image also increases.

Several lossless data hiding methods that work in spatial domain replaces some least significant bits of cover pixels to hide secret data using modulus function

N. Akhtar (✉)
Department of Computer Engineering, Aligarh Muslim University, Aligarh, India
e-mail: nadeemalakhtar@gmail.com

© Springer India 2016   281
A. Nagar et al. (eds.), *Proceedings of 3rd International Conference on Advanced Computing, Networking and Informatics*, Smart Innovation, Systems and Technologies 44, DOI 10.1007/978-81-322-2529-4_29

[3–5]. In other methods [6–9], secret data are hidden in cover image pixel by pixel using modulus function. In [7], secret data compression is also used to improve the stego image quality by reducing the number modified cover pixels. The difference between cover and stego pixel is bounded by (m + 1)/2, where m is the smallest integer greater than all the secret data values.

In this work, an improved scheme using modulus function and compression is proposed and implemented that outperform all the methods that uses modulus function for data hiding. The size of secret image is exactly half of the size of cover image. Cover and secret images are divided into fixed sized non-overlapping blocks. A different value of m is used for each block and all these m values are hidden into unmodified cover pixels. This reduces the number of modified cover pixels but overall stego image quality improves because the value of m is lesser for several blocks.

In [10], authors have used dynamic programming strategy to find the optimal least significant bit replacement. Both the computation time and stego image quality are good in this method.

In [8], Thien and Lin process the image in such a way that each data value to be hidden is less than m. They used Mod m function to hide a secret data value in a cover image pixel, considering each data value separately. Then, they adjusted the stego-pixels to reduce the difference between stego and cover image pixels in such a way that difference between stego and cover image pixel is less than m/2.

To improve the Thien and Lin method, Chen [7] proposed to consider the repetitions of data values to improve the stego-image quality. The heavy repetitions of secret data values (repetitions more than 3) are compressed into a triplet before hiding, which results in less number of cover image pixels modified. In Chen's method too, the difference between stego and cover image pixel is always less than (m + 1)/2. Chen's method outperforms other methods unless secret data is with very strong randomness [7].

The proposed method in this paper further improves Chen's method by dividing the image into fixed sized non-overlapping blocks and a different value of m is selected for each block. In this way, the value of m for some blocks will be smaller than others.

2 Proposed Method-Encoding and Decoding

This section explains proposed method. The size of cover image is exactly double of the size of secret image. Both cover and secret images are grayscales images.

Firstly, each secret image pixel is broken into two by separating the four most significant bits from the four least significant bits, making the size of secret image equal to that of cover image. In the new secret image, all pixel values are less than 16.

Next, cover and secret images are divided into fixed sized non-overlapping blocks of size n × n. Say the number of blocks in the image be N. let m_j be the smallest integer which is larger than all the pixel values of block j of secret image, $1 \leq j \leq N$. Since all pixels values are less than 16, so $m_j \leq 16$.

Each secret pixel of block j is hidden into corresponding cover pixel one by one using modulus function mod m_j. At this stage compression in secret data is considered. The same compression method is used in this work as discussed in [7]. If some secret datum v repeats for r times, $r > 3$, that repetition sequence is compressed into a triplet $<m_j, v, r - 4>$, where m_j is the repetition indicator, v is the secret value being repeated and r is the number of times the value v is being repeated. All the secret data values in block m_j are less than m_j, so m_j can be used as a special value which indicates repetitions of length more than 3. Since only those repetitions are considered, whose length is more than 3, a 0 value can indicate repetition of length 4, so $r - 4$ is stored. Since all values to be hidden are less than m_j, $r - 4$ must also be less than or equal to m_j. This way, the maximum value of repetition length that can be considered for compression in one go is $(m_j + 4)$. If a secret data value repeats for more than $(m_j + 4)$ times, that repetition sequence will be broken into several sequence of length $(m_j + 4)$, except the last and a separate triplet will be stored for each broken such sequence.

Let c_i, h_i and s_i be the cover, secret and stego pixels respectively. Secret pixel h_i belongs to block j. If the secret data value h_i does not repeats or the repetition length is less than 4, the pixel h_i is hidden into corresponding c_i as $s_i = c_i - [c_i \bmod(m_j + 1)] + h_i$. If the repetition length of h_i is $r > 3$, only the triplet $<m_j, h_i, r - 4>$ will be stored in first three corresponding cover pixels and remaining $(r - 3)$ corresponding cover pixels will remain unchanged.

Since the value of flag m_j is different for each block, it is also needed to store the flag m_j for block j and that too at some known location so that data recovery can be done at the decoder side. The flag m_j for secret block j is stored in the first pixel of the cover block j. The first secret data values of all the secret blocks are combined to form a sequence, say fd. The length of sequence fd is equal to the number of blocks. The sequence fd is stored in those cover pixels that are left unchanged by the hiding procedure explained above. Since the length of fd, equal to the number of blocks, is much smaller than the number of unchanged cover pixel, only first few unchanged cover pixels will be used to hide them. Since first cover pixel in the block stores flag m_j, hiding process will start from the second pixel.

For example, let block size n be 4. Secret pixels 2, 2, 2, 2, 2, 2, 2, 2, 6, 5, 7, 3, 4, 2, 6, 3 of block j are to be hidden into cover pixels 65, 67, 63, 64, 68, 70, 68, 69, 72, 67, 69, 63, 64, 64, 63, 69. The smallest integer which is larger than all these secret pixel is $m_j = 8$. The flag m_j will be stored in the first cover pixel as $65 - (65/16) + 8 = 72$. The first secret value 2 is added to the sequence fd. Now, the hiding process start from the second secret pixel. the second pixel is being repeated 7 times, the triplet $<8, 2, 3>$ will be stored in second, third and fourth cover pixels and next four cover pixels will remain unchanged. The next eight secret pixels will be stored according to normal process into the corresponding cover pixels. The sequence fd, which contains the first secret data value of each block, will be stored in the sequence of cover pixels of all the blocks that remain unchanged. The block under consideration will have four unchanged pixels i.e. from fifth to eighth pixel in that sequence. After this process, the stego pixels will be 72, 71, 65, 66, 65, 70, 68, 69, 78, 69, 70, 66, 67, 65, 69, 66. Note that the only pixel in sequence fd is stored in fifth cover pixel. At this stage, the maximum difference between any cover and stego pixel is less than $(m_j + 1) = 9$. Next, adjustment procedure as defined in [9] is followed to make this difference smaller than $(m_j + 1)/2$. Following five cases are considered for reducing the difference further:

Case 1: $(\lfloor (m_j + 1)/2 \rfloor < s_i - c_i \leq m_j$ and $s_t \geq m_j + 1)$ $\quad s_i \leftarrow s_i - m_j - 1$
Case 2: $(-m_j \leq s_i - c_i < - \lfloor (m_j + 1)/2 \rfloor$ and $s_i \leq 254 - m_j) s_i \leftarrow s_i + m_j + 1$
Case 3: $(-\lfloor (m_j + 1)/2 \rfloor \leq s_i - c_i \leq \lfloor (m_j + 1)/2 \rfloor)$ $\quad s_i \leftarrow s_i$
Case 4: $(\lfloor (m_j + 1)/2 \rfloor < s_i - c_i \leq m_j$ and $s_i \leq m_j)$ $\quad s_i \leftarrow s_i$
Case 5: $(-m_j \leq s_i - c_i < - \lfloor (m_j + 1)/2 \rfloor$ and $s_i \geq 255 - m_j)$ $\quad s_i \leftarrow s_i$

After this step, pixel values of stego image become closer to cover image pixel values. After this stage, the modified stego pixels will be **72**, 71, 65, 66, 65, 70, 68, 69, **69**, 69, 70, 66, 67, 65, **60**, 66. The pixels in bold are those which get modified in this process. After this step, the maximum difference between any cover image and stego image pixel values become less than $(m_j + 1)/2 = 5$.

In the decoding process, the stego image is divided into fixed non-overlapping blocks of size n × n. Each block is processed separately. Firstly, the flag m_j for each block j is found from the first pixel of block j s_{j1} as $m_j = s_{j1} \mod 16$.

Starting from the second pixel, each secret pixel h_{ji} is found from stego pixel s_{ji} as $h_{ji} = s_{ji} \mod (m_j + 1)$. If the value found is less than flag m_j, it is a regular secret pixel. If the value is equal to m_j, it represent repetition of secret data value. Next value identifies the secret value being repeated i.e. v and next to next value identifies the number of times it is being repeated i.e. $r - 4$. Next $(r - 3)$ stego pixels may contain the first data value of blocks i.e. the sequence fd and they will be recovered from those pixels.

3 Experimental Results

The proposed method is compared with simple LSB substitution method, method proposed by Thien and Lin [8] and method proposed by Chen [7]. The size of cover and secret image are 512 × 512 and 512 × 256 respectively. Two standard images baboon and lena are considered as cover images to hide other secret images [11]. The block size n is taken as 4 × 4, 8 × 8 and 16 × 16.

The PSNR for the stego-images is listed in Table 1, for Lena and Baboon as cover images. The results of the proposed method are better than Thein and Lin's and Chen's method.

Table 1 PSNR values for cover images Lena and Baboon

Image	Simple LSB	Thien and Lin's method	Chen's method	Proposed method (16 × 16)	Proposed method (8 × 8)	Proposed method (4 × 4)
Cover image: Lena						
Jet (f16)	31.88	34.84	35.64	35.91	36.00	36.13
Tiffany	31.34	34.85	35.41	35.43	35.42	35.48
Peppers	32.31	34.79	35.41	36.12	36.21	36.36
House	32.48	34.79	36.18	37.11	37.25	37.31
Houses	32.08	34.81	34.96	35.30	35.49	35.78
Man	31.89	34.77	35.06	35.95	36.18	36.43
Tank	32.75	34.72	34.80	36.09	36.14	36.25
Cameraman	32.07	34.76	36.03	36.62	36.80	36.96
Tree	31.93	34.85	31.93	36.18	36.35	36.50
Random	31.82	34.81	34.34	34.35	–	–
Cover image: Baboon						
Jet (f16)	31.93	34.81	35.63	35.88	35.98	36.12
Tiffany	31.41	34.81	35.40	35.40	35.41	35.49
Peppers	32.35	34.80	35.40	36.09	36.19	36.37
House	32.53	34.76	36.20	37.13	37.27	37.33
Houses	32.13	34.81	34.92	35.25	35.46	35.77
Man	31.95	34.80	34.98	35.93	36.16	36.43
Tank	32.83	34.78	34.78	36.10	36.14	36.26
Cameraman	32.09	34.79	35.98	36.61	36.79	36.94
Tree	31.95	34.81	35.68	36.18	36.36	36.52
Lena	32.45	34.78	35.31	36.04	36.17	36.39
Random	31.83	34.80	34.34	34.34	–	–

4 Discussion and Conclusion

From the Table 1, it is clear that performance of proposed scheme is better than the previous methods that use modulus function for data hiding. As the block size is decreased, the improvement in PSNR increases.

In Chen's method, only one flag value m = 16 is used for the entire image. In the proposed method, there are as many flag values as the number of blocks in the image. The flag value for most of blocks are smaller than 16. This reduces the difference between cover and stego pixels. Proposed method need to store several flag values, which are stored in unchanged cover pixels. This increases the difference between cover and stego pixels. But the increase due to storing several flags is much smaller than the decrease due to the reduction in flag values, so overall difference between cover and stego pixels decreases greatly, improving the PSNR for stego image.

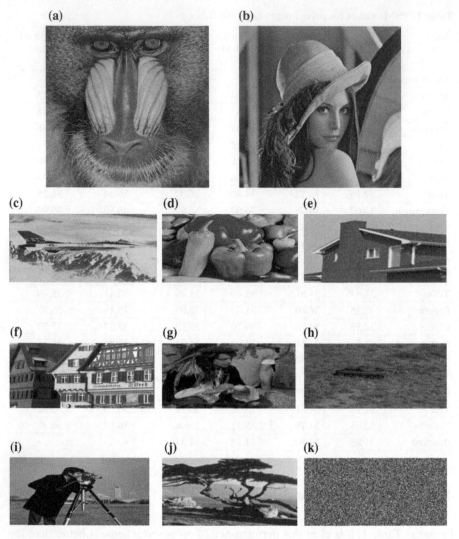

Fig. 1 512 × 512 cover images **a** baboon **b** Lena 512 × 256 secret images **c** f16 **d** pepers **e** house **f** houses **g** man **h** tank **i** cameraman **j** tree **k** random

For random image Fig. 1k, data hiding is not possible using 8 × 8 and 4 × 4 block size because the number of unchanged pixels is smaller than the number of flag values.

The performance of the proposed method is good and suitable to use for data hiding in a wide range of application. Most important features of this scheme is the maximum difference between pixel values of cover and stego image depends on the flag m_j for block j, which can be smaller than 16, unlike other methods.

References

1. Cheddad, A.A., Condell, J., Curran, K., Mc Kevitt, P.: Digital image steganography: survey and analysis of current methods. Signal Process. **90**(3), 727–752 (2010)
2. Atawneh, S., Almomani1, A., Sumari, P.: Steganography in digital images: common approaches and tools. IETE Tech. Rev. **30**(4), 344 (2013)
3. Cheng-Hsing, Y., et al.: Adaptive data hiding in edge areas of images with spatial LSB domain systems. In: IEEE Trans. Inf. Forensics Secur. **3**(3), 488–497 (2008)
4. Jin, H.L., Fujiyoshi, M., Kiya, H.: Lossless data hiding in the spatial domain for high quality images. IEICE Trans. Fundam. Electron. Commun. Comput. Sci. **90**(4), 771–777 (2007)
5. Akhtar, N., Johri, P., Khan, S.: Enhancing the security and quality of LSB based image steganography. In: IEEE International Conference on Computational Intelligence and Computer Networks (CICN), Mathura, India. 27–29 Sept 2013
6. Pan, F., Li, J., Yang, X.: International Conference on Electronics, Communications and Control (ICECC), Zhejiang, pp. 282–284, 9–11 Sept 2011
7. Chen, S.-K.: A module-based LSB substitution method with lossless secret data compression. Comput. Stan. Interfaces **33**, 367–371 (2011)
8. Thien, C.C., Lin, J.C.: A simple and high-hiding capacity method for hiding digit-by digit data in images based on modulus function. Pattern Recogn. **36**, 2875–2881 (2003)
9. Akhtar, N., Bano, A., Islam, F.: An improved module based substitution steganography method. In: Fourth IEEE International Conference on Communication Systems and Network Technologies (CSNT), pp 695–699 (2014)
10. Chang, C.-C., Hsiao, J.-Y., Chan, C.-S.: Finding optimal least-significant-bit substitution in image hiding by dynamic programming strategy. Pattern Recognit. **36**(7), 1583–1595 (2003)
11. http://sipi.usc.edu/database/

On the Use of Gaussian Integers in Public Key Cryptosystems

Aakash Paul, Somjit Datta, Saransh Sharma and Subhashis Majumder

Abstract We present a comparative analysis of the processes of factorization of Gaussian integers and rational integers, with the objective of demonstrating the advantages of using the former instead of the latter in RSA public key cryptosystems. We show that the level of security of a cryptosystem based on the use of the product of two Gaussian primes is much higher than that of one based on the use of the product of two rational primes occupying the same storage space. Consequently, to achieve a certain specific degree of security, the use of complex Gaussian primes would require much less storage space than the use of rational primes, leading to substantial saving of expenditure. We also set forth a scheme in which rings of algebraic integers of progressively higher and higher degrees and class numbers can be used to build cryptosystems that remain secure by forever staying ahead of advances in computing power.

Keywords Gaussian integer · Encryption · Decryption · Algebraic integer

A. Paul (✉) · S. Sharma · S. Majumder
Department of Computer Science and Engineering, Heritage Institute of Technology,
Kolkata, West Bengal, India
e-mail: aakashpaul92@gmail.com

S. Sharma
e-mail: sharma_saransh@hotmail.com

S. Majumder
e-mail: subhashis.majumder@heritageit.edu

S. Datta
Department of Mathematics, Heritage Institute of Technology, Kolkata, West Bengal, India
e-mail: somjitdatta@gmail.com

© Springer India 2016   289
A. Nagar et al. (eds.), *Proceedings of 3rd International Conference
on Advanced Computing, Networking and Informatics*, Smart Innovation,
Systems and Technologies 44, DOI 10.1007/978-81-322-2529-4_30

1 Introduction

The effectiveness of the RSA cryptosystem is well known. The method was pro-
pounded by Adleman et al. [1]; and further elaborations were carried out by Rivest
[2]. The innovations which led to these developments in public key cryptosystems
are due to Diffie and Hellman [3].

We recall the process of encryption and decryption of messages in an RSA
public key cryptosystem. For the results from number theory that we use, we refer
the reader to Hardy and Wright [4], Ireland and Rosen [5] and Davenport [6]. We
use the 26-letter alphabet A–Z with numerical equivalents 11–36. We wish to
encode a message consisting of several words, the total number of letters and spaces
in which is, say, m. If we let 99 denote a space, then the message can be converted
to a string of $2m$ digits. Now we choose two rational prime numbers p and q, each
of at least 100 decimal digits and multiply them to obtain the modulus $n = pq$. Let
$\phi(.)$ denote the Euler ϕ-function. We calculate $\phi(n) = (p - 1)(q - 1)$ and find an
integer j that is relatively prime to $\phi(n)$. The numbers n and j are published but
p and q are kept secret. Next we divide the string of $2\,m$ digits obtained above into
numbers that are all less than n, so that we have a list of numbers $n_1, n_2, n_3 ,...,$
$n_s < n$. We now calculate $n_1^j \equiv c_1 \pmod{n}$, $n_2^j \equiv c_2 \pmod{n}$, $n_3^j \equiv c_3 \pmod{n}...,$
$n_s^j \equiv c_s \pmod{n}$. The encoded message that is transmitted is $\{c_1, c_2, c_3,..., c_s\}$.

In order to decode it, the recipient of the message must find the numbers $n_1, n_2,$
$n_3,..., n_s$ and convert them to letters or spaces by the scheme mentioned above. The
numbers $n_1, n_2, n_3,..., n_s$ can be found by solving the congruences $x^j \equiv c_1 \pmod{n}$,
$x^j \equiv c_2 \pmod{n}$, $x^j \equiv c_s \pmod{n},..., x^j \equiv c_s \pmod{n}$. These solutions can be carried
out by a standard procedure, provided $\phi(n)$ can be found. To find $\phi(n)$, a knowledge
of p and q is necessary. Since p and q are each of at least 100 decimal digits, $n = pq$
is of at least 200 digits, and in the current state of knowledge of factorization
algorithms, it is impossible to factorize such a number. If in future it becomes
possible to factorize such a number, we merely need to choose p and q sufficiently
large, say each of 400 digits, such that it is impossible to factorize $n = pq$.
Therefore, unless p and q are known to a person, he cannot decode the message.

This cryptosystem depends for its success on the fact that given the product of
two large and unknown rational primes it is so difficult to factorize it as to be
effectively impossible.

However, given the speed with which the computing power of computers is
increasing, this cryptosystem would need larger and larger primes p and q, thus
rapidly augmenting the necessary storage space. It would certainly be advantageous
to have a cryptosystem based on numbers for which the method of factorization is
far more complicated than that for the rational integers. Note that the factorization
of Gaussian integers is substantially more complicated, both theoretically and
computationally, than that of rational integers. In fact, to understand the process of
factorization of Gaussian integers, the Law of Quadratic Reciprocity and certain
other concepts of number theory must be known. We discuss below the method of

factorization of Gaussian integers and demonstrate the advantages of their use in cryptosystems.

2 Encryption and Decryption Using Gaussian Integers

For the principles and processes of number theory that are used in what follows, we refer the reader to Apostol [7], Cormen et al. [8], Davenport [6], Knuth [9], Koblitz [10], Niven et al. [11] Rosen [12], Silverman [13]. We recall that if $s + it = (a + ib)$ $(c + id)$ then $N(s + it) = N(a + ib)N(c + id)$, where $N(.)$ denotes the norm. Now suppose $a + ib$ and $c + id$ are split Gaussian primes. Since the ring of Gaussian integers has class number 1, the factorization of $s + it$ given above is unique. If $a + ib$ and $c + id$ are not known to us, and we are asked to factorize $s + it$, then we have to first find the norm of $s + it$ and factorize it: Let $N(s + it) = n = pq$ where p and q are rational primes; then $N(a + ib) = a^2 + b^2 = p$ and $N(c + id) = c^2 + d^2 = q$. Our task now is to find a, b, c and d. In other words, we have to find the representations of p and q as sums two squares.

Since $a + ib$ and $c + id$ are split, $p \equiv 1 \pmod 4$ and $q \equiv 1 \pmod 4$. It is well-known that every rational prime $\equiv 1 \pmod 4$ can be expressed as the sum of two squares. To find this representation of such a prime as the sum of two squares, a method of descent is necessary, which we now describe.

We intend to express p as a sum of two squares. Since $p \equiv 1 \pmod 4$, the Law of Quadratic Reciprocity implies that the congruence $x^2 \equiv -1 \pmod p$ can be solved. Let $x = C$ be a solution. Then $C^2 + 1^2 = Mp$ for some integer M. It can be shown that $M < p$. We now choose an integer $R \equiv C \pmod M$ such that $-\frac{1}{2}M \le R \le \frac{1}{2}M$. The $R^2 + 1^2 \equiv C^2 + 1^2 \equiv 0 \pmod M$. Hence, $R^2 + 1^2 = MK$ for some integer K. It can be shown that $1 \le K \le M/2$. Now a well-known algebraic identity gives us $(R^2 + 1^2)$ $(C^2 + 1^2) = (RC + 1)^2 + (C - R)^2$. Since $(R^2 + 1^2)(C^2 + 1^2) = M^2Kp$, we have $(RC + 1)^2 + (C - R)^2 = M^2Kp$. It can be shown that both $RC + 1$ and $C - R$ are divisible by M. Dividing the result last displayed, we obtain

$$\left(\frac{RC + 1}{M}\right)^2 + \left(\frac{C - R}{M}\right)^2 = Kp.$$

Therefore, we have expressed Kp, a smaller multiple of p than Mp, as a sum of two squares. By the same procedure, we can express Lp, a smaller multiple of p than Kp, as a sum of two squares. Hence we can express smaller and smaller multiples of p as sums of two squares, until finally we obtain an expression of p itself as the sum of two squares. Observe that, since Mp, Kp, Lp, Tp... are the successive multiples of p obtained in this descent, we have $K \le \frac{M}{2}$, $L \le \frac{K}{2}$, $T \le \frac{L}{2}$.... Hence $L \le \frac{M}{4}$, $T \le \frac{M}{8}$....

Suppose we have found, by the descent described above, that $p = a^2 + b^2$ and $q = c^2 + d^2$. Then $s + it = (a + ib)(c + id)$ and the desired factorization is complete.

Since the ring of Gaussian integers possesses unique factorization, the processes of encoding and decoding of messages involving rational integers that we have recalled above are valid for the Gaussian integers as well, $s + it$, $a + ib$ and $c + id$ taking the places of n, p and q respectively. Therefore, any person who wishes to break a code based on Gaussian integers will have to go through the lengthy and laborious process of factorizing $s + it$ described above.

It can be shown that in the descent procedure described above, the multiple of p is at least halved at every stage. We recall that $2^{10} = 1024$ and note that if p is of more than 100 digits, then it is greater than 2^{300}. As a very rough estimate, we can say that probabilistically approximately $(300 + 0)/2 = 150$ iterations would be necessary in the descent procedure. By a judicious choice of the prime p, the number of necessary iterations can be made as high as possible.

Moreover, it is a computationally non-trivial task to solve $x^2 \equiv -1 \pmod{p}$ and $x^2 \equiv -1 \pmod{q}$ when p and q are each of 100 digits. Lagrange found a method of solving this type of congruence when the prime modulus is the form $4k + 1$, which is precisely the case we are interested in. He proved, using Wilson's Theorem, that the solutions is $x \equiv \pm(2k)! \pmod{p}$. But since $p = 4k + 1$ is of 100 digits, $2k$ would be of close to 100 digits too, and $(2k)!$ would be such an enormous number that finding it and reducing it mod p to obtain the solution would pose serious computational difficulties. Therefore, for solving this congruence, trial and error is as good as any other method, and that would require a substantial amount of computation.

2.1 A Typical Example

We discuss an example in order to illustrate the proposed method discussed above. For the sake of simplicity, we choose a very short message. Actual messages are, of course, likely to be much longer.

Suppose we wish to convey the message "GO AHEAD" in an encrypted form. As mentioned above, in our scheme, A is denoted by 11, B by 12, C by 13 and so on till Z, and a space is denoted by 99. Then the message is converted to the string 1715991118151114. Now, by standard methods of primality testing, we find two Gaussian primes $a + ib$ and $c + id$ whose norms, p and q respectively are each of at least some desired number of digits, and calculate their product $s + it = (a + ib)$ $(c + id)$ We also calculate $\phi(s + it)$. We then choose an integer j that is relatively prime to $\phi(s + it)$.

The string 1715991118151114 is broken up into the 2-tuples (17,15), (99,11), (18,15) and (11,14). Consider now the Gaussian integers $17 + 15i$, $99 + 11i$, $18 + 15i$, and $11 + 14i$. Let these be denoted by n_1, n_2, n_3, and n_4, the notation being the same as that of the discussion above. In this case, the string has been broken up into 2-tuples of 2-digit entries for the sake of simplicity. All we have to ensure is that their norms are all less than $N(s + it) = pq$, which is of at least 200 digits.

We now calculate $n_1^j \equiv c_1 \pmod{s + it}$, $n_2^j \equiv c_2 \pmod{s + it}$, $n_3^j \equiv c_3 \pmod{s + it}$, and $n_4^j \equiv c_3 \pmod{s + it}$, where c_1, c_2, c_3 and c_4 are Gaussian integers. Let us suppose, for the sake of definiteness, that $c_1 = 2 + 3i$, $c_2 = 4 + 7i$, $c_3 = 9 + 17i$, and $= 11 + 19i$. Then the encoded message that is transmitted will be (2, 3), (4, 7), (9, 17), (11, 19).

The decoding procedure is the same as for rational integers. We have to solve the congruences $x^j \equiv 2 + 3i \pmod{s + it}$, $x^j \equiv 4 + 7i \pmod{s + it}$, $x^j \equiv 9 + 17i \pmod{s + it}$, and $x^j \equiv 11 + 19i \pmod{s = it}$. The solutions will be $17 + 15i$, $99 + 11i$, $18 + 15i$, and $11 + 14i$ respectively, from which we reconstruct the string 1715991118151114, and elicit from it the message "GO AHEAD".

Algorithm 1: Method of Descent

Input : A prime p, such that $p\%4 = 1$
Output: Integers x and y, such that $x^2 + y^2 = p$

1: Set z equal to $\lfloor p/2 \rfloor$.
2: Calculate $\alpha = z^2 + 1$.
3: Check if α is a multiple of p.
4: If α is a multiple of p, GOTO step 5. Else reduce z by 1 and GOTO step 2.
5: Set m equal to p/α.
6: Set x equal to z, and y equal to 1.
7: Calculate u by adding a suitable multiple of m to x, such that the resulting sum lies between $-m/2$ and $m/2$. The multiple of m can be negative. Similarly calculate v by using y in place of x.
8: Set r equal to $(u^2 + v^2)/m$, a equal to $(x*u + y*v)$ and b equal to $(x*v - y*u)$.
9: Set x equal to a/m, y equal to b/m and m equal to r.
10: Perform steps 7 to 9, until m equals 1.

---

3 Decryption Without Private Keys

If someone does not possess the private keys, then in order to decrypt the message he has to follow the method of descent. The method of descent is used to express any given prime of the form $4k + 1$, where k is an integer, as the sum of two perfect squares. First, a multiple of the prime is expressed as the sum of two perfect squares. Then, the multiple is reduced in successive iterations, and finally brought down to one. Algorithm 2.1 gives the steps for the descent procedure.

We also give below the pseudo-code for a procedure that a hacker may possibly run in order to break the encrypted code.

```
input : prime; output :x,y
for z = −prime/2 to 0
      u = z * z + 1
      if(u%prime = 0)
            m = u/prime; x = z
            if(z < 0)
                  x = x * (−1)
            y = 1
end for

while(m! = 1)
      u = rangeModulo(x,m); v = rangeModulo(y,m)
      if(u > m/2)
            u = u − m
      if(v > m/2)
            v = v − m
      r = (u² + v²)/m; a = x * u + y * v; b = x * v − y * u
      x = a/m; y = b/m; m = r
end while

procedure rangeModulo(a,b)
if(a >= 0)
      return (a%b)
else
      x = (−1) * (a/b); a = a + (x * b)
      if(a < 0)
            a = a + b
return (a%b)
```

3.1   A Small Example of Descent

An illustration of the steps in the process is as follows. Let 9999999937 be the
prime number supplied as input to the program. Since $9999999937 = 1 \pmod 4$, it
can be expressed as the sum of two perfect squares. Initially, $z = -1432580440$,
$m = 205228673$. The values of the variables m, z, x and y in the above algorithm in
successive iterations are as follows.

Iteration 1: $X = 28063413$, $y = 7$, $m = 78754$
Iteration 2: $X = 952019$, $y = -2492$, $m = 9065$
Iteration 3: $X = 2919487$, $y = -2616600$, $m = 1537$
Iteration 4: $X = 2440924$, $y = 43474$, $m = 596$
Iteration 5: $X = -1198369$, $y = -117948$, $m = 145$
Iteration 6: $X = -411572$, $y = 566223$, $m = 49$
Iteration 7: $X = -66279$, $y = 419055$, $m = 18$
Iteration 8: $X = -58796$, $y = 80889$, $m = 1$

3.2 Experimental Results

The Method of Descent Algorithm was implemented in the Java programming language (OS—Fedora 15), in a 32-bit machine equipped with 2 GB RAM and Intel Core i3-2100 CPU, with a processor speed of 3.10 GHz X 4.

Consider Table 1. It is evident from it that the running time for the descent procedure increases at a rapid rate as we apply it to primes with progressively higher number of digits. It can be seen that on an average the running time for $n + 1$ digits is about 4 times that for n digits, though there are exceptions. In fact, the time for 13 digits is about 1.3 times that of 12 digits, however, the time for 12 digits is more than 35 times that for 11 digits. Therefore, the factor of 4 may also be a very conservative estimate of the growth of the runtime; actually from 7 to 13 digits, as per Table 1, the factor is about 7.16. But even with this conservative factor of 4, if we consider the time for 7 digits as approximately 1 s, then a rough estimate of the time for 50 digits is 4^{43} = 7737125245533626726718119526 s, which is approximately 2.45×10^{18} years or close to two and a half billion billion years. It is unlikely that any hacker would ever live that long. In fact, the runtime for merely 24 digits is more than 540 years. Hence, this degree of security can be achieved if we can find suitable Gaussian primes $a + bi$ and $c + di$, in which each of a, b, c and d is merely of 12 digits. We recall that in a cryptosystem based on rational integers two primes each of at least 100 digits must be used, whereas each of $a + ib$ and $c + id$

Table 1 Running times for descent procedure	Number of digits	Primes	Time
	6	999961	0.199 s
		823541	0.209 s
		826669	0.216 s
	7	9999973	0.967 s
		9980869	0.848 s
		9446989	0.996 s
	8	99999989	12.927 s
		94008721	12.471 s
		91744613	15.074 s
	9	987653201	111.77 s
		999999937	132.65 s
		999319777	136.29 s
	10	9848868889	10 min 57 s
		5513600773	15 min 55 s
		9000000001	17 min 45 s
	11	16831813861	41 min 08 s
		99999999977	31 min 43 s
	12	816425361649	28 h 58 min 15 s
	13	9740985827329	37 h 33 min 54 s

can be denoted by the use of only 24 digits. Therefore, the saving of storage space by the use of Gaussian integers instead of rational integers is sufficiently substantial to vindicate our proposed adoption of the former in preference to the latter.

4 An Outline of Further Developments

It is evident that the RSA cryptosystem requires continual upgradation, because of the rapidity with which the computing power of computers is increasing. With advances in quantum computing, it is likely that within a few years or decades it would be easy to factorize the product of two primes each of, say, 200 decimal digits. It would then be necessary to use primes, of say, 250 or 300 decimal digits in order to achieve an acceptable level of security. With further advances in computing, progressively larger and larger primes would be necessary, requiring major advances in primality testing. This process would necessitate continual augmentation of storage capacities. Indeed, storage may soon become so prohibitively difficult, that the RSA cryptosystem may become ineffectual.

Therefore, it is necessary to devise a sophisticated and efficient scheme whereby in the race between the difficulty of factorizing the product of two large primes and the rapid advancement of computing power, the former can stay perpetually ahead of the latter. The present scheme of choosing progressively larger and larger primes may indeed prove to be unsophisticated, crude and eventually impracticable.

We suggest the following scheme. In this paper, we have set forth the advantages of the use of Gaussian integers over that of rational integers in possible future cryptosystems. Again with advances in computing power, if a system based on Gaussian integers becomes insecure, then, instead of simply choosing Gaussian primes of larger and larger norms, a new cryptosystem based on some other ring of algebraic integers, of degree higher than two but having the unique factorization property, can be devised. If, in due course of time, that too becomes insecure, a system based on algebraic integers of still higher degree having the unique factorization property can be devised, and so on and so forth.

If with some unforeseen and dramatic advance in computing power, even the scheme described above is rendered ineffectual, then the following more sophisticated scheme can be pursued; hitherto we have suggested the use of rings of algebraic integers of class number 1. As and when necessary, a cryptosystem based on rings of integers of class number 2 can be devised, using the arithmetic of ideals. Assuming that there is no upper limit to the progressive increase in computing power, cryptosystems based on rings of algebraic integers of class numbers 3, 4, 5, 6,…etc. can be successively developed, the process being illimitable, since it is known that algebraic integers of arbitrarily large class numbers exist.

By the process suggested above, the difficulty of factorization of integers of sufficiently large norm can forever stay ahead of the advances in computing power. In this paper, we have discussed the first step of this prospectively interminable process of devising cryptosystems of progressively greater and greater security.

5 Conclusion

The procedure for factorizing the product of two large Gaussian primes involves advanced concepts of algebraic number theory. A prospective breaker has to deal with the extremely laborious process of factorization that we have discussed above. We have empirically shown that such a laborious process is going to be extremely time consuming and will most likely remain infeasible for performance in real time for several years to come. We also discussed a scheme in which rings of algebraic integers of progressively higher and higher degrees and class numbers can be used to build cryptosystems that remain secure by forever staying ahead of advances in computing power.

References

1. Adleman, L.M., Rivest, R.L., Shamir, A.: A method for obtaining digital signatures and public-key cryptosystems. Commun. ACM **21**, 12–126 (1978)
2. Rivest, R.L.: RSA chips (past/present/future). In: Advances in Cryptography, Proceedings of Eurocrypt, vol. 84, pp. 159–165. Springer, New York (1985)
3. Diffie, W., Hellman, M.E.: New directions in cryptography. IEEE Trans. Inf. Theory IT **22**, 644–654 (1976)
4. Hardy, G.H., Wright, E.M. (Revised by Heath-Brown, D.R., Silverman, J.H.): *An Introduction to the Theory of Numbers*, 6th edn. Oxford University Press, Oxford (2008)
5. Ireland, K., Rosen, M.: A Classical Introduction to Modern Number Theory, 2nd edn. Springer, New York (1990)
6. Davenport, H.: The Higher Arithmetic, 7th edn. Cambridge University Press, Cambridge (1999)
7. Apostol, T.M.: Introduction to Analytic Number Theory. Springer, New York (2010)
8. Cormen, T.H., Leiserson, C.E., Rivest, R.L., Stein, C.: Introduction to Algorithms, 3rd edn. PHI Learning Pvt. Ltd., New York (2011)
9. Knuth, D.E.: The Art of Computer Programming, vols. I and II. Addison-Wesley, Reading (1973)
10. Koblitz, N.: A Course in Number Theory and Cryptography, 2nd edn. Springer, Berlin (1994)
11. Niven, I., Zuckerman, H.S., Montgomery, H.L.: An Introduction to the Theory of Numbers, 5th edn. Wiley, New York (2006)
12. Rosen, K.H.: Discrete Mathematics and Its Application, 4th edn. Tata McGraw-Hill Publishing Company Limited, New Delhi (1999)
13. Silverman, J.H.: A Friendly Introduction to Number Theory, 3rd edn. Prentice Hall, Upper Saddle River (2011)

Stack Overflow Based Defense for IPv6 Router Advertisement Flooding (DoS) Attack

Jai Narayan Goel and B.M. Mehtre

Abstract Internet Protocol version 6 (IPv6) is the future for Internet. But still it has some serious security problems e.g., IPv6 Router advertisement (RA) flood. IPv6 Router advertisement (RA) flooding is a severe Denial of Service attack. In this attack using only single computer, attacker can attack all the computers connected to the Local area Network. As a result all victim machines get frozen and become unresponsive. In this chapter, we have described IPv6 RA flood attack in a test environment and its effects on victims. We proposed an effective solution to counter IPv6 RA flood attack by using a stack. The proposed system would use a stack at victim's machine. All the incoming RA packets would be sent to this stack before they are processed by the system. After regular interval, RA packet would Pop up from the stack and would be sent for processing. The proposed approach would protect from the IPv6 RA flood, detect the attack and also raise an alarm to alert the user. This proposed approach can also be used to counter other DoS attacks with some modification. We have also provided an algorithm and experimental results in this chapter.

Keywords IPv6 · Router advertisement flooding · Stack overflow · Cyber defense · Network security · DoS attack · Flood_router26

J.N. Goel
School of Computer and Information Sciences, University of Hyderabad,
Hyderabad 500046, India
e-mail: jainarayangoel@gmail.com

J.N. Goel · B.M. Mehtre (✉)
Center for Information Assurance and Management, Institute for Development
and Research in Banking Technology, Hyderabad 500057, India
e-mail: bmmehtre@idrbt.ac.in

© Springer India 2016   299
A. Nagar et al. (eds.), *Proceedings of 3rd International Conference
on Advanced Computing, Networking and Informatics*, Smart Innovation,
Systems and Technologies 44, DOI 10.1007/978-81-322-2529-4_31

1 Introduction

The use of IPv6 is increasing rapidly. With 32 bit address format IPv4 is not capable to fulfil all the future demand for IP addresses. With 128 bit format IPv6 provide 2^{128} bit addresses that are enough for future need. In IPv4 Dynamic Host Configuration Protocol [1] (DHCP) is used to provide router address and other network configuration to hosts. IPv6 [2] provide auto configuration [3] service to provide IP address and other network configuration information to hosts. Auto configuration service remove the need for manually configure network configuration in computers connected to that LAN and also provide more mobility. Two ICMPv6 messages are used for this purpose (RFC 4861 [4], RFC 1256 [5]). They are:—*Router Solicitation* message and *Router Advertisement* message. When a host, inside a Local Area Network need address of default router of LAN, It broadcasts *Router Solicitation* message in that LAN. Every host connected to that LAN (including router) get the *Router Solicitation* message. As soon as router get the *Router Solicitation* message, it broadcasts a *Router Advertisement* message in that LAN. When a host gets the *Router Advertisement* message, it takes the IP address of the sender of *Router Advertisement* message as its default router and update its other network configuration according to that *Router Advertisement* packet. This process takes lot of computing resources of host.

High resource requirement in Auto configuration [3] in IPv6 is a vulnerability for this protocol. IPv6 router advertisement (RA) flooding exploits this vulnerability of IPv6. The attacker starts flooding of RA packet in the LAN. Every user of that LAN gets flood of IPv6 router advertisements. All these RA packets have different ip address and network information. Hosts try to update their network configuration according to every RA packet. But because every updating process takes lot of computing resources, this flooding of RA packet consume all the resources of hosts. All the hosts connected to that LAN get frozen and unresponsive.

IPv6 RA flooding is very severe DoS attack. With a single system, attacker can attack on entire LAN and bring down entire LAN and the systems connected to that LAN. On the other hand attacker would not be affected by this attack. Defense from this attack is very important because it can make a big Loss to any institute or organization. We proposed a complete solution for this attack in this paper.

The rest of the paper is organized as follows. We describe related work, available solutions and their limitations in Sect. 2. We describe attack model, test bed setup, attack implementation and their results in Sect. 3. We describe our proposed solution for the attack in Sect. 4. In Sect. 5 we analyse our model efficiency and performance. Finally Sect. 6 concludes the paper.

2 Related Work

After IPv6 came into existence much research work have been done to increase its security. Arkko et al. [6] proposed to use Cryptographically Generated Addresses (CGA) and Address Based Keys (ABK) to secure IPv6 Neighbour and Router Discovery. [7–9] discuss various Security Issues and Vulnerabilities, their countermeasures in IPv6. Steffen Hermann et al. [10] evaluated guidelines for secure deployment of IPv6. References [3, 11] discuss the security problems in neighbour discovery and auto configuration in IPv6.

2.1 Available Solutions and Their Limitations

In this section we discuss available solutions for IPv6 RA DoS attack and their Limitations. There are 5 methods to defend from this attack:

RA Guard RA guards [12, 13] are installed on switches to stop Router advertisement flooding. RA guard forwards only those RA packets whose are received from a port known to be connected to an authorized router.

RA guard worked for older version of this attack *flood_router6*. But RA guard can be defeated using an extension header with RA packet. So RA guard does not work for *flood_router26*.

Stop Auto Configuration Stopping auto configuration service [13] for IPv6 is another method to stop this attack. In this we turn off router discovery service in hosts. So system does not send *Router Solicitation* message and does not process any received *Router Advertisement* message. So system does not take care of any *Router Advertisement* message and works normally.

But if we stop auto configuration, any system in the LAN would not be able to configure network settings automatically. Network administrator has to do configuration manually. This is not a feasible solution for large networks and where network configuration changes frequently.

Firewall Firewall [13] can be configured to allow only authorized routers to send RA packets. So unauthorized system cannot send RA packets.

But this method can be defeated easily by spoofing. By spoofing IP address of authorized router, any attacker can also send RA packets. So this approach also not works.

Disable IPv6 Permanently According to this IPv6 should be disabled in system permanently. If IPv6 would not be enabled, RA flood attack would not work.

But we are shifting to IPv6 technology, that is more capable to fulfil IP address demand. So disabling IPv6 permanently in system is also not a feasible solution.

Dynamic IPv6 Activation Deactivation This approach was proposed by us in [14] research paper. This system detects the IPv6 RA flooding attack and raises an alarm to notify the user or network administrator. After detecting the attack it deactivates IPv6 services in that system temporarily. When effect of attack reduces it re activates IPv6 service and user can access Ipv6 services again.

This defending system protects computer to get hanged or frozen from IPv6 RA flooding and provide IPv6 connectivity at regular intervals. Yet our approach was a good solution but still it does not provide uninterrupted IPv6 services.

3 IPv6 RA Flooding

To describe IPv6 RA flood attack we demonstrate it in Virtual LAN. We made a Virtual LAN shown in Fig. 1. We used GNS3 [15] to make this LAN. We made 4 virtual machine using VMware [16]. On first virtual machine we installed KALI Linux [17]. On rest three virtual machine we installed Windows operating system. We connected every virtual machine with GNS3 and made a virtual LAN with these 4 virtual machines.

We used KALI Linux as Attacker machine. Windows machines were Victims. We used *flood_router26* Tool [18] to initiate the IPv6 RA flood attack. First of all we opened terminal in KALI Linux. Then we typed *flood_router26 eth0* command. *Eth0* is used to attack on Ethernet. This started broadcasting flood of *Router Advertisement* messages in the LAN. The screenshot of the terminal of KALI Linux is shown in Fig. 2. Here dots are showing flood of RA packets.

Before starting the attack victim machines were behaving normally. Their CPU utilization was normal. We took screen-shot of resource utilization at one of the victim machines. At that time CPU utilization of that machine was almost 0 %. It is shown in Fig. 3.

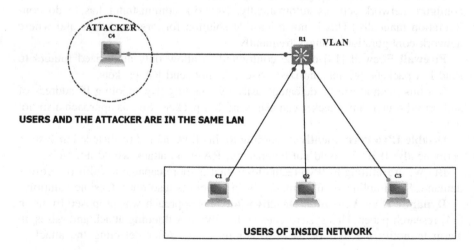

Fig. 1 Virtual LAN

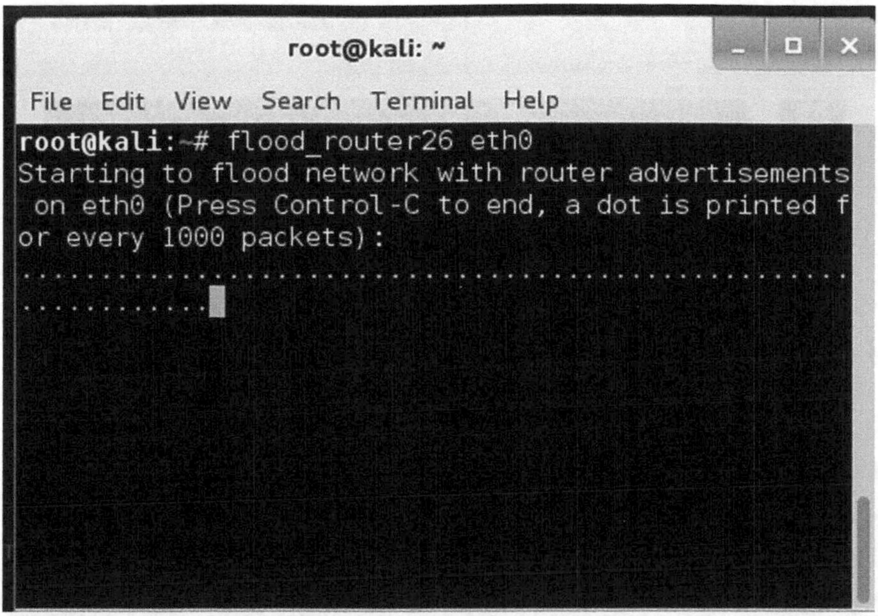

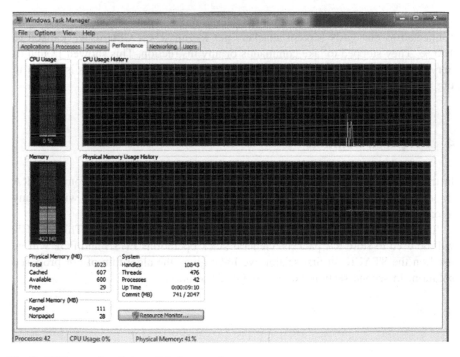

Fig. 2 Flood_router26 attack

Fig. 3 CPU usage in windows 7 before attack

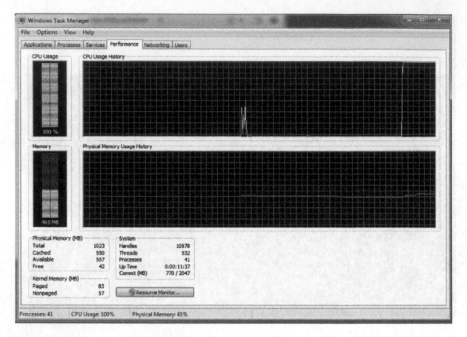

Fig. 4 CPU usage in windows 7 after attack

When we started the attack from KALI Linux on the LAN, all the systems connected to that LAN (except attacker) get frozen. Their CPU usage jumped to 100 %. They stopped to take commands or any instruction. This showed Denial of Service attack. The screenshot of resource utilization of those systems is shown in Fig. 4.

4 Proposed Solution

In this section we have described our proposed approach to defend from IPv6 RA flooding attack. This approach will successfully protect computers without any interruption in any service.

The proposed approach uses a STACK with specific functionality. This STACK will be used to store RA packets. General stack operations PUSH and POP will be used on the STACK. In first section we have given the algorithm of our proposed solution. In second section we explained our solution.

4.1 Algorithm

1: Sender (authorized router/ attacker) sends RA packets.
2: Computer gets the RA packet and PUSH it to the stack.
3: Wait for time t.
4: **if** Stack overflow **then**
5:     Notify user about RA flooding attack by raising alarm and Notification
6: **else**
7:     POP RA packet from top of the stack and send that RA packet to the system.
8: Empty the stack
9: **goto** *Step 3*

4.2 Detailed Solution

In our proposed system STACK would use PUSH and POP operations in general sense. The Fig. 5. describes the stack.

In Fig. 6. we showed incoming RA packets at a victim's machine. When a victim's machine receive a RA packet, the proposed system would PUSH that RA packet in the proposed STACK. Other RA packet would keep on coming to the system and system would keep on PUSH them to the top of STACK.

After a specific time t (decided by the developer of the operating system), the proposed system would POP the RA packet from top of the stack. Because we are using stack so this RA packet would be most recent RA packet. After popping the

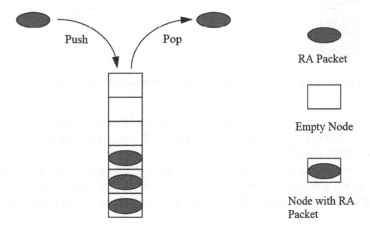

Fig. 5 RA packets are shown by a *blue oval*. Empty node of the STACK is shown by a *rectangle*

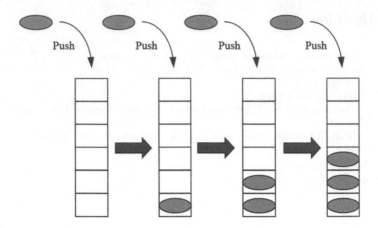

Fig. 6 Incoming RA packets pushed in STACK

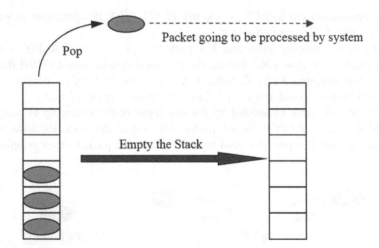

Fig. 7 RA packet popped from STACK and sent to be processed by system

RA packet from the stack the proposed system would send this packet to system to update the network configuration. After this POP operation, the proposed system would EMPTY the STACK. Shown in Fig. 7.

But if within that specific time t the STACK get overflow this would be considered as the signature of IPv6 RA flooding attack. An alarm module would be activated automatically and will notify the user or system administrator. Shown in Fig. 8.

Because in proposed approach RA packets would first go to the STACK so RA flood would not affect the Victim's machine. By changing the size of the STACK and the time t, this solution can adjust itself according to the requirement of the system.

Fig. 8 Attack detection by
stack overflow

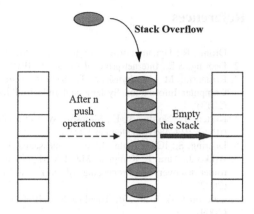

5 Performance Analysis

In the proposed solution RA packets would first go to the STACK before it is
processed by system to update router address and network configuration. Only most
recent RA packet would be sent to system to update router address and other
network configuration. Time t would be decided based on the capacity of the
computer to process RA packets. So core system (victim's machine) would never
be flooded by RA packets and flood would be stopped by our proposed system.
So RA flood would never affect the Victim's machine. Victim's machine would be
protected by the proposed defence system.

No other existing solution is able to provide complete protection from IPv6 RA
flood attack. Our proposed system would provide complete protection with uninterrupted service to the user. So it would be the best solution for this attack.

6 Conclusion

In this chapter we proposed a new and most effective approach against IPv6 Router
advertisement flooding. The proposed system provides full protection to users
against IPv6 flooding. STACK Overflow provides complete attack detection and
defense mechanism against IPv6 RA flooding attack. In Future, researchers can
apply STACK Overflow approach also on other DoS attacks with some modification. If more research is done on this method, this approach can be a holistic
solution for all types of DoS attacks.

References

1. Droms, R.: Dynamic host configuration protocol. RFC1541. March (1997)
2. Deering, S.E.: Internet protocol version 6 (IPv6) specification. RFC2460. December (1998)
3. Rostaaski, M., Mushynskyy, T.: Security issues of IPv6 network autoconfiguration. In: Computer Information Systems and Industrial Management, pp. 218–228. Springer, Berlin (2013)
4. Narten, T., Nordmark, E., Simpson, W., Soliman, H.: Neighbor discovery for IP version 6 (IPv6). RFC4861. September (2007)
5. Deering, S.: ICMP router discovery messages. RFC 1256. September (1991)
6. Arkko, J., Aura, T., Kempf, J., Mntyl, V.M., Nikander, P., Roe, M.: Securing IPv6 neighbor and router discovery. In: Proceedings of the 1st ACM Workshop on Wireless Security, pp. 77–86 (2002)
7. Caicedo, C.E., Joshi, J.B., Tuladhar, S.R.: IPv6 security challenges. IEEE Comput. **42**(2), 36–42 (2009)
8. Lee, J.H., Ernst, T.: IPv6 security issues in cooperative intelligent transportation systems. Comput. J. bxs006 (2012)
9. Ullrich, J., Krombholz, K., Hobel, H., Dabrowski, A., Weippl, E.: IPv6 security: attacks and countermeasures in a nutshell. In: Proceedings of the 8th USENIX conference on Offensive Technologies. USENIX Association 2014, pp. 5–16 (2014)
10. Hermann, S., Fabian, B.: A comparison of internet protocol (IPv6) security guidelines. Future Internet. **6**(1), 1–60 (2014)
11. Nikander, P., Kempf, J., Nordmark, E.: IPv6 neighbor discovery (ND) trust models and threats. RFC 3756, Internet Engineering Task Force. May (2004)
12. Levy-Abegnoli, E., Van de Velde, G., Popoviciu, C., Mohacsi, J.: IPv6 router advertisement guard. RFC 6105, Internet Engineering Task Force. February (2011)
13. Chown, T., Venaas, S.: Rogue IPv6 router advertisement problem statement. RFC6104. February (2011)
14. Goel, J.N., Mehtre, B.M.: Dynamic IPv6 activation based defense for IPv6 router advertisement flooding (DoS) attack. In: 2014 IEEE International Conference on Computational Intelligence and Computing Research (ICCIC), pp. 628–632 (2014)
15. GNS3 Tool. http://www.gns3.com
16. VMware Tool. http://www.vmware.com/in
17. KALI Linux. http://www.kali.org
18. Hauser, V.: A complete tool set to attack the inherent protocol weaknesses of IPv6 and ICMPv6. https://www.thc.org/thc-ipv6/

Ideal Contrast Visual Cryptography for General Access Structures with AND Operation

Kanakkath Praveen and M. Sethumadhavan

Abstract In Visual Cryptographic Scheme (*VCS*) reversing operations (NOT operations) are carried out during reconstruction process to improve the quality of the reconstructed secret image. Many studies were reported on using reversing operation in both perfect and nonperfect black *VCS*. In 2005, Cimato et al. put forward an ideal contrast *VCS* with reversing, which is applicable to any access structures (*IC-GA-VCS*). Here there will not be any change in resolution for reconstructed secret image in comparison with original secret image (*KI*). In this paper a proposal for replacing reversing operations with AND operations during reconstruction in perfect black *VCS* is shown. A comparison of the proposed modification with Cimato et al. construction and other related work are discussed in this paper.

Keywords Visual cryptography · Secret sharing · Perfect reconstruction · Ideal contrast

1 Introduction

Visual cryptography is a technique to split a secret into distinct transparencies, where the sufficient transparencies combine together will reconstruct original secret image (KI). The basic parameters for a *VCS* are pixel expansion (*m*) and contrast (*α*). Naor and Adi Shamir in 1994, developed a deterministic (*k*, *n*)-*VCS* [1] for sharing secret images using OR operation. Droste [3] in 1998, proposed a *VCS* with less pixel expansion than Naor's et al. scheme. A *VCS* for any access structure was

K. Praveen (✉) · M. Sethumadhavan
Amrita Vishwa Vidyapeetham, Amrita Nagar P.O, Ettimadai,
Coimbatore 641112, India
e-mail: praveen.cys@gmail.com

M. Sethumadhavan
e-mail: m_sethu@cb.amrita.edu

© Springer India 2016   309
A. Nagar et al. (eds.), *Proceedings of 3rd International Conference on Advanced Computing, Networking and Informatics*, Smart Innovation, Systems and Technologies 44, DOI 10.1007/978-81-322-2529-4_32

introduced by Ateniese et al. [2] in 1996. Adhikari et al. [4] in 2004, come up with a *VCS* for general access structure. In 2005, Tylus et al. [5] proposed a *VCS* based on XOR operation. In *VCS* after reconstruction, the image becomes blurrier and darker due to graying effect. As a compromise to this graying effect, reversing (NOT) operations are carried out during the reconstruction process, which results in the contrast of the recovered image, ideal. In 2004, Viet et al. [6] first proposed a reversing construction based on customary *VCS*. Cimato et al. [10] in 2005, proposed an ideal contrast general access structure *VCS* with reversing using OR and NOT operations. Ideal contrast means the reconstructed secret image have same resolution as that of original secret image. Hu et al. [7] and Yang et al. [8, 9] in subsequent years proposed a *VCS* with reversing. But the resolution of the reconstructed image in both Hu et al. [7] and Yang et al. [8, 9] are not same as that of original secret image. Yang et al. outlined solutions for both perfect and non perfect black schemes. In 2009, Zhang et al. [11] come up with a solution which encodes the pixels block by block, were the encoding capacity is high when compared to all the above schemes.

In this paper, we propose an IC-GA-VCS with AND operation. The reversing operations of Cimato et al. scheme are replaced with AND operations in the proposed scheme. This will in turn reduce the number of operations used during reconstruction. The proposed scheme will work only for schemes which perfectly reconstruct the black pixels. The proposed scheme and the comparison with other related schemes are given in the Sects. 2, 3 and 4. The following paragraph shows some background definitions and notations of *VCS*.

Let $PT = \{PT_1, PT_2, \ldots, PT_n\}$ be an n participant set and PS be the power set of PT. Let us denote Γ_Q as qualified set and Γ_F as forbidden set. Let $\Gamma_Q \in PS$ and $\Gamma_F \in PS$ where $\Gamma_Q \cap \Gamma_F = \varphi$. Any set $A \in \Gamma_Q$ can recover the secret image, whereas any set $A \in \Gamma_F$ cannot leak out any secret information. The minimal qualified subset of PT is denoted as $\Gamma_0 = \{A \in \Gamma_Q : A' \notin \Gamma_Q \text{ for all } A' \subseteq A, A' \neq A\}$. The pair $\Gamma = (\Gamma_Q, \Gamma_F)$ is called the access structure of *VCS*. Let S be a $n \times m$ Boolean matrix and $A \subseteq PT$, the vector attained by applying the Boolean OR operation to the rows of S corresponding to the elements in A is denoted by S_A. Let $w(S_A)$ denotes the Hamming weight of vector S_A.

Definition [2] Let (Γ_Q, Γ_F) be an access structure on PT. Two collections (multisets) of $n \times m$ Boolean matrices S^0 and S^1 constitute a (Γ_Q, Γ_F, m)-*VCS* if there exists a positive real number α and the set of thresholds $\{t_A \mid A \in \Gamma_Q\}$ satisfying the two conditions:

1. Any (qualified) set $A = \{i_1, i_2, \ldots, i_q\} \in \Gamma_Q$ can recover the shared image by stacking their transparencies. Formally, $w(S_A^0) \leq t_A - \alpha.m$, whereas $w(S_A^1) \geq t_A$.
2. Any (forbidden) set $A = \{i_1, i_2, \ldots, i_q\} \in \Gamma_F$ has no information on the secret. Formally the two collections of $q \times m$ matrices D_t, where $t \in \{0, 1\}$ obtained by restricting each $n \times m$ matrix in S^t to rows $i_1, i_2, \ldots, i_q$ are indistinguishable, in the sense that they contain the same matrices with the same frequencies.

2 Cimato's IC-GA-VCS

Let $PT = \{PT_1, PT_2, \ldots, PT_n\}$ is the participant set and $A = \{i_1, i_2, i_3, \ldots, i_q\} \in \Gamma_Q$ is the qualified set. The share generation and decryption phase is given below.

1. Let KI be the secret binary image of size $(p \times r)$. For each participant PT_u, $1 \leq u \leq n$ the share transparencies is constructed as follows;

$$Sh_{(u,j)}(l,h) = \begin{cases} (u,j)\text{th element of } S^0 & \text{if } KI(l,h) = 0 \\ (u,j)\text{th element of } S^1 & \text{if } KI(l,h) = 1 \end{cases};$$

   where
   $1 \leq l \leq p$, $1 \leq h \leq r$ and $1 \leq j \leq m$. S^0 and S^1 are the basis matrices which are constructed using perfect black scheme given in paper [2].

2. Let us define a function $f(s) = \begin{cases} 1 & \text{if } s == 0 \\ 0 & \text{if } s == 1 \end{cases}$.

3. In the decryption phase apply steps from (a) to (c)

   a. Let $\lambda_j(l,h) = \text{OR-ing } (Sh_{\langle i_1, j \rangle}, Sh_{\langle i_2, j \rangle}, Sh_{\langle i_3, j \rangle}, \ldots, Sh_{\langle i_q, j \rangle})$ for all $j = 1, \ldots, m$.
   b. $\sigma(l,h) = \text{OR-ing all } f(\lambda_j(l,h))$ for all $j = 1, \ldots, m$.
   c. Find $KI = f(\sigma(l,h))$ for all $1 \leq l \leq p$, $1 \leq h \leq r$.

3 Proposed IC-GA-VCS

Let $PT = \{PT_1, PT_2, \ldots, PT_n\}$ is the participant set and $A = \{i_1, i_2, i_3, \ldots, i_q\} \in \Gamma_Q$ is the qualified set. The share generation and decryption phase is given below.

1. Let KI be the secret binary image of size $(p \times r)$. For each participant PT_u, $1 \leq u \leq n$ the share transparencies construction is given as.

$$Sh_{(u,j)}(l,h) = \begin{cases} (u,j)\text{th element of } S^0 & \text{if } KI(l,h) = 0 \\ (u,j)\text{th element of } S^1 & \text{if } KI(l,h) = 1 \end{cases};$$

   where
   $1 \leq l \leq p$, $1 \leq h \leq r$ and $1 \leq j \leq m$. S^0 and S^1 are the basis matrices which are constructed using perfect black scheme given in paper [2].

2. In the decryption phase apply the following steps

   a. Let $\lambda_j(l,h) = \text{OR-ing } (Sh_{\langle i_1, j \rangle}, Sh_{\langle i_2, j \rangle}, Sh_{\langle i_3, j \rangle}, \ldots, Sh_{\langle i_q, j \rangle})$ for all $j = 1, \ldots, m$.
   b. $KI = \text{AND-ing all } \lambda_j(l,h)$ for all $j = 1, \ldots, m$.

Example Let $PT = \{PT_1, PT_2, \ldots, PT_n\}$ is the participant set and the minimal qualified set is given by $\Gamma_0 = \{\{PT_1, PT_2\}, \{PT_2, PT_3\}, \{PT_3, PT_4\}\}$. The matrices S^0 and S^1 corresponding to Γ_0 are

$$S^0 = \begin{bmatrix} 0 & 1 & 1 & 0 \\ 0 & 1 & 1 & 1 \\ 0 & 1 & 1 & 1 \\ 0 & 1 & 0 & 1 \end{bmatrix} \text{ and } S^1 = \begin{bmatrix} 1 & 0 & 0 & 1 \\ 1 & 1 & 1 & 0 \\ 1 & 1 & 0 & 1 \\ 1 & 0 & 1 & 0 \end{bmatrix} \text{ respectively.}$$

Let $KI = \begin{bmatrix} 1 & 0 \\ 0 & 1 \end{bmatrix}$, then the share transparencies for each participant are given in the Table 1. During the decryption phase, λ_j's for reconstructing KI is shown in Table 2.

4 Comparison with Related Work

This paper mainly focused on the related work of ideal contrast *VCS* for black and white images. In paper Viet et al. [6], it is given that a near ideal contrast is achieved by applying the customary *VCS* infinite times. The secret splitting process of Cimato et al. [10] shows that, the total number of runs required for reconstruction is m. In

Table 1 Share transparencies of each participant

Participant	Share transparencies
PT_1	$Sh_{(1,1)} = \begin{bmatrix} 1 & 0 \\ 1 & 0 \end{bmatrix}$; $Sh_{(1,2)} = \begin{bmatrix} 0 & 1 \\ 1 & 1 \end{bmatrix}$; $Sh_{(1,3)} = \begin{bmatrix} 0 & 1 \\ 0 & 1 \end{bmatrix}$; $Sh_{(1,4)} = \begin{bmatrix} 1 & 0 \\ 0 & 0 \end{bmatrix}$
PT_2	$Sh_{(2,1)} = \begin{bmatrix} 1 & 0 \\ 1 & 1 \end{bmatrix}$; $Sh_{(2,2)} = \begin{bmatrix} 1 & 1 \\ 1 & 0 \end{bmatrix}$; $Sh_{(2,3)} = \begin{bmatrix} 1 & 1 \\ 1 & 1 \end{bmatrix}$; $Sh_{(2,4)} = \begin{bmatrix} 0 & 1 \\ 0 & 1 \end{bmatrix}$
PT_3	$Sh_{(3,1)} = \begin{bmatrix} 1 & 0 \\ 1 & 0 \end{bmatrix}$; $Sh_{(3,2)} = \begin{bmatrix} 1 & 1 \\ 1 & 1 \end{bmatrix}$; $Sh_{(3,3)} = \begin{bmatrix} 0 & 1 \\ 1 & 1 \end{bmatrix}$; $Sh_{(3,4)} = \begin{bmatrix} 1 & 1 \\ 0 & 1 \end{bmatrix}$
PT_4	$Sh_{(4,1)} = \begin{bmatrix} 1 & 0 \\ 1 & 1 \end{bmatrix}$; $Sh_{(4,2)} = \begin{bmatrix} 0 & 1 \\ 0 & 0 \end{bmatrix}$; $Sh_{(4,3)} = \begin{bmatrix} 1 & 0 \\ 1 & 1 \end{bmatrix}$; $Sh_{(4,4)} = \begin{bmatrix} 0 & 1 \\ 0 & 0 \end{bmatrix}$

Table 2 λ_j's for reconstructing KI

Qualified set	λ_j, for all $j = 1$ to m
$\{PT_1, PT_2\}$	$\lambda_1 = \begin{bmatrix} 1 & 0 \\ 1 & 1 \end{bmatrix}$; $\lambda_2 = \begin{bmatrix} 1 & 1 \\ 1 & 1 \end{bmatrix}$; $\lambda_3 = \begin{bmatrix} 1 & 1 \\ 1 & 1 \end{bmatrix}$; $\lambda_4 = \begin{bmatrix} 1 & 1 \\ 0 & 1 \end{bmatrix}$
$\{PT_2, PT_3\}$	$\lambda_1 = \begin{bmatrix} 1 & 0 \\ 1 & 1 \end{bmatrix}$; $\lambda_2 = \begin{bmatrix} 1 & 1 \\ 1 & 1 \end{bmatrix}$; $\lambda_3 = \begin{bmatrix} 1 & 1 \\ 1 & 1 \end{bmatrix}$; $\lambda_4 = \begin{bmatrix} 1 & 1 \\ 0 & 1 \end{bmatrix}$
$\{PT_3, PT_4\}$	$\lambda_1 = \begin{bmatrix} 1 & 0 \\ 1 & 1 \end{bmatrix}$; $\lambda_2 = \begin{bmatrix} 1 & 1 \\ 1 & 1 \end{bmatrix}$; $\lambda_3 = \begin{bmatrix} 1 & 1 \\ 1 & 1 \end{bmatrix}$; $\lambda_4 = \begin{bmatrix} 1 & 1 \\ 0 & 1 \end{bmatrix}$

Table 3 Comparison of proposed scheme with Cimato et al. construction

Operation	Cimato et al.	Proposed
OR	$((k - 1) \times m) + (m - 1)$	$(k - 1) \times m$
NOT	$m + 1$	NIL
AND	NIL	$m - 1$

Yang et al. [8, 9] schemes, both reversing and shifting operations are used for reconstruction. The number of runs required in this scheme is less than that of Cimato et al. [10] scheme. In Hu et al. [7] scheme, the size of the participant share transparency was determined based on the cardinality of elements in the minimal qualified set. In the cases of Yang et al. [8, 9], Viet et al. [6] and Hu et al. [7] schemes the resolution of the secret image is changing after recovery process. The scheme proposed by Zhang et al. [11], encodes pixel block in turn improves the encoding capacity, which is superior to Cimato et al. [10]. All the pixels can be encoded at a time in Cimato et al. [10]. But in Zhang et al. [11] scheme parallelism is not possible because of pixel block encoding. The schemes discussed above uses reversing operations (NOT operations) during reconstruction. In the proposed scheme, reversing operations of Cimato et al. scheme are replaced with AND operations, which makes the reconstruction simpler. Table 3 shows the amount of operations done during recovery phase for both Cimato et al. [10] and proposed scheme.

5 Conclusion

In the literature many studies were carried out on *VCS* with reversing operation which reconstructs the secret image without any resolution change compared to original image. Reversing operations are used during reconstruction in both perfect black *VCS* and non perfect black *VCS*. This paper proposes a replacement to reversing operations with AND operations in perfect black *VCS*. A comparison with the related work which makes the reconstruction simpler is shown in this paper. The proposed scheme can also be extended for grey scale images.

References

1. Naor, M., Shamir, A.: Visual cryptography. In: De Santis, A. (ed.) EUROCRYPT 1994. LNCS, vol. 950, pp. 1–12. Springer, Heidelberg (1995)
2. Ateniese, G., Blundo, C., De Santis, A., Stinson, D.R.: Visual cryptography for general access structures. Inf. Comput. **129**(2), 86–106 (1996)
3. Droste, S.: New results on visual cryptography. In: Koblitz, N. (ed.) CRYPTO 1996. LNCS, vol. 1109, pp. 401–415. Springer, Heidelberg (1996)
4. Adhikari, A., Dutta, T.K., Roy, B.: A new black and white visual cryptographic scheme for general access structures. In: Canteaut, A., Viswanathan, K. (eds.) INDOCRYPT 2004. LNCS, vol. 3348, pp. 399–413. Springer, Heidelberg (2004)

5. Tylus, P., Hollman, H.D.L., Lint, J.H.V., Tolhuizen, L.: XOR based visual cryptographic schemes. Design Codes Crypto. **37**(1), 169–186 (2005)
6. Viet, D.Q., Kurosawa, K.: Almost ideal contrast visual cryptography with reversing. In: Okamoto, T. (eds.) CT-RSA 2004. LNCS, vol. 2964, pp. 353–365 (2004)
7. Hu, C.M., Tzeng, W.G.: Compatible ideal contrast visual cryptography schemes with reversing. In: Zhou, J. (ed.) ISC 2005. LNCS, vol. 3650, pp. 300–313. Springer, Heidelberg (2005)
8. Yang, C.N., Wang, C.C, Chen, T.S.: Real perfect contrast visual secret sharing schemes with reversing. In: Zhou, J., Yung, M., Bao, F. (ed.): ACNS 2006. LNCS, vol. 3989, pp. 433–447. Springer, Heidelberg (2006)
9. Yang, C.N., Wang, C.C., Chen, T.S.: Visual cryptography schemes with reversing. Comput. J. **51**(6), 710–722 (2008)
10. Cimato, S., De Santis, A., Ferrara, A.L., Masucci, B.: Ideal contrast visual cryptography schemes with reversing. Inf. Process. Lett. **93**(4), 199–206 (2005)
11. Zhang, H., Wang, X., Huang, Y.: A novel ideal contrast visual secret sharing scheme with reversing. J. Multimedia **4**(3), 104–111 (2009)

Extending Attack Graph-Based Metrics for Enterprise Network Security Management

Ghanshyam S. Bopche and Babu M. Mehtre

Abstract Measurement of enterprise network security is a long standing challenge to the research community. However, practical security metrics are vital for securing enterprise networks. With the constant change in the size and complexity of enterprise networks, and application portfolios as well, network attack surface keeps changing and hence monitoring of security performance is increasingly difficult and challenging problem. Existing attack graph-based security metrics are inefficient in capturing change in the network attack surface. In this paper, we have explored the possible use of graph-based distance metrics for capturing the change in the security level of dynamically evolving enterprise networks. We used classical graph similarity measures such as *Maximum Common Subgraph (MCS)*, and *Graph Edit Distance (GED)* as an indicator of change in the enterprise network security. Our experimental results shows that graph similarity measures are efficient and capable of capturing changing network attack surface in dynamic (i.e. time varying) enterprise networks.

Keywords Graph similarity measures · Attack surface · Attack graph · Network security and protection · Security metric

G.S. Bopche (✉) · B.M. Mehtre
Center for Information Assurance & Management (CIAM), Institute for Development and Research in Banking Technology (IDRBT), Castle Hills, Masab Tank, Hyderabad 500057, India
e-mail: ghanshyambopche.mca@gmail.com

B.M. Mehtre
e-mail: mehtre@gmail.com

G.S. Bopche
School of Computer and Information Sciences (SCIS), University of Hyderabad, Gachibowli, Hyderabad 500046, India

© Springer India 2016
A. Nagar et al. (eds.), *Proceedings of 3rd International Conference on Advanced Computing, Networking and Informatics*, Smart Innovation, Systems and Technologies 44, DOI 10.1007/978-81-322-2529-4_33

315

1 Introduction

With various choices of using network devices and creating network topologies, increasingly complex application portfolios, flexibility in installing variety of applications by an end user's, frequent change in hosts or network configurations, today's network infrastructure undergoes continuous evolution and hence there is always a possibility of information exposure to a larger threat landscape. As an attack surface in dynamic network keeps changing, temporal aspect of enterprise network security needs to be taken into account and hence an anomaly detection technique in network security need to be utilized regularly. Early detection of security anomalies such as unauthorized access to the web server or database server can greatly enhance the security administrator's preparedness in the wake of emerging cyber threats or cyber attacks. Again, early detection of change in network attack surface is essential for building capability like an early warning system for possible intrusion. In order to achieve this, it is vital that network security behavior should be quantified in some manner.

An attack graph, a graphical formalism and respective graph-assisted metrics were proposed in literature for enterprise network security management. These metrics measure the size of an attack graph (in terms of number of attack paths, length of shortest attack path), as graph size is used as a prime indicator of the amount of security risk an enterprise network may face in the near future from adversaries. As we know, no single metric is capable of identifying all aspects of network security, numerous attack graph-based metrics were proposed in the literature for example *shortest path (SP)* [1], *number of paths (NP)* [2], *mean of path lengths (MPL)* [3], *suite of the new complementary set of metrics based on attack path lengths* [4], *attack vector and reachability machines* [5]. These metrics measure different aspect of network security, computed by different mathematical operations viz. mean, standard deviation (SD), median, and mode etc. Such severity values do not have great significance, rarely support decision-making, and hence rarely used in enterprise network security management. Further, these metrics are not sensitive enough to capture change in network attack surface. In the event of dynamically changing enterprise network and hence the threat landscape, new metrics need to be introduced that are sensitive to such change. New metrics should identify problems or security events proactively so that necessary corrective actions can be taken before security attack becomes reality.

The metrics we proposed in this paper are inspired from the *computer vision and pattern recognition (CVPR)* discipline. In pattern recognition, different graphs are evaluated for their structural similarity using a kind of graph similarity measures such as maximum common subgraph-based distance measures *(MCS)* [6], graph edit distance-based measures *(GED)* [7] etc. Measures based on *MCS* and *GED* are very flexible compared to other graph matching methods because of their ability to manage with random/unpredictable structured graphs with unconstrained and unique alphabetic labels for both nodes and edges [8]. Hence, we have adopted the classical algorithmic framework for *MCS* and *GED—based* distance computation to our specific problem of security management in dynamic enterprise networks. We proposed metrics to monitor change in the network security over a time through the identification of large relative change in the nodes and edges of the representative attack graph.

The rest of the article is organized as follows: Sect. 2 first reviews the attack graph model. Section 3 defines the proposed metrics, and present an algorithm for efficient computations of the metrics. Experimental system, and initial results are discussed in Sect. 4. Sensitivity analysis of the proposed metrics is done in Sect. 5. Section 6 closes with conclusions and suggestions for future work.

2 Preliminaries

In general, attack graph is a graphical formalism that depicts prior knowledge about the system vulnerabilities, their interdependencies, and network reachability information [9]. Attack graph for the enterprise network consists of two types of nodes, namely, *exploits* and *security conditions* [10, 11]. An *exploit* represents an adversarial action on the network host in order to take advantage of the vulnerability. *Security conditions* represent properties of system/network vital for successful execution of an exploit. The existence of a host vulnerability, network reachability, and trust relationship between the hosts are the kind of security conditions required for successful exploitation of vulnerability on a remote host.

Directed edges interconnect exploits with conditions. No two *exploits* or two *security conditions* are directly connected. A directed edge from security condition to an exploit represent the *require relation* and it states that for successful execution of an exploit all the security conditions need to be satisfied. A directed edge from an exploit to the *security condition* indicates the imply relation [10]. Once an exploit is successfully executed, there will be generation of few more security conditions called *postconditions*. These newly created postcondition may act as a precondition for other exploit.

With the perception of an attack graph discussed above, Wang et al. [11] formally defines exploit-dependency graph as follows:

Definition 1 (*Attack Graph Model*) "Given a set of exploits e, a set of conditions c, a require relation $R_r \subseteq c \times e$, and an imply relation $R_i \subseteq e \times c$, an attack graph G is the directed graph $G(e \cup c, R_r \cup R_i)$, where $(e \cup c)$ is the vertex set and $(R_r \cup R_i)$ is the edge set" [11].

It is clear from the above definition, that an attack graph G is a directed bipartite graph with two disjoint sets of vertices, namely, *exploit* and *security condition*. The edge set consist of two types of edges, namely, *require edge* and *imply edge* [12]. As an important feature of an attack graph, require relation (R_r) should be always conjunctive, and the imply relation (R_i) should be always disjunctive [11]. It means require edge should capture the conjunctive nature of the preconditions in order to successfully execute an exploit. In other words, an exploit cannot be executed until all of its preconditions have been satisfied. An imply edge should identify those conditions which are generated after the successful execution of an exploit.

3 Proposed Metrics

In this section, we explored classical graph similarity measures, namely, maximum common subgraph (*MCS*) [6] and graph edit distance (*GED*) [7]. These measures we used in our study for the detection of change in the network attack surface. The algorithm 3 presented in this paper for metric computation works on labeled attack graphs. Let L_V and L_E denote the finite set of vertex and edge labels in an attack graph, respectively.

Definition 2 An attack graph G is a four-tuple $G = (V, E, \rho, \mu)$, where

- V is a finite set of vertices, i.e. $V = (e \cup c)$.
- $E \subseteq V \times V$ is a finite set of Edges, i.e. $E = (R_r \cup R_i)$
- $\rho : V \rightarrow L_V$ is a function that assign unique label to all the vertices $v \in V$
- $\mu : E \rightarrow L_E$ is a function that assign unique label to all the edges in E

Here, L_V represents the set of symbolic labels uniquely identifying each node/vertex in the attack graph G.

Definition 3 (*Subgraph*) Let $G = (V, E, \rho, \mu)$ and $G' = (V', E', \rho', \mu')$ be two attack graphs; G' is a subgraph of G, i.e. $G' \subseteq G$, if

- $V' \subseteq V$
- $\rho'(x) = \rho(x), \ \forall x \in V'$
- $E' \subseteq E$ i.e. $E' = E \cap (V' \times V')$
- $\mu'(x, y) = \mu(x, y), \ \forall (x, y) \in V' \times V'$

From Definition 3, it follows that, given an attack graph $G = (V, E, \rho, \mu)$, any subset V' of it's vertices uniquely defines a subgraph of G. This subgraph is said to be induced by V'.

Definition 4 (*Common Subgraph*) Let $G = (V, E, \rho, \mu)$ and $G' = (V', E', \rho', \mu')$ be two attack graphs. A graph G'' is called common subgraph of G and G' if

- $G'' \subseteq G$ and $G'' \subseteq G'$.

Definition 5 (*Maximum Common Subgraph (mcs)*) Graph G'' is called maximum common subgraph (mcs) of G and G' if G'' is common subgraph of both G and G' and there exist no other common subgraph of G and G' that has more nodes and, for a given number of nodes, more edges than G'' [13].

Definition 6 (*Maximum Common Subgraph-based Distance (MCS)*) The distance between two non-empty attack graphs $G = (V, E, \rho, \mu)$ and $G' = (V', E', \rho', \mu')$ is defined by:

$$MCS(G, G') = 1 - \frac{|mcs(G, G')|}{\max(|G|, |G'|)} \tag{1}$$

Here $mcs(G, G')$ is is computed by counting either the number of common edges or common nodes in G and G'.

Definition 7 (*Graph Edit Distance (GED)*) Let the graph $G = (V, E, \rho, \mu)$ represents the attack graph generated from the enterprise network observed at time t, and let $G' = (V', E', \rho', \mu')$ describes the attack graph of the same network observed at time t' where $t' = t + \Delta t$. Then the graph edit distance $GED(G, G')$ is defined by:

$$GED(G, G') = |V| + |V'| - 2|V \cap V'| + |E| + |E'| - 2|E \cap E'| \tag{2}$$

Clearly, here the *GED*, as a measure of change in an attack surface, increases with an increase in the amount of change experienced by an enterprise network over sampling interval Δt. *GED* is bounded below by $GED(G, G') = 0$ when G and G' are isomorphic (i.e. there is no change in the network attack surface), and above by $GED(G, G') = |V| + |V'| + |E| + |E'|$, when the attack surface is completely changed.

Algorithm 1 Algorithm for metrics computation (*MCS* and *GED* metric)

Input:
$\langle G_1, G_2 \rangle$ a pair of an attack graphs
$G_1 \rightarrow$ attack graph for the enterprise network at time t
$G_2 \rightarrow$ attack graph for the enterprise network at time $t + \Delta t$
$\Delta t \rightarrow$ a sampling interval
Output:
MCS and *GED*-based distance {For a successive graphs G_1 and G_2 in an attack graph sequence generated for the same network at regular interval Δt over the period W}

1: $\langle V_1, E_1 \rangle \leftarrow$ MST (G_1) ▷ Apply minimum spanning tree algorithm (MST) on G_1 to identify all the unique nodes and edges in G_1

2: $\langle V_2, E_2 \rangle \leftarrow$ MST (G_2) ▷ Apply minimum spanning tree algorithm (MST) on G_2 to identify all the unique nodes and edges in G_2

3: $\langle V_3, E_3 \rangle \leftarrow G_1 \cap G_2$ ▷ Compute $G_1 \cap G_2$ in terms of either number of common nodes or common edges. Number of nodes i.e., $V_1 \cap V_2$ or edges i.e., $E_1 \cap E_2$ common to both G_1 and G_2 can be considered here.

4: Compute MCS-based distance metric

$$MCS(G_1, G_2) = \frac{|mcs(G_1, G_2)|}{max(|G_1|, |G_2|)}$$

5: Compute GED-based metric

$$GED(G_1, G_2) = |V_1| + |V_2| - 2|V_1 \cap V_2| + |E_1| + |E_2| - 2|E_1 \cap E_2|$$

---

4 Experimental Setup and Results

We conducted a simulation study similar to [4] in order to assess the behavior of the proposed metrics and to provide further evidence to support our claim. A simulated test bed is implemented using multiple virtual machines (VMs) with the help of VMware Workstation. We have chosen a set of base network models such as flat network, external-internal network, and DMZ network. For each network model as shown in Fig. 1, there is an attacker host who is capable of exploiting all discovered vulnerabilities on a network and whose goal is to gain root-level access on the target/victim host V. The network models being used dictate the connectivity between an attacker and other network hosts. Hosts B through D are the intermediate hosts between an attacker and victim V and each of these machines is accessible to an attacker under various vulnerability densities. In a flat network, since each host can connect to any other host, an attacker can have direct access to the victim host V. For external-internal network, a filtering device (i.e., firewall) is placed to disallow the connectivity to a subset of addresses and ports in either direction. Here, an attacker can directly connect herself to host B only. Other hosts, i.e. host B, C, D, and V can connect to each other. Finally, in external-DMZ-internal network model, two filtering devices viz. a DMZ filtering device (to filter external

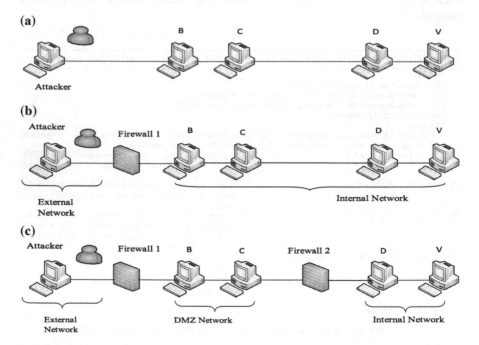

Fig. 1 Network models [4]. **a** Flat network model. **b** External-internal network model. **c** External-DMZ-internal network model

network connections that are destined for DMZ network) and an internal filtering device (to filter DMZ connections, which are destined for internal networks) are used. Here, an attacker can directly connect herself to host B only. Host B is directly connected to host C. Host C is directly connected to host B, D, and V. Host D can directly connect to host V. Host V can connect directly to host C and D.

Hosts with vulnerability are added to each of the network models in incremental fashion and then a goal oriented exploit-dependency attack graph with unique node and edge labels is generated using SGPlan [14]. Here the nodes correspond to either exploits or security conditions (pre-conditions or post-conditions) of vulnerabilities [15]. Hosts can connect to each other on a single port (i.e., port 80) and the number of vulnerabilities considered, are modified for this port only. Here we assume that each vulnerability is remotely exploitable and reachability is the precondition for vulnerability exploitation. Successful exploitation of remote vulnerability gives an attacker a capability to execute/run arbitrary program on a compromised host. The issue of locally exploitable vulnerabilities is not considered here and we left this as a part of future work.

The enterprise network at time t is represented by an attack graph G_1. An attack graph for the same network is generated again at time $t + \Delta t$. This newly generated attack graph i.e. G_2 represents the network attack surface of an enterprise network at time $t + \Delta t$. Here Δt is some arbitrary sampling interval. It is an important parameter in enterprise network security monitoring and it defines how often an attack graph is constructed for the given enterprise network and hence how often network security measurement is taken. It determines the length of the time required for gathering vulnerability information, network configuration details and time required for construction of an attack graph as well. Δt governs the type of attacks can be detected. Δt could be static or variable, slow or fast. It is ideal to generate the attack graph for an enterprise network when new host joined the network or new applications installed over the networked host or some services enabled or disabled. The selection of Δt should be carefully done and it relies on the expertise of the security administrator. It should be selected in such a way that any security change that may occur will be detected within an acceptable period.

For a given window of time W, various attack graphs are generated (at discrete instant of time based on Δt) for a given enterprise network. This generates a sequence of attack graphs. Each graph in a series is compared with immediate predecessor using graph similarity measures proposed in this paper. Results obtained from similarity measures indicate the amount of change occurred in the attack surface of an enterprise network over the time interval Δt. The graph features like, nodes, edges that successfully characterize the variation or change in the network attack surface can be used as a measurement variable. Such variables are sensitive to changes in an attack surface and can be mapped very easily to the external causes and hence leads to an event detection capability [16]. The external causes may be the introduction of vulnerable host, installation of vulnerable software, disabling or enabling of services, failure of security countermeasures such as firewall, IDPS etc. We have combined these variables in our proposed metrics so that single series or sequence analysis can be utilized.

Proposed graph similarity measures such as *MCS* and *GED* are applied to each graph in a sequence for each network model. This results in a new series or sequence of numbers whose value represents a measure of change in the network attack surface during the period Δt. We used number of common nodes and common edges in a successive attack graphs in a graph sequence as a feature for measuring graph dissimilarity. By observing changes in the attack surface continuously over successive time interval, the graph-similarity measures (i.e., *MCS* and *GED*) provide a trend in the dynamic security behavior af the enterprise network as it evolves over time [16]. If an attack graph with larger change in network attack surface is detected, then administrator has to examine each node of an attack graph in order to find external causes of the problem.

5 Sensitivity Analysis

As shown in Fig. 2, along the X-axis there is an incremental assignment of remotely exploitable vulnerabilities to the different hosts. The vulnerability assignment to the host is given in the following format: jH, where j represents the remotely exploitable vulnerabilities on host H. We have taken an incremental approach of network security evaluation. In each step, a vulnerable host or vulnerability on host is added to the each network model and goal-oriented attack graph is generated and evaluated for all proposed metrics. Figure 1 shows the scenario when all hosts have joined the network in each network model. As shown in Fig. 2, the abrupt change in

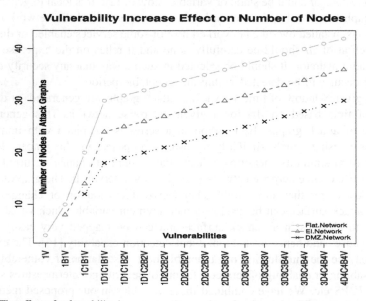

Fig. 2 The effect of vulnerability increase on the number of nodes under different network models

the parameter of the series i.e., number of nodes in an attack graphs, occurs on the order of sampling interval Δt.

The use of firewall(s) before the target host stops most of the attacks and hence results in an attack graph of smaller size (in terms of nodes and edges). It means the proposed metrics provides a quantitative support for the well-known defense in depth network protection strategy. Furthermore, by looking at the values for External-Internal and DMZ network models in Fig. 2, the values are not defined until the attacker can exploit enough vulnerabilities to reach the target. Such kind of observation suggests another strategy for protecting vulnerable hosts. The strategy says, if a host in an enterprise network is known to be vulnerable or more vulnerable than other hosts then access to the machine should be reduced. Such a countermeasure is applied by closing the ports or reconfiguring the service so that the effort required to access the service is increased.

Figures 3 and 4 show the mcs-based distance metric and graph edit distance metric, respectively for different vulnerability distributions in the network. Irrespective of the network model, the above metrics show sudden increase in the size of an attack graph when new host joins the network. This is because of the hosts being in the same protection domain and having the ability to start an attack from any other host. The proposed metrics are sensitive enough to detect the introduction of new exploitable vulnerabilities or vulnerable hosts in the network. And hence, the proposed metrics can be used as a useful tools in monitoring security of an enterprise network.

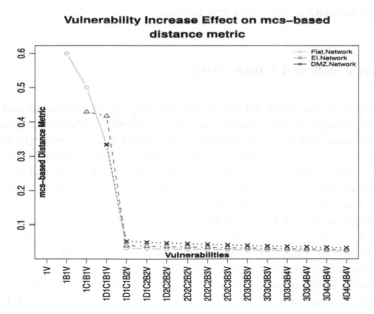

Fig. 3 The effect of vulnerability increase on the maximum common subgraph-based distance metric (*MCS*) under different network models

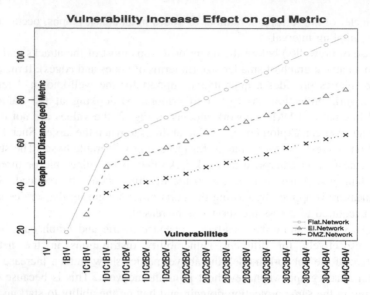

Fig. 4 The effect of vulnerability increase on the graph edit distance (*GED*) metric under different network models

From the temporal analysis standpoint, each network model in Figs. 2, 3, and 4 shows typical behavior. An analyst has to identity the external causes of the change in network attack surface, and then redimensioning of the network is done for providing adequate security when it is needed.

6 Conclusion and Future Work

In this paper, we have explored the possible use of graph-based distance metrics in measuring changes in the security level of dynamically evolving enterprise networks. We used maximum common subgraph (*MCS*) [6], and graph edit distance (*GED*) [7], two classical graph similarity measures as an indicator of enterprise network security changing over time. Experimental result show that the graph similarity measures are efficient and capable of detecting change in network attack surface in dynamic enterprise networks.

Our future work will be on producing graph-based metrics that can detect the regions of change in an attack graphs as a result of change in computer network. Such metrics will draw administrator's attention to the region of a graph that changed most and hence to the more significant network events that hurts security of a network. Such metrics will improve the administrator's capability in evaluating a network's security more efficiently and hence network hardening as well. Finally, our future work is to develop efficient means of computing the proposed metrics and to investigate complexity analysis of the proposed algorithm rigorously.

References

1. Phillips, C., Swiler, L.: A graph-based system for network vulnerability analysis. In: Proceedings of the 1998 workshop on New Security Paradigms (NSPW '98), pp. 71–79. ACM, New York, NY, USA (1998)
2. Ortalo, R., Deswarte, Y., Kaaniche, M.: Experimenting with quantitative evaluation tools for monitoring operational security. IEEE Trans. Softw. Eng. **25**, 633–650 (1999)
3. Li, W., Vaughn, B.: Cluster security research involving the modeling of network exploitations using exploitation graphs. In: Sixth IEEE International Symposium on Cluster Computing and the Grid (CCGRID''06). IEEE Computer Society, Washington, DC, USA (2006)
4. Idika, N., Bhargava, B.: Extending attack graph-based security metrics and aggregating their application. IEEE Trans. Dependable Secure Comput. **9**(1), 75–85 (2012)
5. Noel, S., Jajodia, S.: Metrics suite for network attack graph analytics. In: 9*th* Annual Cyber and Information Security Research Conference (CISRC). Oak Ridge National Laboratory, Tennessee (2014)
6. Bunke, H., Messmer, B.: A graph distance metric based on the maximal common subgraph. Pattern Recogn. Lett. **19**, 255–259 (1998)
7. Messmer, B., Bunke, H.: An algorithm for error-tolerant sub-graph isomorphism detection. IEEE Trans. Pattern Anal. Mach. Intell. **20**(5), 492–504 (1998)
8. Rinsen, K., Emmenegger, S., Bunke, H.: A novel software toolkit for graph edit distance computation. In: Kropatsch, W. et al. (eds.) GbRPR 2013. LNCS, vol. 7877, pp. 142–151. Springer, Berlin (2013)
9. Wang, L., Singhal, A., Jajodia, S.: Measuring the overall security of network configurations using attack graphs. In: 21st Annual IFIP WG 11.3 Working Conference on Data and Applications Security, vol. 4602, pp. 98–112. Springer, Berlin (2007)
10. http://www.facweb.iitkgp.ernet.in
11. Wang, L., Noel, S., Jajodia, S.: Minimum-cost network hardening using attack graphs. J. Comp. Comm. **29**(18), 3812–3824 (2006)
12. Kundu, A., Ghosh N., Chokshi I., Ghosh S.: Analysis of attack graph-based metrics for quantification of network security. In: 2012 Annual IEEE India Conference (INDICON) (2012)
13. Dickinson, P.J.: Matching graphs with unique node labels. J. Pattern Anal. Appl. **7**(3), 243–254 (2004) (Springer)
14. SGPlan 5. http://wah.cse.cuhk.edu.hk/wah/programs/SGPlan/
15. Beytullah, Y., Gurkan, G., Fatih, A.: Cost-aware network hardening with limited budget using compact attack graphs. In: 2014 IEEE Military Communications Conference (MILCOM), pp. 152–157, 6–8 Oct 2014
16. Showbridge, P., Kraetzl, M., Ray, D.: Detection of abnormal change in dynamic networks. In: Information, Proceedings of Decision and Control, 1999, (IDC'99), pp. 557–562 (1999)

A Graph-Based Chameleon Signature Scheme

P. Thanalakshmi and R. Anitha

Abstract Ordinary digital signatures are universally verifiable but they are unsuitable for certain applications that are personally or commercially sensitive. Chameleon signature introduced by Krawczyk and Rabin is non interactive and prevents the receiver of a signed document from disclosing its contents to third parties. Most of the efficient chameleon signature schemes are based on some known hard problems such as integer factorization and discrete logarithm. In this paper, a chameleon signature scheme is proposed and to the best of our knowledge it is the first such scheme using graph theory. It is a collision-resistant scheme that also satisfies the security requirements such as message-hiding, key exposure freeness and semantic security.

Keywords Directed graphs · Chameleon hashing · Chameleon signature

1 Introduction

Digital signature is a cryptographic primitive employed to provide document signer authentication, integrity and non-repudiation. It allows anyone in the system to verify the signature by using signer's public key. The universal verifiable property of digital signature is desirable for certain applications. But they are unsuitable for applications that are personally or commercially sensitive, as self authenticating signatures designed for such applications make it valuable to the industrial spy or extortion.

Suppose that an institution invites companies to propose authentic quotations for a project. No company involved in this process wants to reveal its quotation to other

P. Thanalakshmi (✉) · R. Anitha
PSG College of Technology, Coimbatore, India
e-mail: ptl@amc.psgtech.ac.in

R. Anitha
e-mail: anitha_nadarajan@mail.psgtech.ac.in

© Springer India 2016
A. Nagar et al. (eds.), *Proceedings of 3rd International Conference on Advanced Computing, Networking and Informatics*, Smart Innovation, Systems and Technologies 44, DOI 10.1007/978-81-322-2529-4_34

327

companies. So the quotations and the signatures of the companies has to be encrypted such that they can be read and verified only by the institution. Since the institution's goal is to obtain a low price, it could show some decrypted signed quotations to other companies to influence them in making low quotations. Here arises a conflict between authenticity and privacy. In the above scenario and in many other cases, there is a need to prevent the disclosure of contents to the third parties and provide non-repudiation in the event of legal disputes. Krawczyk and Rabin introduced chameleon signature in [9] that satisfies the contradictory requirements such as non-repudiation and non-transferability. The signature verification process is non interactive with less complexity compared to the undeniable signature in [3]. Hash-and-sign model is used to construct chameleon signatures. The signer authenticates a message by first computing the message digest using a chameleon hash function and then the hash value is signed by using a standard signing algorithm. Finding hash collisions in chameleon hash is computationally infeasible, but the recipient of the signature is able to find collisions by using his private key. Due to this situation, no third party would be ready to accept the signed information exposed by the recipient. In case of disputes, the judge can easily ascertain whether the proposed message signature pair is actually the original one committed by the signer or not. When another message is authenticated with the same signature of the original one, the contested signature is considered as a forged signature because such collision can be produced only by the recipient.

2 Related Work

The problem of key exposure limits the original chameleon signature scheme proposed by Krawczyk and Rabin [9]. It creates a hindrance for the recipient to forge signatures. Hence in [1] Ateniese and de Medeiros stated that the key exposure problem due to forgery threatens the claim of non transferability. As non-transferability is required in many applications, it is essential to build a chameleon signature scheme using a hash function which is free from key exposure. To address this type of problem, Ateniese et al. proposed an identity based chameleon hash function in [1]. In this scheme, the signature forgery by the recipient results the signer to recover the recipient's trapdoor informations related to that particular transaction. The informations obtained are not sufficient to deny signatures on any message in other transactions. Thus they provide a solution but do not solve the problem of key exposure completely. To overcome this drawback, Chen et al. in [4] proposed the first complete chameleon hash function which is key exposure free in the gap Diffie-Hellman group with bilinear pairings. Ateniese et al. in [2] proposed three non-pairing based key exposure free schemes out of which two schemes are based on RSA and factoring, and one is based on Strong Diffie-Hellman and Discrete Logarithm Problem(DLP). In [7], Wei Gao proposed a chameleon hash function based on factoring and proved its security by introducing a variant Rabin signature scheme and in [6] they proposed a DLP based chameleon hash function with Schnor signature and it is key exposure free. In [5] Chen et al. proposed a key exposure free chameleon hash and signature scheme based on DLP, without using the gap Diffie Hellman groups. In [10] Pan et al. proposed a family of chameleon hash functions and strongly unforgeable one-time signature schemes based on DLP over inner automorphism groups.

In this paper, we propose a graph based chameleon signature scheme. The unforgeability property of the proposed scheme relies completely on the unforgeability of the underlying digital signature scheme. The additional features of chameleon signature such as non-transferability and non-repudiation are derived from the underlying chameleon hash function. The proposed chameleon hash is an efficient key exposure free function constructed using graphs. Its security relies on a collision resistant one-way hash function and a pseudo random generator. A distinctive advantage of the resulting chameleon hashing scheme is in case of disputes the signer can efficiently prove whether the message was the original one or not. This property is known as message hiding or message recovery.

The rest of the paper is organized as follows. Section 3 contains the preliminaries needed. The proposed chameleon hashing scheme and chameleon signature scheme based on graph theory are given in Sect. 4 and the conclusion is given in Sect. 5.

3 Preliminaries

This section recalls the standard definitions for the collision resistant one-way hash function, pseudorandom generator and the graph terminology as in [8]. Also it introduces the definitions and chameleon hashing's security requirements in [1, 9].

3.1 Cryptographic Primitives

Collision-Resistant Hash Functions Let $\Gamma = \{H : R^2 \to R\}$ be a family of functions. If for a randomly chosen H in Γ, any adversary with computationally bounded resources (collision-finder) cannot find two messages m and m' such that $m \neq m'$ and $H(m) = H(m')$ except with negligible probabilities, then the family Γ is collision-resistant.

Pseudorandom Generators Consider a deterministic function $PG : R \to R^2$. If on a random input x, no adversary with computationally bounded resources (distinguisher) can distinguish a truly random value on R^2 from $PG(x)$ with non-negligible probability, then PG is said to be a pseudorandom Generator.

3.2   Graphs

A pair $G = (V, E)$ is said to be a directed graph, if V is a finite set of vertices and E is a set of ordered pairs of vertices representing directed edges. In G, a sequence of vertices $p = (v_0, \ldots, v_l)$ starting from v_0 and ending at v_l such that there is an edge between v_{j-1} to v_j for all $j = 1, \ldots, l$ is called a path of length $l \geq 0$. A directed graph without cycles is called a directed acyclic graph (*DAG*). In this paper for simplicity, only *DAG* with a single source $v_\perp$ and a single sink v^τ is considered. Let $n > 0$ be the number of interior vertices other than source and sink vertices. Throughout this work, a special type of *DAG* which has $v_\perp$ with outdegree 1, v^τ with indegree 1 and two types of interior vertices called expansion vertices and compression vertices. Expansion vertices are vertices with one incoming edge and 2 outgoing edges and compression vertices are vertices with 2 incoming edges and one outgoing edge. A cut in a graph G is a nontrivial partition $C = (S, V - S)$ of the vertices such that $v_\perp \in S$ and $v^\tau \in V - S$. For easiness, a cut can be represented by a single set of vertices (S) with the convention that if $v_\perp \in S$ then (S) represents $(S, V - S)$, while if $v^\tau \in S$ then (S) represents $(V - S, S)$. An edge $e = (u, v)$ crosses a cut $C = (S, V - S)$ if $u \in S$ and $v \in V - S$. The collection of edges crossing C is represented by $Edges(C)$. We label the edges in G from the set Z_n. A labeling is a partial function λ from E to Z_n i.e., a function $\lambda : T \to Z_n$ where $T \subseteq E$. The domain T of the labeling is denoted as $dom(\lambda)$. Let PG be a pseudo-random generator and H be a collision resistant hash function defined by $PG : Z_n \to Z_n^2$ and $H : Z_n^2 \to Z_n$ respectively. The labelling λ is said to be consistent, if the labelling of the edges satisfy the following conditions: (i) if $e_0 \in dom(\lambda)$ be an incoming edge to any expansion vertex and $e_1, e_2 \in dom(\lambda)$ be the outgoing edges from that vertex, then $PG(\lambda(e_0)) = (\lambda(e_1), \lambda(e_2))$ and (ii) if $e_0, e_1 \in dom(\lambda)$ be the incoming edges to any compression vertex and $e_2 \in dom(\lambda)$ be the outgoing edge from that vertex, then $H(\lambda(e_0), \lambda(e_1)) = \lambda(e_2)$. Independently we can label the edges of the cut C. $\sigma : C$ represents a labelling function σ defined on the $Edges(C)$. The set of all such labellings is denoted by $\{\sigma : C\}$.

For any ordered pair of cuts $C_1 \subseteq C_2$ and labeling function $\sigma : Edges(C_1) \to Z_n$, a unique extension labeling denoted by $Ext_{C_2}(\sigma : C_1)$ is defined as the labelling on the cut $\sigma : C_1$ is uniquely extended to $Edges(C_2)$ through a consistent labelling.

3.3   Chameleon Hashing and Its Security Requirements

A chameleon hash scheme is defined by a quadruplet of polynomial-time algorithms (*KG, Hash, UForge, IForge*).

KG (1^k): When the security parameter k is given as input to a probabilistic polynomial time algorithm, it produces a secret and a public key pair (sk, pk).

Hash: When a public key pk, a message m and an auxiliary random parameter r are given as input to a probabilistic polynomial-time algorithm, it outputs the hash value $h = Hash(pk, m, r)$.

UForge (universal forge): When the secret key sk, a message m and an auxiliary random parameter r are given as input to a deterministic polynomial-time algorithm, it outputs a pair (m', r') a collision for (m, r) i.e. $Hash(pk, m, r) = h = Hash(pk, m', r')$.

IForge (instance forge): When a tuple (pk, m, r, m', r') is given as input to a probabilistic polynomial time algorithm, it outputs another collision pair (m'', r'') that also satisfies $h = Hash(pk, m'', r'')$.

3.4 The Security Requirements of a Chameleon Hash Includes

Collision-Resistance With the only input (pk, m, r) there exists no efficient algorithm with non negligible probability can find a second pair (m', r') such that $Hash(pk, m, r) = h = Hash(pk, m', r')$.

Semantic Security Let $H[X]$ denote the entropy of a random variable X, and $H[X|Y]$ the entropy of the variable X given the value of a random function Y of X. Semantic security means that the conditional entropy $H[m|h]$ of the message m given its chameleon hash value h equals the total entropy $H[m]$.

Key Exposure Freeness If a recipient with the public key has never computed a collision, then given $h = Hash(pk, m, r)$ there exists no efficient algorithm in polynomial time can find a collision. It remains to be true even if the adversary has oracle access to *UForge* (sk,..) and allowed to make polynomially many queries on (m_i, r_i) of his choice, except the challenge query.

Message Hiding Using *UForge* algorithm suppose that a recipient has computed a collision pair (m', r'), then the signer without revealing any information of the original message can argue successfully that the pair generated by the recipient is invalid by releasing another collision pair (m'', r''). Moreover, the entropy of the original value (m, r) is unaffected by the revelation of the pairs $(m', r'), (m'', r'')$ and any further collisions $H[(m, r)|h, (m', r'), (m'', r'')] = H[(m, r)|h]$.

4 Proposed Graph Based Chameleon Hash and Signature Construction

Let G be a publically known *DAG* with a cut C. Consider a collision resistant hash function h defined by $h : M \rightarrow (Z_n - \{0\}, *)$ where M is the message space and $*$ be a multiplicative binary operation defined on $Z_n - \{0\}$. The proposed scheme is described below by the following polynomial time algorithms.

4.1 Chameleon Hash Construction

KG (1^k) : select a collision resistant hash function $H \in \Gamma$ and randomly choose $\sigma_s \in \{\sigma : \{\upsilon_\perp\}\}$ as the private key. Compute the public key
$\sigma_p = Ext_{\{\upsilon^\tau\}}(\sigma_s : \{\upsilon_\perp\})modn$ and output $pk = (H, \sigma_p), sk = (H, \sigma_s)$.

Hash: It is a protocol which runs between the signer and the designated receiver. For a message m, Alice randomly chooses a cut $C \in G, \sigma_{R_2} \in Z_n - \{0\}$, $\sigma_{R_{1(C)}} \in Z_n^l - \{0\}$(where l is the no. of edges crossing C) and computes $\sigma = Ext_{\{\upsilon^i\}}\left((h(m) + \sigma_{R_{1(C)}}) : C\right)modn$. Using the public key σ_p of Bob she computes σ_{CH} the chameleon hash value as $\sigma_{CH} = (\sigma + \sigma_p) * \sigma_{R_2}modn$ and outputs along $(m, \sigma_{R_{1(C)}}, \sigma_{R_2})$.

UForge: Using the secret key σ_s, the tuple $(m, \sigma_{R_{1(C)}}, \sigma_{R_2})$ and the message m' the verifier computes $\sigma_{s(C)} = Ext_{\{C\}}(\sigma_s : \{\upsilon_\perp\})modn$ and get the labelling for every edge i in the *Edges(C)* by computing $\sigma_{h_i} = \left(\sigma_{s(C)_i} + h(m) + \sigma_{R_{1(C)_i}}\right) * \sigma_{R_2}modn$. The verfier using $\sigma_{h_i}, \forall i \in C$ gets σ_h and verifies $\sigma_{CH} = Ext_{\upsilon^\tau}(\sigma_h : C)$ for $(m, \sigma_{R_{1(C)}}, \sigma_{R_2})$. With the known values $\sigma_{s(C)_i}$ and σ_{h_i}, for the message m' the verfier chooses σ'_{R_2} such that $gcd(\sigma'_{R_2}, n) = 1$ so that he can compute $\sigma'_{R_{1(C)_i}}$ by solving the linear congruence equation as $\sigma'_{R_{1(C)_i}} * \sigma'_{R_2} = \sigma_{h_i} - (\sigma_{s(C)_i} + h(m')) * \sigma'_{R_2}modn$ outputs a collision $(m', \sigma'_{R_{1(C)}}, \sigma'_{R_2})$ for the chameleon hash value σ_{CH}.

IForge: On input a pair of collision $(m, \sigma_{R_{1(C)}}, \sigma_{R_2}), (m', \sigma'_{R_{1(C)}}, \sigma'_{R_2})$, the signer first recover $\sigma_{s(C)_i}$ the secret labelling on the ith edge of the cut by solving the linear congruence equation

$$(\sigma_{s(C)_i} + h(m) + \sigma_{R_{1(C)_i}}) * \sigma_{R_2} \equiv (\sigma_{s(C)_i} + h(m') + \sigma'_{R_{1(C)_i}}) * \sigma'_{R_2}modn$$

Similarly solving l linear congruence equations, $\sigma_{s(C)}$ is computed. Next, as in *UForge*, with $\sigma_{s(C)}$ one can compute another pair $(m'', \sigma''_{R_{1(C)}}, \sigma''_{R_2})$ which has the same hash value σ_h which in turn gives σ_{CH}.

An example of *DAG* is depicted in Fig. 1 with the source vertex $\upsilon_\perp = 1$, the sink vertex $\upsilon^\tau = 16, V_H = \{2, 3, 4, 5, 6, 7, 8\}$ is the set of compression vertices and $V_G = \{9, 10, 11, 12, 13, 14, 15\}$ is the set of expansion vertices. Figure 1 depicts an example cut $S(C1) = \{0, 1, 2, 3, 4, 5, 6, 7, 8, 11, 12\}$ with $Edges(C1) = \{(5, 9), (8, 9), (5, 10), (6, 10), (11, 14), (12, 14)\}$. Consider $n = 10, H(x, y) \overset{def}{\leftarrow} x + y$, and $PG(x) \overset{def}{\leftarrow} (x, x)$. Let $\sigma_s = 2$ then $\sigma_p = 6$ and choose $h(m) = 4, \sigma_{R_2} = 7$, $\sigma_{R_{1(C)}} = ((5, 9), 3), ((8, 9), 3), ((5, 10), 3), ((6, 10), 3), ((11, 14), 6), ((12, 14), 6)$ then $\sigma = Ext_{\{vt\}}\left((h(m) + \sigma_{R_{1(C)}}) : C\right)mod10 = 8$ and $\sigma_{CH} = \left((\sigma + \sigma_p) * \sigma_{R_2}\right)mod10 = ((8 + 6) * 7)mod10 = 8$.

Fig. 1 *DAG*

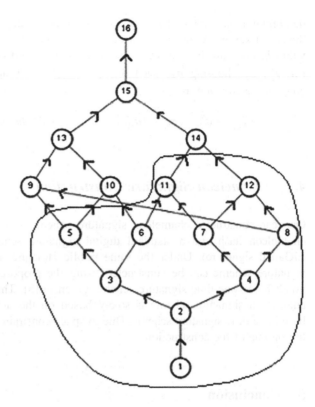

The security of the above chameleon hash scheme is discussed below:

Theorem 1 The chameleon hash scheme proposed satisfies collision-resistant, semantic security, message hiding and key exposure free properties.

Proof **Collision-resistance and key-exposure-freeness**: Exposing a collision allows any body to extract the trapdoor labeling $\sigma_{s(C)} : C$, the extended labeling of the secret key σ_s. As $\sigma_{s(C)}$ is the preimage of σ_p, computing collisions results a contravention to the preimage resistance property of the hash function H. Hence we conclude that finding collisions is hard without the knowledge of the trap-door labeling $\sigma_{s(C)}$. Thus the scheme is secure from key exposure as one-way hash function is a preimage resistant.

Semantic security: The message m and the the values $\sigma_{R_{1(C)}}, \sigma_{R_2}$ uniquely determines the hash value as $h = Hash(m, \sigma_{R_{1(C)}}, \sigma_{R_2})$. As m and $\sigma_{R_{1(C)}}, \sigma_{R_2}$ are independent variables, we obtained the conditional probability as $c(m|h) = c(m|\sigma_{R_{1(C)}}, \sigma_{R_2})$ and $c(m|\sigma_{R_{1(C)}}, \sigma_{R_2}) = c(m)$. The conditional probability $c(m|h) = c(m)$ infers that the chameleon hash value h discloses no information about the message m.

Message hiding: Suppose a pair $(m', \sigma'_{R_{1(C)}}, \sigma'_{R_2})$ that forms a collision for the hash value h is exposed. It allows anyone in the system to compute a de-commitment to h under any message m'' of her choice, without the knowledge of the secret key σ_s. As in IForge, using (m', r') and (m, r) that results the same hash value h, one can find the trapdoor labeling $\sigma_{s(C)}$ and compute another collision (m'', r'') by choosing σ_{R_2}'' such that $gcd(\sigma_{R_2}'', n) = 1$ and by solving the linear congruence equation as

$$\sigma_{R_{1(C)_i}}'' * \sigma_{R_2}'' = \sigma_{h_i} - (\sigma_{s(C)_i} + h(m'')) * \sigma_{R_2}'' \bmod n \text{ to obtain } \sigma_{R_{1(C)_i}}''.$$

$\square$

4.2 Chameleon Signature Construction

One can construct a chameleon signature scheme as in [9] by using the above chameleon hash and a standard digital signature scheme(such as RSA, DSS, ElGamal signature). Under the same public structure a graph based chameleon signature scheme can be constructed using the proposed chameleon hash and a graph based one-time signature scheme given in [8]. The security of graph based chameleon signature scheme is solely based on the unforgeability of the graph based one-time signature scheme. Due to space constraint, we mentioned here only the outline of the construction.

5 Conclusion

In this paper a new graph based chameleon hash scheme is proposed. The desired security requirements of chameleon hash such as collision-resistant, semantic security, message hiding and key exposure freeness are proved. Also a graph based unforgeable chameleon signature scheme can be constructed with a graph based chameleon hash and a graph based one-time signature scheme.

References

1. Ateniese, G., de Medeiros, B.: Identity-based chameleon hash and applications. In: Financial Cryptography, pp. 164–180. Springer (2004)
2. Ateniese, G., de Medeiros, B.: On the key exposure problem in chameleon hashes. In: Security in Communication Networks, pp. 165–179. Springer (2005)
3. Chaum, D., Van Antwerpen, H.: Undeniable signatures. In: Advances in Cryptology—CRYPTO'89 Proceedings, pp. 212–216. Springer (1990)
4. Chen, X., Zhang, F., Kim, K.: Chameleon hashing without key exposure. In: Information Security, pp. 87–98. Springer (2004)
5. Chen, X., Zhang, F., Tian, H., Wei, B., Kim, K.: Discrete logarithm based chameleon hashing and signatures without key exposure. Comput. Electr. Eng. 37(4), 614–623 (2011)

6. Gao, W., Li, F., Wang, X.: Chameleon hash without key exposure based on schnorr signature. Comput. Stand. Interfaces **31**(2), 282–285 (2009)
7. Gao, W., Wang, X.L., Xie, D.Q.: Chameleon hashes without key exposure based on factoring. J. Comput. Sci. Technol. **22**(1), 109–113 (2007)
8. Hevia, A., Micciancio, D.: The provable security of graph-based one-time signatures and extensions to algebraic signature schemes. In: Advances in Cryptology—ASIACRYPT, pp. 379–396. Springer (2002)
9. Krawczyk, H.M., Rabin, T.D.: Chameleon hashing and signatures (2000), US Patent 6,108,783
10. Pan, P., Wang, L., Yang, Y., Gan, Y., Wang, L., Xu, C.: Chameleon hash functions and one-time signature schemes from inner automorphism groups. Fundamenta Informaticae **126**(1), 103–119 (2013)

Design of ECSEPP: Elliptic Curve Based Secure E-cash Payment Protocol

Aditya Bhattacharyya and S.K. Setua

Abstract The present scenario in the area of e-commerce the most popular term is E-cash. E-cash is developed to allow fully anonymous secure electronic cash transfer to support online trading between buyers and sellers. E-cash transfer system has helped us to make transaction electronically. In this paper we propose an elliptic curve based secure e-cash payment protocol. The proposed system secures the transactions not only by the nature of the curve but also makes use of the hash function to enhance the desired security measure. It also ensures mutual authentication, anonymity, non-repudiation and traceability of the users.

Keywords Elliptic curve cryptography (ECC) · Payment gateway · E-commerce · E-cash

1 Introduction

E-commerce has become the current trends of internet technology where customer privacy is the main concern. To maintain customer's privacy Blind signature schemes are used in handling e-cash. If we want to prevent the unauthorized access of e-cash, we need a scheme which can trace out the pretender.

We have used the concept of Abad-Peiro's [1] general payment model where the entities are

Customer (C): A user who will buy the products.
Merchant (M): An entity who sells the products.

A. Bhattacharyya (✉)
Vidyasagar University, Midnapore, West Bengal, India
e-mail: adityamcavu@gmail.com

S.K. Setua
University of Calcutta, Kolkata, West Bengal, India
e-mail: sksetua@gmail.com

© Springer India 2016   337
A. Nagar et al. (eds.), *Proceedings of 3rd International Conference on Advanced Computing, Networking and Informatics*, Smart Innovation, Systems and Technologies 44, DOI 10.1007/978-81-322-2529-4_35

Issuer (I): Bank in which the customer has an account
Acquirer (A): Bank in which the merchant has an account

Payment Gateway (PG): Acts as an interface between the acquirer and issuer for the purpose of payment clearance with reliability, speed and security.

To make e-commerce more secured and fast, numerous protocols have been proposed in this field. Chaum et al. [2] proposed the first untraceable electronic system in the year 1982. After that, various technologies and protocols have been proposed. Bellare et al. [3] proposed iKP secure electronic payment system. Increasing the value of i from 1 to 3, the security requirements met by iKP increases. Kungspisdan's contribution in the area of mobile payment protocol revolutionizes the area [4]. They proposed a new secure lightweight mobile bill payment protocol in simpler way. Tellez contributed a client centric anonymous payment model PCMS [5]. They implemented their protocol on NOKIA N95 through PCMS wallet and they claimed that the protocol preserves the security. Beside this Xing et al. [6], Liu and Huang [7], Raulynaitis et al. [8] have enriched the research area.

For its shorter key length and higher security in compare to other public key cryptosystem, ECC plays an important role. As online e-cash systems demands huge involvement of bank in respect to the time and space, it is best option to implement ECC in e-cash.

Again a cryptographic hash function is a function that makes impossible to get back the original message from digest. In security aspect, it supports digital signature, message authentication code (MAC). Beside this, the characteristics of hash (impossible to modify the message without changing the hash, infeasible to find two different messages with the same hash) makes it popular among the cryptographers. It is very easy to hash deterministically and efficiently into an elliptic curve. With the hardness of discrete logarithm problem of elliptic curve and hash function, any system seems to be infeasible to attack.

In this paper, we propose an efficient payment protocol. Our protocol satisfies the requirement for a secure e-cash transfer system like integrity, authentication, confidentiality, authorization and non-repudiation. Side by side the lower memory consumption and higher computational performance of Elliptic curve cryptography makes our protocol efficient and useful too.

Rest of the paper is organized as follows. In the next section a brief introduction on the elliptic curve cryptography and one way hash function is given. Description of our model is given in the Sect. 3. Section 4 analyzes security and performance of the proposed protocol. In Sect. 5, we conclude our works.

2 Elliptic Curve Cryptography and One Way Hash

ECC was introduced independently by Miller and Koblitz [9]. Since then many implementation of it have been proposed. The discrete logarithm problem is the basis for security for EC cryptosystem. The Weirstrass equation of an elliptic curve

Fig. 1 Elliptic curve point addition

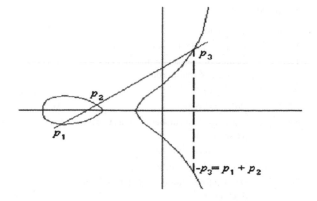

be $y^2 \bmod p = x^3 + ax + b \bmod p$ where $4a^3 + 27b^2 \neq 0$, where O is the point at infinity.

There are two basic operations performed on the curve-Point Addition and Point Doubling. Different values for a and b taking from finite field, Fp, can construct different elliptic curve. As for example, let the two distinct points be $P_1 (X_1, Y_1)$ and $P_2 (X_2, Y_2)$ where $P_1 \neq \pm P_2$ (Fig. 1). The addition of two points according to chord-and-tangent rule will give the following results:

Let $P_1 + P_2 = P_3 ((X_1 + Y_1) + (X_2 + Y_2) = (X_3, Y_3))$
where

$S = (Y_1 - Y_2)/(X_1 - X_2) \pmod{p}$
$X_3 = S^2 - X_1 - X_2 \pmod{p}$ and
$Y_3 = -Y_1 + S (X_1 - X_3) \pmod{p}$

Again let P ($P \in E(F_p)$) be a point of order n. Given $Q \in E(F_p)$, then for any integer l, discrete logarithm problem can be formed where $0 \le l \le n - 1$ such that $Q = l\,P$. To secure our system we not only depend upon Elliptic Curve, we considered Hash onto the Elliptic Curve. Let a function into an E, we can define L as the maximum size of $f^{-1}(P)$ where P is the point on E. Then $L = \max_{P \leftarrow E} (|f^{-1}(P)|)$ for any encoding function $f_{a,b}$, we can show that $L \le 4$. In our construction, we set up a one way hash function H(m) into the curve $E_{a,b} (F_q)$ as $H(m) = f(h(m))$ where $h:\{0,1\}^* \rightarrow Fq$.

Notations: The symbols used in our protocol are given in the following:

Notations	Description	Notations	Description
P	Point on the curve	r,s	Session keys used to encrypt the message
kP	Multiplication of point P with scalar k	PRequest	Payment request
p	A large prime number	E	Encrypted form of message
q	A large prime number such that $q\|\#E(Fp)$	Pack	Payment acknowledge

(continued)

(continued)

Notations	Description	Notations	Description
T_c	Temporary ID of customer/client	ASrequest	Amount subtraction request
TID	Identity of transaction that includes time and date of transaction	ASack	Amount subtraction acknowledge
ID_x	Identity of X	ACresponse	Amount claim response
OD	Order description	ACack	Amount claim acknowledge
n	Nonce/random number to protect against replay attack	PG	Payment gateway
Price	Cost/payment amount	TC	Type of card (debit/credit)
h()	One way hash function	CKx	Time at the clock used by X
OI	Order information contain TID, OD, h (OD, price)	$KS_{A\text{-}B\ t}$	Shared the key between A to B for t bit shifting
$K_{C\text{-}M}$	Secret share between client and merchant	TS	Transaction status (whether accepted or rejected)

3 Our Protocol

I. Registration and Payment Initialization

The registration phase involves between the customer and payment gateway. For a particular payment system apart from customer and payment gateway, the presence of merchant M is important. The client and merchant exchanges the messages through PG.

I. C sends the TID with its identity, a shared key $K_{C\text{-}M}$, a nonce n with a sessional key r to encrypt the total message towards the PG.

II. The PG then sends the message to M. M receives the request with the session key S and sends back its own identity ID_M with the hashed message to PG.

III. The PG forwards the message to C.

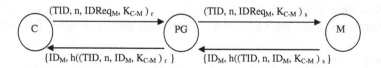

II. Payment Request and Amount Subtraction Request

Client C creates payment request with OI to the I through an encrypted text by which only I can identify the customer and his/her transaction.

The amount subtraction request bears the customer's identity i.e., details of his/her account. The request is executed by PG which is directly linked with issuer bank. The AS request includes price, h(OI), CK, TC and $h(KS_{C-IZ})$ in encrypted manner with the key KS_{C-I}. Now the request is sent to the PG.

III. Payment Request Forwarding to Merchant

A payment confirmation message is then send to M by PG for getting back the confirmation. The shared key helps to encrypt the message.

$$PG \xrightarrow{\quad E\{(T_c,TID,ID_M,n),\ Price,\ h(ID_M,OI),\ CK_c,\ KS_{PG-M}\}\quad} M$$

IV Acknowledge the Request

In this phase, M validate the payment request message by decrypting it with the shared key for validating the data OI, CK_c and AS request.

The message is then send by M to PG with ACrequest. It contains CK_M, h(OI), order's amounts, I's identity and TC with encrypted manner to the PG with key shared KS_{M-PG}.

$$M \xrightarrow{\quad ACrequest = E\{TS,T_c,\ n,\ CK_M,\ KS_{M-PG},\ Price,\ h(OI)\}\quad} PG$$

V. Inter Banking Transfer

On receiving the $AC_{request}$, PG checks whether the response is rejected or accepted with n and TID. Here timeliness is also verified. Then ASrequest is prepared which contains the ID_M, TID, OI, T_c and shared key KS_{C-I} and forwarded to the I. Beside this a message is sent to A with necessary information, OI, ID_M and the price that will be transferred to M's account.

When A receives the price, it calculates price ∥nonce ∥accept/reject and checks it with PRequest. If it is true, a AC_{ack} message is sent to PG.

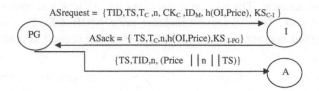

VI. Payment Response

PG generates the ACresponse which contains TS and. the response is encrypted with KS_{PG-M} and this response is sent to M.

VII. Value Subtraction Response

PG generates the payment acknowledgement (Pack) and sent it to I. The Pack is encrypted with KS_{PG-I}. On confirming the integrity, I transfers the ASack to C. After the completion of transaction, KS_{C-M}, KS_{C-I}, KS_{PG-M}, KS_{PG-C} are kept in the revocation list to prevent the replay attack.

4 Security Analysis

We analyze the efficiency and different aspects of security of our proposed protocol in this section.

4.1 Mutual Authentication

In the inter-banking transfer phase authentication has been done in between the customer and the issuer. If an adversary wants to steal the information, it is quite impossible to him/her to solve k from the equation $Q = kPh(M)$, where M is the message, because the discrete logarithm problem needs to be solved.

4.2 Anonymity

In our protocol, the client use a nickname i.e., temporary id, Tc instead of his/her real identity. So neither PG nor M can map the true identity of client.

4.3 Non-repudiation

Since the session KS_{C-I}, KS_{C-M} is only generated by client or issuer in our protocol, not by merchant, the merchant can provide a non-repudiable evidence that the client has send a request to M during payment initialization and payment request phase.

4.4 Replay Attack

The usage of different shared session key for every transaction prevents the replay attack. The attacker can't replay the messages without the shared key. Again, the timestamp which is generated in message ensures that replay attack is infeasible to it.

4.5 Password Guessing Attack

An attacker intercepts the payment request and amount claim response and tries to analyze it. He/she may guess h(OI). But the real problem is that the adversary can't compute k from the equation of the elliptic curve Q = kP h(OI), because he/she needs to solve the ECDLP which is impossible to guess.

4.6 Traceability/Double Spending

If any customer illegally performs the same transaction twice through the PG, this illegal transaction can be traced out by verifying T_c, TID.

Table 1 Storage space comparison for different e-cash system (in bits)

Phase	Proposed	Au et al. [11]	Chou et al. [10]	Canard [12]
Payment	800	5188	1184	30,740
Withdrawal/amount subtraction	1060	8160	2208	6420

5 Analysis

We know that each pairing takes five 512-bit multi-based exponentiations [10]. We compare our protocol with different e-cash system with respect to the storage space for different operations. Let the size of the point P be 160 bits, p be 160 bits, q of 160 bits and the size of each party's ID be 20 bits, then our protocol costs only 800 bits for total size of message which is shown in Table 1. For example, the customer sends the following message to the merchant in payment phase:

$$\text{TID}, \text{ID}_M, \text{Price}, n, h(\text{OD}, \text{Price}) = 160 + 160 + 80 + 160 + (160 + 80)$$
$$= 800 \text{ bits.}$$

6 Conclusion

In this paper we proposed a secured hash integrated elliptic curve based e-cash payment protocol. The following security features: mutual authentication, anonymity, non-repudiation, traceability are satisfied successfully by our system. Beside this, we compared our protocol with different existing e-cash payment systems. The result demands that our protocol is more efficient than others.

References

1. Abad-Peiro, J.L., Asokan, N., Steiner, M., Waidner, M.: Designing a generic payment service. IBM Syst. Res. J. **37**(1), 72–88 (1998)
2. Chaum, D.: Blind signature for untraceable payments. In: Proceedings of Eurocrypt'82, Plenum, New York, pp. 199–203 (1983)
3. Bellare, M., Garay, J.A., Hauser, R., Herzberg, A., Krawczyk, H., Steiner, M., Tsudik, G., Herreweghen, E.V., Waidner, M.: Design, implementation, and deployment of the ikp secure electronic payment system. IEEE J. Sel. Areas Commun. **18**(4), 611–627 (2000)
4. Kungpisdan, S.: Design and analysis of secure mobile payment systems. PhD thesis, Monash University (2005)
5. Tellez, J., Sierra, J.: Anonymous payment in a client centric model for digital ecosystem. In: IEEE DEST, pp. 422–427 (2007)

6. Xing, H., Li, F., Qin, Z.: A provably secure proxy signature scheme in certificate less cryptography. Informatica **21**(2), 277–294 (2010)
7. Liu, J., Huang, S.: Identity-based threshold proxy signature from bilinear pairings. Informatica **21**(1), 41–56 (2010)
8. Raulynaitis, A., Sakalauskas, E., Japertas, S.: Security analysis of asymmetric cipher protocol based on matrix decomposition problem. Informatica **21**(2), 215–228 (2010)
9. Koblitz, N.: Elliptic curve cryptosystems. Math. Comput. **48**, 203–209 (1987)
10. Chou, J.S., Chen, Y.-L., Cho, M.-H., Sun, H.-M.: A novel ID-based electronic cash system from pairings. In: Cryptology ePrint Archive, Report 2009/339 (2009). http://eprint.iacr.org/
11. Au, M., Susilo, W., Mu, Y.: Practical anonymous divisible e-cash from bounded accumulators. In: Proceedings of Financial Cryptography and Data Security, LNCS, vol. 5143, pp. 287–301. Berlin (2008)
12. Canard, S., Gouget, A.: Divisible e-cash systems can be truly anonymous. In: Proceedings of EUROCRYPT 2007, Lecture Notes in Computer Science, vol. 4515, , pp. 482–497. Springer, Berlin (2007)

Security Improvement of One-Time Password Using Crypto-Biometric Model

Dindayal Mahto and Dilip Kumar Yadav

Abstract In many e-commerce systems, to counter network eavesdropping/replay attack, OTP concept has been used. However if the OTP itself gets attacked and then there might be possibility of attacking the account of the legitimate client too. This paper proposes a model for improving the security of OTP using ECC with iris biometric for an e-commerce transaction. This model also offers improve security with shorter key length than the RSA and also avoids to remember the private keys as the private keys are generated dynamically as and when required.

Keywords One-time password (OTP) · Elliptic curve cryptography (ECC) · Biometrics · Iris

1 Introduction

ECC being an asymmetric cryptography has a limitation of managing the private keys. A private key of the client must be confidential, non-sharable, and safe. Due to rapid breakthrough in cryptanalysis, there is a demand of bigger cryptographic key length that again leads to a difficulty of memorize such cryptographic keys. Hence such keys must be stored somewhere, which may become vulnerable for security of keys. In order to overcome the above difficulty of managing keys, this paper suggests to generate the keys using iris biometric templates, such keys are generated dynamically as and when client needs. Finally the generated cryptographic keys are used to implement ECC for OTP security.

D. Mahto (✉) · D.K. Yadav
Department of Computer Applications, National Institute of Technology,
Jamshedpur 831014, Jharkhand, India
e-mail: dindayal.mahto@gmail.com

D.K. Yadav
e-mail: dkyadav.ca@nitjsr.ac.in

© Springer India 2016   347
A. Nagar et al. (eds.), *Proceedings of 3rd International Conference
on Advanced Computing, Networking and Informatics*, Smart Innovation,
Systems and Technologies 44, DOI 10.1007/978-81-322-2529-4_36

This paper is organized as follows. Section 2 describes the related works and literature reviews. Section 3 describes OTP. Section 4 describes ECC. Section 5 describes iris biometric. Section 6 explains the proposed model., and Sect. 7 describes conclusion.

2 Related Works and Literature Reviews

Many researchers have illustrated the implementation of cryptographic models using keys generated from biometric traits. A brief review of few popular papers is presented here. Hao et al. [6] have illustrated the implementation 128-bit AES cryptography model using iris based biometric cryptographic keys. This research paper generates genuine iris codes first, and then a re-generate-able binary digits known as biometric key gets created of up to 140 bits. Janbandhu et al. [8] suggested a technique to generate private key for RSA based digital signature model using a 512 byte iris template. This paper generates a larger number, based on iris template till the number becomes eligible for co-prime with Euler Totient Function. Once the number gets generated, then the number becomes private key for the client. Some of the other suggested approaches related to the crypto-biometrics are given by [3, 10, 11, 13].

3 One-Time Password (OTP)

OTP is a mechanism to counter eavesdropping/replay attack occurs during data communication between two parties in a network [5]. However an OTP itself needs its security during transmission in Client/Server architecture.

4 ECC

ECC was proposed by two different researchers (Koblitz [9] and Miller [12]) in late 1985. ECC is considered to be a competitor for RSA algorithm. Due to ECC's complex numerical calculation, it provides better security per bit with smaller key length than RSA. The security level of ECC-160 and ECC-224 are equivalent to the security level of of RSA-1024 and RSA-2048 respectively [4].

5   Iris Biometric

Iris of an eyeball is a colored circular boundary that appears in the outer part of the pupil. Due to its distinctiveness, large amount and non-counterfeiting [6] texture pattern [2], iris compared to other biometrics traits provides highly reliable and accurate user identification method. Iris code is generated after finding the surrounded boundary between the iris and pupil portions and outer boundary between the iris and sclera portions of the eyeball's image. Here iris is localized as done by [1, 6, 7] and then the localized feature in turn generates the iris code.

6   Proposed Model

The Fig. 1 shows the proposed model. This model uses iris traits of bank clients for generating their cryptographic keys, then the keys are used in ECC to provide data communication security while sending the OTP from Bank Transaction Server to client. At the client-end, when client decrypts and gets plain-OTP, then the he enters Plain-OTP in OTP Input Screen, which in turn forwards the Plain-OTP to Encryption Module, which generates server's temporary public key of desired length based on Plain-OTP. This Encryption Module encrypts the Plain-OTP using ECC with recently generated server's public key. At the server-end, server

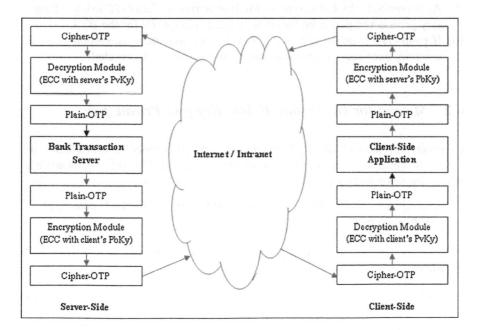

Fig. 1 Architecture of the proposed model

generates its own temporary private key of desired length based on original Plain-OTP. If Decryption Module able to decrypt the incoming Cipher-OPT using ECC with its recently generated private key and matches the decrypted Plain-OTP with transaction server's original Plain-OTP, then monetary transaction gets successfully executed, otherwise monetary transaction gets canceled.

6.1 Steps of the Proposed Methodology

1. Bank Transaction Server (BTS) generates plain-OTP.
2. ECC Encryption Module receives plain-OTP and generates cipher-OTP with of client's public key (based on iris code).
3. Cipher-OTP gets forwarded over Internet/Intranet to the client's mobile.
4. At Client-Side, Client mobile gets and forwards the cipher-OTP to ECC Decryption Module.
5. The ECC Decryption Module generates plain-OTP with of client's private key (based on iris code).
6. Client enters the recently retrieved plain-OTP in the OTP Input Screen on Client-Side Application (CSA), which in turn forwards the plain-OTP to ECC Encryption Module.
7. ECC Encryption Module generates cipher-OTP using a temporary public key generated based on entered plain-OTP for the BTS.
8. The cipher-OTP gets forwarded through Internet/Intranet Server-Side.
9. At Server-Side, ECC Decryption Module retrieves plain-OTP using a temporary private key generated based on original plain-OTP for the BTS.
10. If plain-OTP matches with original-OTP, then monetary transaction gets successfully executed, otherwise monetary transaction gets canceled.

6.2 Method for Generating Public Key and Private Key

Cryptographic keys of client get generated using hash function with client's iris code. The generated hash value becomes private key (d_A) for the client known as Alice. Steps for Alice's public key generation is as follows:

1. Both Alice and BTS select a big prime number 'p' and the ECC parameters 'a' and 'b' such that

$$y^2 \ mod \ p = (x^3 + ax + b) \ mod \ p; \tag{1}$$

where,

$$4a^3 + 27b^2 \neq 0 \tag{2}$$

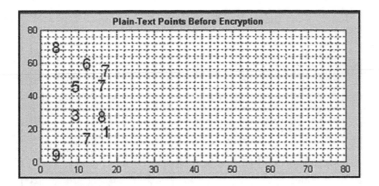

Fig. 2 Sample data points of the input message [11]

2. Base point: G(x, y) gets selected from the elliptic curve.
3. Calculate client's public key: $P_A = d_A * G(x, y)$

6.3 OTP Message Encryption

BTS generates plain-OTP as msg to be sent as points P_{msg} (x, y) as shown in the Fig. 2. These points are encrypted as a cipher-OTP and later same are decrypted. Steps for encryption are given below:

1. ECC Encryption Module generates $P_{msg} = (x, y)$ from msg.
2. This Module selects a global variable: h = 20, Calculate, X = msg $*$ h $+$ i; takes the value of i from 1 to h $-$ 1 and keeps on trying to get an integral value of y. Thus, msg is generated as the form of point(x, y). The degeneration of msg = floor((x $-$ 1)/h). The cipher-OTP is a collection of two points: $C_{msg} = (h * G, P_{msg} + h * P_A)$ ECC Encryption Module forwards this cipher-OTP to Alice as shown in the Fig. 3.

6.4 OTP Message Decryption

1. Alice multiplies the point-1 with its private key and then subtract the resultant point from point-2:

$$= (P_{msg} + h * P_A - d_A * h * G) = (P_{msg} + h * (d_A * G) - d_A * (h * G))$$
$$= P_{msg}$$

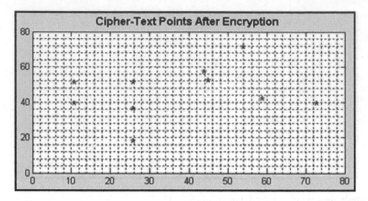

Fig. 3 Encrypted data points [11]

2. Finally the decrypted-OTP is P_{msg} as shown in the Fig. 2.

The procedure for encryption and decryption of Client-Side's plain-OTP is performed as mentioned in proposed model section.

7 Conclusion

This paper proposed an improvement of OTP security using ECC with iris biometric. ECC offers better security with lesser key length. Iris biometric offers highly reliable and accurate user identification method due to its distinctiveness, large amount and non-counterfeiting texture pattern. Now a days many e-commerce firms doing monetary transaction with OTP. The OTP is not secure and completely dependent on the SMS providers. The proposed model mitigates the demerits of the present OTP based e-commerce transaction.

References

1. Bakshi, S., Mehrotra, H., Majhi, B.: Real-time iris segmentation based on image morphology. In: Proceedings of the International Conference on Communication, Computing and Security. ACM, pp. 335–338, (2011)
2. Daugman, J.: New methods in iris recognition. IEEE Trans. Syst. Man Cybern. Part B Cybern. **37**(5), 1167–1175 (2007)
3. Dodis, Y., Reyzin, L., Smith, A.: Fuzzy extractors: how to generate strong keys from biometrics and other noisy data. In: Cachin, C., Camenisch, J. (eds.) Advances in Cryptology —EUROCRYPT, Lecture Notes in Computer Science, Springer Heidelberg, vol. 3027, pp. 523–540 (2004)
4. Gura, N., Patel, A., Wander, A., Eberle, H., Shantz, S.: Comparing elliptic curve cryptography and RSA on 8-bit cpus. In: Joye, M., Quisquater, J. (eds.) Cryptographic Hardware and

Embedded Systems—CHES, Lecture Notes in Computer Science, Springer Heidelberg, vol. 3156, pp. 119–132 (2004)

5. Haller, N.: The s/key one-time password system. Network Working Group (1995)
6. Hao, F., Anderson, R., Daugman, J.: Combining crypto with biometrics effectively. IEEE Trans. Comput. **55**(9), 1081–1088 (2006)
7. Hollingsworth, K., Bowyer, K., Flynn, P.: The best bits in an iris code. IEEE Trans. Pattern Anal. Mach. Intell. **31**(6), 964–973 (2009)
8. Janbandhu, P.K., Siyal, M.Y.: Novel biometric digital signatures for internet-based applications. Inf. Manage. Comput. Secur. **9**(5), 205–212 (2001)
9. Koblitz, N.: Elliptic curve cryptosystems. Math. Comput. **48**(177), 203–209 (1987)
10. Mahto, D., Yadav, D.K.: Network security using ECC with Biometric. In: Singh, K., Awasthi, A.K. (eds.) QSHINE, LNICS-SITE, vol. 115, pp. 842–853. Springer, Heidelberg (2013)
11. Mahto, D., Yadav, D.K.: Enhancing security of one-time password using elliptic curve cryptography with biometrics for e-commerce applications. In: Proceedings of the Third International Conference on Computer, Communication, Control and Information Technology (C3IT), pp. 1–6. IEEE (2015)
12. Miller, V.S.: Use of elliptic curves in cryptography. In: Williams, H. (ed.) Advances in Cryptology CRYPTO 85 Proc., Lecture Notes in Computer Science, Springer Heidelberg, vol. 218, pp. 417–426 (1986)
13. Zhang, L., Sun, Z., Tan, T., Hu, S.: Robust biometric key extraction based on iris cryptosystem. In: Tistarelli, M., Nixon, M. (eds.) Advances in Biometrics, Lecture Notes in Computer Science, Springer Heidelberg, vol. 5558, pp. 1060–1069 (2009)

An Image Encryption Technique Using Orthonormal Matrices and Chaotic Maps

Animesh Chhotaray, Soumojit Biswas, Sukant Kumar Chhotaray and Girija Sankar Rath

Abstract Many image encryption techniques have been designed based on block based transformation techniques or chaos based algorithms. In this paper, a two stage image encryption technique is proposed. In the first stage, a new method of generating orthonormal matrices is presented and used along with sparse matrices to generate key matrices which are used to transform the original image. Since, chaos based encryption techniques are very efficient due to sensitivity to initial conditions and system parameters, this paper makes an analytical study of some recent chaos based image encryption techniques. In the second stage, the image is encrypted using one of the studied algorithms which has an optimal balance between robustness and execution time. This transformation is very fast and the overall cryptosystem is very robust which can be observed from entropy analysis, resistance to statistical and differential attacks and large key space.

Keywords Image encryption · Orthonormal matrices · Chaotic maps · Asymmetric cryptosystem · Robust · Fast · Statistical and differential attacks

A. Chhotaray (✉)
School of Computer Engineering, Kalinga Institute of Industrial Technology, Bhubaneswar, Odisha, India
e-mail: animesh.chhotaray@gmail.com

S. Biswas
School of Electronics Engineering, Kalinga Institute of Industrial Technology, Bhubaneswar, Odisha, India
e-mail: sbdisaster40@gmail.com

S.K. Chhotaray
Department of Electronics and Communication Engineering, Sardar Vallabhbhai Patel Institute of Technology, Vasad, Gujarat, India
e-mail: sukantchhotaray@gmail.com

G.S. Rath
Department of Electronics and Communication Engineering, C. V. Raman College of Engineering, Bhubaneswar, Odisha, India
e-mail: gsrath2011@gmail.com

© Springer India 2016   355
A. Nagar et al. (eds.), *Proceedings of 3rd International Conference on Advanced Computing, Networking and Informatics*, Smart Innovation, Systems and Technologies 44, DOI 10.1007/978-81-322-2529-4_37

1 Introduction

Transmission of digital images has found its application in military image databases, confidential video conferencing, online personal photograph album etc., that require robust, fast and reliable security system to transmit digital images [1]. Hence, prevention of this image from unauthorized access has become a great security concern.

In recent years, many encryption methods have been proposed. Among them chaotic based encryption methods are considered very efficient due to their sensitivity to initial values and control parameters. Some of the commonly used chaotic maps are logistic map, tent map, sine map, piecewise linear chaotic map etc.

A matrix of integers can be used to represent an image or a part of an image. Transform matrices are used to transform the whole image or a part of it [2–5]. Let $m \times n$ represents an image I and R and S are the transform matrices of dimension $m \times m$ and $n \times n$ respectively. Then a transformed image TI of dimension $m \times n$ can be obtained by the following relation.

$$TI = RIS \tag{1}$$

The original image I can be obtained from T provided inverse of P and Q exists i.e. E can be recovered using the following relation

$$E = (R^{-1})(TI)(S^{-1}) \tag{2}$$

A matrix A is said to be orthogonal if the product of the matrix and the transpose of the matrix results in any scalar multiple of identity matrix I i.e. $A^T \cdot A = kI$, where k is any integer. A will be an orthonormal matrix if $k = 1$. Since, we are using an asymmetric cryptosystem to encrypt the original image, the public key is a product of orthonormal matrix and a sparse matrix whose inverse can be easily calculated [2].

2 Encryption

In the first stage of encryption, the cipher image is obtained using standard block based transformation algorithms. A product of orthonormal matrix and sparse matrix is used as the key matrix. In this stage, all operations are carried out in GF (p).

Sukant et al. proposed an algorithm to generate orthonormal matrices and sparse matrices and their usage in designing an asymmetric block based image encryption technique. In this paper, another algorithm for generating orthonormal matrices is presented and is multiplied with the sparse matrix S obtained using Sukant et al.'s algorithm to get the public key.

2.1 Generation of Orthonormal Matrix

A matrix 'X' generated by the following relation will be orthonormal,

$$X = \begin{bmatrix} n_1 x_{11} & n_2 x_{12} \\ n_2 x_{21} & n_1 x_{22} \end{bmatrix} \tag{3}$$

if x_{11}, x_{12} and x_{22} are orthonormal matrices, $x_{21} = -x_{22} x_{12}^T x_{11}$, $n_1^2 + n_2^2 = 1$. Here, n_1 and n_2 are integers.

With the help of this method, generation of higher order orthonormal matrices becomes very easy. Value of n_1 and n_2 can be generated with help of two integers m_1 and m_2 and using the following relations.

$$n_1 = \frac{m_1^2 - m_2^2}{m_1^2 + m_2^2}, \, n_2 = \frac{2m_1 m_2}{m_1^2 + m_2^2}$$

The 8×8 orthonormal matrix (X) used to generate the public key of second stage encryption is mentioned below.

$$X = \begin{bmatrix} 9 & 12 & 234 & 175 & 44 & 102 & 137 & 250 \\ 12 & 242 & 175 & 17 & 102 & 207 & 250 & 114 \\ 151 & 9 & 205 & 165 & 213 & 209 & 188 & 222 \\ 9 & 100 & 165 & 46 & 209 & 38 & 222 & 63 \\ 40 & 24 & 129 & 112 & 177 & 32 & 10 & 106 \\ 24 & 211 & 112 & 122 & 32 & 74 & 106 & 241 \\ 179 & 174 & 46 & 173 & 222 & 188 & 89 & 33 \\ 174 & 72 & 173 & 205 & 188 & 29 & 33 & 162 \end{bmatrix} \tag{4}$$

The public key is generated by

$$key = X \cdot S \cdot X'$$

and its value is given by

$$key = \begin{bmatrix} 188 & 247 & 7 & 239 & 6 & 52 & 201 & 220 \\ 48 & 77 & 154 & 120 & 3 & 250 & 234 & 196 \\ 27 & 58 & 61 & 191 & 15 & 40 & 239 & 14 \\ 81 & 144 & 63 & 136 & 226 & 130 & 129 & 29 \\ 43 & 110 & 221 & 241 & 110 & 166 & 182 & 214 \\ 130 & 81 & 119 & 231 & 199 & 157 & 60 & 215 \\ 84 & 158 & 71 & 105 & 248 & 5 & 169 & 171 \\ 53 & 98 & 170 & 37 & 39 & 90 & 215 & 6 \end{bmatrix} \tag{5}$$

The 8×8 sparse matrix (S) generated by using [2] is given as

$$S = \begin{bmatrix} 31 & 0 & 0 & 0 & 0 & 0 & 0 & 0 \\ 0 & 7 & 0 & 0 & 0 & 0 & 0 & 0 \\ 0 & 0 & 191 & 0 & 0 & 0 & 0 & 0 \\ 0 & 0 & 0 & -1 & 1 & 0 & 0 & 0 \\ 0 & 0 & 0 & -1 & 0 & 1 & 0 & 0 \\ 0 & 0 & 0 & 8 & 0 & 0 & 0 & 0 \\ 0 & 0 & 0 & 0 & 0 & 0 & 47 & 0 \\ 0 & 0 & 0 & 0 & 0 & 0 & 0 & 127 \end{bmatrix} \tag{6}$$

The cipher image is further encrypted in the second stage using an efficient chaos based algorithm.

3 Study of Recent Chaos Based Image Encryption Techniques

Khanzadi et al. proposed an encryption algorithm based on Random Bit Sequence Generator (RBSG) using logistic and tent maps. In this algorithm, random ergodic matrices (REM) and random number matrices (RNM) are used to permute the pixels and resulting bitmaps. Random bit matrices (RBM) are used to obtain the substituted bit maps (SBM). The final encrypted image is obtained by combining the SBM's [1].

Zhou et al. proposed an chaotic encryption method using a combination of two chaotic maps. The chaotic system can be a logistic-tent or logistic-sine or tent-sine system. Random pixel insertion is done at the beginning of each row followed by pixel substitution of sub images. Each round is completed with rotation of combined sub-images. The encrypted image is obtained after four such rounds [6].

Fouda et al. proposed an image encryption technique based on piece wise linear chaotic maps (PWLCM). The solutions of Linear Diophantine Equation (LDE) are used to generate the permutation key (PK) and substitution key (SK) which in turn are used to permute and mask the image. PK and SK has new set of initial conditions whose value depend on total encrypted image. The resultant is encrypted image at the end of first round. The whole encryption process is carried out n times [7].

The above algorithms were simulated in Matlab with standard images of different resolutions under identical conditions. Cameraman (C) and lena (L) images of 256×256 resolution and baboon (B) image of 512×512 resolution with initial entropy values 7.5683, 7.0097 and 7.3579 respectively were used for simulation. The simulation results are shown in Tables 1, 2 and 3. The various metrics used for comparison are entropy (E), 2D correlation coefficient (C), NPCR, UACI, Correlation Coefficients (Horizontal (H), Vertical (V), Diagonal (D), Average (A)) and time (T) in seconds. From the above analysis, it is observed that the degree of randomness of the pixels in the encrypted image is similar in all the three methods

Table 1 Image metrics for cameraman 256 × 256 image

Method	E	C	NPCR	UACI	H	V	D	A	T (s)
Khanzadi	7.9599	0.0051	99.4858	30.4345	0.0385	0.0553	−0.0779	0.0053	156.103
Zhou	7.9970	−0.0041	99.5865	31.1852	−0.0431	0.0225	0.0581	0.0125	2.98
Fouda	7.9968	0.0061	99.5956	31.1804	0.0090	0.0078	0.0279	0.0149	25.875

Table 2 Image metrics for lena 256 × 256 image

Method	E	C	NPCR	UACI	H	V	D	A	T (s)
Khanzadi	7.9877	0.0013	99.6048	29.6489	0.0577	0.0357	0.0489	0.0474	147.48
Zhou	7.9972	0.0054	99.6017	30.4247	−0.0227	−0.0156	−0.0069	−0.0151	2.92
Fouda	7.9971	0.0061	99.6017	30.2863	0.0026	0.0023	−0.0049	0.00004	23.448

Table 3 Image metrics for baboon 512 × 512 image

Method	E	C	NPCR	UACI	H	V	D	A	T (s)
Khanzadi	–	–	–	–	–	–	–	–	Very high
Zhou	7.9993	0.0027	99.5979	27.8237	0.0257	0.0307	−0.0478	0.0029	8.00
Fouda	7.9992	0.0015	99.6124	27.8334	0.0156	−0.0393	0.0804	0.0189	252.388

and all other parameters have nearly similar values except execution time. Since the execution time of method 1 and method 3 is very high compared to method 2, Zhou's method will be used to perform second stage encryption.

4 Results and Security Analysis

The two stage encryption algorithm is tested with the same set of metrics and images used previously to analyse the three chaotic based image encryption algorithms and the results are presented below.

From Tables 1, 2, 3 and 4, it can be observed that the proposed technique's execution time is still substantially less than Khanzadi and Fouda's methods. The image metrics do not show significant deviation.

Table 4 Image metrics for proposed algorithm

Image	E	H	V	D	A	UACI	NPCR	C	T (s)
Cameraman (256)	7.9972	0.0431	0.0147	0.0189	0.0256	31.0643	99.5514	0.0047	2.71
Lena (256)	7.9973	−0.0086	−0.0339	−0.0186	−0.0204	30.5578	99.5880	0.0028	2.76
Baboon (512)	7.9992	−0.0076	−0.0439	0.0292	−0.0074	27.8128	99.6033	0.0022	7.76

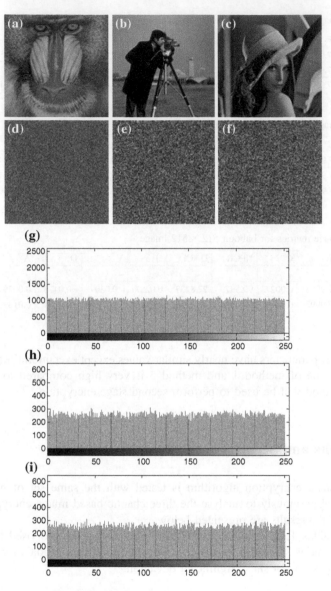

Fig. 1 Simulation results of proposed algorithm. **a** Baboon. **b** Cameraman. **c** Lena. **d** Encrypted baboon. **e** Encrypted cameraman. **f** Encrypted Lena. **g** Encrypted baboon histogram. **h** Encrypted cameraman histogram. **i** Encrypted Lena histogram

4.1 Keyspace

Due to two-stage encryption, brute force attack becomes very difficult as the number of keys becomes very large. In the second stage encryption there are six parameters whose initial value can be between 0 and 4. If 14 decimal places are considered, then number of keys equals 10^{84} [6]. For $p = 251$, number of 2×2 orthonormal matrices are 124,749 [2]. With increase in dimension of matrix and change in value of p the number of orthonormal matrices increases by significant magnitude. For $p = 251$, around 251^n n × n sparse matrices can be generated. Similar to orthonormal matrices, the number of sparse matrices also increases as value of p changes and alternate representation of sparse matrices are considered. Hence, the key space of the proposed algorithm is huge and can easily thwart any brute force attack.

4.2 Statistical and Differential Analysis

From Fig. 1 it is evident that all the pixels of the final encrypted image are uniformly distributed and hence prevent chosen-cipher text attack. Also, the ciphered images are completely scrambled leaving no trace of original image. The correlation between the pixels of encrypted image and original image is very low which is evident from the correlation coefficients and high entropy value. Differential attack is difficult due to large NPCR and low UACI values. Some of the figures of simulation are shown in the following figure.

4.3 Comparison

The proposed algorithm is compared with AES (128) and algorithms by Khanzadi and Fouda. The former comparison can be done using Tables 4 and 5 whereas the latter can be done using Tables 1, 2, 3 and 4. The proposed algorithm has very small execution time whereas the NPCR, UACI, entropy and correlation coefficient values vary by a non-significant magnitude.

Table 5 Performance metrics with AES (128)

Image	AES (128)				
	E	C	NPCR	UACI	T (s)
Lena (256)	7.9971	0.0023	99.6185	30.5049	69.31
Cameraman (256)	7.9992	0.0077	99.63	31.03	69.93
Baboon (512)	7.9992	0.0037	99.5975	27.8054	283.65

5 Conclusion

In this paper, a two-stage encryption algorithm is proposed. In the first stage, an asymmetric block based encryption technique is used in which the public key is a product of an orthonormal matrix and a sparse matrix. A new method of generating orthonormal matrices is also introduced. The cipher image is further encrypted by a chaos based algorithm proposed by Zhou et al. Since, all operations in the first stage are carried in GF (251), matrix inversion becomes very difficult and hence finding the inverse of the public key is cumbersome. The total number of keys for the entire algorithm is huge which makes brute force attack very difficult. Statistical attack is very difficult since the encrypted images are completely scrambled and correlation coefficients are low. Differential attack is unlikely which is evident from the high NPCR value and low UACI value. Apart from the security aspect, the proposed algorithm has low computational complexity which is evident from the small execution time. Hence, the proposed algorithm is very efficient.

References

1. Khanzadi, H., Eshgi, M., Borujeni, S.E.: Image encryption using random bit sequence based on chaotic maps. Arab. J. Sci. Eng. **39**, 1039–1047 (2014)
2. Chhotaray, S.K., Chhotaray A., Rath, G.S.: Orthonormal matrices and image encryption. In: International Conference on Devices, Circuits and Communication (ICDCCOM), pp. 1–5, BIT Mesra, Ranchi, India (2014). doi:10.1109/ICDCCom.2014.7024732
3. Cui, D., Shu, L., Chen, Y., Wu, X.: Image encryption using block-based transformation with fractional Fourier transform. In: 8th International ICST Conference on Communications and Networking, pp. 552–556 (2013)
4. Younes, M.A.B., Jantan, A.: Image encryption using block-based transformation algorithm. IAENG Int. J. Comput. Sci. **35**, 1 (2008)
5. Karagodin, M.A., Osokin, A.N.: Image compression by means of Walsh transforms. In: IEEE Transactions on Modern Techniques and Technologies (MTT), pp. 173–175 (2002)
6. Zhou, Y., Bao, L., Chen, C.L.P.: A new 1D chaotic system for image encryption. Sig. Process. **97**, 172–182 (2014)
7. Fouda, J.S.A.E., Effa, J.Y., Sabat, S.L., Ali, M.: A fast chaotic block cipher for image encryption. Commun. Nonlinear Sci. Numer. Simul. **19**, 578–588 (2014)

Learning Probe Attack Patterns
with Honeypots

Kanchan Shendre, Santosh Kumar Sahu, Ratnakar Dash
and Sanjay Kumar Jena

Abstract The rapid growth of internet and internet based applications has given
rise to the number of attacks on the network. The way the attacker attacks the
system differs from one attacker to the other. The sequence of attack or the signature of an attacker should be stored, analyzed and used to generate rules for
mitigating future attack attempts. In this paper, we have deployed honeypot to
record the activities of the attacker. While the attacker prepares for an attack, the
IDS redirects him to the honeypot. We make the attacker to believe that he is
working with the actual system. The activities related to the attack are recorded by
the honeypot by interacting with the intruder. The recorded activities are analyzed
by the network administrator and the rule database is updated. As a result, we
improve the detection accuracy and security of the system using honeypot without
any loss or damage to the original system.

Keywords Honeypot · Virtual honeypot · Intrusion detection system · Honeyd

1 Introduction

Honeypot is the system to deceive the attacker by providing the decoy system
which seems to be highly valuable, but badly secured so that the attacker can
interact with that system. The administrator is able to analyze the attacker's

K. Shendre (✉) · S.K. Sahu · R. Dash · S.K. Jena
National Institute of Technology, Rourkela, India
e-mail: kanchanshendre19@gmail.com

S.K. Sahu
e-mail: santoshsahu@hotmail.co.in

R. Dash
e-mail: ratnakar@nitrkl.ac.in

S.K. Jena
e-mail: skjena@nitrkl.ac.in

© Springer India 2016   363
A. Nagar et al. (eds.), *Proceedings of 3rd International Conference
on Advanced Computing, Networking and Informatics*, Smart Innovation,
Systems and Technologies 44, DOI 10.1007/978-81-322-2529-4_38

interaction with the system and categorize that attack by which the intent of the attackers can be known as discussed in [1, 2]. If a honeypot successfully interacts with the intruder, the intruder will never know that she/he is being monitored and tricked. Most of the honeypots are installed inside firewalls through which it can be controlled in a better way, although it can also be installed outside the firewalls. A firewall restricts the traffic coming from the Internet, whereas honeypot allows the traffic from the Internet, and restricts the traffic sent back from the system [1].

The parameters that are used to know the value fetched from a honeypot are given by [3]: (i) Type of deployment of honeypot and (ii) Scenario of deployment (location of deployment i.e. behind firewall inside DMZ, in front of firewall etc.). On the basis of these parameters a honeypot can act in the same way as bulgur alarm for detection of attacks, Prevention of attacks by deception and deterrence, responding to attacks by providing valuable logs regarding attack [3].

1.1 Areas of Deployment

There are two areas of deployment of honeypot: physical honeypots and virtual honeypots. In case of physical honeypots, the original system is allowed to completely compromise by the intruder. There is a risk to the system to be damaged by the intruder. So, another approach called as a virtual honeypot which provides the attacker with a vulnerable system which is not actually the real system is used, but the attacker never knows that he is dealing with the virtual system.

1.2 Types of Honeypot

There are two types of honeypot: High Interaction honeypot and Low Interaction honeypot. In a high-interaction honeypot the attacker can interact with a real system. While a low-interaction honeypots provides only some parts such as the network stack. The high interaction honeypot allows the adversary to fully compromise the system to launch the network attack. There is a higher risk in deploying high interaction honeypot. It takes more time for analyzing the events; it may take several days to know the intent of the attacker. It needs high maintenance so it is very hard to deploy. These are the drawbacks of high interaction honeypot.

Due to the drawbacks and risk in deployment of high-interaction honeypot, we have used the low interaction honeypot. Low-interaction honeypots are used to collect the statistical data and high-level information about attack patterns. Since an attacker interacts just with a simulation, he cannot fully compromise the system. A controlled environment is constructed by Low-interaction honeypots and thus the limited risk is involved: As the attacker cannot completely compromise the system, we do not need to worry about abuses of our low-interaction honeypots.

2   Related Work

The different types of honeypot can be used to detect different types attack by using different honeypot tools. Some previously known attacks and work done in honeypot is summarized as shown below (Table 1).

3   Objective

The objective of this paper is to learn the Probe attack patterns and generate rules for unknown probe attacks and update new rules into snort rule set. We not only trap the attacker but also try to know the motives and tactics used by the attacker.

4   Proposed Work

The honeypot is configured on the virtual system like Vmware. In low interaction honeypot, there are certain fingerprint files which contain the information about how the particular operating system will respond. For example, if we want to show the attacker that we are running Windows XP operating system, it will respond with certain characteristics, which will be used by the honeypot to respond to the attacker. The attacker will think that he is actually working with the Windows XP operating system but he will never know that he is actually dealing with the virtual operating system. The few of the important features of honeyd are creation, setting, binding and adding. In the creation process, we are going to create a template with some name or default. The structure of the template is as follows:

```
create<template-name>
create default
dynamic<template-name>
{Then we set the personality of the honeypot, i.e, the
operating system and mention certain protocol or action
such as reset, block or open.}
set<template name>personality<personality-name>
set<template name>default<proto>action<action>
We are adding the particular template along with protocol
name, port number and action.
add<template-name><proto>port<port-number><action>
```

Figure 1 shows the working model of IDS and honeypot together. The intrusion detection system redirects the attacker to the honeypot, when the malicious activity

Table 1 Related works in honeypot

Year	Author	Type	Attack type	Work done
2011	Saurabh et al. [3]	*l*	NA	Due to the lack of capabilities of existing security devices, there is a need to study honeypot deployment and analyze tools, methods and targets of the attacker
2006	Nguyen et al. [2]	*h*	NA	The purpose of this paper is to deploy a honeypot in such a way that it is well concealed from the intruder. The honeypot is deployed on Xen virtual machine with system Xebek
2004	Kuwatly et al. [4]	*h* + 1	NA	The dynamic honeypot approach integrates passive or active probing and virtual honeypot
2006	Alata et al. [5]	*h* + 1	U2R	The results based on a 6 months period using high interaction honeypot concludes that if the password is found to be weak it is replaced by strong one
2009	Vinu et al. [6]	NA	DOS	This paper has proposed the effective honeypot model for secured communication of authorized client and server
2009	Shujun et al. [7]	NA	Phishing	Honeypot is used to collect important information regarding attackers' activity
2009	Jianwei et al. [8]	*h*	malware	This paper has introduced a high interaction toolkit called HoneyBow containing three tools MwFetcher, MwWatcher, MwHunter
2003	Lance et al. [9]		The advance insider	This paper detects the threats done by the authorised insider
2009	Almotairi et al. [10]	*l*	NA	The technique for detecting new attacks using PCA with low interaction honeypot is presented in this paper

Note l Low interaction honeypot and *h* High interaction honeypot

Fig. 1 The working model of
IDS and honeypot

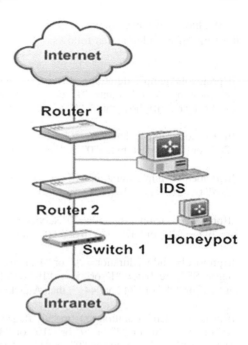

is detected. The intruder interacts with the honeypot and tries to know its vulnerabilities and open ports. The honeypot allows to gain access to the limited resources of the system so that it should not make any harm to the important files and resources. The attack activities of the particular intruder is logged by the honeypot. This log file is then used to create new rules which are further added to the list of already generated rules. Once this is done, when the same type of behavior occurs next time, this is directly considered as attack and there is no need to redirect that intruder to the honeypot. In this way, the novel attacks can be detected by the intrusion detection system.

5 Result and Discussion

We have studied the probe attack patterns and represented the number of instances of each type as follows (Table 2):

Table 2 Number of instances
for each type of probe attack

Sl. no.	Name of attack	No. of instances
1	nmap	11,609
2	portsweep	1,915
3	ipsweep	2,177
4	satan	2,013

We have estimated some snort rules by using honeypots and represented them in the form of pseudocode as follows:

If(protocol="icmp", duration="0",service="eco_i" or "ecr_i", flag="SF", src_byte= "8", dest_byte= "0" count= "1" or "2" or "46", srv_diff_host_rate = "1" to "17", dst_host_srv_diff_host_rate = "0" to "0.4") then Attack= "nmap"

If(protocol= "udp", duration= "0",service="private", flag= "SF", src_byte="100" or "207" or "215", dest_byte="0" or "100" or "207") then Attack="nmap"

If(protocol= "udp", duration="0",service="private", flag= "SF", src_byte= "100" or "207" or "215", dest_byte= "0" or "100" or "207") then Attack= "nmap"

If(protocol= "icmp", duration="0",service="eco_i" or "ecr_i" or "urp_i", flag="SF", src_byte= "20" or "37", dest_byte= "0") then Attack= "satan"

If(protocol= "udp", duration= "0" or "4", service="domain_u" or "other" or "private", flag= "SF", src_byte= "1" or "5" or "19" or "40", dest_byte="0" or "1" or "4" or "5" or "26" or "28" or "74" or "444") then Attack= "satan"

If(protocol="tcp", duration="0" to "9", flag=all except "SF" and "OTH", src_byte= "0" or "5" or "6" or "7" or "9" or "10" or "30" or "31" or "39" or "44" or "54" or "103" or "1710" dest_byte= "0" or "4" or "15" or "19" or "23" or "25" or "26" or "28" or "31" or "32" or "34" or "35" or "40" or "43" or "44" or "53" or "54" or "60" or "75" or "77" or "109" or "112" or "114" or "121" or "131" or "143" or "144" or "147" or "151" or "164" or "178" or "186" or "192" or "196" or "292" or "375" or "536" or "556" or "672" or "1405" or "1886" or "18056") then Attack= "satan"

If (protocol= "tcp", duration="0" to "7" or "12743", service="ctf" or "domain" or "ftp_data" or "gopher" or "http" or "link" or "mtp" or "name" or "private" or "remote_job" or "rje" or "smtp" or "ssh" or "telnet" or "time" or "whois", flag= "REJ" or "RSTO" or "SF", src_byte="0" or "4113", dest_byte= "0" or "3" or "4" or "12" or "15" or "61" or "77" or "79" or "82" or "83" or "84" or "85" or "89" or "90" or "91" or "96" or "132" or "133" or "142" or "51633", dst_host_count= "1" to "72", dst_host_srv_count= "1" to "194", dst_host_diff_srv_rate= "0" or "0.99" or "1", dst_host_serror_rate=   "0",   dst_host_rerror_rate=   "0.5"   to   "1", dst_host_srv_rerror_rate= "0.01" to "0.07" or "0.13" or "0.25" or "0.29" or "0.5" "0.67" or "1") then Attack= "ipsweep"

If(protocol= "icmp", duration="0",service="eco_i" or "ecr_i" or "urp_i", flag= "SF", src_byte= "8" or "18", dest_byte= "0") then Attack= "ipsweep"

If the attacker sends probe requests to multiple hosts using a specific port, then this attempt recorded as portsweep attack.

6 Conclusion

The primary objective of the honeypot is to collect intense attack patterns and decode it into human understandable format. In this paper, we have implemented a virtual honeypot using honeyd which is installed on Ubuntu 14 machine and the attack patterns are captured whenever recommended by the IDS. The well-known probe attacking tools are used for attacking the system by us. The packets captured by the honeypot is decoded and converted into csv format for subsequent analysis. Finally, the patterns are processed and the snort rule set is updated to detect these type of attacks that may take place in future. It helps the administrator to protect the system from probe attacks and to analyze the signatures of the attacks.

References

1. Provos, N., Holz, T.: Virtual Honeypots: from Botnet Tracking to Intrusion Detection. Pearson Education, New Delhi (2007)
2. Quynh, N.A., Takefuji, Y.: Towards an Invisible Honeypot Monitoring System, Information Security and Privacy. Springer, Berlin (2006)
3. Chamotra, S., et al.: Deployment of a low interaction honey pot in an organizational private network. In: International Conference on Emerging Trends in Networks and Computer Communications (ETNCC), IEEE, 2011
4. Kuwatly, I., et al.: A dynamic honeypot design for intrusion detection. In: International Conference on Pervasive Services, ICPS 2004. IEEE/ACS, IEEE (2004)
5. Alata, E., et al.: Lessons learned from the deployment of a high-interaction honeypot. arXiv preprint arXiv:0704.0858 (2007)
6. Das, V.V.: Honeypot scheme for distributed denial-of-service. In: International Conference on Advanced Computer Control, ICACC'09. IEEE (2009)
7. Li, S., Schmitz, R.: A novel anti-phishing framework based on honeypots. IEEE (2009)
8. Zhuge, J., et al.: Collecting autonomous spreading malware using high-interaction honeypots. In: Information and Communications Security. Springer, Berlin, pp. 438–451 (2007)
9. Spitzner, L.: Honeypots: Catching the insider threat. In: Proceedings of 19th Annual Computer Security Applications Conference, IEEE (2003)
10. Almotairi, S., et al.: A technique for detecting new attacks in low-interaction honeypot traffic. In: Fourth International Conference on Internet Monitoring and Protection, ICIMP'09. IEEE (2009)

Part V
Operating System and Software Analysis

Fuzzy Based Multilevel Feedback Queue Scheduler

Supriya Raheja, Reena Dadhich and Smita Rajpal

Abstract In multilevel feedback queue scheduling algorithm the major concern is to improve the turnaround time by keeping the system responsive to the user. Presence of vagueness in a system can further affect these performance metrics. With this intent, we attempt to propose a fuzzy based multilevel feedback queue scheduler which deals with the vagueness of parameters associated with tasks as well as to improve the performance of system by reducing the waiting time, response time, turnaround time and normalized turnaround time. Performance analysis shows that our methodology performs better than the multilevel feedback scheduling approach.

Keywords Multilevel feedback queue (MLFQ) scheduling algorithm · Scheduler · Fuzzy set · Fuzzy inference system (FIS) · Fuzzy based multilevel feedback queue scheduling algorithm

1 Introduction

Multilevel Feedback Queue scheduling (MLFQ) is the preferable choice of operating system designer's for schedulers. However, no algorithm is exception from the problems and issues [1, 2]. The issues with MLFQ scheduling algorithm are threefold: to assign the optimum length of time quantum to each queue, to minimize

S. Raheja (✉)
Department of Computer Science & Engineering, ITM University, Gurgaon, India
e-mail: supriya.raheja@gmail.com

R. Dadhich
Department of Computer Science & Informatics, University of Kota, Kota, India
e-mail: reena.dadhich@gmail.com

S. Rajpal
Alpha Global IT, Toronto, Canada
e-mail: smita_rajpal@yahoo.co.in

© Springer India 2016
A. Nagar et al. (eds.), *Proceedings of 3rd International Conference on Advanced Computing, Networking and Informatics*, Smart Innovation, Systems and Technologies 44, DOI 10.1007/978-81-322-2529-4_39

373

the response time to improve the user interactivity and to improve the turnaround time of the tasks [3, 4]. With all these issues one major concern is how to handle the vagueness of the tasks in a system.

In this paper we are designing a fuzzy based multilevel feedback queue scheduler (FMLFQ) which considers all the issues with the MLFQ scheduling algorithm. It dynamically generates the time quantum based on the present state of ready queue and also reduces the response time and turnaround time by not affecting the other factors.

This paper is organized as follows. Section 2 gives the brief explanation of the multilevel feedback queue scheduling. This section also discusses the related work of MLFQ. Section 3 provides the reader with the background information on fuzzy set theory. Section 4 describes FMLFQ scheduler in detail. Section 5 discusses the performance analysis with the help of example task set and results. Finally, Sect. 6 concludes the work.

2 MLFQ Scheduling Algorithm and Related Work

MLFQ scheduling is one of the widely known scheduling approaches for interactive systems. It contains multiple queues and each queue has a different priority. The task from the top priority queue is scheduled first with CPU. Let us consider an example of MLFQ where ready queue is divided into three queues Q0, Q1 and Q2 where Q0 has higher priority than Q1, and Q2 has the lowest priority. Each queue has its own scheduling algorithm. The scheduler follows RR scheduling approach for queues Q0 and Q1 whereas for Q2, it follows FCFS approach. When the task enters in the system, firstly it is added at the end of Q0 and then system allots a fixed single time quantum. This scheduling algorithm provides the facility to move the tasks from one queue to another queue. If the task consumes more CPU time, the task is moved to the lower priority queue Q1 and is allotted double time quantum [2, 5].

Since MLFQ is mainly preferable algorithm for interactive tasks, Parvar et al. have utilized the recurrent neural network to optimize the number of queues [6]. Kenneth Hoganson has pointed the performance of MLFQ scheduler in terms of task's starvation [7]. He has presented an approach that extenuates the MLFQ starvation problem. Ayan Bhunia has also given a solution for the MLFQ scheduler for the tasks which get starved in the lower priority queues for waiting CPU. The main issue with the MLFQ scheduling algorithm is to provide the optimum value to queues and to improve the performance [8]. Authors have focused on issues with MLFQ but in literature no author has given solution to handle the vagueness of MLFQ scheduling algorithm.

3 Fuzzy Set Theory

Real world objects are imprecise in nature, since human reasoning and judgment are the major sources of impreciseness and fuzziness. For example, "set of tall men" cannot be precisely defined in mathematical terms. With this concern in 1965, Prof. Zadeh has given a new theory of logic i.e. fuzzy logic. Fuzzy logic is a mathematical and methodological logic which can extract the uncertainties and imprecise information related to human thinking.

Fuzzy set is a class of objects that contains the continuous degrees of membership. In fuzzy set, each element is assigned a single membership value in between 0 and 1. It permits the gradual assessment of the membership of elements in a set. Let $X = \{x_1, x_2...x_n\}$ be the universe of discourse. The membership function $\mu_A(x)$ of a fuzzy set A is a function where $\mu_A(x): X \in [0, 1]$ [9]. The greater value of $\mu_A(x)$ illustrates more belongingness of element x to the set A.

A fuzzy inference system (FIS) is the major unit of fuzzy logic system. FIS is used to take the decisions out of knowledge base through fuzzy inference engine [10, 11]. The fuzzy inference engine provides the methods of reasoning which extracts the information from the knowledge base and makes the decision [12, 13]. Therefore, it is considered as the brain of an expert system.

4 FMLFQ Scheduler

FMLFQ scheduler works in two phases. In the first phase it deals with the vagueness and in the second phase it schedules the tasks based on the result received from the first phase as shown in Fig. 1.

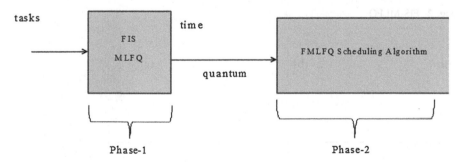

Fig. 1 FMLFQ scheduler

4.1 FIS-MLFQ

In the first phase, FMLFQ scheduler uses a FIS called FIS-MLFQ. It's a Mamdani type FIS which handles the vagueness present with the active tasks of a system. FIS-MLFQ maps the two inputs number of active tasks N and average burst time BT of these tasks into a single output time quantum TQ as shown in Fig. 2.

We have assigned three membership functions LW, MD and HG to both input variables BT and N as well as to output variable TQ as illustrated in Fig. 3. Rule base provides a relationship between input variables and output variable used in real world. In rule base, we have defined total nine rules (3 × 3 matrix) as shown in Fig. 4. Our scheduler will take the decisions based on the rules defined in the rule base.

Through the rule viewer tool, we can compute the varying length of time quantum based on the current state of active tasks which is generated automatically by the system without any interference of user. For example, if N is 6 and BT is 46, the system will generate TQ as 9.8, whereas if N is 5 and BT is 32, the TQ is 8.2 and when we change the BT to 20 and N to 2, system will change the TQ as 2.45. These all different combinations can be viewed through the surface view as shown in Fig. 5.

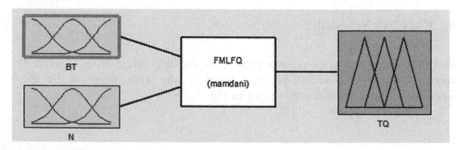

Fig. 2 FIS-MLFQ

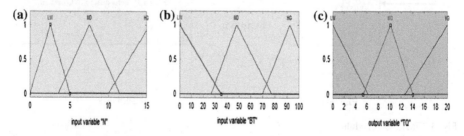

Fig. 3 a Membership functions 'N' **b** membership functions 'BT' **c** membership functions 'TQ'

Fig. 4 Rule
base-FIS-FMLFQ

1. If (BT is LW) and (N is LW) then (TQ is LW) (1)
2. If (BT is MD) and (N is LW) then (TQ is MD) (1)
3. If (BT is HG) and (N is LW) then (TQ is HG) (1)
4. If (BT is LW) and (N is MD) then (TQ is LW) (1)
5. If (BT is MD) and (N is MD) then (TQ is MD) (1)
6. If (BT is HG) and (N is MD) then (TQ is MD) (1)
7. If (BT is HG) and (N is HG) then (TQ is MD) (1)
8. If (BT is MD) and (N is HG) then (TQ is MD) (1)
9. If (BT is LW) and (N is HG) then (TQ is LW) (1)

Fig. 5 Surface view

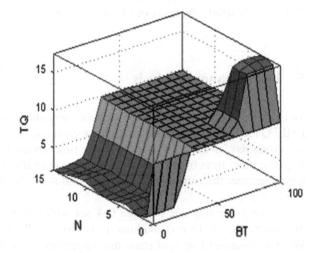

4.2 FMLFQ Scheduling Algorithm

In the second phase, FMLFQ scheduler uses the scheduling algorithm to decide the next task to execute with CPU. These are the steps followed by scheduler

Step 1: Initialize the variables

$$B = \text{burst time of task}$$
$$A = \text{arrival time of task}$$
$$T_S = \text{static time quantum}$$

Step 2: Calculate the number of active tasks at time t ($t \geq 0$) and assign to N.

Step 3: Calculate the average burst time BT.

$$BT = \left(\sum B_i\right)\Big/N \text{ where i is 1 to N.}$$

Step 4: Calculate the length of time quantum TQ_0 using FIS-MLFQ.

Step 5: Assign the TQ_0 to queue Q0 and schedule the tasks in round robin fashion.

Step 6: if $(B_i > TQ_0)$ shift the ith task from Q0 to Q1.

Step 7: Calculate the remaining burst time of task RBT.

Step 8: Calculate the length of time quantum TQ_1.

$$TQ_1 = TQ_0 + T_S$$

Step 9: Assign the TQ_1 to queue Q1 and schedule the tasks in round robin fashion.

Step 10: if $(RBT_i > TQ_1)$ shift the ith task from Q1 to Q2.

Step 11: Schedule the tasks of Q2 in first come first serve manner.

5 Performance Analysis

To analyze the performance of our proposed scheduler, we have applied MLFQ and FMLFQ methodologies on the multiple random task sets. These two methodologies are compared in terms of waiting time: time spend by a task in ready queue, response time: time from entering into a system to first response, turnaround time: total time from the entry time to exit time and normalized turnaround time: relative delay of a task.

We are presenting one example task set with 6 tasks (T1 to T6) having arrival time (0, 5, 10, 12, 18 ms) and burst time (40, 50, 70, 60, 30, 45 ms) respectively. We have assumed length of static time quantum T_S as 10 ms. In phase-1 FMLFQ scheduler uses the FIS-MLFQ which returns the value of time quantum for Q0 as 9.8 ms and FMLFQ scheduling algorithm returns the length of time quantum for Q1 as 19.8 ms.

However, MFLQ scheduling algorithm uses 10 ms for Q0 and 20 ms for Q1. We have scheduled the sample task set using MLFQ scheduling algorithm and FMLFQ scheduler. After scheduling, we have compared the two methodologies in terms of performance metrics as shown in Fig. 6. Then we have calculated the average waiting time, average response time, average turnaround time and average normalized turnaround time. Figure 7 illustrates the performance of 5 different task sets. From the overall results we can analyze that proposed FMLFQ scheduler performs better than the MLFQ scheduler. It reduces both turnaround time as well as response time which are the major concerns of this work along with the handling of vagueness using fuzzy set.

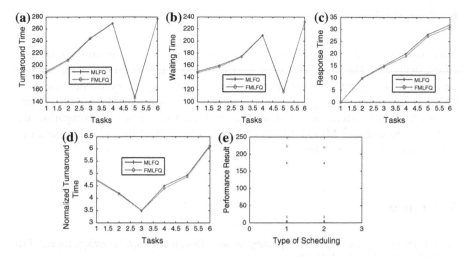

Fig. 6 **a** Waiting time **b** response time **c** turnaround time **d** normalized turnaround time **e** overall performance

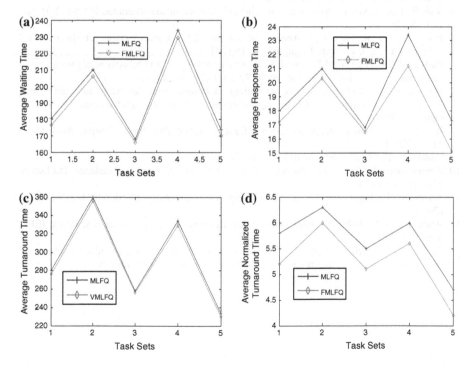

Fig. 7 **a** Average waiting time **b** average response time **c** average turnaround time **d** average normalized turnaround time

6   Conclusion

This paper proposes a fuzzy based approach to a multilevel feedback scheduler which deals with the vagueness associated with tasks. It also improves the average turnaround time of tasks by further minimizing the average response time of tasks. FMLFQ scheduler follows a dynamic approach by assigning the length of time quantum to each queue at run time. We evaluated our work by comparing the performance with the MLFQ scheduler and results proved that FMLFQ performs better than MLFQ scheduler.

References

1. Tanenbaum, A., Woodfhull, A.: Operating Systems Design and Implementation, 3rd edn. PHI publication, Netherlands (2006)
2. Silberschatz, G.Gagne: Operating Systems Concepts, 8th edn. Wiley, Hoboken (2009)
3. Remzi, H., Arpaci-Dusseau, Andrea, C.: Operating Systems: Three Easy Pieces, Scheduling: Multilevel Feedback Queue. version 0.80 (2014)
4. Rao, P.: Complexity analysis of new multilevel Feedback queue scheduler. JCCST. 2(3), 86–105 (2012)
5. Panduranga, R.M., Shet, K.C.: Analysis of new multilevel feedback queue scheduler for real time kernel. Int. J. Comput. Cogn. 8(3), 5–16 (2010)
6. Parvar, M.R.E., Parvar, M.E., Saeed, S.: A starvation free IMLFQ scheduling algorithm based on neural network. Int. J. Comput. Intell. Res. 4(1),27–36 (2008)
7. Hoganson, K.: Reducing MLFQ scheduling starvation with feedback and exponential averaging. Consortium for Computing Sciences in Colleges. Southeastern Conference, Georgia (2009)
8. Bhunia, A.: Enhancing the performance of feedback scheduling. Int. J. Comput. Appl. 18(4), 11–16 (2011)
9. Zadeh, L.: Fuzzy sets. Inf. Control 8, 338–356 (1965)
10. Zimmerman, J.: Fuzzy Set Theory and Its Applications. Kluwer Academic Publishers, Massachusetts (2001)
11. Atanassov, K.: Intuitionistic Fuzzy Sets: Theory and Applications. Physica-Verlag, New York (2000)
12. Abdurazzag, A., Miho, V.: Fuzzy logic based algorithm for uniprocessor scheduling. IEEE. 499–504 (2008)
13. Raheja, S., Dadhich, R., Rajpal, S.: An optimum time quantum using linguistic synthesis for round robin scheduling algorithm. Int. J. Soft Comput. 3(1), 57–66 (2012)

Dynamic Slicing of Feature-Oriented Programs

Madhusmita Sahu and Durga Prasad Mohapatra

Abstract We intend to suggest a dynamic slicing algorithm for feature-oriented programs. We have named our algorithm *Execution Trace file Based Dynamic Slicing* (ETBDS) algorithm. The ETBDS algorithm constructs an intermediate program representation known as *Dynamic Feature-Oriented Dependence Graph* (DFDG) based on various dependences exist amongst the program statements. We use an execution trace file to keep the execution history of the program. The dynamic slice is computed by first performing breadth-first or depth-first traversal on the DFDG and then mapping out the resultant nodes to the program statements.

Keywords Feature-oriented programming (FOP) · Dynamic feature-oriented dependence graph (DFDG) · FeatureC++ · Mixin layer

1 Introduction

Feature-Oriented Programming (FOP) is the study of feature modularity in program families and programming models supporting it. The key idea behind FOP is to build software by the composition of features. *Features* are the characteristics of software that distinguish members of a program family. The FOP paradigm is concerned with identifying functionality in the form of features.

The rest of the paper is organized as follows. Section 2 highlights an overview of some previous works. Section 3 presents a brief idea of Feature-Oriented

M. Sahu (✉) · D.P. Mohapatra
Department of Computer Science & Engineering,
National Institute of Technology, Rourkela 769008, Odisha, India
e-mail: 513CS8041@nitrkl.ac.in

D.P. Mohapatra
e-mail: durga@nitrkl.ac.in

© Springer India 2016   381
A. Nagar et al. (eds.), *Proceedings of 3rd International Conference on Advanced Computing, Networking and Informatics*, Smart Innovation, Systems and Technologies 44, DOI 10.1007/978-81-322-2529-4_40

Programming. Section 4 describes our proposed work on dynamic slicing of feature-oriented programs. Section 5 concludes the paper. We use node and vertex interchangeably in this paper.

2 Overview of Previous Work

Apel et al. [1] presented FeatureC++ with an additional adoptation to Aspect-Oriented Programming (AOP) concepts. Apel et al. [2] presented a novel language for FOP in C++ namely *FeatureC++*. They showed few problems of FOP languages during implementation of program families and proposed ways to solve them by the combination of AOP and FOP features. We find no work discussing the slicing of FOP. We present an approach for dynamic slicing of FOPs using FeatureC++ as a FOP language. We have extended the work of Mohapatra et al. [3] to incorporate feature-oriented features.

3 Feature-Oriented Programming (FOP)

The term Feature-Oriented Programming (FOP) was coined by Christian Prehofer in 1997 [4]. FOP is a vision of programming in which individual features can be defined separately and then can be composed to build a wide variety of particular products. The *step-wise refinement*, where features are incrementally refined by other features, results in a layered stack of features. A suitable technique for the implementation of features is the use of *Mixin Layers*. A Mixin Layer is a static component that encapsulates fragments of several different classes (Mixins) to compose all fragments consistently.

FeatureC++ is an extension to C++ language supporting Feature-Oriented Programming (FOP). Figure 3 shows an example FeatureC++ program. This program checks the primeness of a number. Figure 1 shows different features supported by our prime number checking problem and Fig. 2 shows the corresponding stack of Mixin Layers. Details of FeatureC++ can be found in [1, 2].

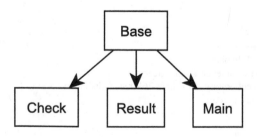

Fig. 1 Features supported by prime number checking problem

Fig. 2 Stack of mixin layers in prime number checking problem

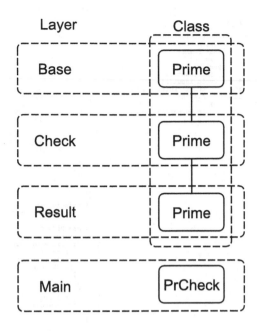

4 Proposed Work

This section describes our *Execution Trace file Based Dynamic Slicing* (ETBDS) algorithm to compute dynamic slices of FOPs alongwith few definitions.

4.1 Definitions

Definition 1: Digraph A *directed graph* or *digraph* is defined as a collection of a finite set, V, of elements called *vertices* or *nodes* and the set, E, of ordered pairs of elements of V called *edges* or *arcs*.

Definition 2: Arc-classified digraph An *arc-classified digraph* is defined as a digraph where more than one type of edges exist between two vertices and the direction of edges between two vertices are not same.

Definition 3: Path A *path* in a digraph is a sequence of edges or arcs connecting a sequence of vertices and the edges are directed in the same direction.

Definition 4: Execution trace The path that an input data to a program actually executes is referred to an *execution trace*. For example, Fig. 4 shows the execution trace of the program given in Fig. 3 for the input data $n = 5$.

(a)
```
       #include<iostream>
       using namespace std;
       class Prime{
          int n;
       public:
          int check();
9         void input(){
10           cout<<"Enter a number: ";
11           cin>>n;
          }
          void output();
       };
```

(b)
```
       #include<iostream>
       using namespace std;

       refines class Prime{
       public:
12        int check(){
13           int i;
13           i=2;
14           while(i<n){
15              if(n%i==0)
16                 break;
17              i++;
          }
18           if(i==n)
19              return 1;
             else
20              return 0;
          }
       };
```

(c)
```
       #include<iostream>
       using namespace std;

       refines class Prime{
       public:
21        void output(){
22           if(check())
23              cout<<"Number is prime";
             else
24              cout<<"Number is not prime";
          }
       };
```

(d)
```
       #include<iostream>
       using namespace std;

       class PrCheck{
       public:
5         static void main(){
6            Prime p;
7            p.input();
8            p.output();
          }
       };
1      int main(){
2         PrCheck::main();
3         int c=getchar();
4         return 0;
       }
```

(e)
```
       Base
       Check
       Result
       Main
```

Fig. 3 A FeatureC++ program to check the primeness of a number. **a** Base/Prime.h.
b Check/Prime.h. **c** Result/Prime.h. **d** Main/PrCheck.h. **e** Test-Prime.equation

4.2 The Dynamic Feature-Oriented Dependence Graph (DFDG)

The DFDG is an *arc-classified digraph* consisting of vertices corresponding to the statements and edges showing dynamic dependence relationships exist amongst statements. The following types of dependence edges exist in the DFDG of a feature-oriented program:

Control dependence edge: *Control dependences* represent the control conditions or predicates on which the execution of a statement or an expression depends. For example, in Fig. 5, the edge between nodes 21 and 22 indicates that node 21 controls the execution of node 22.

Data dependence edge: *Data dependences* represent the flow of data amongst the statements and expressions. For example, in Fig. 5, the edge between nodes 13 and 14 indicates that node 14 uses the value of *i* defined at node 13.

Mixin call edge: *Mixin call* edges denote the entry of a function in a mixin layer in response to a function call in another mixin layer. For example, in Fig. 5, the edge from node 12 to node 22 indicates that node 22 in one mixin layer calls a function check() that is defined in another mixin layer at node 12.

Mixin data dependence edge: *Mixin data dependences* represents the flow of data amongst the statements and expressions in different mixin layers. For example, in

Fig. 5, the edge from node 14 to node 11 indicates that node 14 in one mixin layer uses the value of n and n is defined in another mixin layer. Similarly, node 22 in one mixin layer uses the value returned by node 19 and node 19 exist in another mixin layer.

Call edge: *Call* edges are used to reflect the entry of a function in response to a function call. For example, in Fig. 5, there is an edge from node 5 to node 2 since node 2 calls a function main defined at node 5 in the same mixin layer.

4.3   Computation of Dynamic Slices

Let *FP* be a feature-oriented program and $G = (V, E)$ be the DFDG of *FP*. We use a slicing criterion with respect to which the dynamic slice of FOP is to be computed. A *dynamic slicing criterion* for G has the form $<x, y, e, i>$, where $x \in V$ represents an occurrence of a statement for an execution trace e with input i and y is the variable used at x. A *dynamic slice* DS_G of G on a given slicing criterion $<x, y, e, i>$ is a subset of vertices of G such that for any $x' \in V$, $x' \in DS_G(x, y, e, i)$ if and only if there exists a path from x' to x in G.

Algorithm 1 gives our proposed *ETBDS* algorithm.

Fig. 4 Execution trace of the program given in Fig. 3 for $n = 5$

```
1(1)    int main()
2(1)    PrCheck::main();
5(1)    static void main()
6(1)    Prime p;
7(1)    p.input();
9(1)    void input()
10(1)   cout<<"Enter a number: ";
11(1)   cin>>n;
8(1)    p.output();
21(1)   void output()
22(1)   if(check())
12(1)   int check()
13(1)   i=2;
14(1)   while(i<n)
15(1)   if(n%i==0)
17(1)   i++;
14(2)   while(i<n)
15(2)   if(n%i==0)
17(2)   i++;
14(3)   while(i<n)
15(3)   if(n%i==0)
17(3)   i++;
14(4)   while(i<n)
18(1)   if(i==n)
19(1)   return 1;
23(1)   cout<<"Number is prime";
3(1)    int c=getchar();
4(1)    return 0;
```

Working of ETBDS Algorithm

The working of our ETBDS algorithm is exemplified with an example. Consider the example FeatureC++ program given in Fig. 3. The program executes the statements 1, 2, 5, 6, 7, 9, 10, 11, 8, 21, 22, 12, 13, 14, 15, 17, 14, 1, 17, 14, 15, 17, 14, 18, 19, 23, 3, 4 in order for the input data $n = 5$. The execution trace file, shown in Fig. 4, kept these executed statements. Then, applying Steps 6 to 23 of ETBDS algorithm

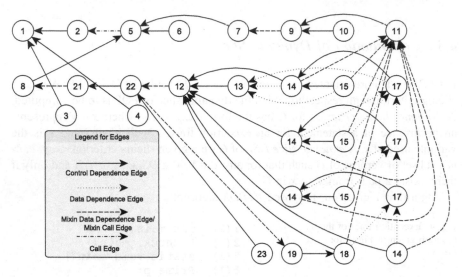

Fig. 5 Dynamic feature-oriented dependence graph (DFDG) for the execution trace given Fig. 4

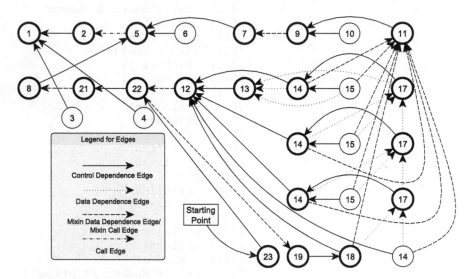

Fig. 6 Dynamic feature-oriented dependence graph (DFDG) for the execution trace given Fig. 4 and node 23 as the starting point

and using the trace file shown in Fig. 4, the *Dynamic Feature-Oriented Dependence Graph* (DFDG) is constructed. The DFDG of the example program given in Fig. 3 is shown in Fig. 5 with input data $n = 5$. Suppose, we want to find the dynamic slice of the output statement at node 23. Then, starting from node 23 and applying the Step 25 of the ETBDS algorithm, the breadth first traversal yields the vertices 23, 22, 21, 19, 8, 5, 2, 1, 18, 12, 11, 17, 9, 7, 14, 17, 14, 17, 14, 13 and the depth first traversal yields the vertices 23, 22, 21, 19, 18, 12, 11, 17, 14, 17, 14, 17, 14, 13, 9, 7, 5, 2, 1, 8. Both the traversals yields the same result i.e., the same set of vertices in the slice. These vertices are shown bold in Fig. 6. The statements corresponding to these vertices are found using the Steps 26 to 27 of the ETBDS algorithm. This produces the required dynamic slice consisting of statements numbered 1, 2, 5, 7, 8, 9, 11, 12, 13, 14, 17, 18, 19, 21, 22, 23.

Algorithm 1 ETBDS Algorithm

1: Execute the program for a given input.
2: Keep each executed statement in a trace file in the order of execution.
3: **if** the program contains loops **then**
4:     Keep each executed statement inside the loop in a trace file after each time it has been executed.
5: **end if**
6: Make a vertex in the DFDG for each statement in the trace file.
7: **for** each occurrence of a statement in the trace file **do**
8:     Make a separate vertex in the DFDG.
9: **end for**
10: **if** if node y controls the execution of node x **then**
11:     Insert control dependence edge from x to y.
12: **end if**
13: **if** if node x uses a variable defined at node y **then**
14:     Insert data dependence edge from x to y.
15: **end if**
16: **if** if node y calls a function defined at node x in the same mixin layer **then**
17:     Insert call edge from x to y.
18: **end if**
19: **if** if node x in one mixin layer uses a variable or value defined at or returned from node y in another mixin layer **then**
20:     Insert mixin data dependence edge from x to y.
21: **end if**
22: **if** if node y in one mixin layer calls a function defined at node x in another mixin layer **then**
23:     Insert mixin call edge from x to y.
24: **end if**
25: Do the breadth-first or depth-first traversal throughout the DFDG taking any vertex x as the starting point of traversal where x corresponds to the statement of interest.
26: Define a mapping function $g : DS_G(x, y, e, i) \rightarrow$ FP.
27: Map out the yielded slice obtained in Step 25 throughout the DFDG to the source program FP using g.

---

5   Conclusion

We presented an algorithm for dynamic slicing of feature-oriented programs. First, we executed the program for a given input and stored the execution history in an execution trace file. Then, we constructed an intermediate representation called *Dynamic Feature-Oriented Dependence Graph* (DFDG) based on various dependences. The DFDG was traversed in breadth-first and depth-first manner and the resultant nodes were mapped to the program statements to compute the dynamic slice.

References

1. Apel, S., Leich, T., Rosenmuller, M., Saake, G.: FeatureC++: on the symbiosis of feature-oriented and aspect-oriented programming. In: Proceedings of the International Conference on Generative Programming and Component Engineering (GPCE'05), pp. 125–140. Springer, Berlin (2005)
2. Apel, S., Leich, T., Rosenmuller, M., Saake, G.: FeatureC++: feature-oriented and aspect-oriented programming in C++. Technical report (2005)
3. Mohapatra, D.P., Sahu, M., Mall, R., Kumar, R.: Dynamic slicing of aspect-oriented programs. Informatica **32**(3), 261–274 (2008)
4. Prehofer, C.: Feature-oriented programming: a fresh look at objects. In: Proceedings of 11th ECOOP, Lecture Notes in Computer Science, pp. 419–443. Springer, Berlin, Heidelberg (1997)

Mathematical Model to Predict IO Performance Based on Drive Workload Parameters

Taranisen Mohanta, Leena Muddi, Narendra Chirumamilla
and Aravinda Babu Revuri

Abstract Disk drive technologies have evolved rapidly over the last decade to address the needs of big data. Due to rapid growth in social media, data availability and data protection has become an essence. The availability or protection of the data ideally depends on the reliability of the disk drive. The disk drive speed and performance with minimum cost still plays a vital role as compared to other faster storage devices such as NVRAM, SSD and so forth in the current data storage industry. The disk drive performance model plays a critical role to size the application, to cater the performance based on the business needs. The proposed performance model of disk drives predict how well any application will perform on the selected disk drive based on performance indices such as response time, MBPS, IOPS etc., when the disk performs intended workload.

Keywords Drive performance model · Linear polynomial method · IO performance prediction · Drive workload parameters

1 Introduction

The rapid growth of social media in the last decade has changed the electronic data storage. The data storage essentially takes place on the disk drives. The disk drive technology also rapidly evolved to cater the need for the Big Data, Data revival,

T. Mohanta (✉) · L. Muddi · N. Chirumamilla · A.B. Revuri
HP India Software Operations Pvt. Ltd, Bangalore, India
e-mail: taranisen_mohanta@yahoo.com

L. Muddi
e-mail: leena.muddi@hp.com

N. Chirumamilla
e-mail: narendra.chirumamilla@gmail.com

A.B. Revuri
e-mail: abr@hp.com

© Springer India 2016   389
A. Nagar et al. (eds.), *Proceedings of 3rd International Conference
on Advanced Computing, Networking and Informatics*, Smart Innovation,
Systems and Technologies 44, DOI 10.1007/978-81-322-2529-4_41

Data storing and Data Mining Purpose. The protection of the data essentially depends on the reliability of the disk drive. The disk drive speed and performance with minimum cost still plays the vital role as compared to other faster storage devices such as NVRAM, SSD and so forth in the data storage industry. The disk drive performance model plays a critical role to size the application, to cater the performance based on the business need.

This paper has made use of different mathematical models and compared them in order to predict the performance model of the disk drive based on the real time disk drive performance data. It compares the real time performance data of the disk drive with different kind of workload along with attributes and predicts the performance for user required workload based on the proposed model. The goal and use is to size the application based on the disk drive performance to meet the application performance for the business. This model can be used by the hard disk pre-sales team or the marketing teams to actually predict the IO performance of the storage systems running with different applications.

The proposed disk drive performance model predict how well any application will perform on the selected disk drive based on performance indices such as response time, MBPS, IOPS etc. when the disk performs intended workload. The experimental results and the model used in this paper to validate the efficiency or accuracy of proposed models with an error bound of 5 % using the real-time collected performance data. This paper work compares performance prediction with two different models and suggests linear polynomial method is the better model as it shows least deviation from the actual performance data.

2 Background and Related Work

The Data Storage Industry uses different storage technologies such as DAS (Direct Attached Storage), NAS (Network Attached Storage) and SAN (Storage Area Network). These data storage techniques are used in the modern datacenters which essentially use the disk drives. The performance of the disk plays a major role in order to meet the need of the users by depending on their type usage and different applications. The performance of the storage system depends on the performance of the hard disk. There are different kinds of storage system to cater the need of high performance of the application. There are storage systems such as NVRAM, SSD, FC, SCSI, and SATA et al. [1–3].

Many attempts have been made to compute or analyze the performance of the disk drives. In order to setup the storage system the performance prediction for different kind of workload plays a major role. There are several attempts made to predict the performance of different kinds of hard disk drives et al. [1–4]. Many research activities are done on the performance model of the hard disk in both analytical model as well as simulation way. But the deployment of the model is a big challenge in terms of time, expertise, complexity and the kind of resources required for the predictive model to run. Others such as authors in Ref. [2]

have proposed different approaches on the hard disk performance prediction mechanism using the machine language tools such as CHART model and artificial neural network model. Similarly, other authors in [5] work proposed a different approach that is based on Adaptive Neuro Fuzzy Inference Systems.

In this situation, it is highly desirable to have a black box model for disk drive performance prediction with simple and accurate algorithm. Although there are research on different black box model for the disk drive model [6–8] has done but the efficient, simple and improved model is highly desirable. The goal is to be able to device a method to find out the performance model without any prior detail of complex design and algorithm of the disk drive 1. In order to achieve the performance model of the disk drive the working storage setup with access to the disk drive is required. With the different kinds of workload inputs which potentially affect the disk drive performance has to be trained using the efficient mathematical equation the data generation system will learn the disk drive behavior, functionality etc. using the different scenarios.

2.1 Polynomial Model for Prediction

The polynomial model is considered to be the simplest one that analyzes the data in a very effective manner. Although the polynomials are of different order, but the order of the polynomial should be as low as possible. The high-order polynomials should be avoided unless they can be justified for reasons outside the data. Therefore, in order to predict the performance of the hard disk, here we consider a linear polynomial model that analyzes the data and predict the performance of the disk drive. A linear polynomial is any polynomial defined by an equation which is of the form

$$p(x) = ax + b \tag{1}$$

where a, b are real numbers and $a! = 0$.

2.2 Radical Model for Prediction

In order to analyze the data and predict the performance of the disk drives, a radical equation can also be used which is derived out of the experimental result pattern.

In order to show the superiority of the linear polynomial model, we have compared the prediction of disk performance through this model with radical equation model.

A radical equation is one that includes a radical sign, which includes square root $\sqrt{x}$, cube root $\sqrt[3]{x}$, and nth root $\sqrt[n]{x}$. A common form of a radical equation is

$$a = \sqrt[n]{x^m} \tag{2}$$

(Equivalent to $a = x^{\frac{m}{n}}$) where m and n are integers.

3 Approach and Uniqueness

The main goal, and also the primary difference compared to other competitive approaches available for the performance prediction is to design a disk drive performance model that has no prior knowledge about the disk drive functional design and its implementation details. This will enable us to use mathematical tools to implement the methodology and, the same can be used over a wide variety of storage systems with minimal (hopefully none) additional effort to predict the performance.

3.1 Workload Representation in the Model

As it is already mentioned, our performance model uses the real performance data and validates against the proposed models for the prediction functionality. Particularly in this case we have identified the different parameters which influence the performance of the disk drive when they vary. In any disk drive based storage devices, the major components that influence the change in its performance are different modes/types of IOs performed by applications on the Host. The typical working model is as mentioned in the Fig. 1.

Fig. 1 Working model

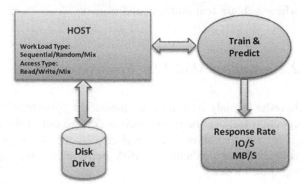

To come out with a best performance model, the following parameters are considered for the study.

- Data access pattern (random, sequential, segmented in this case the values considered as 1, 2, and 3 respectively)
- Types of IO requests (reads, writes and mix of reads and writes).
- Data transfer size (KB).
- Q-depth: number of processes simultaneously used to issue IO requests on to the disk drive.

The expected output performance of the disk drive that has to be predicted is

1. IO's per second.
2. MB's per second.
3. Response time in milliseconds.

Input data and output data that has to be predicted are used as Int-input data and Out-actual output to be obtained.

4 Results and Contributions

The Sample data is collected from a test setup as shown below in Fig. 2. The setup has a Windows host server connected to a storage system using FC SAN switched infrastructure. At the backend of the storage system, a series of daisy chained disk enclosures consisting of 72 GB size disks are used to pool the disk space.

Fig. 2 Test setup

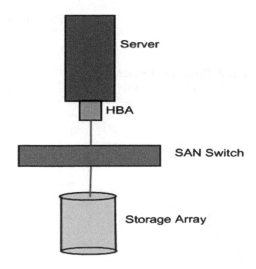

To observe the behavior and result of the different approaches, we present a set of comparison graph representations. The percentage of relative error is shown as

$$R_e = \left|T_{op}(Model) - T_{op}(Real)\right| * 100/T_{op}(Real) \tag{3}$$

where,

op   read or write,
$T_{op}(Real)$ real throughput,
$T_{op}(Model)$ throughput predicted by the model.

In Fig. 3, the graph represents the comparison between the linear polynomial method, the Radical method and the actual experimentation for 5K reads.

Polynomial derived from the sample data points

$$f(x) = \frac{(b * x) + a}{(x + c)}. \tag{4}$$

where $a = 40.58$, $b = 2.12$ and $c = 7.024$ are the constants derived from the regression analysis out of the sample data.

Estimated Response time (RT) is calculated as

$$RT = f(x) * x \tag{5}$$

where x is queue depth, outstanding commands in the queue.

The estimated IOPs is calculated as

$$IOPs = 1000/f(x) \tag{6}$$

Fig. 3 Comparison between various performance models

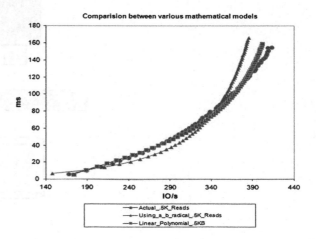

The Radical equation derived from the sample data points

$$f(x) = (a + b)/\sqrt{(1 + x)} \qquad (7)$$

where $a = 2$ and $b = 4.75$.

Estimated Response time (RT) is calculated as

$$RT = f(x) * x \qquad (8)$$

where x is queue depth, outstanding commands in the queue.

The estimated IOPs is calculated as

$$IOPs = 1000/f(x) \qquad (9)$$

4.1 Comparison Between Actual, Radical, and Linear Polynomial Model for Different IO Rate

The graph shows that the polynomial curve is very close to the actual IOPS for a 5K read performance. Hence the application can very well predict the IOPS that can be achieved for a given input x just by applying the mathematical Linear Polynomial.

4.2 Comparison for Random Reads and Random Writes

From Fig. 4, the graphs shows the similar evidences when the IO throughout versus Response times are plotted using Liner Polynomial method for different samples of 8K reads and 8K writes. Actual graph is pretty much close to the modeled graph.

Fig. 4 Random writes measured versus estimated

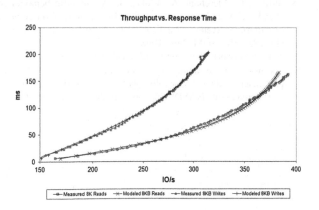

5 Conclusions

This paper confirms an innovative approach for performance modeling of disk drives using the linear polynomial model. The essential objective of this model is to design a self-managed storage system for the different kinds of application considering the disk drive performance. Another possibility of the model is to define the performance of a given storage device. If manufacturers published this model for the different storage devices as part of the specification or as a generic tool, potential buyers could trace their applications performance, and feed them in the models for different storage devices to see which one is better for them before they actually buy the storage device. Based on the results obtained by applying this linear polynomial model, it has been observed that the proposed model performs better as compare to the counterpart. The scope for future research includes advanced model to accommodate different disk sizes and vendor requirements, and come up with a performance model for the disk drive based storage arrays, considering the Array firmware overhead on the disk drive firmware.

References

1. Anderson, E.: Simple table based modeling of storage devices. Technical Report HPL-SSP-2001-04, HP Laboratories (2001). http://www.hpl.hp.com/SSP/papers/
2. Hassoun, M.H.: Fundamentals of Artificial Neural Networks. MIT Press, Cambridge (1995)
3. Wang, M., Au, K., Ailamaki, A., Brockwell, A., Faloutsos, C., Ganger, G.R.: Storage Device Performance Prediction with CART Models, pp. 588–595. MASCOTS (2004)
4. Marquardt, D.W.: An algorithm for least-squares estimation of nonlinear parameters. SIAM J. Appl. Math. 11(2), 431–441 (1963)
5. Jang, J.S.R.: ANFIS: Adaptive-network-based fuzzy inference system. IEEE Trans. Syst. Man, Cybern. 23(5/6), 665–685 (1993)
6. Ruemmler, C., Wilkes, J.: An introduction to disk drive modeling. IEEE Comput. 27(3), 17–28 (1994)
7. Schindler, J., Ganger, G.R.: Automated disk drive characterization. In: Proceedings of the Sigmetrics 2000, pp. 109–126. ACM Press (2000)
8. Shriver, E., Merchant, A., Wilkes, J.: An analytic behavior model for disk drives with read ahead caches and request reordering. International Conference on Measurement and Modeling of Computer Systems. Madison, WI, 22–26 June 1998. Published as Perform. Eval. Rev. 26(1), 182–191. ACM, June 1998
9. Neural Network Toolbox: http://www.mathworks.com
10. Taranisen, M., Srikanth, A.: A method to predict hard disk failures using SMART monitored parameters. Recent Developments in National Seminar on Devices, Circuits and Communication (NASDEC2—06), pp. 243–246, Nov 2006

Measure of Complexity for Object-Oriented Programs: A Cognitive Approach

Amit Kumar Jakhar and Kumar Rajnish

Abstract A new cognitive metric named "Propose Object-Oriented Cognitive Complexity" (POOCC) is proposed to measure the complexity of Object-Oriented (OO) programs. The proposed metric is based on some fundamental factors like: number of operands, operators available in a method of an OO class, cognitive weight of different basic control structures and ratio of accessing similar parameters of an OO class. Evaluation of propose cognitive metric is done with the help of 16 C++ programs and Weyuker's property is used to validate it, seven out of nine properties are satisfied by the proposed metric. Furthermore, this work is also present the relationship between POOCC and Lines of Code (LOC) to examine the density of the code.

Keywords Lines of code · Classes · Object oriented · Cognitive · Complexity

1 Introduction

According to IEEE's definition [1], complexity is "the degree to which a system or component has a design or implementation that is difficult to understand and verify". Software industry peoples are struggling for a certain method which can measure the complexity of the software more precisely. Therefore, measuring the required effort to develop, test and maintain the software become effective. But in the software system all things like un-measurable because software cannot touch and visualize.

Cognitive Informatics (CI) can be utilized in several research arenas to search the solution to a given problem such as software engineering, cognitive sciences and

A.K. Jakhar (✉) · K. Rajnish
Department of Computer Science & Engineering, Mesra, Ranchi, Jharkhand, India
e-mail: amitjakhar69@gmail.com

K. Rajnish
e-mail: krajnish@bitmesra.ac.in

© Springer India 2016
A. Nagar et al. (eds.), *Proceedings of 3rd International Conference on Advanced Computing, Networking and Informatics*, Smart Innovation, Systems and Technologies 44, DOI 10.1007/978-81-322-2529-4_42

artificial intelligence [2–4]. Cognitive informatics performs a very significant role to measure the important characteristics of the software. To comprehend the software in several aspects like: input/output, identifiers, branches, loops, functions, the cognitive weights reflect difficulty level of individual structure of the source code for developers to develop and understand the software. This can further be used to measure the effort required test and maintain the software. It has been found that the functional complexity of the source code is relying on three factors: inputs, outputs, and the architectural flow [4]. The cognitive complexity focuses on the both internal and control structure of the software system. Some of the related work regarding cognitive complexity measurement for OO systems is presented in [5–9]. A very traditional and common method to measure the physical size of a software system is Lines of code (LOC) [10]. This measuring technique is not so much effective because it is user dependent and also due to the software flexibility.

The rest of the paper is organized as follows: Sect. 2 presents the Weyuker's property. Section 3 presents the proposed cognitive metric along with the analytical evaluation and empirical results. Section 4 presents the conclusion and future scope of this work.

2　Weyuker's Property

Weyuker's property [11] are described in Table 1. The notations utilized in this work are described as follows: P, Q and R signify the classes; P + Q indicates the grouping of class P and class Q; M denotes the selected metric and M (P) denotes the value of the selected metric for class P; and $P \equiv Q$ (both P and Q classes are equivalent), in the other means that two class designs deliver the same functionality to each other. The definition of two united classes into one is taken as the same as recommended by [12], i.e. the grouping of both the classes outcome in another third class whose properties (methods/functions and variables) are the union of the properties of both the component classes. Weyuker properties are utilized to validate a new complexity measure. A good measure satisfies the maximum properties.

3　Propose Cognitive Complexity Metric

Mainly a program consists of three structures: Sequential, Branch, and Iteration. Wang et al. [7] describes the cognitive weight of each BCSs.

This section presents the definition of propose cognitive complexity metric along with the analytical and empirical results.

Table 1 Weyuker property

Property no.	Property name	Description
1	Non-coarseness	Given a class P and a metric M, another class Q can always be found such that, M (P) ≠ M (Q)
2	Granularity	There is a finite number of classes having the same metric value. This property will be met by any metric measured at the class level
3	Non-uniqueness (notion of equivalence)	There can exist distinct classes P and Q such that, M (P) = M (Q)
4	Design details are important	For two class designs, P and Q, which provide the same functionality, it does not imply that the metric values for P and Q will be the same
5	Monotonicity	For all classes P and Q the following must hold: M (P) ≤ M (P + Q) and M (Q) ≤ M (P + Q) where P + Q implies the combination of P and Q
6	Non-equivalence of interaction	∃ P, ∃ Q, ∃ R such that M (P) = M (Q) does not imply that M (P + R) = M (Q + R)
7	Interaction among statements	Not considered for OO metrics
8	No change on renaming	If P is renaming of Q then M (P) = M (Q)
9	Interaction increases complexity	∃ P and ∃ Q such that: M (P) + M (Q) < M (P + Q)

3.1 Definition of Proposed Object-Oriented Cognitive Complexity (POOCC)

The proposed cognitive complexity metric for OO program is defined as:

$$POOCC = N_{operands} + N_{operators} + W_c + RASP$$

where,

$N_{operands}$ is the number of operands available in a method,

$N_{operators}$ is the number of operators available in a method,

W_c is the cognitive weight of all basic control structures, and

RASP is the Ratio of Accessing Similar Parameters of a class.

RASP is calculated by the intersection of the methods on the basis of used parameters of the concerned class. The same is applied to all the methods and the resulted sum is divided by the number of parameters of the class.

$$RASP = \sum M_i \cap M_j / na$$

where M is the method and na indicate the number of instance variables in a particular class.

$$AMP = \left(\sum_{i=1}^{n} M_i \right) / n$$

where, AMP represents Average Method Parameters, M is the method and n represent the number of methods/functions in a class.

As the number of instance variables, number of operators and W_c increases POOCC also increases as a result program complexity increases.

AMP provides how each method of the class accessing n parameters on an average.

3.2 Evaluation of POOCC Against Weyuker Properties

Property 1 and Property 3: $(\exists P)(\exists Q)(|P| \neq |Q|$ and $|P| = |Q|)$. Where P and Q are the two different classes. These properties state that there are some classes whose cognitive values are equally complex or different respectively. Refer to Table 2, program 1 and 2 satisfies property 1 and program 2 and 4 satisfies property 3.

Table 2 Analysis of metric values for 16 C++ classes

Prog. No	POOCC	RASP	AMP	Wc	LOC	CE	Ref. code
1	85	3	0.75	28	48	1.77	216–218
2	43	1	2.00	19	32	1.34	224–225
3	23	3	1.00	7	36	0.64	231–232
4	43	3	1.20	11	52	0.83	233–234
5	84	1	1.50	21	60	1.40	235–236
6	16	1	1.00	3	16	1.00	238–239
7	37	1	3.00	9	26	1.42	265–266
8	60.33	3.33	2.00	13	45	1.34	267–269
9	25	3	1.50	5	24	1.04	280–281
10	51	1	1.33	13	32	1.59	306–307
11	50	3	4.00	9	34	1.47	307–308
12	34	6	1.00	13	24	1.42	312–313
13	58	3	4.00	13	44	1.32	314–316
14	54	3	4.00	9	40	1.35	316–317
15	49	6	2.00	9	34	1.44	318–319
16	37	3	2.00	9	29	1.28	320–321

Total POOCC = 749.33 Total LOC = 576

Property 2: Granularity: There are a finite number of classes having the same metric value. This property will be met by any Cognitive Metric which is measured at the class level.

Property 4: Design details are important: For any two classes of design say (P and Q), which provides the similar functionalities; it does not imply that cognitive metric values for class P and class Q will be the same. It is because of the number of operands and operators used in a method is design implementation dependent. Hence, it is satisfied.

Property 5: $(\forall P)(\forall Q)(|P| \leq |P; Q|$ and $|Q| \leq |P; Q|)$.

The program 4 and program 6 are considered. The cognitive complexity of these two programs is '43' and '16'. Both programs have a sequential structure. A program is given in Appendix. This program combines the functionalities of program 4 and program 6. And the cognitive complexity of this program is '52'. Since, it is clear from this example that the proposed approach holds this property.

Property 6: Non-equivalence of interaction: program 2 (P) and program 4 (Q) are used which have same cognitive complexity '43'. Let program 3 (R) is incorporated in P and Q. The resulted program gives the cognitive complexity of both the program is '65' and '56' respectively. This is due to the similar function in program Q as in program R. Hence, this property is satisfied by the proposed cognitive complexity approach.

Property 7: There are program bodies P and Q such that Q is formed by permuting the order of the statement of P and $(|P| \neq |Q|)$. The proposed cognitive measured utilized the operands, operators, basic control structures. No matter where they are located in the program. Hence, this property is not satisfied by the proposed approach.

Property 8: If P is renaming of Q, then $|P| = |Q|$

Proposed cognitive complexity measure uses different parameters of the program, and it has no relation with the name of a program. So, this property is clearly satisfied by the proposed approach.

Property 9: $(\exists P) (\exists Q) (|P| + |Q|) < (|P;Q|)$

Consider the program 4 and program 6 which have '43' and '16' cognitive complexity. When the functionality of both the program is added into one, then the cognitive complexity of the third developed program shown in Appendix, is '52'. Total complexity of program P and Q is 43 + 16 = 59, and 59 > 52. Hence, the proposed measure does not hold this property.

3.3 Empirical Results

This section presents the summary of the results that is based on data collected from [13].

3.4 Comparative Study of POOCC and LOC

This subsection presents the evaluation of POOCC and Lines of Codes (LOC) by using 16 C++ classes which are collected from [13]. The paper has also attempted to present the relationship between POOCC versus LOC for analyzing the nature of coding efficiency.

Each concerned programs of this work is examined the POOCC in terms of a cognitive weight unit (CWU) which indicates the complexity level of individual program, as CWU increases means the complexity increases and vice versa (the results is given in Table 2 and shown in Figs. 1 and 2).

LOC measure the software as physical size or length. This measure can be used to estimate the characteristics of the program such as required effort and difficulty level of the source code. However, the LOC illustrates only one particular aspect that is the static length of software. Because, it does not consider the functionality of the software. An interesting relationship between POOCC and LOC is known as Coding Efficiency (CE) can be measured as follows:

$$CE = POOCC/LOC[CWU/LOC]$$

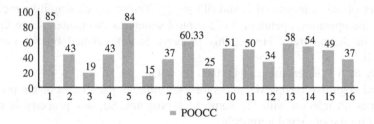

Fig. 1 POOCC values for 16 C++ classes

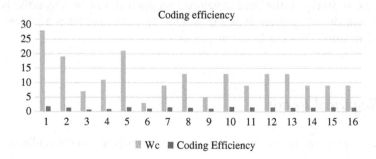

Fig. 2 Coding efficiency between LOC versus POOCC

Table 2 provides the coding efficiency of all the 16 programs according to the study of this paper and the average coding efficiency (CE) for the 16 classes is given below:

$$CE = POOCC/LOC$$
$$= 749.33/576 \approx 1.3009$$

The throughout coding efficiency of the classes is ranging from 0.64 to 1.77 CWU/LOC, i.e. 1.0 LOC = 1.3009 CWU on an average. The value obtained for CE indicates that as CWU increases per LOC, classes will become more complex to understand as it includes more number of operands and methods that increase the entire complexity of a class. Certain interesting observation has been made from the result of Table 2, Figs. 1 and 2 which are as follows:

- From Table 2, it is found that the LOC and POOCC follow the same pattern while increasing and decreasing the complexity. As the value of LOC increases, so does the corresponding POOCC and vice versa. It is significant to note that there are two points for which the POOCC grows up (see painted with bold and italics row which is mentioned for POOCC and LOC, given in Table 2). This discussion indicates that both the two classes have greater coding efficiency. Fewer LOC implements more complex code as mentioned by BCSs weight $W_c = 28$ with LOC = 48 and POOCC = 85 for program 1 and BCSs weight $W_c = 21$ with LOC = 60 and POOCC = 84 for program 5.
- Figure 1 indicates the actual cognitive complexity values for all classes. High POOCC values indicates more complexity, as it involves more number of operands, operators, BCSs weight and ratio of accessing similar parameters of the classes and vice versa with low POOCC.
- From Fig. 2, it is found that as BCSs weight W_c increases, CE of a class is also increase as it predicting more complex classes with fewer or more LOCs.

4 Conclusion and Future Scope

This paper presents a new cognitive complexity metric for OO programs. The new metric is evaluated analytically with the help of Weyuker property and validating empirically against a set of 16 C++ classes. Cognitive weights have been introduced for different control structures to measure the complexity of all the 16 C++ programs. An experiment has done to examine the correlation between POOCC and LOC of all the classes. From the results and comparative study mentioned above in Sects. 3.3 and 3.4, it has been observed that POOCC is a good metric for OO program to measure the cognitive complexity based on the cognitive weights. It also observed from analytical evaluation mentioned in Sect. 3.2 property 9 is not satisfied. The reason for POOCC metric not satisfying the property 9 of Weyuker's, because by decomposing a class, there is an overall increase in the $N_{operands}$, $N_{methods}$, W_c and RASP values for all the generated subclasses. In other means, the complexity will be increased.

The future scope focuses on some other essential factors: (1) Programs utilized in this study are very small as compared to the huge software system. The extension of this work can be carried out with large systems with automated tool besides with some other existing cognitive complexity measures to evaluate the effect on OO programs. (2) The proposed work can be further extended to estimate the required effort, cost and time to develop the software projects. (3) This work can also be extended to estimate the required effort to test and maintain the software.

Appendix: Combination of Program 4 and Program 6

```
class first{
int j;
public:
void input(void);
void show(void);
friend void change (first &, second &);
friend void add(first &, second &);
};
void item : : input(){
cout<<"enter the value";
cin<<j;}
void item : : show (){
cout<<j;
}
class second{
int k;
public:
void input(void);
void show(void);
friend void change (first &, second &);
friend void add(first &, second &);
};
void item : : input(){
cout<<"enter the value";
cin<<k;}
void item : : show (){
cout<<k;
}
void change(first &x, first &y){
int temp;
```

```
temp=x.j;
x.j=y.k;
y.k=temp;
}
void add(first &x, second &y){
int c=x.j+x.k;
cout<<"\nthe addition is="<<c;
}
int main(){
first f;
second s;
f.input();
s.input();
change(s,f);
cout<<"\nafter change";
f.show();
s.show();
f.input();
s.input();
cout<<"\n the addition =";
add(s,f);
return 0;
}
```

References

1. IEEE CS: IEEE Standard Glossary of Software Engineering Terminology, IEEE Standard 610.12 (1990)
2. Wang, Y.: On cognitive informatics. In: Proceedings of IEEE (ICCI'2002), pp. 34–42 (2002)
3. Wang, Y.: On the informatics laws of software. In: Proceedings of 1st IEEE International Conference on Cognitive Informatics (2002)
4. Wang, Y.: On the cognitive informatics foundations of software engineering. In: Proceedings of 3rd IEEE International Conference on Cognitive Informatics (2004)
5. Mishra, S., Akman, I.K.: Weighted class complexity: a measure of complexity for object-oriented design. J. Inf. Sci. Eng. 24, 1689–1708 (2008)
6. Mishra, S., Akman, I.K., Koyuncu, M.: An inheritance complexity metric for object-oriented code: a cognitive approach. Indian Acad. Sci. 36(3), 317–337 (2011)
7. Wang, Y., Shao, J.: A new measure of software complexity based on cognitive weights. Can. J. Electr. Comput. Eng. 28(2), 69–74 (2003)
8. Aloysius, A., Arockiam, L.: Coupling complexity metric: a cognitive approach. IJITCS, MECS Publisher, 9, pp. 29–35 (2012)
9. Gupta, V., Chabbra, K.J.: Object-oriented cognitive spatial complexity measures. Int. J. Electr. Electr. Eng. 3(6), 370–377 (2009)
10. Kearney, J.K., Sedlmeyer, R.L., Thompson, W.B., Gary, M.A., Adler, M.A.: Software Complexity Measurement. ACM Press, New York, vol. 28, pp. 1044–1050 (1986)

11. Weyuker, J.E.: Evaluating software complexity measures. IEEE Trans. Softw. Eng. **14**, 1357–1365 (1998)
12. Chidamber, S.R., Kemerer, C.F.: A metrics suite for object oriented design. IEEE Trans. Softw. Eng. **20**(6), 476–493 (1994)
13. Kamthane, A.N.: Object-Oriented Programming with ANSI & Turbo C++, 4th edn. Pearson Education

Measurement of Semantic Similarity: A Concept Hierarchy Based Approach

Shrutilipi Bhattacharjee and Soumya K. Ghosh

Abstract Resolving semantic heterogeneity is one of the major issues in many fields, namely, natural language processing, search engine development, document clustering, geospatial information retrieval and knowledge discovery, etc. Semantic heterogeneity is often considered as an obstacle for realizing full interoperability among diverse datasets. Appropriate measurement metric is essential to properly understand the extent of similarity between concepts. The proposed approach is based on the notion of concept hierarchy which is built using a lexical database. The *WordNet*, a semantic lexical database, is used here to build the semantic hierarchy. A measurement metric is also proposed to quantify the extent of similarity between a pair of concepts. The work is compared with existing methodologies on *Miller-Charles* benchmark dataset using three correlation coefficients (*Pearson's*, *Spearman's* and *Kendall Tau rank* correlation coefficients). The proposed approach is found to yield better results than most of the existing techniques.

Keywords Semantic heterogeneity · Concept hierarchy · *Wordnet* · Correlation coefficient

1 Introduction

Semantic similarity is a context dependent and dynamic phenomenon [1, 2]. Proper analysis of semantic relationship is required to identify the association between two concepts. Appropriate measurement is also needed to rank the concepts according to the extent of similarity. Different lexical databases like *WordNet* [3], *Nexus* [4],

S. Bhattacharjee · S.K. Ghosh (✉)
School of Information Technology, Indian Institute of Technology,
Kharagpur 721302, West Bengal, India
e-mail: skg@iitkgp.ac.in

S. Bhattacharjee
e-mail: shrutilipi.2007@gmail.com

© Springer India 2016
A. Nagar et al. (eds.), *Proceedings of 3rd International Conference on Advanced Computing, Networking and Informatics*, Smart Innovation, Systems and Technologies 44, DOI 10.1007/978-81-322-2529-4_43

407

etc. have been established to analyze the semantic relationship between a pair of concepts. Some important semantic relations are, *synonym* (different concepts with almost identical or similar meanings), *hyponym* or *is-a* (concept or phrase whose semantic field is included within that of another concept), *hypernym* (concept that stand when their extensions stand in the relation of class to subclass) [3] etc.

One of the basic methods to find semantic similarity is to calculate the number of hops (concepts) in the shortest path [5] connecting two concepts in the semantic databases/taxonomy. But, with the same semantic distance, any two pairs of concepts may not be equally similar conceptually. Li et al. [6] have proposed a non-linear model which combines structural semantic information (from a lexical taxonomy) and information content (from a corpus). Cilibrasi et al. [7] have proposed a distance metric, *normalized Google distance* (NGD), between concepts using page-counts retrieved from a web search engine. It is defined as the normalized information distance between two strings. However, the *homonym* relationship has not been considered here. Thus, for non-independent, hierarchical taxonomies, it produces suboptimal results. Sahami et al. [8] have proposed a method that measures semantic similarity between two queries using snippets returned by the search engine. In [9], a feature vector is formed using frequencies of two thousand lexical patterns in snippets and four co-occurrence measures are considered, namely, *dice coefficient, overlap coefficient, jaccard coefficient* and *point-wise mutual information*. In our previous work [2, 10], we have identified the requirements of semantic resolution of query in the field of spatial analysis. However, proper quantification of the semantic extent between concepts is still lacking. In case of hierarchical distance based similarity analysis, the general *context resemblance* method [11] has been used extensively in many applications. However, this method is constrained by some criteria which are not always applicable in every domain, for example spatial decision making [10].

Most of the approaches stated above, have not considered the context and the domain in which the concepts are related. Proper analysis of semantic relations between concept pairs is still a challenge. The proposed approach considers the context of retrieval considering the shortest path (if there exists multiple paths) between two concepts. Domain expert can choose the domain of retrieval, thus avoiding the problem of handling *homonyms*. The proposed approach is compared with other existing schemes of similarity measurement on *Miller-Charles* (*MC*) benchmark dataset [12]. The *MC* dataset has been frequently used to benchmark any semantic similarity measures. The concept pairs are rated on a scale from 0 (no similarity) to 4 (perfect *synonymy*). The degree of correlation between the human ratings in the *MC* dataset and the similarity scores produced by different approaches (including the proposed approach) will show the efficacy of the proposed scheme. The contributions of the proposed work are as follows:

- Proposing a similarity measurement metric between any pair of concepts, considering the context and the domain of retrieval

- Comparison of the proposed metric with existing methodologies on *Miller-Charles* benchmark dataset in terms of correlation coefficients
- Implementation of the proposed methodology using Protégé and the measurement of similarity between concept-pairs

2 SHSM: Semantic Hierarchy Based Similarity Measurement

The proposed *SHSM* scheme aims at measuring the semantic similarity between concepts based on concept hierarchy. The hierarchy is derived from *WordNet* lexical database. Firstly, a concept hierarchy is formed using all the concepts in the domain of interest. This concept hierarchy contains the most general concept, *Entity* (of *WordNet*) as its *root*. For example, two concepts are considered, namely, *food* and *shore* in Fig. 1. The concept hierarchy is converted to *concept-vector* hierarchy by converting each concept (in concept hierarchy) to a *concept-vector*.

The *concept-vector* tuple is represented by <*name of the concept, synonym set of the concept, the hierarchy-vector of the concept*>. Algorithm 1 constructs the *hierarchy-vector*s of the concepts (in concept hierarchy) and the *query concept*. The semantic similarity score between query concept and each concept in the hierarchy is calculated using Algorithm 2.

2.1 Representation of Concept-Vectors

The semantic concept hierarchy, *H* is formed by using all the concepts under consideration. It is a sub-tree of *WordNet* hierarchy. Each concept in *H* is converted to its corresponding *concept-vector*.

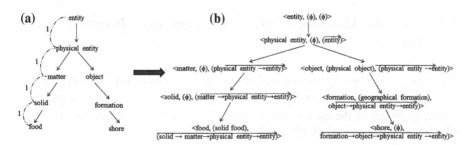

Fig. 1 *Concept-vector* representation of any hierarchy. **a** Concept hierarchy. **b** Concept-vector hierarchy

Concept-vector $\vec{C}$: $<N, S, \vec{V}>$ where, N: Name of the concept; S: Synset (set of all the *synonyms*) of the concept; V: Hierarchy vector of the concept.

Thus, a *concept* can be represented as $<N, (S_1, S_2, \ldots, S_m), \overrightarrow{(V_l \Rightarrow V_{l-1} \Rightarrow \ldots \Rightarrow V_1)}>$; where, *m*: number of *synonyms* in the synset; *l*: (number of concepts from *Entity* (most general concept of *WordNet*) to N) $- 1$; "$\Rightarrow$": navigation from specific concept to more general concept.

2.2   Generation of Concept-Vectors from Concept Hierarchy

The *concept-vector* can be represented as the full path from the most general concept of the word relator to the target concept. For any *hierarchy-vector* $\overrightarrow{(V_l \Rightarrow V_{l-1} \Rightarrow \ldots \Rightarrow V_1)}$, of the *concept-vector* $\vec{C}$, V_1 is the root element (*Entity*, in case of *WordNet*) and V_l is the direct *hypernym* of the concept C (with reference to *WordNet*); i.e., $\forall V_l$, $\exists V_{l+1}$, such that, $V_{l+1} = \vec{C}.Name$.

Algorithm 1 provides the procedure to convert each *concept* into its corresponding *concept-vector*. For example, the semantic hierarchy depicted in Fig. 1(a), is converted to *concept-vector* hierarchy by replacing each concept to its corresponding *concept-vector*. It is shown in the Fig. 1(b). Each *concept-vector* now consists of three fields: its name, *synonym* set and the *hierarchy-vector*. In Algorithm 1, the *for* loop of the line number 3–14 repeats for each concept in the hierarchy. Line 4 locates a particular concept N_i in *WordNet*. First the temporary concept variable W_1 is initialized as *Entity*, i as 1. The while loop in line number 7–9 backtracks the all possible paths in *WordNet* hierarchy until it reaches N_i. It stores the whole path from *Entity* to N_i in array $W[1..l]$. Line 12 forms the *hierarchy-vector* $\overrightarrow{N_i.hierarchy\text{-}vector}$ by processing the $W[1..l]$ backward from index $l-1$ to 1 in reverse order, i.e., $\overrightarrow{W_{l-1} \Rightarrow W_{l-2} \Rightarrow \ldots \Rightarrow W_1}$. Finally, it forms the *concept-vector* N_i with the triple, $<N_i.name, (N_i.synsets), \overrightarrow{(N_i.hierarchy - vector)}>$.

2.3   Semantic Matching of Query Concept with Hierarchy Concepts

The query concept Q is converted to its corresponding concept vector $\vec{Q} = <Q, QS, \overrightarrow{QV}> \equiv <Q, (QS_1, QS_2, \ldots, QS_m), \overrightarrow{(QV_l \Rightarrow QV_{l-1} \Rightarrow \ldots \Rightarrow QV_1)}>$; where, Q: Name of the query concept, QS: Synset (set of all the *synonyms*) of Q, QV: *Hierarchy-vector* of Q; QV_1 is *Entity*.

To check the extent of similarity between query concept Q, with any concept in H, it is compared with the one or more than one concept in the *concept-vector* hierarchy using Algorithm 2. In Algorithm 2, the *for loop* (line number 3–29) repeats for each concept in the hierarchy, if target concepts in the hierarchy is not known. Line 4–7 check whether *concept-vector* N_i and the query-vector Q are same or *synonymous*, by comparing the name field and the synset field of both N_i and Q. If same, Algorithm 2 returns the similarity value as 1 and both are said to be *synonymous*. If false, the *hierarchy-vector* of both N_i and Q are compared by processing the $W[1..l]$ array. The *while loop* (line number 10–12) checks how many concepts in both the *hierarchy-vectors* are matching; variable j returns the number of matched concept. Average similarity can be represented by the average of the fraction of matched concepts of both N_i and Q (ref. to line 13). The amount of dissimilarity between the concepts are calculated in different manner i.e., how much each concept contributes to amount of dissimilarity. If j is not equal to either of the length of the *hierarchy-vectors*, then dissimilarity between them is the mean of $\left(\frac{|Q.QV|-j}{|Q.QV|} + \frac{|N_i.V|-j}{|N_i.V|}\right)$ (line 15). So, both are *hyponym* of $N_i.V_j$. If, j matches with any one the *hierarchy-vector*'s length, then the latter vector is the *hyponym* of the former. Both of them contribute to the amount of dissimilarity equal to the number non-matching concepts in the *hierarchy-vector* of the latter concept (so, twice the amount of dissimilarity) (ref. to line 19). Finally, the similarity score is calculated by subtracting amount of dissimilarity from similarity, thus the absolute similarity score is found. Negative scores indicate no similarity, i.e., similarity score is 0.

> **Input:** Concept hierarchy H
> **Output:** *Concept-vector* $\overrightarrow{N_i}$
> 1　Generate_Concept_Vector(N_i)
> 2　{
> 3　**foreach** *concept N_i in H* **do**
> 4　　　Search *WordNet* with $N_i.name$;
> 5　　　i=1;
> 6　　　*WordNet* Concept, $W_i = Entity$;
> 7　　　**while** $N_i.name \neq W_i$ **do**
> 8　　　　　i++;
> 9　　　　　$W_i = W_{i-1}.hyponym$
> 10　　　**end**
> 11　　　l = i;
> 12　　　Form the *hierarchy-vector* as $\overrightarrow{N_i.hierarchy\text{-}vector} \equiv$
> 　　　　$W_{i-1} \Rightarrow W_{i-2} \Rightarrow ... \Rightarrow W_1$
> 13　　　Modify node N_i as $\overrightarrow{N_i} \equiv\ < N_i.name, (N_i.synsets), (\overrightarrow{N_i.hierarchy\text{-}vector}) >$
> 14　**end**
> 15　}

Algorithm 1: Conversion of concept hierarchy to *concept-vector* hierarchy

Table 1 Calculation of the similarity scores using proposed *SHSM* scheme and other existing methodologies on *Millar-Charles* standard datasets

Concept-pair	MC	Jaccard	Dice	Overlap	PMI	NGD	SH	CODC	SVM	RMSS	SHSM
Automobile-car	3.920	0.650	0.664	0.831	0.427	0.466	0.225	0.008	0.980	0.918	1
Journey-voyage	3.840	0.408	0.424	0.164	0.468	0.556	0.121	0.005	0.996	1.000	0.864
Gem-jewel	3.840	0.287	0.300	0.075	0.688	0.566	0.052	0.012	0.686	0.817	1
Boy-lad	3.760	0.177	0.186	0.593	0.632	0.456	0.109	0	0.974	0.958	0.850
Coast-shore	3.700	0.783	0.794	0.510	0.561	0.603	0.089	0.006	0.945	0.975	0.750
Asylum-madhouse	3.610	0.013	0.014	0.082	0.813	0.782	0.052	0	0.773	0.794	0.864
Magician-wizard	3.500	0.287	0.301	0.370	0.863	0.572	0.057	0.008	1	0.997	1
Midday-noon	3.420	0.096	0.101	0.116	0.586	0.687	0.069	0.010	0.819	0.987	1
Furnace-stove	3.110	0.395	0.410	0.099	1	0.638	0.074	0.011	0.889	0.878	0.600
Food-fruit	3.080	0.751	0.763	1	0.449	0.616	0.045	0.004	0.998	0.940	0.571
Bird-cock	3.050	0.143	0.151	0.144	0.428	0.562	0.018	0.006	0.593	0.867	0.864
Bird-crane	2.970	0.227	0.238	0.209	0.516	0.563	0.055	0	0.879	0.846	0.654
Implement-tool	2.950	1	1	0.507	0.297	0.750	0.098	0.005	0.684	0.496	0.813
Brother-monk	2.820	0.253	0.265	0.326	0.623	0.495	0.064	0.007	0.377	0.265	0.414
Crane-implement	1.680	0.061	0.065	0.100	0.194	0.559	0.039	0	0.133	0.056	0.524
Brother-lad	1.660	0.179	0.189	0.356	0.645	0.505	0.058	0.005	0.344	0.132	0.414
Car-journey	1.160	0.438	0.454	0.365	0.205	0.41	0.047	0.004	0.286	0.165	0
Monk-oracle	1.100	0.004	0.005	0.002	0	0.579	0.015	0	0.328	0.798	0.505
Food-rooster	0.890	0.001	0.001	0.412	0.207	0.568	0.022	0	0.060	0.018	0.181
Coast-hill	0.870	0.963	0.965	0.263	0.350	0.669	0.070	0	0.874	0.356	0.333
Forest-graveyard	0.840	0.057	0.061	0.230	0.495	0.612	0.006	0	0.547	0.442	0
Monk-slave	0.550	0.172	0.181	0.047	0.611	0.698	0.026	0	0.375	0.243	0.678
Coast-forest	0.420	0.861	0.869	0.295	0.417	0.545	0.060	0	0.405	0.150	0.100
Lad-wizard	0.420	0.062	0.065	0.050	0.426	0.657	0.038	0	0.220	0.231	0.478

(continued)

Table 1 (continued)

Concept-pair	MC	Jaccard	Dice	Overlap	PMI	NGD	SH	CODC	SVM	RMSS	SHSM
Cord-smile	0.130	0.092	0.097	0.015	0.208	0.460	0.025	0	0	0.006	0
Glass-magician	0.110	0.107	0.113	0.396	0.598	0.488	0.037	0	0.18	0.050	0
Rooster-voyage	0.080	0	0	0	0.228	0.487	0.049	0	0.017	0.052	0
Noon-string	0.080	0.116	0.123	0.040	0.102	0.488	0.024	0	0.018	0.000	0
Pearson's correlation coefficient		0.260	0.267	0.382	0.549	0.205	0.580	0.694	0.834	0.867	0.872
Spearman's rank correlation coefficient		0.395	0.396	0.403	0.518	0.148	0.619	0.702	0.818	0.849	0.869
Kendall tau rank correlation coefficient		0.266	0.266	0.274	0.332	0.112	0.447	0.544	0.635	0.667	0.712

3 Comparative Study

A comparative study of the proposed *SHSM* with other existing methodologies is presented in Table 1. The *Miller-Charles* dataset and ratings (*MC*) have been considered to evaluate the performance of the algorithms. It contains 30 word pairs rated by a group of 38 human subjects. The word pairs are rated on a scale from 0 (no similarity) to 4 (perfect *synonymy*). But most researchers have used only 28 pairs for evaluations, as other two word pairs are missing from earlier version of *WordNet*. The considered semantic similarity measures in Table 1 are: *jaccard, dice, overlap, PMI* [13], *normalized Google distance* (NGD) [7], *Sahami and Heilman* (SH) [8], *co-occurrence double checking model* (CODC) [14], *support vector machine-based approach* (SVM) [13], and *relational model based similarity measurement approach* (RMSS) [9]. The similarity scores, evaluated by all the similarity measures are shown in Table 1. The above mentioned approaches (along with the *SHSM* technique) are evaluated on *MC* dataset using three correlation coefficients [15], namely, *Pearson's correlation coefficient, Spearman's rank correlation coefficient* and *Kendall tau rank correlation coefficient* (shown in the last three rows in Table 1). All the similarity scores (from column 3–12 of Table 1), except for the human-ratings in *Miller-Charles* dataset (*MC*) (column 1 of Table 1), are normalized to [0, 1] range for the ease of comparison. The correlation coefficient of a particular method (column-wise) is measured by taking the column against the respective scheme and *MC* data.

It may be observed from Table 1 that the proposed *SHSM* gives the highest correlation scores over other existing methodologies, for all the three correlation coefficient measurements. As *SHSM* similarity measure uses the contextual and domain information, it gives better performance by differentiating between *homonym*s.

Input: Hierarchy H, Query concept Q
Output: Semantic relation, Similarity score

1 Call Generate_Concept_Vector(Q); //Call Algotithm 1

2 Query vector $\overline{Q} =< Q.name, (Q.synsets), (\overrightarrow{Q.QV}) >$ **foreach** *concept-vector* $\overrightarrow{N_i}$ *in H* **do**

3   **if** $Q.name == N_i.Name \| Q.name == N_i.synsets \| Q.synsets == N_i.Name \| Q.synsets == N_i.synsets$ **then**

4   Similarity score $= 1$;

5   set Q as the *synonym* of N_i;

6   **end**

7   **else**

8          $j = 1$;

9   **while** $Q.QV_j = N_i.V_j$ **do**

10   j++;

11   **end**

12   Avg.Similarity $= \frac{\frac{j}{|Q.QV|} + \frac{j}{|N_i.V|}}{2}$;

13   **if** $j \neq |Q.QV| \; \&\& \; j \neq |N_i.V|$ **then**

14   Avg.Dissimilarity $= \frac{\frac{|Q.QV|-j}{|Q.QV|} + \frac{|N_i.V|-j}{|N_i.V|}}{2}$

15   set Q and N_i as *hyponyms* of $N_i.V_j$;

16   **end**

17   **if** $j \neq |Q.QV| \oplus j \neq |N_i.V|$ **then**

18   Avg.Dissimilarity $= 2 * \frac{\max\{\frac{|Q.QV|-j}{|Q.QV|}, \frac{|N_i.V|-j}{|N_i.V|}\}}{2}$

19   **if** $j = |Q.QV_j|$ **then**

20   set N_i as the *hyponym* of Q;

21   **end**

22   **else**

23   set N_i as the *hypernym* of Q;

24   **end**

25   **end**

26   Similarity score $= \max\{0,$ Avg.Similarity - Avg.Dissimilarity$\}$

27   **end**

28 **end**

Algorithm 2: Semantic matching of concepts

4 Conclusion

In this paper, an approach for semantic hierarchy based similarity measurement between a pair of concepts has been proposed. The proposed metric has been tested with the *Millar-Charles* benchmark dataset and found to produce better results than most of the existing methodologies. This work is being extended further for the applications like information retrieval in geospatial domain. Measuring the semantic similarity between different geospatial concepts for various applications is considered as the future prospect of this work.

References

1. Shvaiko, P., Euzenat, J.: A survey of schema-based matching approaches. In: Spaccapietra, S. ed.: Journal on Data Semantics IV. Volume 3730 of Lecture Notes in Computer Science, pp. 146–171. Springer, Berlin Heidelberg (2005)
2. Bhattacharjee, S., Ghosh, S.K.: Automatic resolution of semantic heterogeneity in GIS: an ontology based approach. In: Advanced Computing, Networking and Informatics, vol. 1, pp. 585–591. Springer (2014)
3. Miller, G.A.: WordNet: a lexical database for English. Commun. ACM **38**(11), 39–41 (1995)
4. Jannink, J.F.: A word nexus for systematic interoperation of semantically heterogeneous data sources. PhD Thesis, Stanford University (2001)
5. Rada, R., Mili, H., Bicknell, E., Blettner, M.: Development and application of a metric on semantic nets. IEEE Trans. on Syst. Man Cybern. **19**(1), 17–30 (1989)
6. Li, Y., Bandar, Z.A., McLean, D.: An approach for measuring semantic similarity between words using multiple information sources. IEEE Trans. Knowl. Data Eng. **15**(4), 871–882 (2003)
7. Cilibrasi, R.L., Vitanyi, P.M.: The Google similarity distance. IEEE Trans. Knowl. Data Eng. **19**(3), 370–383 (2007)
8. Sahami, M., Heilman, T.D.: A web-based kernel function for measuring the similarity of short text snippets. In: Proceedings of the 15th International Conference on World Wide Web, pp. 377–386. ACM (2006)
9. Bollegala, D., Matsuo, Y., Ishizuka, M.: A relational model of semantic similarity between words using automatically extracted lexical pattern clusters from the web. In: Proceedings of the 2009 Conference on Empirical Methods in Natural Language Processing, vol. 2-vol. 2, pp. 803–812. Association for Computational Linguistics (2009)
10. Bhattacharjee, S., Mitra, P., Ghosh, S.K.: Spatial interpolation to predict missing attributes in GIS using semantic kriging. IEEE Trans. Geosci. Remote Sens. **52**(8), 4771–4780 (2014). doi:10.1109/TGRS.2013.2284489
11. Manning, C.D., Raghavan, P., Schütze, H.: Introduction to Information Retrieval, vol. 1. Cambridge University Press, Cambridge (2008)
12. Miller, G.A., Charles, W.G.: Contextual correlates of semantic similarity. Lang. Cogn. Proc. **6**(1), 1–28 (1991)
13. Bollegala, D., Matsuo, Y., Ishizuka, M.: Measuring semantic similarity between words using web search engines. WWW **7**, 757–766 (2007)
14. Chen, H.H., Lin, M.S., Wei, Y.C.: Novel association measures using web search with double checking. In: Proceedings of the 21st International Conference on Computational Linguistics and the 44th Annual Meeting of the Association for Computational Linguistics, Association for Computational Linguistics, pp. 1009–1016 (2006)
15. Chen, V.Y.J., Chinchilli, V.M., Donald St, P.R.: Robustness and monotonicity properties of generalized correlation coefficients. J. Stat. Planning Infer. **141**(2), 924–936 (2011)

Evaluating the Effectiveness of Conventional Fixes for SQL Injection Vulnerability

Swathy Joseph and K.P. Jevitha

Abstract The computer world is definitely familiar with SQL as it plays a major role in the development of web applications. Almost all applications have data to be stored for future reference and most of them use RDBMS. Many applications choose its backend from the SQL variants. Large and important applications like the bank and credit-cards will have highly sensitive data in their databases. With the incredible advancement in technology, almost no data can survive the omniscient eyes of the attackers. The only thing that can be done is to make the attackers work difficult. The conventional fixes help in the prevention of attacks to an extent. However, there is a need for some authentic work about the effectiveness of these fixes. In this paper, we present a study of the popular SQL Injection Attack (SQLIA) techniques and the effectiveness of conventional fixes in reducing them. For addressing the SQLIA's in depth, a thorough background study was done and the mitigation techniques were evaluated using both automated and manual testing. We took the help of a renowned penetration testing tool, SQLMap, for the auto-mated testing. The results indicate the importance of incorporating these mitigation techniques in the code apart from going for complex fixes that require both effort and time.

Keywords Web-attacks · SQLIA · SQL injection

S. Joseph (✉) · K.P. Jevitha
Department of Computer Science and Engineering,
Amrita Vishwa Vidyapeetham, Coimbatore, India
e-mail: swathyjoseph90@gmail.com

K.P. Jevitha
e-mail: kp_jevitha@cb.amrita.edu

© Springer India 2016
A. Nagar et al. (eds.), *Proceedings of 3rd International Conference on Advanced Computing, Networking and Informatics*, Smart Innovation, Systems and Technologies 44, DOI 10.1007/978-81-322-2529-4_44

417

1 Introduction

SQL Injection is defined by OWASP as "an attack that consists of insertion or injection of a SQL query via input data from client to the application" [1]. SQLIA has probably existed for ages, dating back to when SQL databases were first linked to web applications. Despite its age, it still persists. SQLIA tops the recent list of OWASP's Top 10 web-attacks [2]. The evidences that these attacks still exist are the Lizamoon [3] and the Lilupophilupop [4] attacks that compromised almost a million websites.

The key objective of this work is to evaluate the effectiveness of the conventional techniques in reducing the SQLIA. It was seen that the use of legacy code in application development is one of the main reasons for this attack. The other reasons include lack of input sanitization, architectural issues etc. We hope that our work throws some light on the effectiveness of the code level defense techniques. The important SQLIA's considered here are the boolean-based blind, stacked queries, time-based blind attack and the error-based injection attacks [5]. The conventional fixes for SQLIA evaluated in this work include the use of parameterized queries, whitelist validation and stored procedures.

The rest of the paper is structured as follows: In Sect. 2, we present the related work in this domain. In Sect. 3, we give a briefing about the different SQLIA techniques and the intent for using them. In Sects. 4 and 5, we describe the use of SQLMap tool and the testbed setup. In Sect. 6, we discuss about the different defensive coding techniques and the implementation of code level mitigations. In Sect. 7, we present the evaluation results and we conclude in Sect. 8.

2 Related Works

Diallo et al. [6] has a survey on the various dimensions of SQLIAs that includes the different classes of attacks and the available approaches against them. Bono and Domangue [7] describe the application of attack ideologies and the mitigations. Input sanitization, prepared statements, stored procedures, principle of least privileges and security audits are listed out in this paper. Shar and Tan [8] conclude that the best strategy against the injection attacks is the integration of defensive coding with runtime prevention methods. This work also gives a summary of the various detection methods and run-time prevention methods. Ahmad et al. [9] has a detailed study of each attack for the purpose of categorization of SQLIAs. This paper categorizes the SQLIA into order-wise attacks, those against the database and finally as blind attacks. They conclude that this classification can help in reducing the possibility of occurrence of the vulnerability. Bisht et al. [10] throw some light onto the prevention of SQLIA. Jane and Chaudhari [11] propose an approach which is related to the inference of the intended query structure within an application. This paper states that, for an input to be a useful candidate, it should be benign and the program's control must flow in the same path. The programmer intention is mined out and hence the approach is concluded to be quite promising. Halfond and Osro [12] propose a technique that uses finite state machines for the detection of SQLIAs and thereby preventing them. This technique finds the hotspots and creates automaton for every hotspot. When a query is issued, this is verified against its corresponding automata model and only if they match the user is allowed to continue.

3 Important SQLIA Techniques

In this section, we present and discuss the important kinds of SQLIAs known to date [13]. These techniques are used either individually or a few of them together to perform the attack. Out of the described techniques, time-based and boolean based attacks are inferential attacks.

Boolean-based blind technique: The attacker has no direct knowledge about the authentication he has to obtain. This can be done by attaching true or false statements along with legitimate SQL queries and observing the responses.

Time-based technique: The intent of the attacker is to verify the precision of his guesses based on the time injection queries used. SQL statements used in this type of injection attack holds the back-end database for a certain number of seconds. The attacker notes the response time to infer if the injection is successful or not.

Error-based technique: This kind of attack exploits the error messages returned by the database management software. If these are observed carefully, database fingerprinting might be possible. Once the attacker gets to know the details of the backend, he can use technology-specific methods for the attack. But this kind of attack works only for the applications that are configured to disclose back-end dbms error messages.

Piggy-backed technique: This technique is also known as the stacked queries method. This attack uses multiple queries separated by a semicolon which can be applied for data manipulations.

These are the first-order injection techniques. There are second-order SQLIAs that take in the attack-intended input normally and the effect of that input will be noticeable only from the next access to the database.

4 SQLMap

The SQLMap [5] tool, is an open source penetration testing tool developed in Python. It automates the detection and exploitation of SQL injection flaws and takes advantage of the vulnerabilities to gain access to the contents in the backend database system. It supports the exploitation of 5 different injection types and provides support for a number of databases. The tool uses the bisection algorithm

```
python sqlmap.py -u "http://localhost:8080/bookstore_current-0.1-dev/Login.jsp"
POST --data="FormName=Login&Login=admin&Password=admin&FormAction=login&
ret_page=&querystring=" -p "Login" --proxy "http://127.0.0.1:8081" --beep --risk=1
--level=2
```

Fig. 1 Sample SQLMap usage

for the implementation of boolean-based and time-based injections. Figure 1 provides a sample usage of the SQLMap. The tool was used to test all the applications by changing the level and risk options available within the tool. The risk argument specifies the risk of tests that were to be done like the default tests or heavy query tests. The level argument specifies the level of tests to be performed. The number of tests performed increases as the level value is increased. The http request is given as a parameter and p indicates the injectable parameter.

5 Testbed Description

This section gives a description of the testbed used for evaluating the different SQLIA prevention techniques. For the purpose of comparison, we have used four different test applications from the Amnesia [12] Testbed. The subjects are of varying sizes. A brief description of what the applications dealt with, is given below:

- Bookstore: online purchasing of books.
- Classifieds: advertise items and pets to be sold.
- Employee directory: details of the employees in a company.
- Events: details of various events to be held/hosted.

The applications were Java based with MySQL 5.6 as their back-end. These applications were deployed using tomcat on windows 7. The SQLMap gives the penetration tester, a broad range of options which eases his work. Initially, on the raw code, with no fixes implemented, the SQLMap tool was run. It was seen that 3 types of injections were possible for all the applications. The injections possible were error-based, time-based and boolean-based blind [5]. The manual testing on these applications also showed the same result. Table 1 gives the vulnerability of each application

Table 1 Number of injectable pages per application

Application	Number of web-pages per application	Number of pages injectable
Bookstore	19	2
Classifieds	20	3
Employee directory	14	2
Events	13	2

6 Mitigation Techniques

Three mitigation techniques were tried out—the parameterized queries, the stored procedures and the whitelist validation.

Parameterized queries: This is language dependent. Java's JDBC provides the use of prepared statement class, while PHP provides the PHP Data Object (PDO) package. This supports the placeholders and named parameters. Since our applications are Java-based we go for the usage of prepared statements. Parameterized queries provide the benefit that they take the user input as such.

```
E.g.: PreparedStatement prepStat = con.prepareStatement("select mem_id,
mem_lvl from membrs where mem_login =? and mem_passwd=?");
```

can be used instead of:

```
rs = openrs( stat, "select mem_id, mem_lvl from membrs where
mem_login = " + toSQL(sLogin, adText) + " and mem_passwd="
+ toSQL(sPassword, adText));
```

Whitelist validation: The Blacklist validation is popular. However, the whitelist validation is more effective than the other one. Here we accept only what is known to be true. We took the help of regular expressions for implementing this and used the java.util.regex package. Table 2 shows the expressions used for the whitelist validation. Data type, data size, range etc. are to be kept in mind while designing a regular expression. So the bottom line is that, the applications would accept only the input in the mentioned formats.

Stored procedures: This makes the task easier when repetitive tasks are to be used. The SQL code is predefined in the database and then accessed from the application. Stored procedure method is also said to give the effect of prepared statements mentioned above, if used safely.

```
E.g.: begin select mem_id, mem_level into res1, res2 from membrs
      where mem_login = plog and mem_passwd = ppas;
```

Table 2 Sample regular expressions used for whitelist validation

Regular expression	Purpose
"\ \w*@\ \w*[.]\ \w{3}"	For accepting an email_id
"\ \d{13}"	For accepting credit card number
"\ \d{7}"	Phone number
"\ \w*"	First and last name
"\ \w{3,15}"	Login and password

We used this as the SQL procedure for accepting credentials for a login page in the application which was stored in the database and later accessed from application using:

```
String simpleProc = "{ call pbookstre2 (?,?,?,?) }";
```

Hence the test applications were re-coded using these three conventional fixes separately.

7 Evaluation and Results

In this section, we describe how the effectiveness of the above mentioned techniques were evaluated. The injection points were found out and then the tool was run over the re-coded applications. The results were taken in the same manner varying the risk and level parameters. Similarly the results were taken for every injection point. Studying the queries that caused the injections even after the implementation of the mitigation techniques revealed that they were complex and nested queries. An example of a http request which caused error-based injections possible bypassing the whitelist validation technique is:

```
Payload: FormName=Login&Login=admin' AND (SELECT 6410 FROM
(SELECT COUNT(*),CONCAT(CHAR (58, 103, 112, 97, 58), (SELECT
(CASE WHEN (6410=6410) THEN 1 ELSE 0 AND =login&
ret_page=&querystring=
```

Table 3 shows a sample reading taken for the bookstore application when prepared statements completely secured the application from injection attacks. Table 4 shows a sample reading for classifieds application. It was seen that the injections were possible even after the use of prepared statements. Most of the queries found its way to the data stored in the database through the virtual database of MySQL called information_schema.

The injections possible in the testbed applications were found to be of three major types: boolean-based blind, time-based blind and error-based injections [5]. The frequency of the injection attacks are depicted in Fig. 2. The time-based boolean attack seems to be used heavily for the exploitation purpose, followed closely by the error-based injection techniques.

The pie-chart in Fig. 3 shows the analysis of the fixes after recoding the applications. Out of all the injections possible after the application of the fixes, the analysis showed that 64 % of the queries bypassed the prepared statement fix. The queries were mostly of second-order. 22 % of the queries bypassed stored procedures and 14 % bypassed the whitelist validation. Our analysis leads to a conclusion that the prepared statements prevent the first-order SQL injections to an extent but not the second order injections. There are a few loopholes mentioned in [14] with the use of stored procedures. A code which uses a procedure which has

Table 3 A sample reading with zero injections possible when prepared statement is used

Risk	Level	Number of injections before applying fix	After applying fix 1: prepared statements	After applying fix 2: stored procedures	After applying fix 3: whitelist validation
1	1	3	0	0	0
1	2	3	0	0	0
1	3	3	0	0	0
1	4	3	0	0	0
1	5	3	0	0	0
2	1	3	0	0	0
2	2	3	0	0	0
2	3	3	0	0	0
2	4	3	0	0	0
2	5	3	0	0	0
3	1	3	0	1	3
3	2	3	0	1	3
3	3	3	0	1	3
3	4	3	0	1	3
3	5	3	0	1	3

Table 4 A sample reading for classifieds application which shows injections are possible even when prepared statement is used

Risk	Level	Number of injections before applying fix	After applying fix 1: prepared statements	After applying fix 2: stored procedures	After applying fix 3: whitelist validation
1	1	3	2	0	0
1	2	3	2	0	2
1	3	3	2	0	2
1	4	3	2	0	2
1	5	3	2	0	2
2	1	3	2	0	2
2	2	3	2	0	2
2	3	3	2	0	2
2	4	3	2	0	2
2	5	3	2	0	2
3	1	3	2	1	2
3	2	3	2	1	2
3	3	3	2	1	2
3	4	3	2	1	2
3	5	3	2	1	2

Fig. 2 Frequency graph of
the injection attacks possible

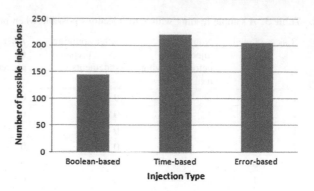

Fig. 3 Analysis of
effectiveness of the fixes

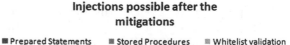

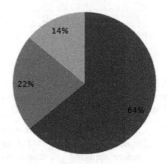

exec(@input) stored in its backend database will execute the input given by the user
regardless of what it is. While designing the stored procedures, the design of
parameters with larger sizes will make the application prone to SQL injections. The
designer should take ample care to allocate only limited sizes to the parameters in
the stored procedures. Given that the developer pays attention to the above criteria,
the stored procedures can effectively work against the first and second order
injections. The regular expressions used proved to be efficient. However, the one
used for accepting the user's name was seen to be injectable as we did not mention
any particular constraints for it. Integrating validation with both the techniques
mentioned above is almost unbreakable for the attackers. The use of regular
expressions for whitelist validation is the most effective way to do it. There is
nothing as good as a properly crafted regular expression for the validation. The
entire data in the database could be dumped if injection was possible, through the
injection points. Our analysis shows that the whitelist validation can control the first
and second order injections to a good extent by efficiently utilizing the power of
regular expressions.

Fig. 4 Response time

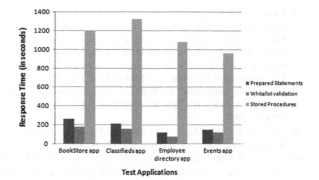

The Fig. 4 shows the response time of different applications to the SQLMap tool. The execution took approximately equal time for both prepared statements and whitelist validation. The time taken by the tool was very high for applications using stored procedures compared to the ones using the other two fixes. We believe that the difference in time for execution of applications that use stored procedure, is due to the fact that the actual functionality is stored in the database and hence access time is quite large.

8 Conclusion

The conventional mitigations techniques were evaluated based on parameters like the response time and decrease in the number of injections possible. The work converged to a conclusion that by the usage of the conventional mitigation techniques, we can reduce the number of injections considerably. Techniques which focus on prevention of the attack should incorporate defensive coding. Dynamically built SQL statements should be created properly with the conventional mitigation techniques mentioned rather than using the simple concatenation technique. This will make sure that only the legitimate queries are passed to the database server. Using validation along with the other two techniques proved to be a reasonable solution for this attack. The study of effectiveness of the conventional fixes showed their inability for ultimate eradication and this opens the scope for further research work.

References

1. OWASP: https://www.owasp.org/index.php/SQLInjection
2. OWASP Top 10 list: https://www.owasp.org/index.php/Top_10_2013-Top_10
3. LizaMoon the Latest SQL-Injection Attack: http://blogs.mcafee.com/mcafee-labs/lizamoon-the-latest-sql-injection-attack

4. Lilupophilupop: Tongue-twister SQL injection attacks pass one million mark: http://www. infosecurity-magazine.com/news/lilupophilupop-tongue-wister-sql-injection/
5. SQLMap: https://github.com/sqlmapproject/sqlmap/wiki
6. Kindy, D.A., Pathan, A.K.: A Detailed survey on various aspects of SQL injection in web applications: vulnerabilities, innovative attacks and remedies. In: International Journal of Communication Networks and Information Security, vol. 5, no. 2, pp. 80–92 August 2013
7. Bono, S.C., Domangue, E.: SQL Injection: A Case Study, Whitepaper Oct 2012
8. Shar, L.K., Beng, H., Tan, K.: Defeating SQL Injection. IEEE Comput. Soc. **46**(3), 69–77 (2013) (IEEE)
9. Ahmad, K., Shekhar, J., Yadav, K.P.: Classification of SQL injection attacks. In: VSRD-TNTJ, vol. 1, no. (4), pp. 235–242(2010)
10. Bisht, P., Madhusudan, P., Venkatakrishnan, V.N.: CANDID: Dynamic candidate evaluations for automatic prevention of SQL injection attacks. In: ACM Transactions on Information and System Security, vol. 13, no. 2, p. 139. ACM (2010)
11. Jane, P.Y., Chaudhari, M.S.: SQLIA: Detection and prevention techniques: a survey. IOSR J. Comput. Eng. **2**, 56–60. IOSR J. (2013)
12. Halfond, W.G.J., Orso, A.: AMNESIA: analysis and monitoring for neutralizing SQL injection attacks. In: Proceedings of the 20th IEEE/ACM International Conference on Automated Software Engineering, pp. 174–183. ACM, New York (2005)
13. Clarke, J.: SQL Injection Attacks and Defense. Elsevier Inc (2009)
14. Howard, M., LeBlanc, D.: Writing Secure Code, 2nd edn. Microsoft Press, Washington (2003)

Salt and Pepper Noise Reduction Schemes Using Cellular Automata

Deepak Ranjan Nayak, Ratnakar Dash and Banshidhar Majhi

Abstract Two filters in the light of two-dimensional Cellular Automata (CA) are presented in this paper for salt and pepper noise reduction of an image. The design of a parallel algorithm to remove noise from corrupted images is a demanded approach now, so we utilize the idea of cellular automata to cater this need. The filters are mainly designed according to the neighborhood structure of a cell with different boundary conditions. The performances of the proposed filters with that of existing filters are evaluated in terms of peak signal-to-noise ratio (PSNR) values and it has been observed that the proposed filters are extremely promising for noise reduction of an image contaminated by salt and pepper noise. The primary point of interest in utilizing these proposed filters is; it preserves more image details in expense of noise suppression.

Keywords Cellular automata (CA) · Boundary condition · Impulse noise · Peak signal-to-noise ratio · NNBCA · TFNBCA

1 Introduction

Image Noise Filtering is one of the most challenging approaches in digital image processing. Impulsive noise is one sort of noise, which is an often encountered problem in acquisition, transmission and processing of images. This noise is brought out by malfunctioning pixels in camera sensors, faulty memory locations in hardware, transmission in a noisy channel [1]. Impulsive noise is of two types: Salt

D.R. Nayak (✉) · R. Dash · B. Majhi
Department of Computer Science and Engineering, National Institute of Techonoloy,
Rourkela 769008, India
e-mail: depakranjannayak@gmail.com

R. Dash
e-mail: ratnakar.dash@gmail.com

B. Majhi
e-mail: bmajhi@nitrkl.ac.in

© Springer India 2016
A. Nagar et al. (eds.), *Proceedings of 3rd International Conference on Advanced Computing, Networking and Informatics*, Smart Innovation, Systems and Technologies 44, DOI 10.1007/978-81-322-2529-4_45

427

and Pepper Noise or Random Valued Impulsive Noise. In this paper, we limit our work to Salt and Pepper Noise. However, adequately addressing to this noise is still a pending research assignment.

It has been observed that the linear filtering technique is not desirable to remove impulse noise from the image. On account of good denoising power and computational efficiency, median filter was considered as the most popular non-linear filter for removing salt and pepper noise [2]. However, it destroys a few subtle elements like edges of the image when the noise density is more than 30 %. Subsequently, variations of median filters such as weighted median filter [3], center weighted median filter [4] have been proposed to enhance the result. But these methods do not differentiate noisy and non-noisy pixels and, therefore, do not preserve more details. Later adaptive median filter [5], progressive-switching median filter [6], and adaptive center weighted median filter [7] are suggested. These methods work in two stages; in the first stage noise detection is carried out and in the second stage filtering is performed. They are good at detecting noise, even at high noise density. A decision-based impulse noise filter for restoring images tainted by impulse noise is presented in [8]. This algorithm produces significantly better image quality over standard median filters. Many soft computing approaches have also been applied to restore images corrupted by impulse noise. In [9], a noise adaptive fuzzy switching median filter is proposed, which functions admirably at high noise level. However, most of the soft computing approaches suffer from limitations like lack of robustness, high costs of computation, high complexity and tuning of parameters [10]. By looking at these constraints, we introduce two novel filters based on cellular automata to address all the above issues.

Cellular automata have been effectively utilized as a part of the area of image processing in the last few decades due to its parallel computational behavior [11]. Cellular automata are composed of a regular grid of cells, where every cell can have a limited number of conceivable states. In every discrete time step, the state of a cell is determined by a transition rule that is designed by the states of its neighbors in the past time step. The state of every cell in the grid is updated in parallel. So, the rules of cellular automata are local and uniform [12]. If the cells are arranged in a linear array, it is called as a one-dimensional CA. In two-dimensional CA, the cells are organized in a 2D lattice. An image can be mapped as a two-dimensional CA where every cell relates to a pixel in the image, and the possible states are the gray values or colors [13]. Since the states of every cell are updated synchronously at a discrete time step, the time expected to solve a particular image processing errand is the minimum. Few researchers have proposed filters taking into account cellular automata and its varieties. For instance, in [10], a hybrid method based on cellular automata (CA) and fuzzy logic called Fuzzy Cellular Automata (FCA) is proposed to eliminate impulse noises from noisy images. Rosin in [11] has used the sequential floating forward search method for feature selection to select good rule sets for noise filtering.

In this paper, we introduce two filters based on CA for removing of salt and pepper noise from the images. The filters are outlined with the assistance of different neighborhood and boundary conditions. We utilize the statistical properties to develop the transition function of the proposed filters. These filters do not require any parameters to tune, and hence no training stage is needed. The proposed methods are tested on different images and found to produce better results in terms of the qualitative and quantitative measures of the image.

The paper is organized as follows. In Sect. 2, we introduce mathematical preliminaries of cellular automata. The proposed scheme is discussed in Sect. 3. Section 4 presents the experimental results followed by the conclusion in Sect. 5.

2 CA Preliminaries

CA can be represented analytically with five-tuples, $CA = \{L, N, Q, \delta, q_0\}$; where L represents the regular lattice of cells, Q is the finite set of states, q_0 is known as the initial state and $q_0 \in Q$, N is the neighborhood (of size $n = |N|$) and $\delta : Q^n \to Q$ is the transition function or rule of CA [13].

In case of 3-neighborhood one dimensional CA, the transition rule can be defined as

$$q_i^{t+1} = \delta(q_i^t, q_{i-1}^t, q_{i+1}^t) \tag{1}$$

where q_i^{t+1} and q_i^t represent the states of the ith cell at time instant $t + 1$ and t respectively, q_{i-1}^t and q_{i+1}^t denote the states of the left and right neighbors of the ith cell at time instant t, and δ is the transition rule. To tackle an issue utilizing CA, we need to characterize essential terms like the neighborhood structures, boundary conditions, and the initial conditions.

2.1 Neighborhood Structure

The neighborhood of a cell, known as center cell, comprises of the center cell and its encompassing cells whose states determine the next state of the core cell. Cellular automata utilize various neighborhood models for diverse sort of uses. But the models employed as a part of this paper are Moore and extended Moore neighborhood that is demonstrated in Fig. 1. The transition function of Moore neighborhood is represented as

Fig. 1 Neighborhood models. **a** Moore. **b** Extended Moore

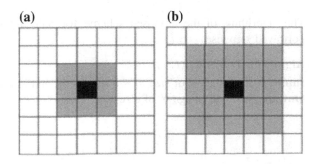
(a)   (b)

$$q_{i,j}^{t+1} = \delta\left(q_{i,j}^t, q_{i,j+1}^t, q_{i+1,j+1}^t, q_{i+1,j}^t, q_{i+1,j-1}^t, q_{i,j-1}^t, q_{i-1,j-1}^t, q_{i-1,j}^t, q_{i-1,j+1}^t\right) \quad (2)$$

Similarly, it can be defined for extended Moore neighborhood model.

2.2 Boundary Conditions

Boundary conditions are required when a transition rule is applied to the boundary cells of the CA. A CA is said to be a null boundary CA (NBCA) if the endmost cells in a grid are connected to logic-0 state and a periodic boundary CA (PBCA) if the endmost cells are neighboring one another [14]. In a fixed boundary CA (FBCA) the endmost cells are connected to any fixed state value. With the adiabatic boundary condition, the endmost cells are linked to its repeat state and in reflexive boundary condition mirror states replace the endmost cells [15].

3 Proposed Work

Two filters based on nine neighborhood CA (NNBCA) and twenty-five neighborhood CA (TFNBCA) under different boundary conditions are presented in this section. These filters are simple but efficient.

3.1 Methodology

The proposed methodology works in two steps; at first, the noisy pixels are detected by NNBCA or TFNBCA; in the second step, the detected noise pixels are replaced by NNBCA or TFNBCA. The architecture of the proposed scheme is indicated in Fig. 2.

3.2 Algorithms

The algorithms of the proposed scheme for Salt and Pepper noise reduction is portrayed stepwise.

Algorithm 1: $NNBCA(I_{m \times n})$
Input: Noisy image I of size $m \times n$ that is the initial configuration of *NNBCA*.
Output: Restored image Y of size $m \times n$.

Fig. 2 The architecture of the
proposed methodology

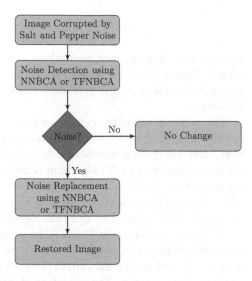

Step 1: Read the input image I corrupted by Salt and Pepper noise.
Step 2: Apply the periodic/adiabatic/reflexive boundary CA to the image I.

$$B'_{m+2 \times n+2} = PBCA[I_{m \times n}] \text{ or } ABCA[I_{m \times n}] \text{ or } RBCA[I_{m \times n}]$$

where $PBCA[.]$, $ABCA[.]$, and $RBCA[.]$ are the method for adding periodic, adiabatic and reflexive boundary to the noisy image I respectively.
Step 3: (**Noise Detection using NNBCA**) Let q_{min}, q_{max}, q_{med} and q_{std} are the minimum, maximum, median and standard deviation values of the nine state values (the pixel values) in the neighborhood of a cell. Then the CA rule for impulse detection is defined by the following pseudo code:

if $q_{i,j} < q_{min}$ *or* $q_{i,j} > q_{max}$
then $N_{i,j} = 1$;
go to step 4;
else
$N_{i,j} = 0$;
$q_{i,j}^{t+1} = q_{i,j}^{t}$;
end
$N_{i,j} = 1$ means the pixel being considered is a noisy pixel, so it needs to be replaced. $N_{i,j} = 0$ means the pixel is not a noisy pixel that is the pixel left unchanged in the next time step.

Step 4: (**Noise Estimation using NNBCA**) For noise replacement, apply following CA rule to the cell (pixel) where $N_{i,j}$ has value 1.

$$q_{i,j}^{t+1} = (N_{i,j} \times q_{med} + (1 - N_{i,j}) \times q_{i,j}^t) + (\alpha \times q_{std}) \qquad (3)$$

where α is a adjusting parameter whose value is 0.01 to 0.2.

Step 5: Repeat step 3 and 4 for every pixels of the noisy image. Finally, a restored image Y of size $m \times n$ is produced and it preserves thin lines and other detail features.

The proposed algorithm is basic and simple to execute. The choice of boundary conditions used in step 2 will not affect the result. The transition function used in step 3 and 4 is applied simultaneously to all the pixels of the image. Consequently, the time complexity of this algorithm is the least. But the algorithm needs very few iterations from step 2–4 when the noise density will be more than 20 %, (i.e. for 30 % it needs 2 iterations, for 40 % it needs 3 iterations and so on) which is the downside of this algorithm. Our next algorithm TFNBCA handle this problem.

TFNBCA is same as the NNBCA, however, varies in step 3 and step 4 that is as opposed to taking nine neighborhood we take twenty-five neighborhood to detect noisy pixels and replace the current pixel. Due to the paucity of space we have not described the algorithm. The advantage of this algorithm is unlike NNBCA, it does not require any iteration when the noise density is more.

4 Experimental Results and Discussions

Experiments are carried on three familiar images lena, pepper and baboon. The performance of proposed schemes is compared against traditional methods in terms of peak signal to noise ratio (*PSNR*) values. *PSNR* is the measure of quality in an image. *PSNR* (*dB*) is characterized by the mean squared error (*MSE*) as

$$PSNR = 10 \log_{10} \left(\frac{MAX_I^2}{MSE} \right) \qquad (4)$$

and

$$MSE = \frac{\sum_{i=1}^{M} \sum_{j=1}^{N} (X(i,j) - Y(i,j))^2}{M \times N} \qquad (5)$$

where MAX_I is the maximum possible pixel value of the original image *I*. $X(i, j)$ is the pixel of the original noisy image and $Y(i, j)$ is the pixel of the restored image at (i,j)th position. *M* and *N* are the number of rows and columns of the image.

First, we take a pepper image corrupted with Salt and Pepper Noise of 10–50 % noise densities. The noisy images are subjected to filtering by the two proposed algorithms NNBCA and TFNBCA along with the existing schemes discussed in Sect. 1. The *PSNR* (in dB) values obtained are presented in Table 1. Similarly, simulations are conducted with other images. For lena and baboon image, the

Table 1 PSNR (dB) values of different filters for pepper test image corrupted by salt and pepper noise

Filters	Noise ratio				
	10 %	20 %	30 %	40 %	50 %
SMF3 [2]	34.70	32.67	31.37	30.45	29.77
SMF5 [2]	33.66	32.08	30.94	30.10	29.29
WMF [3]	34.46	32.60	31.31	30.32	29.63
CWMF [4]	34.55	32.68	31.38	30.46	29.69
PSMF [6]	36.48	33.78	32.06	30.99	30.01
ACWMF [7]	36.74	33.92	32.22	31.01	29.99
DBAIN [8]	36.79	33.92	32.17	31.07	30.04
NAFSM [9]	37.02	34.04	32.29	31.04	30.13
Proposed NNBCA	37.15	34.15	32.49	31.37	30.60
Proposed TFNBCA	37.08	34.09	32.32	31.10	30.13

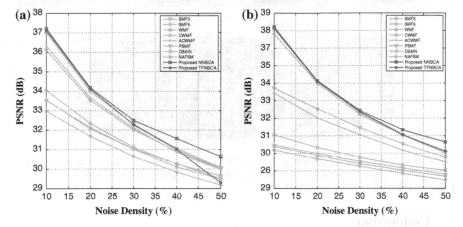

Fig. 3 PSNR plot of restored lena and baboon image of varying densities. **a** Lena image. **b** Baboon image

PSNR values are plotted in Fig. 3a, b respectively. One subjective comparison of pepper image is shown in Fig. 4 that indicates the restored results of pepper image corrupted with a noise density of 50 %.

From the quantitative and qualitative results, it is evident that the performances of the proposed schemes are superior to other filters used here for comparison. Figure 4 exhibits the proposed schemes preserve more edge details than others. However, the PSNR value of TFNBCA is quite less than NNBCA. All the filters are simulated in MatLab 2013 a, Microsoft Windows 7 Operating System and Intel Core 2 Duo 2.10 GHz CPU with 4 GB of RAM.

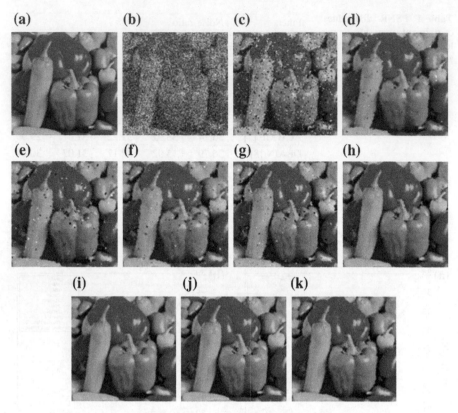

Fig. 4 Noise filtering of pepper image corrupted with 50 % of noise by different filters. **a** Original image. **b** 50 % noisy image. **c** SMF3. **d** SMF5. **e** WMF. **f** CWMF. **g** PSMF. **h** DBAIN. **i** NAFSM. **j** Proposed NNBCA. **k** Proposed TFNBCA

5 Conclusion

Two efficient cellular automata based filters for restoring images corrupted by salt and pepper noise are discussed in this paper. The results show that the performance of these schemes is better than other existing schemes. Simplicity, robustness and parallel behaviors of the cellular automata are the main advantages of these filters. We can use other varieties of cellular automata to upgrade the performance further.

References

1. Selvapeter, P.J., Hordijk, W.: Cellular automata for image noise filtering. IEEE Conf. (NaBIC), 193–197 (2009)
2. Gonzalez, R.C., Woods, R.E.: Digital Image Processing, 2nd edn edn. Prentice-Hall, New Delhi (2002)

3. Yin, L., Yang, R., Gabbouj, M., Neuvo, Y.: Weighted median filters: a tutorial. IEEE Trans. Circ. Syst. II: Anal Digit Signal Process. **43**, 157Â–92 (1996)
4. Ko, S.J., Lee, Y.H.: Center weighted median filters and their applications to image enhancement. IEEE Trans. Circ. Syst. **38**, 984–993 (1991)
5. Akkoul, S., Ledee, R., Leconge, R., Harba, R.: A new adaptive switching median filter. IEEE Signal Process. Lett. **17**, 587–590 (2010)
6. Wang, Z., Zhang, D.: Progressive switching median filter for the removal of impulse noise from highly corrupted images. IEEE Trans Circ. Syst II: Anal Dig Signal Process. **46**, 78–80 (1999)
7. Chen, T., Wu, H.R.: Adaptive impulse detection using center-weighted median filters. IEEE Signal Process. Lett. **8**, 1–3 (2001)
8. Srivasan, K.S., Ebenezer, D.:A new fast and efficient decision-based algorithm for removal of high-density impulse noises. IEEE Signal Process. Lett. **14**(3) (2007)
9. Toh, K., Isa, N.: Noise adaptive fuzzy switching median filter for salt-and-pepper noise reduction. Signal Process Lett. IEEE. **17**, 281–284 (2010)
10. Sadeghi, S., Rezvanian, A., Kamrani, E.: An efficient method for impulse reduction from images using fuzzy cellular automata. International J. Elec. Comm. 772–779 (2012)
11. Rosin, P.L.: Training cellular automata for image processing. IEEE Trans. Image Process. **15** (7), 2076–2087 (2006)
12. Wolfram, S.:Computation theory of cellular automata. Commun. Math. Phys., 14–57 (1984)
13. Chang, C., Zhang, Y., Gdong, Y. :Cellular automata for edge detection of images. IEEE Proc. Mach. Learn. Cybern. 26–29 (2004)
14. Sahoo, S., Choudhury, P.P., Pal, A., Nayak, B.K.: Solutions on 1-D and 2-D density classification problem using programmable cellular automata. J. Cell. Automata **9**(1), 59–88 (2014)
15. Nayak, D.R., Patra, P.K., Mahapatra, A.: A survey on two dimensional cellular automata and its application in image processing. arXiv: 1407.7626 [cs.CV] (2014)

Part VI
Internet, Web Technology, and Web Security

Optimizing the Performance in Web Browsers Through Data Compression: A Study

S. Aswini, G. ShanmugaSundaram and P. Iyappan

Abstract Web browsers are used on daily basis, where web compression plays an important role in compressing the web pages content and thereby, providing the users in launching the pages much faster. The study is carried out with the concept of web compression and its existing techniques. Outcome of the study on existing research contributions would present the various challenges in web compression. Other problem issues related to this work are discussed and it will be taken further in the future works.

Keywords Web performance · Data compression · Web browsers · Web caching · Web compression

1 Introduction

Compression is the encoding of data or information which can be done in the form of texts, images, audios or videos [1]. The data compression [2] is defined as the reduction of the data size. Lossless Compression is generally termed as reconstruction of the original message from the compressed data [1]. Lossy Compression is the reconstruction of the approximate message from compression, which can't be able to reproduce the original message. The lossless compression is better approach because it can exactly produce the original content. Whereas in lossy compression, it will result in losing some originality of exact content or pixel [2]. There are

S. Aswini (✉) · G. ShanmugaSundaram · P. Iyappan
Sri Manakula Vinayagar Engineering College
(Affiliated to Pondicherry University), Puducherry, India
e-mail: aswini1405@outlook.com

G. ShanmugaSundaram
e-mail: shanmugasundaram@smvec.ac.in

P. Iyappan
e-mail: iyappan_be@yahoo.co.in

© Springer India 2016
A. Nagar et al. (eds.), *Proceedings of 3rd International Conference on Advanced Computing, Networking and Informatics*, Smart Innovation, Systems and Technologies 44, DOI 10.1007/978-81-322-2529-4_46

different types of lossless compression algorithms used in the data compression namely Lempel Ziv (LZ77/78), Lempel Ziv Storer Szymanski (LZSS) and Lempel Ziv Markov chain Algorithm (LZMA) [1]. There are two types of coders namely statistical coder and dictionary coder. Both the statistical coders and dictionary coders [3] are mandatory, because the statistical coders provides better compression, whereas the dictionary coders helps in enhancing the performance.

Maintaining higher compression performed in higher speed is one of the most challenging task. If the task is accomplished in pushing and embedding it to the web browsers, then there will be 3× increase over the web performance. Enhancing the web performance will improve the internet speed to a higher level, where less memory will be cached temporary in physical memory, where it will also enhance the overall performance of our systems too [4].

The rest of the paper as organized follows: Sect. 2 explains the background of this study which focuses on the algorithmic techniques used previously along with drawbacks. Section 3 discusses about the related study where it briefly explains the limitations. Section 4 discusses further on the problem issues in detail with respect to Sect. 3. Section 5 concludes the paper.

2 Background

2.1 Aim of the Study

- To enhance the performance of the web browsers through compression, which optimizes and thereby reducing the traffic and latency.
- To study about the lossless compression algorithms relevant to web compression.

2.2 Algorithmic Techniques

2.2.1 LZ77

LZ77 was a sliding window based technique in which the dictionary consists of a set of fixed length phrases into previously processed text. This algorithm makes use of both lookahead buffer and search buffer. The lookahead buffer consist of the first symbols of the data that needs to be encoded, whereas the search buffer consists of the pre-defined symbols with an input alphabet. The process initiates from the search buffer and then, it continues towards the lookahead buffer [5]. LZ77 had drawbacks in strings replacement.

2.2.2 LZSS

(LZSS) Lempel Ziv Storer Szymanski was a dictionary encoding technique which replaced a string of symbols with the reference towards the location of the dictionary that had the same strings [2].

There were some drawbacks as follows [1, 2]:

- The LZSS gives poor compression since it didn't have any matches for data in text window.
- LZSS also had troubles with the pre-loading, where it had no idea what sort of data would come towards the input streams.
- It faced issues in the aspect of compressed speed when the window gets increased due to the additional work performed while navigating and maintaining the large binary tree structures.
- While compressing the tiny files in LZSS, the files gets affected due to the increased time for building a complete dictionary that resulted in delays.
- Also, the LZSS had inefficiency over some strings. Perhaps, the side effect was due to the decompression that kept track of the duplicate strings that resulted in a significant cutoff in the expansion speed [2].

2.2.3 LZ78

LZ78 was an improved version of LZ77 and LZSS that helped in building up phrases into one symbol at a time through addition of a new symbol over an existing phrase whenever a match was occured [6, 7].

Also, Few drawbacks was found in the LZ78 algorithm [8]:

- The related data or information that had been crossing the phrase boundaries were lost during the parsing of input strings. Also, there would be many patterns crossing over the phrase boundaries and those patterns would probably affect the next symbol in that respective sequence set.
- In optimal predictability, the convergence rate of LZ78 is slow.

2.2.4 LZMA

The Lempel Ziv Markov chain Algorithm (LZMA) was preferably used in the reduction of data size that restores back to the original one without the loss of data [9] which comes under the lossless compression. The LZMA came from the dictionary scheme and the markov model. The range encoder [10] was used in LZMA which was the only difference. LZMA was the combination of dictionary based scheme and Markov chains. Also, LZMA used range encoding [11] instead of the Huffman coding. In range encoding, all the symbols in a message was encoded into a number. But, in Huffman encoding, every symbol was assigned with a bit pattern.

After the assignment of those bit patterns, then those assigned bit patterns was linked together. Therefore, the better compression was experienced in range encoding.

Markov chain was termed as a mathematical system which undergoes transition from one state to the another on a finite state manner. The Markov chain was represented in the form of probabilistic state diagrams where the transition occur in between the finite states which was labelled with the set of possible occurrences [12]. While parsing the source data, the Markov chain was used in LZMA for encoding and also in the form of finite state machine known as the Finite State Compressor [2]. The Markov chains was related with LZMA through variable to variable length coding that uses the number of encoded source symbols and the number of encoded bits which were done as per the codewords [9, 13]. The Markov chain is not recommended when compared towards the Markov Chain Monte Carlo (MCMC) as well as the Hidden Markov methods. The range encoding was one of the entropy coding method used in producing a stream of bits for representing a stream of symbols and their possibilities [14]. Then, the range decoding decodes the encoded data. The range coding was found to be effective when compared with the arithmetic coding [15]. Range coding does faster arithmetic coding where it re-normalizes the process into one byte when compared with one bit. Thus, range coding runs nearly 2× faster than the arithmetic coding where it uses bytes as the encoding digits instead of the bits. LZMA has issues in the areas of both statistical coding and dictionary coding where the performance level was highly minimal [11].

3   Related Works

In this section the existing works that used the lossless compression algorithms are presented.

According to Hoobin et al. [3]: The LZ77 was adopted, where Gzip had been used along with the adoption [7]. Gzip mechanism works under the Deflate algorithm and Huffman coding. The range coding is better than Huffman coding since range coding uses fast arithmetic coding. Hence, thereby it requires improvement.

According to Parekar et al. [1]: The lossless data compression algorithms were addressed such as the Run-Length Encoding (RLE) algorithm, LZ77 [5], LZMA and LZW. The concept of LZMA does not satisfy speed in terms of compression.

The web prefetching mechanism in cache was proposed by Seema et al. [14] for improving cache size, where it gradually provided in improving the performance. Yet, it failed to address data compression which would have been one of the best measure for enhancing web performance.

According to Sudha et al. [16]: The LZMA mechanism is used in successfully accomplishing the compression with reference to [2]. But, [16] failed to show the performance values and also failed to prove it is better than others through comparison. The LZMA has drawbacks in sliding window dictionary which is possibility of overhead when heavy process is performed. So, it requires improvements upon all basis.

According to GopalRatnam and Cook [17]: An optimized LZ77 tool was done with reference to [5]. But, experiencing some drawbacks in LZ77, then [17] proceeded to improvise it using LZMA mechanism tool [10] including block sorting data compression [18], where Static HTML Transform (SHT) and Semi-Dynamic HTML Transform (SDHT) was proposed. Yet, it failed to satisfy the compression in CSS page bloats which includes various fonts and styles, which led those CSS pages getting uncompressed instead of compressing them.

The attempt by Mahad and Wan-Kadir [19] to improve web server performance was proposed inorder to reduce the overloading of network traffic and network congestion. But, it failed in terms of memory compression.

From studying the above related works, the drawbacks has shown that the web browser performance still requires better improvements.

4 Problem Issues

From the background and related works, a lot of problem issues are being analyzed in detail. Coming from Sect. 3, the related works were discussed regarding the recent works as well as the limitations faced in [1–3, 5, 10, 14, 16–19] The issues regarding the compression were discussed towards the performance improvement by means of lossless algorithms [1].

The lossless algorithms such as LZ77/78 [5, 7] were discussed, where it led to the birth DEFLATE mechanism [20, 21] in which the GZip and ZLib [22] were developed. Also, LZMA [1, 5, 9–11, 16] was being discussed which is a part of lossless algorithm. The difference between Deflate and LZMA is that LZMA uses faster arithmetic coding known as range coding [14, 15], which is better than the Huffman coding used. Both LZMA and GZip along with block sorting [18] was used [17]. Yet, it failed where it had flaws in CSS resulting in page bloats, which downgraded the web optimization level especially while loading the websites and during transferring files, the uncompressed data will also be added due to the bloats. In [19], the two-tier cache where it used SQUID cache, which is good to improve performance in web servers. But, [19] failed to focus more on cache storage, where heavy logs will gradually lead to overhead since it hasn't followed memory compression by compressing the cache.

After a depth study, a list of redundancy issues are being analyzed namely, (i) Spatial redundancy, (ii) Coding redundancy, (iii) Spectral redundancy, (iv) Psycho visual redundancy and (v) Temporal redundancy.

The spatial redundancy is related to predictive issues, where this problem occurs when too many nodes are present and this causes confusion in finding nearest neighbor node. But, this could be solved through our new parameter hidden markov web.

The coding redundancy was one of the major issues that can be solved by introducing range coding (encoding/decoding), which is a faster arithmetic coding mechanism.

The spectral redundancy had an issue where the colours gets affected during the compression process. During reconstruction (un-compression), it results in losing some originality.

The Psycho-Visual redundancy had an issue that occurs while compressing fonts, images and videos. Also, when we zoom-in or full-screen-view, then the view gets blurred especially affecting colour view and pixels too. Temporal redundancy is a common issue that happens in "Live Video Streaming".

5 Conclusion

From this research study, the web caching, compression and algorithms such as LZ77, LZSS, LZ78 and LZMA were being addressed along with the issues and analysis that were discussed in the aspect of performance, where the solutions will be used as a part in upcoming future works.

References

1. Parekar, P.M., Thakare, S.S.: Lossless data compression—a review. Int. J. Comput. Sci. Inf. Technol. 5(1), 276–278 (2014)
2. Nelson, M., Loup Gail, J.: The data compression book, 2nd edn
3. Hoobin, C., Puglisi, S.J., Zobel, J.: Relative Lempel Ziv factorization for efficient storage and retrieval of web collections. In: 38th International Conference on Very Large Databases, Proceedings of the VLDB Endowment, vol. 5, no. 11 (2011)
4. Gavaletz, E., Hamon, D., Kaur, J.: In-browser network performance measurement. IEEE Trans. Parallel Distrib. Comput. Syst. 2009
5. Ziv, J., Lempel, A.: A universal algorithm for sequential data compression. IEEE Trans. Inf. Theory 23(3), 337–342 (1977)
6. Jebamalar Leavline, E., Gnana Singh, A.A.: Hardware implementation of LZMA data compression algorithm. Int. J. Appl. Inf. Syst. 5(4) (2013)
7. Ziv, J., Lempel, A.: Compression of individual sequences via variable rate coding. IEEE Trans. Inf. Theor. 24(5), 530–536 1978
8. Skibiński, P.: Improving HTML compression. In: IEEE Data Compression Conference 2008
9. Farina, A., Navarro, G., Parama, J.R.: Word-based statistical compressors as natural language compression boosters. IEEE Trans. Data Compr. (2008)
10. Igor Pavlov: 7-zip compression utility. http://www.7-zip.org (2009)
11. Ferragina, P., Manzini, G.: On compressing the textual web. ACM (2010)
12. Veerarajan, T.: Probability, Statistics and Random Processes, 3rd edn
13. Claude, F., Farina, A., Martinez-Prieto, M.A.: Indexes for highly repetitive document collections. ACM Trans. Oct 2011
14. Priyanka Makkar, S.: An approach to improve the web performance by prefetching the frequently access pages. Int. J. Adv. Res. Comput. Eng. Technol. 1(4), 215–219 (2012)

15. Langdon, G.G.: An introduction to arithmetic coding. International Business Machines Corporation (IBM) Trans. **28**(2), 135–149 (1984)
16. Sudha, M., Dr. Palani, S.: A novel implementation of wavelet transform and Lzma for compression and decompression of document images. Int. J. Innov. Eng. Technol. (IJIET) **2**(3) (2013)
17. GopalRatnam, K., Cook, D.J.: Active Lezi: An incremental parsin algorithm for sequential prediction. Int. J. Artificial Intell. Tools (2004)
18. Burrows, M., Wheeler, D.J.: A block-sorting data compression algorithm. SRC Research Report 124. Digital Equipment Corporation (1994)
19. Mahad, F.S., Wan-Kadir, W.M.N.: Improving web server performance using two-tier web caching. J. Theor. Appl. Inf. Technol. **52**(3) (2013)
20. Deflate Compression Algorithm: http://www.gzip.org/algorithm.txt
21. Deutsch, P.: RFC 1951: Deflate compressed data format specification version 1.3, May 1996
22. Deutsch, P., Gailly, J.L.: RFC 1950: ZLib compressed data format specification version 3.3. May 1996

Architectural Characterization of Web Service Interaction Verification

Gopal N. Rai and G.R. Gangadharan

Abstract Web service interaction utilizes disparate models as it still does not have its own model for verification process. Adaptation of a different model is not always beneficial as it may prune several significant characteristics that worth considering in verification. The primary reason behind this adaptation is that the primitive characteristics are not well identified, standardized, and established for Web service interaction model. In this article, we therefore investigate the primitive characteristics of Web service interaction model that need to be well considered in verification. Further, we study the appropriateness and effectiveness of two modeling and verification phenomena namely *model checking* and *module checking* with respect to investigated primitive characteristics.

Keywords Web services composition · Formal methods · Modeling · Model checking · Module checking

1 Introduction

Web services are self contained, self-describing modular applications that can be published, located, and invoked across the web [1]. Since Web services interact with each other through messages (synchronous and asynchronous), concurrency related bugs and/or inconsistency problems are possible in the interaction patterns (communication patterns). In order to overcome the problem, Web service interaction verification is required.

G.N. Rai (✉) · G.R. Gangadharan
IDRBT, Castle Hills, Masab Tank, Hyderabad 500 057, India
e-mail: gopalnrai@gmail.com

G.R. Gangadharan
e-mail: geeyaar@gmail.com

G.N. Rai
SCIS, University of Hyderabad, Gachibowli, Hyderabad 500 046, India

© Springer India 2016   447
A. Nagar et al. (eds.), *Proceedings of 3rd International Conference
on Advanced Computing, Networking and Informatics*, Smart Innovation,
Systems and Technologies 44, DOI 10.1007/978-81-322-2529-4_47

Classical testing techniques are inadequate to verify the interaction among services [2, 3]. A Web service interaction scenario resembles with reactive and concurrent systems, with component-based systems [4, 5], and with multi-agent systems [6] up to a certain extent. Therefore, various Web service interaction modeling and verification techniques are proposed on the basis of these approaches [6, 7]. However, composition among services tangles the task of verification as it poses unique primitive verification requirements that are not covered by either classical tests or other verification approaches.

Many efficient and industry wide accepted formal verification tools are available that could verify Web service interaction. However, transformation of a service interaction scenario into an input model for any existing verification tool may loose the architectural originality and significant native primitive notions, thus, restricting the verifiable properties (requirements). In this article, we characterize the model of Web service interaction for verification, a novel effort towards establishing *Web service interaction verification model*.

The rest of the article is structured as follows. Section 2 investigates fundamental characteristics of a Web service interaction model. We study the appropriateness and effectiveness of model checking and module checking in Sect. 3. Section 4 provides conclusion and future directions.

2 Primitive Characteristics of Web Service Interaction Model

In this section, primitive characteristics of the Web service interaction model are discussed as follows.

2.1 Modular and Hierarchical Architecture

All Web services basically fall into two categories: either basic or composite. A basic Web service does not take help of other Web services to accomplish the job whereas a composite Web service does.

Modular architecture of Web services interaction model refers to the design of a system composed of independent Web services that could interact with each other. The advantage of the modular architecture is that a module could be added or substituted with another suitable module without affecting the rest of the system.

Hierarchical architecture of Web services interaction model comes in the existence only when a composite service comes in the existence. A composite service depends on the other basic or composite services. A composite service is a higher level abstraction and basic Web services are lowest level abstraction. Web service interaction possess hierarchical architecture but nesting of Web services are not allowed. Indeed, nesting of Web services does not make any sense as basic Web services work as independent modules.

Therefore, Web service interaction model has both shades of hierarchical and modular architectural archetype. It requires modeling a Web service in such a way that it preserves its identity and supports composite services at a time. Consequently, a Web service can serve as an independent module and can serve as an building block of a composite service.

2.2 Open System

Each basic Web service works as an independent module that fundamentally yields an open system from a verification perspective. An open system is one which is designed to interact with an environment [8]. A web service cannot anticipate all behaviors of its interacting partner in advance, thus enabling the open system phenomenon. In order to be able to model the open system, *environment modeling* is inevitable. Classical modeling approaches and many other new verification techniques does not support environment modeling. They typically apply to closed systems whose behavior is fully specified.

2.3 Trace Modeling

Informally, a trace (in the context of Web services) is a linear, unidirectional Web service composition workflow path in which each node represents a Web service and the directed edges indicate the flow from one service to another. Trace modeling and computation are among the most primitive modeling requirements for Web service interaction as the computation of trace related phenomenons (such as trace inclusion, trace crossing, and trace merging) reduce the time and space complexity for verification.

Formal definition of a Web service trace is given as follows.

Definition 1 (*Trace*) A trace is a tuple $T = (\mathcal{W}, I, w_i, w_n)$, where $\mathcal{W} = \{w_1, \cdots, w_m\}$ is a finite set of the Web services, $I : \mathcal{W} \to \mathcal{W}$ is an invocation function such that $w_i I w_j$ if and only if w_i invokes w_j, $w_i \in \mathcal{W}$ is a service from which trace generation begin and $\not\exists w_j \in \mathcal{W} : w_j I w_i$, and $w_n \in \mathcal{W}$ is a service on which trace ends up and $\not\exists w_j \in \mathcal{W} : w_n I w_j$.

Let w_i be a Web service then T_{w_i} represents a set which contains all the traces generated by the service w_i. The concept of trace is utilized mainly while studying behavioral equivalence of services.

2.4 Asynchronous Messaging

A Web service consists ports and a port consists sets of input and output messages. Messages consisting activities are unit of interaction. There are two principal messaging models used in Web services namely synchronous and asynchronous model. The two Web service messaging models are distinguished by their way of request-response operation handling mechanism.

Synchronous messaging and asynchronous messaging among less number of services can be handled in a fair manner without much complexity. If involved number of services are high, asynchronous messaging complicates the verification process.

2.5 Recursive Composition

Composition and recursive composition are fundamental characteristics of Web service interaction model. Composition of services is aggregation of facilities provided by services. Recursive composition refers recursive aggregation of services. The difference between recursive composition and non-recursive composition is explained as follows.

Let $\langle A, B \rangle$ represents the knowledge that a Web service A is composed of a Web service B. Let $\{\langle A, B \rangle, \langle B, C \rangle, \langle C, D \rangle\}$ represents a composition scenario. If a composition dependency holds among tuples such as $\langle A, B \rangle$ depends upon $\langle B, C \rangle$ and $\langle B, C \rangle$ depends upon $\langle C, D \rangle$ then it forms a recursive composition scenario otherwise non-recursive composition.

2.6 Dynamic Reconfiguration

There are two crucial factors that make consideration of dynamic reconfiguration in the case of Web service inevitable. First, a Web service resembles a module (a basic Web service resembles an independent module whereas a composite Web service resembles a dependent module). Second, dynamic availability of services. Web services are accessible through the Web and a Web service could become unavailable/removed at any time or a new Web service could be introduced at any time. Ethically, a composition designer or verifier must be ready for substitution, replacement, and introduction of services. Introduction or unavailability of a service could make complete chaos if not handled properly. Automatic dynamic service composition is a rapidly emerging paradigm and research topic that is based on dynamic reconfiguration phenomenon.

2.7 Hierarchical Concurrency

Classical model-based verification approaches do not consider hierarchy among Web services while verifying the interaction. They keep transmitting all variables that are being considered for verification through every state in their state transition diagram. All variables need not be considered at a time. A hierarchy must be found among services and must be considered in verification process. We explain a hierarchical concurrency scenario among Web services as follows.

Let $\langle A, B \rangle$ represents the knowledge that A is composed of B and $\langle A, B \rangle \rightarrow \langle B, C \rangle$ represents the knowledge that $\langle A, B \rangle$ depends upon $\langle B, C \rangle$. Let us consider the following scenario

$$
\begin{array}{ccccc}
\langle A, B \rangle & & & & \langle I, J \rangle \\
\downarrow & & & & \downarrow \\
\langle B, C \rangle & \leftarrow & (B, J) & \rightarrow & \langle J, K \rangle \\
\downarrow & & & & \downarrow \\
\langle C, D \rangle & & & & \langle K, L \rangle
\end{array}
$$

Consider that concurrency has to be resolved between services B and J for a given specification. If $\langle B, C \rangle$ and $\langle J, K \rangle$ do not affect $\langle B, J \rangle$ regarding concurrency then concurrency will be resolved between B and J only (B and J are first level services). No need to involve the services C and K (C and K are second level services). If the concurrency is not resolved at first level then only second level services will be considered. Again, if second level services are also not sufficient to resolve the concurrency, third level services (D and L are third level services) will be considered.

We introduce a conceptual term *sphere of influence* to ease the process of hierarchical concurrency verification in the context of Web service interaction. Sphere of influence is a set of Web services computed for an Web service such that either each member of the set is directly invokable by the center service or invokes the center service. Figure 1 is a pictorial representation of the sphere of influence for a Web service namely WS1. This diagram infers that services WS2, WS3, WS4, and WS5 constitutes the sphere of influence set for the service WS1. Figure 2 is an

Fig. 1 Sphere of influence for a Web service WS1

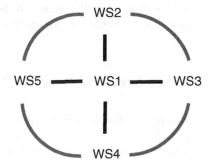

Fig. 2 Evolved version of
sphere of influence depicted
in Fig. 1

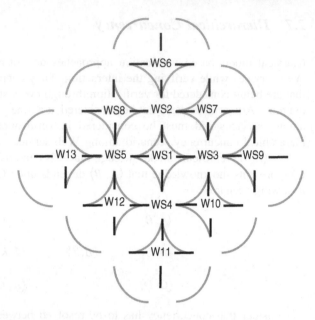

evolved version of sphere of influence depicted in Fig. 1. Sphere of influence serves
as a basic model to verify hierarchical concurrency.

2.8   Verification of Adversarial Specification

Classical temporal logics such as linear temporal logic (LTL) and computation tree
logic (CTL) are unable to specify collaborative as well as adversarial interactions
among different Web services. Alternating Temporal Logic (ATL) [8] is designed
to write requirements of open system [8] and is able to express collaborative as well
as adversarial interaction specifications. ATL models each Web service as an agent.
Let Σ be a set of agents corresponding to different Web services, one of which may
correspond to the external environment. Then, the logic ATL admits formulas of the
form $\langle\langle A \rangle\rangle\Diamond p$, where p is a state predicate and A is a subset of agents. The formula
$\langle\langle A \rangle\rangle\Diamond p$ means that the agents in the set A can cooperate to reach a p-state no
matter how the remaining agents resolve their choices.

3   Web Service Interaction Verification

In this section, we study and compare two different model-based formal verification
approaches that are suitable for service interaction verification: model checking and
module checking. Further, we study the working mechanisms of their representative

tools and its input languages whether there are native language constructs or not to support the desired characteristics discussed earlier in this paper.

Model Checking Model checking has been widely used as one of the formal techniques for Web service composition and interaction verification [7, 9–12]. In order to be able to model check a Web service interaction scenario, the scenario has to be modeled as an input model of model checking. The transformation of service interaction into an input model of model checking yields a monolithic structure. This transformed structure collapses several significant characteristics such as modular cum hierarchical architecture. Further, monolithic structure increases computation overhead and is not adequate for reasoning. The biggest drawback one faces with model checking is that it does not support open system modeling that is primary requirement for modeling of Web service interaction. Model checking does not have any provision for *interface* and *private* variable types. Consequently, a chaos develops among variables in verification.

SPIN [13] and NuSMV [14] are representative tools of model checking. SPIN supports only linear-time temporal logic whereas NuSMV supports both linear-time temporal logic as well as branching-time temporal logic. We study NuSMV (NuSMV 5.0) as a representative tool for model checking.

Module Checking Module checking is also an model-based formal verification technique. It could be considered as an alternative of model checking. Module checking differs from model checking in various means as follows. From modeling point of view, module checking supports the heterogeneous modeling framework of reactive modules whereas model checking supports unstructured state transition graphs. From specification writing point of view, module checking employs ATL to write specification about the system whereas model checking verifies the specifications written in LTL/CTL. From architecture point of view, module checking facilitates with hierarchical design and verification along with modular design whereas model checking provides only modular designing and verification.

We study MOCHA [15] (jMOCHA 2.0) as a representative tool for module checking. MOCHA is an interactive verification environment that supports a range of compositional and hierarchical verification methodologies.

A comparative analysis between NuSMV 5.0 and jMOCHA 2.0 with respect to investigated characteristics is shown in Table 1.

4 Related Work

To the best of our knowledge, a paper merely focusing on architectural characterization of Web service interaction is not available in literature. However, some of the studied characteristics are discussed individually in articles. Hierarchical

Table 1 Verification requirements versus verification approaches

Verification requirements	Model checking (NuSMV)	Module checking (jMOCHA)	Property type
Modular architecture	Yes	Yes	Modeling
Hierarchical architecture	No	Yes	Modeling
Variables support	Global	External, interface, private	Modeling
Lazy evaluation	No	Yes	Modeling
Dynamic reconfiguration	No	No	Modeling
Composition	No	Yes	Modeling
Recursive composition	No	No	Modeling
Open system modeling	No	Yes	Modeling
Trace containment verification	No	Yes	Modeling
Adversarial specification	No	Yes	Specification

architecture of Web services is discussed in [16]. Hnetynka et al. [4] discusses dynamic reconfiguration of Web services and collaboration among them. Roglinger [17] presents a requirements framework for verification. However, the proposed requirements in [17] corresponds to model checking and not characterizing the Web service interaction. Pistore et al. [18] present a requirement model for verification of Web services. However, the authors focus on business requirements not architectural characterization in the paper [18].

We do not find any evidence of Web service interaction verification using module checking tool (jMOCHA) in literature. However, ATL is used for Web service verification along with Petri net in [19].

5   Conclusion and Future Work

On the basis of our study, we conclude that Web service interaction is a modular cum hierarchical architecture. Each Web service works as an open system. Recursive composition is a fundamental characteristic of Web service interaction. And a verification approach for Web services must support hierarchical concurrency verification with adversarial specification facility.

Though module checking is an efficient verification technique for reactive, concurrent, and open systems, it is not widely accepted among industry and academia. Reasons might be lack of popularity and complex specification writing scheme. However, on the basis of our study, we advocate use of module checking rather than model checking for verification of Web services interaction.

There are many improvement suggestions that could be considered as parts of future work. First, we want to investigate hierarchical structure of Web service interaction in more depth. Both verification approaches model checking and module checking do not support recursive composition and dynamic introduction or removal of a service. Therefore, realization of a verification approach based on investigated characteristics and sphere of influence will be our second future work. Discovering all fundamental and advanced verification requirements for Web service interaction is also a part of our future work.

References

1. Wang, H., Huang, J.Z., Qu, Y., Xie, J.: Web services: problems and future directions. Web Semantics: Sci. Serv. Agents World Wide Web 1(3), 309–320 (2004)
2. Bozkurt, M., Harman, M., Hassoun, Y.: Testing Web Services: A survey. Department of Computer Science, King's College London, Tech. Rep. TR-10–01 (2010)
3. Mei, L., Chan, W., Tse, T., Jiang, B., Zhai, K.: Preemptive regression testing of workflow-based web services. IEEE Trans. Services Comput. (2014)
4. Hnetynka, P., Plášil, F.: Dynamic reconfiguration and access to services in hierarchical component models. In: Component-Based Software Engineering, pp. 352–359. Springer (2006)
5. Lau, K.K., Wang, Z.: Software component models. IEEE Trans. Softw. Eng. 33(10), 709–724 (2007)
6. Walton, C.: Model checking multi agent web services. In: AAAI Symposium of Semantic Web Services (2004)
7. El Kholy, W., Bentahar, J., El Menshawy, M., Qu, H., Dssouli, R.: Modeling and verifying choreographed multi-agent-based web service compositions regulated by commitment protocols. Expert Syst. Appl. 41(16), 7478–7494 (2014)
8. Alur, R., Henzinger, T.A., Kupferman, O.: Alternating-time temporal logic. J. ACM (JACM) 49(5), 672–713 (2002)
9. Bentahar, J., Yahyaoui, H., Kova, M., Maamar, Z.: Symbolic model checking composite web services using operational and control behaviors. Expert Syst. Appl. 40(2), 508–522 (2013)
10. Boaro, L., Glorio, E., Pagliarecci, F., Spalazzi, L.: Semantic model checking security requirements for web services. In: 2010 International Conference on High Performance Computing & Simulation, pp. 283–290 June 2010
11. Marques, A.P., Ravn, A.P., Srba, J., Vighio, S.: Model-checking web services business activity protocols. Int. J. Softw. Tools Technol. Transf. 15(2), 125–147 (2013)
12. Sheng, Q.Z., Maamar, Z., Yao, L., Szabo, C., Bourne, S.: Behavior modeling and automated verification of Web services. Inf. Sci. 258, 416–433 (2014)
13. Holzmann, G.J.: The SPIN model checker: Primer and reference manual, vol. 1003. Addison-Wesley Reading (2004)
14. Cimatti, A., Clarke, E., Giunchiglia, E., Giunchiglia, F., Pistore, M., Roveri, M., Sebastiani, R., Tacchella, A.: Nusmv 2: An opensource tool for symbolic model checking. In: Computer Aided Verification, pp. 359–364. Springer (2002)
15. De Alfaro, L., Alur, R., Grosu, R., Henzinger, T., Kang, M., Majumdar, R., Mang, F., Meyer-Kirsch, C., Wang, B.: Mocha: Exploiting Modularity in Model Checking. Tech. Rep, DTIC Document (2000)
16. Kreger, H.: Web services conceptual architecture (wsca 1.0). IBM Software Group 5 (2001)

17. Rglinger, M.: Verification of web service compositions: an operationalization of correctness and a requirements framework for service-oriented modeling techniques. Business & Inform. Syst. Eng. **1**(6), 429–437 (2009)
18. Pistore, M., Roveri, M., Busetta, P.: Requirements-driven verification of web services. Electronic Notes Theor. Comput. Sci. **105**, 95–108 (2004)
19. Schlingloff, H., Martens, A., Schmidt, K.: Modeling and model checking web services. Electron. Notes Theoret. Comput. Sci **126**, 3–26 (2005)

Revamping Optimal Cloud Storage System

Shraddha Ghogare, Ambika Pawar and Ajay Dani

Abstract With the prosperous growth of cloud computing, it is being used widely in business as well as in researches today, considering its numerous advantages over traditional approaches. However, security and privacy concerns are ascending day by days. Cloud is being utilized for not only to use software and platform over the Internet, but also for storing confidential data. This paper presents a novel approach wherein multi-cloud environment is used to mitigate data privacy apprehensions. The viability and design of this method are described in the proposed paper.

Keywords Cloud computing · Cloud storage · Flat files · Privacy · Multi-cloud environment

1 Introduction

In an always changing economic and business world, businesses strive to achieve more profit and enhance future growth. This can be effectuated by using cloud computing, wherein businesses can perform computing on the cloud. The cloud is essentially a pool of hardware, network, software resources, storage, interfaces and services that leads to delivery of computing as a service. Figure 1 shows services offered by cloud.

S. Ghogare (✉) · A. Pawar
CSE Department, SIT, Symbiosis International University (SIU), Pune, India
e-mail: shraddha.ghogare@sitpune.edu.in

A. Pawar
e-mail: ambikap@sitpune.edu.in

A. Dani
Indian Business School, Hadapsar, Pune, India
e-mail: ardani_123@rediffmail.com

© Springer India 2016   457
A. Nagar et al. (eds.), *Proceedings of 3rd International Conference on Advanced Computing, Networking and Informatics*, Smart Innovation, Systems and Technologies 44, DOI 10.1007/978-81-322-2529-4_48

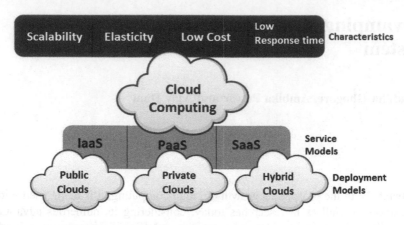

Fig. 1 Cloud computing: an overview

Cloud computing has several advantages over traditional computing systems such as on-demand service, elasticity, scalability and reduced response time. Even so, there exist certain concerns related to security of data. Many researchers and professionals have focused on security of the data deployed on cloud. However, data privacy has not gained much attention. These data privacy issues on service provider's sides are handled using SLAs that are agreed by end users and service provider in the negotiation phase. Furthermore, monitors are used to keep a check up on insiders. Privacy issues on a client's application server's side still persist.

Users are skeptical about their data stored on cloud data storage being mined or used for advertising and marketing purpose. Existing cryptographic techniques can be used for data security; but protecting data privacy needs significant focus. Who should have how much access right is another challenge to data security. In Google's approach to IT Security [1], Google's security strategy that provides control at multiple levels of data storage, transfer and access have been addressed. Security concerns have been addressed precisely; nevertheless, data privacy concerns still need to be stressed. Another approach to solve data privacy issues is to have a role-based access control wherein permissions and users are assigned to roles. Any authenticated user may misuse the access right assigned thereby compromising data privacy.

Moreover, in a cloud based environment, the data are stacked away at several geographic locations. Laws and Regulations vary for each country that defines the way confidential information is collected and stored and the access rights for that data. Therefore, protecting private data on public cloud is a major concern. Use of several clouds solves this problem as files are divided into parts and are uploaded on multiple clouds. This paper extends this approach. It provides a method in that at files are divided into chunks and are uploaded on different cloud service providers (CSPs).

2 Related Work

Maintaining privacy of data on public clouds is a core concern. Once a user deploys his data-processing applications on the cloud, the cloud provider is granted full control over the processes. Hence, trust and governance are two major issues. [2, 3] suggested an approach wherein the multi-cloud concept had been proposed, which includes three patterns: replication of application, partition of an application system into layers and small partitions of the application tier into fragments. All afore-mentioned approaches have been implemented in a multi-cloud environment. Moreover, assessment and comparison of this approach in terms of security, feasibility and regulation have been presented in [4]. However, switching time from one cloud to another and cost of each cloud are two important factors to be addressed to [5].

Many organizations now have moved to cloud storage to store data. These providers either use it to store archive data (passive data) or to store data on the social network that is sensitive. Although there are many advantages of using cloud storage, data privacy and losing control over data are major issues [6]. [7] proposes that the user would trust small part of memory on their system wherein they would encrypt the data and then store on the cloud. The pitfall is that not all users are aware of how to encrypt or secure this little part. However, this would be beneficial rather than uploading plain text.

[8] have proposed a mechanism which addresses Privacy concerns in a cloud environment by suggesting an optimal cloud storage management system wherein multiple cloud storages are used to store the data. The application developed offers the user to select cloud storage(s) out of 67 providers who have been analysed. The feature of this application is that the user can customize sign-up privacy, throughput, response time, capacity. Furthermore, the user has an option to select storage strategy as UseAllInParallel or RoundRobin. However, if the user is not aware of what these techniques are, he might keep the default one as selected. Also, the file splitter does not run concurrently as of now, which may hamper application's overall performance and file splitting operation does not consider a file type so this may incur extra processing overhead on the application server. Users have to sign-up manually to various cloud storage providers at their website; no single sign-on in presence as yet.

It has also been depicted how cloud storage services take no liability of any loss of data or corruption and declare rights to read, write, modify or even delete data. Furthermore, they can sell the data for analysis purpose. [9] proposes a technique by introducing an additional layer of security has been suggested termed as overlay structure here. Cost and switching times are two issues user may face with an increased use of such models. Moreover, performance may degrade due to intro-duction of an additional layer (overlay structure). Key management issues are dealt well in this paper.

Some governing bodies still continue to use flat-files to store their information [10]. In addition, users upload their data to the cloud in the form of files. Even though files are being used widely with growing usage of cloud storage, there has been little work done to provide privacy to files. This scheme can enhance the privacy of such files.

In order to upload files to multiple clouds, file splitting is called for. One of the methods used to develop this scheme is: split the file based on its size in bytes. IDA another algorithm which is inculcated for splitting the file in this model [11, 12]. This system uses IDA algorithm to generate erasure codes by applying either Reed-Solomon or Rabin's IDA wherein simple matrix multiplication of bytes of file and coefficient of a matrix which are computed by using a prime number or Fermat's number system is done [13]. In future, file splitting will be done based on file type, various methods as discussed in [14] will be utilized.

3 Our Approach

This paper considers all the pitfalls of existing methods and brings out a proficiency to enhance privacy protection of cloud storage by using file splitter, encryption and multiple cloud storage.

3.1 Proposed Design

There have been tremendous efforts taken to protect data uploaded on a cloud. The approaches discussed and implemented earlier were primarily engineered for exclusive cloud environments wherein data on singular cloud was provided with enhanced privacy protection by applying certain models. Figure 2 shows how users can upload their private files to the cloud by using client application server with the help of privacy algorithms.

Although this approach may solve some of the concerns related to data privacy, it may fall through in cases where cloud itself goes wrong or is attacked. Moreover, threat of insiders still needs to be addressed effectually.

Considering aforementioned issues, this model makes use of multiple cloud environments for storing the data. Figure 2 shows the flow of operations.

The detailed description of the system is as given below:

As illustrated in Fig. 3, when the user selects a file to be uploaded, it first is divided into m chunks by using a simple file splitter. After this, IDA (Information Dispersal Algorithm) is applied on every chuck of the original file to ensure availability and integrity. Moreover, to achieve confidentiality, encryption algorithm is applied on these n fragments of chunks. And then, these encrypted fragments of chunks are uploaded on multiple clouds. To begin with, the project is using Dropbox accounts to store these fragments. On the other hand, to download a file user has uploaded via this system, first decryption is done, after that out of n

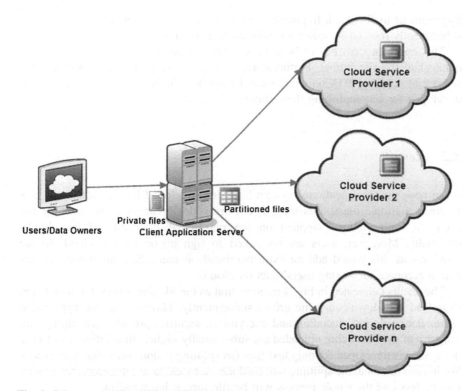

Fig. 2 Privacy protection with multiple clouds

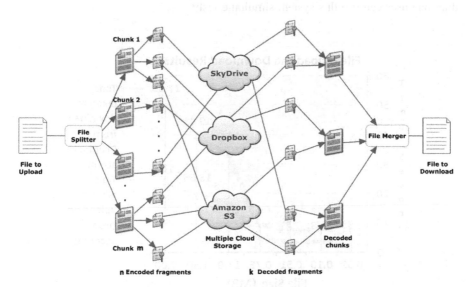

Fig. 3 Working principle of the system

fragments of m chunks, k fragments are retrieved from multiple clouds, and file is subsequently merged by using a simple merger program.

This entire process has to be done on byte stream to make the system independent of the file format. Furthermore, to improve response time, parallel processing of applying IDA on chunks can be applied. In future, the system shall be developed for accomplishing these things.

3.2   Results

As of now, the aforementioned system has been developed for uploading text files wherein, a simple file splitter-merger, erasure codes' generator, and database are used. AES algorithm is applied on the partitioned fragments of the chunks optionally. Moreover, users are not asked to sign up on various cloud storage providers as this would add an extra overhead on them. So, this model utilizes admin accounts for storing users' data on clouds.

The results delineated in Fig. 4 indicate that as the file size grows, the time taken to upload and download a file grows subsequently. However, as the application server does splitting, encoding and encryption; security, privacy, availability and integrity of the files being uploaded are substantially higher. In addition, as of now the system is developed for only text files (as splitting is done on a character array). So, in time to come, the splitting will be done on bytes' array; therefore, time taken will be less and the whole process will be file format independent.

There is no single-point failure here, as IDA takes care of it [12, 13]. Also, more than one user can use this system simultaneously.

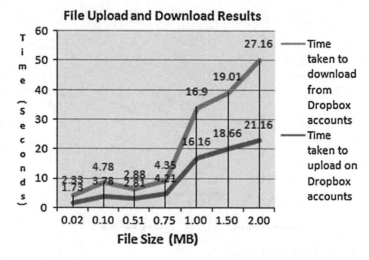

Fig. 4 Results for text files

3.3 Conclusion

This paper addressed core concerns regarding privacy of data stored on the cloud by applying multi-cloud environment and encryption. This approach would reduce privacy issues in cloud. Furthermore, as this model would be developed as a web-based application, number of users can use it simultaneously for storing the data in cloud storage. In the future, the utilization of parallel algorithms or Hadoop map-reduce has been planned in order to reduce processing overhead from the application server and time needed to store/retrieve data on the cloud. Moreover, for partitioning files into pieces, dynamic selection of the algorithm according to file format will be done.

References

1. Kincaid, J.: Google privacy blunder shares your docs without permission. In: TechCrunch (2009)
2. Bohli, J.M., Jensen, M., Gruschka, N., Schwenk, J., Iacono, L.L.L.: Security prospects through cloud computing by adopting multiple clouds. In: Proceedings of IEEE Fourth International Conference on Cloud Computing (CLOUD) (2011)
3. Iacono, L.L., Marnau, N.: Security and privacy-enhancing multicloud architectures
4. Hubbard, D., Sutton, M.: Top threats to cloud computing v1. 0. Cloud Security Alliance (2010)
5. Kumar, G., Shrivastava, N.: A survey on cost effective multi-cloud storage in cloud computing
6. Yao, A.C.: Protocols for secure computations. In: Proceedings of the 23rd Annual IEEE Symposium on Foundations of Computer Science, FOCS82, p. 160164 (1982)
7. Cachin, C., Keidar, I., Shraer, A.: Trusting the cloud. ACM Sigact. News **40**(2), 81–86 (2009)
8. Muller, J., Spillner, J., Schill, A.: Creating optimal cloud storage systems. In: Future Generation Computer Systems, p. 10621072 (2013)
9. Svenn, M.: Secure data management for cloud-based storage solutions. Cloud-Based Softw. Eng. 59, 2013
10. Chickowski, E.: Flat-file databases often overlooked in security schemes (2010). http://www.darkreading.com/risk/flat-file-databases-often-overlooked-in-security-schemes/d/d-id/1133363?
11. Béguin, P., Cresti, A.: General information dispersal algorithms. Theoret. Comput. Sci. **209** (1), 87–105 (1998)
12. Rabin, M.O.: Efficient dispersal of information for security, load balancing, and fault tolerance. J. ACM (JACM) 36(2), 335–348 (1989)
13. Lin, S.-J., Chung, W.-H.: An efficient (n, k) information dispersal algorithm for high code rate system over fermat fields. IEEE Commun. Lett. 16(12), 2036–2039 (2012)
14. Bian, J.: JIGDFS: A secure distributed file system and its applications. PhD thesis, University of Arkansas (2007)

Performance Improvement of MapReduce Framework by Identifying Slow TaskTrackers in Heterogeneous Hadoop Cluster

Nenavath Srinivas Naik, Atul Negi and V.N. Sastry

Abstract MapReduce is presently recognized as a significant parallel and distributed programming model with wide acclaim for large scale computing. MapReduce framework divides a job into *map, reduce* tasks and schedules these tasks in a distributed manner across the cluster. Scheduling of tasks and identification of "slow TaskTrackers" in heterogeneous Hadoop clusters is the focus of recent research. MapReduce performance is currently limited by its default scheduler, which does not adapt well in heterogeneous environments. In this paper, we propose a scheduling method to identify "slow TaskTrackers" in a heterogeneous Hadoop cluster and implement the proposed method by integrating it with the Hadoop default scheduling algorithm. The performance of this method is compared with the Hadoop default scheduler. We observe that the proposed approach shows modest but consistent improvement against the default Hadoop scheduler in heterogeneous environments. We see that it improves by minimizing the overall job execution time.

Keywords Hadoop · MapReduce · Job scheduling · TaskTracker · Heterogeneous environments

N.S. Naik (✉) · A. Negi
School of Computer and Information Sciences, University of Hyderabad,
Hyderabad 500046, India
e-mail: srinuphdcs@gmail.com

A. Negi
e-mail: atulcs@uohyd.ernet.in

V.N. Sastry
Institute for Development and Research in Banking Technology,
Hyderabad 500057, India
e-mail: vnsastry@idrbt.ac.in

© Springer India 2016
A. Nagar et al. (eds.), *Proceedings of 3rd International Conference
on Advanced Computing, Networking and Informatics*, Smart Innovation,
Systems and Technologies 44, DOI 10.1007/978-81-322-2529-4_49

465

1 Introduction

Efficiently storing, querying, analyzing, interpreting, and utilizing these huge data sets presents one of the impressive challenges to the computing industry and the research community [1]. A large number of organizations across the world use Apache Hadoop, created by Doug Cutting, which is an open source implementation of the MapReduce framework and processes massive amounts of data in-parallel on large clusters of commodity systems. Take Yahoo, for example. Uses a Hadoop cluster of 4,000 nodes, having 30,000 CPU cores, and 17 petabytes of disk space [2]. The structure of MapReduce is based on the master-slave architecture [3]. A single master node monitors the status of all slave nodes in the cluster and allocates jobs to them. The benefits of MapReduce framework are the capability of fault tolerance and appropriate distribution of tasks to multiple processing nodes in the cluster [4].

The basic assumption of Hadoop framework is that the nodes of the cluster are homogeneous [5]. Several issues which will directly affect the performance of MapReduce framework are node heterogeneity, stragglers, data locality and "slow TaskTrackers" [6]. These issues have been undervalued by researchers in most of the proposed MapReduce scheduling algorithms, which leads to poor performance of Hadoop [7]. Minimizing the execution time of a job by appropriately assigning tasks to the available nodes is a common goal of the MapReduce schedulers and it is likewise a significant research topic because it betters the performance of MapReduce framework [8].

In this research work, we address the problem of identifying "slow TaskTrackers" in the heterogeneous Hadoop cluster by integrating it with the Hadoop default scheduler. The proposed work helps the JobTracker not to schedule any task on these identified "slow TaskTrackers" instead schedule on the remaining TaskTrackers, which minimizes the job execution time and certainly improves the overall performance of the MapReduce framework in heterogeneous environments. Throughout this paper by "slow TaskTracker" we are referring to a TaskTracker which has some tasks under it that are running slower relative to other tasks.

The rest of the paper is structured as follows. A background of the MapReduce framework and the Hadoop's default scheduler as related work is given in Sect. 2. Procedure for identifying "slow TaskTrackers" in the heterogeneous Hadoop cluster is given in Sect. 3 and Sect. 4 conducts a performance evaluation of the proposed work. Finally, we conclude the paper and give some outlines of our future research work in Sect. 5.

2 Related Work

This section provides a brief view of the MapReduce framework and explains about the Hadoop default scheduling algorithm with its limitations.

2.1 Basic Concepts in MapReduce

In Hadoop cluster, HDFS (Hadoop Distributed file system) contains one single NameNode called master node and a number of DataNodes called worker nodes [9]. NameNode maintains the meta-data information about the locations of data chunks and DataNode stores the chunks of data in the cluster. For running a job in the cluster, MapReduce component is used, which contains one JobTracker and a series of TaskTrackers [10]. JobTracker manages the jobs and assigns tasks to the TaskTrackers and TaskTracker processes the tasks on the corresponding node in the cluster [11].

Scheduling of the MapReduce system has following stages while scheduling a job in the cluster [12].

1. The Hadoop framework first breaks the input data file into *M* pieces of identical data sizes and then distributed in the cluster.
2. The master node will pick up the idle worker nodes and allocates them *M map* tasks. After intermediate output is produced by *map* tasks, the master node will allocates *R reduce* tasks to the worker nodes which are idle.
3. The intermediate (key, value) pairs from the *map* function are buffered to local disks at regular intervals.
4. The above buffered pairs are split into *R* regions by (*map*) worker using a partition function (default is *hash* (intermediate key) mod *R*), so that same intermediate (key, value) pairs go to one partition.
5. Reducers will read the data from the *map* workers using remote procedure calls, then it sorts and groups the data by intermediate key so that all values of the same key are collected together.
6. After complete execution of the *map* and *reduce* tasks, the outcomes will be fed back to the user by the master node.

2.2 Hadoop Default Scheduling Algorithm

The progress score (*PS*) of a task *t* is denoted by PS_t, which is calculated using (1) for *map* tasks and (2) for *reduce* tasks [13].

$$PS_t = M/N \tag{1}$$

$$PS_t = (1/3)(K + M/N) \tag{2}$$

where, *M* is the number of (key, value) pairs that have been processed successfully, *N* is the overall number of (key, value) pairs and *K* is the stage (shuffle, sort and merge) value in a *reduce* phase.

The average progress score of a job PS_{avg} is calculated using (3), $PS[i]$ is the progress score of a task t_i and n is the number of executable tasks in a job.

$$PS_{avg} = \sum_{i=1}^{n} PS[i]/n \qquad (3)$$

Limitations of Hadoop Default Scheduler [13]

1. The *map* and *reduce* task weights in different stages are $M_1 = 1$, $M_2 = 0$ and $(R_1 = R_2 = R_3 = 1/3)$ but these weights will change when tasks run in a heterogeneous environment.
2. Default scheduler cannot identify the "slow TaskTrackers" in a heterogeneous Hadoop cluster.
3. Default scheduler unobserved the accurate straggler tasks which need to be re-executed in the cluster.

3 Proposed Method for Identifying Slow TaskTrackers in Heterogeneous Hadoop Cluster

The performance of distributed and parallel systems like MapReduce is closely related to its Task scheduler. If a task is scheduled on a "slow TaskTracker" then the overall execution time of a job will be increased. Finding "slow TaskTrackers" in heterogeneous Hadoop cluster is an interesting research problem because the efficient way of finding it can significantly *reduce* the overall job execution time and thus improves the performance of the MapReduce framework in heterogeneous environments.

The Progress score of a TaskTracker in the cluster is calculated using (4)

$$PSTT_i = \sum_{j=1}^{t} PS_j/t \qquad (4)$$

Here, the progress score of ith TaskTracker is $PSTT_i$, PS_j is the progress score of a task calculated based on how much a task's (key, value) pairs have been finished per second, which is calculated as in Hadoop default scheduler and t is the number of tasks on the ith TaskTracker in the cluster.

The average progress score of all TaskTrackers in the Hadoop cluster for a given job is calculated using (5)

$$APSTT = \sum_{i=1}^{T} PSTT_i/T \qquad (5)$$

Here, *APSTT* is the average progress score of all TaskTrackers in the cluster and T is the number of TaskTrackers present in the Hadoop cluster.

We can find the "slow TaskTrackers" present in the cluster using (6)

$$PSTT_i > APSTT(TTTh + 1) \tag{6}$$

For the ith TaskTracker, if it satisfies the above equation, then we can say that particular TaskTracker is a "slow TaskTracker" otherwise it is the fast TaskTracker in the heterogeneous Hadoop cluster.

TaskTracker Threshold (*TTTh*) is in the range [0,1] is used to categorize the TaskTrackers in the Hadoop cluster into slow and fast. According to (6), if *TTTh* is too small then it will categorize some fast TaskTrackers to be "slow TaskTrackers" and if *TTTh* is too large then it will categorize some "slow TaskTrackers" to be fast TaskTrackers. Thus, we have chosen 0.5 as an appropriate value for *TTTh* in our experiments.

Input: The set of TaskTrackers present in the heterogeneous Hadoop cluster.
Output: The set of "slow TaskTrackers".

Algorithm 1 Identifying slowTaskTrackers

1: set *slowTaskTrackers*
2: **for** each *TaskTracker* i in the *cluster* **do**
3:     **for** each running *task* j of the *job* **do**
4:         **if** *task* j is a *Map* task **then**
5:             $ProgressScore_j \leftarrow M/N$
6:         **else**
7:             $ProgressScore_j \leftarrow 1/3 * (K + M/N)$
8:         **end if**
9:     **end for**
10:     $PSTT_i = \sum_{j=1}^{t} PS_j/t$
11: **end for**
12: $APSTT = \sum_{i=1}^{T} PSTT_i/T$
13: **for** each running *task* i of the *job* **do**
14:     **if** $PSTT_i > APSTT(TTTh + 1)$ **then**
15:         $slowTaskTrackers.add(i^{th} \ TaskTracker)$
16:     **end if**
17: **end for**
18: return *slowTaskTrackers*

---

4 Evaluation

In this section, we now briefly discuss the experimental environment, workload description and then explains the performance analysis of our proposed method on a heterogeneous Hadoop cluster.

4.1 Experimental Environment

We followed numerous stages to establish the experimental setup required to conduct our experiments and considered heterogeneous nodes in a Hadoop cluster as presented in Table 1, it has different Hadoop cluster hardware environment and configurations. We used Hadoop cluster of five heterogeneous nodes to evaluate our proposed method for finding "slow TaskTrackers". One of the nodes was chosen as a master node which runs the Hadoop distributed file system (NameNode) and MapReduce runtime (JobTracker). The remaining four nodes were worker nodes (DataNodes and TaskTrackers). The nodes were interconnected by Ethernet switch. All systems in the cluster use Ubuntu 14.04 operating system, JDK version 8, and Hadoop 1.2.1 version for performance evaluation.

In our experiments, we evaluate the proposed scheduling method using Hi-Bench benchmark suite [14] because it is a new, realistic and comprehensive benchmark suite for Hadoop.

4.2 Workload Description

We evaluate our proposed method using three different job types: Sort, WordCount, and TeraSort, that simulate micro benchmarks of Hi-Bench benchmark suite. These

Table 1 Hadoop evaluation environment

Node	Hardware configuration	Hadoop configuration
Master node	Intel Xeon(R) CPU E3110 @ 3.00 GHz, 4 GB RAM, 500 GB Disk space	
Slave node 1	Intel core i3-3220 CPU @ 3.30 GHz, 2 GB RAM, 500 GB Disk space	3 *map* and 1 *reduce* slots per node
Slave node 2	Intel core 2 duo CPU E7500 @ 2.93 GHz, 2 GB RAM, 320 GB Disk space	2 *map* and 1 *reduce* slots per node
Slave node 3	Intel Pentium CPU G640 @ 2.80 GHz, 2 GB RAM, 500 GB Disk space	1 *map* and 1 *reduce* slots per node
Slave node 4	Intel Core 2 Duo Processor P8400 @ 2.26 GHz, 3 GB RAM, 250 GB Disk space	2 *map* and 1 *reduce* slots per node

micro benchmarks show the key characteristics of MapReduce clearly and widely used by the Hadoop research community to evaluate the scheduling algorithms in their experiments. We briefly describe the micro-benchmarks as below [14]:

1. The WordCount workload counts the word frequencies from textual data. It is mostly CPU bound (particularly during the *map* phase), causing high CPU usage, light disk or network I/O.
2. The Sort workload depends on the Hadoop framework to sort the final results. It is mostly I/O bound, having moderate CPU usage and heavy disk I/O.
3. The TeraSort workload is very high CPU utilization and moderate disk I/O during the *map* and shuffle phases, and moderate CPU usage and heavy disk I/O during the *reduce* phase.

4.3 Performance Analysis of Our Proposed Method

In order to evaluate the performance, we have integrated our proposed method with the Hadoop default scheduling algorithm to identify the "slow TaskTrackers" in the heterogeneous Hadoop cluster. We compared our proposed method with the Hadoop default scheduler because it is a simple, fast algorithm, extensively used in numerous recent Hadoop clusters and it has no procedure to find the "slow TaskTrackers" and assumes nodes in the cluster as homogeneous. We presented our performance improvement by comparing the proposed method with the Hadoop default scheduler and performed Sort, WordCount, TeraSort benchmarks under heterogeneous environments by considering the Job execution time as a metric for the evaluation.

In our experiments, we presented how "slow TaskTrackers" effect the execution time of a job and performed three micro benchmarks over the MapReduce job execution time metric for performance evaluation in the heterogeneous Hadoop cluster. Figure 1 shows the performance comparison of the Default Hadoop scheduler and Default Hadoop scheduler with the proposed method. In all of these different workloads (Sort, WordCount and TeraSort), our proposed method achieves the best in terms of minimum job execution time compared to the Hadoop default scheduling algorithm in the heterogeneous environments.

Fig. 1 Comparison of job execution time for different workloads

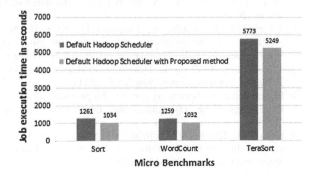

5 Conclusion and Future Work

In this paper, we proposed a scheduling method and integrated it with the Hadoop default scheduler, which aims to find the "slow TaskTrackers" in the heterogeneous Hadoop cluster and it predicts the JobTracker in such a way that it will not schedule any new tasks on the identified "slow TaskTrackers" in the cluster. In this proposed method, when a JobTracker schedules a task on the TaskTracker, first it identifies the "slow TaskTrackers" present in the Hadoop cluster, then it will not schedule the tasks on those particular "slow TaskTrackers" instead schedules on the remaining TaskTrackers in the Hadoop cluster. Our proposed method shows the best performance in terms of job execution time compared to the Hadoop default scheduler when executing the Sort, Word Count, and TeraSort benchmarks and thus it improves the performance of the MapReduce framework in the heterogeneous environments by minimizing the overall job execution time.

As part of the future research work, we would like to further identify the "slow TaskTrackers" in each of the *map* and *reduce* phases of the MapReduce framework in heterogeneous environments.

Acknowledgments Nenavath Srinivas Naik express his gratitude to Prof. P.A. Sastry (Principal), Prof. J. Prasanna Kumar (Head of the CSE Department) and Dr. B. Sandhya, MVSR Engineering College, Hyderabad, India for hosting the experimental test bed.

References

1. Dean, J., Ghemawat, S.: MapReduce: simplified data processing on large clusters. Commun. ACM **51**, 107–113 (2008)
2. Dean, J., Ghemawat, S.: MapReduce: a flexible data processing tool. Commun. ACM **53**(1), 72–77 (2010)
3. Rasooli, A., Down, D.G.: An adaptive scheduling algorithm for dynamic heterogeneous hadoop systems. In: Proceedings of the 2011 Conference of the Center for Advanced Studies on Collaborative Research, pp. 30–44. Canada (2011)
4. Zaharia, M., Borthakur, D., Sarma, J.S., Elmeleegy, K., Shenker, S., Stoica, I.: Job Scheduling for Multi-User MapReduce Clusters. Technical Report, University of California, Berkeley (2009)
5. Dawei, J., Beng, C.O., Lei, S., Sai, W.: The Performance of MapReduce: An In-depth Study. VLDB (2010)
6. Zaharia, M., Konwinski, A., Joseph, A.D., Katz, R., Stoica, I.: Improving mapreduce performance in heterogeneous environments. In: 8th Usenix Symposium on Operating Systems Design and Implementation, pp. 29–42. ACM Press, New York (2008)
7. Tan, J., Meng, X., Zhang, L.: Delay Tails in Mapreduce Scheduling. Technical Report, IBM T. J. Watson Research Center, New York (2011)
8. Ekanayake, J., Pallickara, S., Fox, G.: MapReduce for data intensive scientific analyses. In: Proceedings of the 2008 IEEE Fourth International Conference on eScience, pp. 277–284 (2008)
9. Rasooli, A., Down, D.G.: A hybrid scheduling approach for scalable heterogeneous Hadoop systems. In: Proceeding of the 5th Workshop on Many-Task Computing on Grids and Supercomputers, pp. 1284–1291 (2012)

10. Nanduri, R., Maheshwari, N., Reddyraja, A., Varma, V.: Job aware scheduling algorithm for mapreduce framework. In: Proceedings of the 3rd International Conference on Cloud Computing Technology and Science, pp. 724–729, Washington, USA (2011)
11. Zhenhua, G., Geo, R.F., Zhou, M., Yang, R.: Improving resource utilization in MapReduce. In: IEEE International Conference on Cluster Computing, pp. 402–410 (2012)
12. Rasooli, A., Down, D.G.: COSHH: a classification and optimization based scheduler for heterogeneous Hadoop systems. J. Future Gener. Comput. Syst. 1–15 (2014)
13. Naik, N.S., Negi, A., Sastry, V.N.: A review of adaptive approaches to MapReduce scheduling in heterogeneous environments. In: IEEE International Conference on Advances in Computing, Communications and Informatics, pp. 677–683, Delhi, India (2014)
14. Shengsheng, H., Jie, H., Jinquan, D., Tao, X., Huang, B.: The HiBench benchmark suite: characterization of the MapReduce-based data analysis. In: IEEE 26th International Conference on Data Engineering Workshops, pp. 41–51 (2010)

Router Framework for Secured Network Virtualization in Data Center of IaaS Cloud

Anant V. Nimkar and Soumya K. Ghosh

Abstract Data center exploits network virtualization to fully utilize physical network resources by collocating tenants' virtual networks. The virtual networks consist of sets of virtual routers connected by virtual links. The network virtualization must efficiently embed virtual networks on a physical network of the data center to balance load among physical resources to fully utilize the physical network. The virtual networks must also be securely managed so that they are not compromised by collocated users or a data center network administrator who has direct access to the physical network. In this paper, we propose a router framework in which virtual routers and links can be securely placed on physical router by adding a virtual plane on top of data and control planes, two abstract protocols and an enforcement of Federation Access Control Model (FACM). The two abstract protocols, viz. Secure Virtual Topology Embedding Protocol (SVTEP) and Node-and-Path Label Distribution Protocol (NPLDP) are presented along with a theoretical evaluation of the proposed router framework to fulfill all the aforesaid requirements.

Keywords Router framework · Access control model · Network virtualization · Virtual network embedding · IaaS · Cloud · Data center

1 Introduction

Network Virtualization Environment (NVE) supports virtual networks on top of physical network [1]. Virtual networks consist of virtual nodes and virtual links which are placed on physical routers and physical paths respectively. The virtual

A.V. Nimkar (✉) · S.K. Ghosh
School of Information Technology, Indian Institute of Technology,
Kharagpur 721 302, India
e-mail: anantn@sit.iitkgp.ernet.in

S.K. Ghosh
e-mail: skg@iitkgp.ac.in

© Springer India 2016   475
A. Nagar et al. (eds.), *Proceedings of 3rd International Conference
on Advanced Computing, Networking and Informatics*, Smart Innovation,
Systems and Technologies 44, DOI 10.1007/978-81-322-2529-4_50

resources (i.e. virtual nodes and links) are collectively managed by stakeholders in network virtualization [2]. Network virtualization also needs node and link mapping algorithms so that the loads among physical routers and links are balanced [3]. The main challenge for deploying network virtualization is weaknesses in router virtualization for topology-aware applications [4] and IaaS Cloud Federation [2]. The weaknesses are minimal security provision for management of virtual resources, no load balancing on physical resources and no transparent mapping between physical and virtual resources. Recent router architectures [5–7] install control and data planes, in both centrally or distributively, using container or hypervisor based virtualization. None of the recent architectures e.g. Microsoft SoftRouter [8], OpenVRoute [9] and FIBIUM [10] concentrate on secure management of virtual resources except AVR [11] which allow secure creation and deletion of virtual routers but no secure configuration of virtual routers. Thus, router architectures face three major issues, namely (i) Secure and transparent management of collocated virtual resources, (ii) Load balancing for efficient and secure placement of virtual routers and links on physical router and links respectively, and (iii) Secure mapping between physical and virtual networks.

In this paper, we propose a layered router framework to address the aforementioned issues. The proposed router framework does not significantly change the standard router architecture and it only adds another *virtual plane* on top of control and data plane of existing standard router architecture. The proposed router framework augments a deploy-able unit of Federation Access Control Model (FACM) [12], network embedding algorithm and signaling protocol to provide secure load balancing, management of virtual resources and secure mapping between physical and virtual networks.

The main contributions of the paper has two parts: (i) Router framework for secured multi-tenant network collocation using a security component *Access Control Enforcement Engine* and (ii) Two abstract protocols, viz. *Node-and-Path Label Distribution Protocol* and *Secured Virtual Topology Embedding Protocol*. *Access Control Enforcement Engine* (ACEE) enforces FACM to address the issue of secure management of virtual resources in network virtualization. The abstract *Node-and-Path Label Distribution Protocol* (NPLDP) addresses the issue of mapping tenants' virtual networks to the physical network. The abstract *Secured Virtual Topology Embedding Protocol* (SVTEP) addresses the issue of uniformly distributing virtual nodes and links over the physical network.

The rest of the paper is organized as follows. The related work on architectures of router virtualization is given in Sect. 2. Section 3 presents background works on *Network Virtualization, IaaS Cloud Federation* and *Federation Access Control Model*. Section 4 proposes a router framework by adding *Node-and-Path Label Distribution Protocol* (Sect. 4.1), *Secured Virtual Topology Embedding Protocol* (Sect. 4.2), *Access Control Enforcement Engine* (Sect. 4.3) and the Information Bases (Sect. 4.4) to standard router architecture. Section 4.5 presents the working principle and theoretical evaluation of the proposed router framework. Finally, Sect. 5 concludes this work and gives the future works.

2 Related Work

The router framework for secure network virtualization needs investigation of recent router architectures and virtualization techniques to address the aforesaid three issues. Mattos et al. [7] evaluate three virtualization techniques, namely OpenVZ, Xen and VMware through four metrics namely memory, processor, network and disk performance of virtual routers. The evaluation shows that OpenVZ and Xen return a small performance overhead in terms of memory, processor and disk performance while OpenVZ and Xen return moderate and good network performance respectively. VMware offers fully virtualized solution but introduces significant overhead in handling virtual resources. Autonomic Virtual Routers (AVR) [11] provides automatic virtual router provisioning which serves as a recommendation for the separation of data and control planes in virtual routers. In SoftRouter architecture [8], control planes are far away from data planes. This separation provides easier scalability. SoftRouter does not securely manage virtual resources. FIBIUM [10] is hardware accelerated software router using virtualization on commodity PC and the separation of control and data planes on OpenFlow-based switches. Open Router Virtualization [13] allows data plane and control planes of virtual routers on server and OpenFlow switches respectively. It provides programming interfaces for forwarding plane. OpenVRoute [9] distributes data plane and control plane of virtual routers on server and OpenFlow switches respectively and uses Flow Management Proxy (FMP) to establish communication between data and control planes. Virtual Router as Service (VRS) [14] is a distributed service of forwarding planes of virtual routers and it optimally places forwarding plane and virtual links using some embedding algorithm. VROOM (Virtual ROuters On the Move) [15] provides network-management primitives to freely move virtual routers from one physical router to another physical router.

3 Background

The proposed router framework extends standard router architecture and uses Federation Access Control Model. This section provides fundamentals of Network Virtualization, IaaS Cloud federation and Federation Access Control Model.

3.1 Network Virtualization and IaaS Cloud Federation

Network virtualization environment facilitates collocation of multiple virtual networks on top of physical networks. Virtual network consists of virtual routers and links created on physical nodes and paths respectively. Figure 1 shows two virtual networks collocated on the same physical network. The black plain line polygon

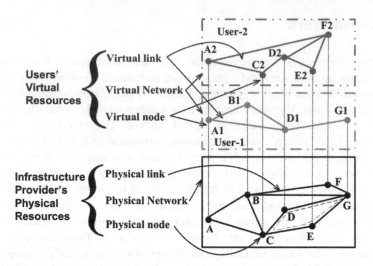

Fig. 1 Network virtualization

shows the physical network infrastructure of cloud provider. The virtual networks of User-1 and User-2 are represented in green dot-dash and red double-dot-dash polygons respectively. The node allocation algorithm map virtual nodes to physical node. Each virtual link is mapped to a physical path using some link allocation algorithm. The virtual nodes A1 and A2 are mapped onto physical node A. Similarly, the virtual link, (D1,G1) can be mapped to the direct edge, D-G or path, D-C-E-G.

IaaS Cloud federation provides virtual nodes, virtual machines and virtual links to get the economies of scale through federation among cloud providers. Figure 2 shows an example of a cloud federation among three IaaS cloud providers for two users. The green dot-dash and red double-dot-dash polygons show the virtual infrastructures of User-1 and User-2 respectively. The virtual nodes e.g. E1, E2, M1 etc. are cooperatively managed by users and corresponding IaaS cloud providers. Similarly, virtual links are managed by user and one/two IaaS cloud provider(s) e.g. the virtual link (H2,Q2) is cooperatively managed by User-2, Cloud Provider-2 and Cloud Provider-3 while the virtual link (L2,Q2) is cooperatively managed by User-2 and Cloud Provider-3. Management of all federated virtual resources among IaaS cloud providers needs a special kind of access control mechanism FACM which handles federated resources.

3.2 Federation Access Control Model

Federation Access Control Model [12] is MAC-and-DAC based access control model to access virtual resources by subjects which are sets of participants from

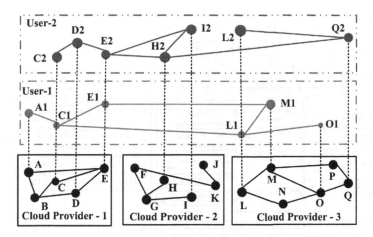

Fig. 2 IaaS Cloud federation

federated infrastructure providers (InPs) and users. Each InP $j \in E$ maintains a part of FACM as $F_j = (S_j, O_j, P_j, F_j, L_j)$ where E is set of all InPs in federated system. S_j is a set of subjects which are collective subset of federating participants. O_j is a set of virtual links, virtual routers and virtual machines. The subjects $s \in S_j$ can perform P_j basic operations (i.e. access rights) on the objects O_j. The set F_j maps security labels of subjects to security labels of objects at particular time instance. L_j is a partially ordered set of security classes of the federations in the system.

FACM employs a different notion of security label which is an ordered tuple. The security label has a form of $\langle SeP_{SL}, U_{SL}, InP_{SL} \rangle$ and; SeP_{SL}, U_{SL} and InP_{SL} are the security labels of service provider, the user from the service provider and a set of federated infrastructure providers respectively. FACM uses two operators namely, the Cartesian equal (i.e. $\equiv$) and Cartesian subset (i.e. $\subseteq$) to check relation between the security class of subjects and objects. FACM also uses Cartesian union (i.e. $\uplus$) to dynamically create the security labels of federated subjects and objects.

4 Router Framework for Secured Network Virtualization

The proposed router framework adds a virtual plane on top of data and control planes of standard router architecture, an deploy-able unit of FACM, a set of information bases to address three aforementioned issues as shown in Fig. 3. The two abstract protocols are *Node-and-Path Label Distribution Protocol* (NPLDP) and *Secured Virtual Topology Embedding Protocol* (SVTEP). The deploy-able unit includes *Access Control Enforcement Engine* (ACEE) and the information bases are *Label Information Base* (LIB), *Forwarding Security Label Information Base* (FSLIB) and *Forwarding Virtual Label Information Base* (FVLIB).

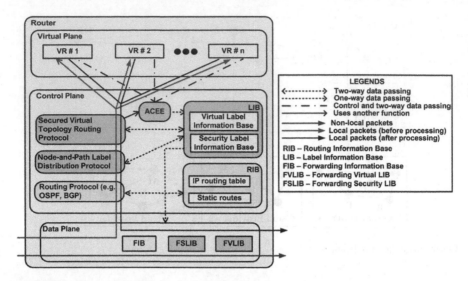

Fig. 3 Router framework for secured network collocation

SVTEP and NPLDP cooperatively run on top of existing routing protocol and provide security through ACEE. ACEE first authorizes the requests received from users/customers for the management of virtual router and links. Then, NPLDP signals physical routers for creation/deletion of virtual resources. NPLDP also distributes node and path labels to provide mapping between physical and virtual networks so that it securely multiplex and de-multiplex data-centric and network-centric traffic. SVTEP securely handles load balancing among physical nodes and links using some virtual node and link allocation algorithms.

LIB consists of (i) Virtual Label Information Base (VLIB) and (ii) Security Label Information Base (SLIB). NPLDP updates VLIB whenever virtual routers and links are created/deleted. ACEE automatically and synchronously updates SLIB to store the security labels of virtual resources whenever the access control services (e.g. create/delete virtual router/link) are executed through a sequence of primitives operations on *Access Control Matrix* (ACM) which is a data structure of FACM. The forwarding plane uses FVLIB and FSLIB which hold subset of virtual and security labels from VLIB and SLIB respectively.

4.1 Node-and-Path Label Distribution Protocol

Node-and-Path Label Distribution Protocol (NPLDP) distributes identities of virtual routers and virtual links as follows. The labels of virtual routers are entered in VLIB whenever FACM successfully executes create-node service. Similarly, the labels of virtual links are entered in VLIB whenever FACM successfully executes create-link service. In case of virtual labels of virtual links, NPLDP has a set of procedures to send virtual labels of source and destination virtual routers along the physical path selected by SVTEP. The deletion of virtual labels from VLIB is performed when delete-node and delete-link services are successfully executed by FACM. The mapping between physical and virtual networks is carried out by NPLDP and it allows multiplexing and de-multiplexing of all users' traffics.

4.2 Secured Virtual Topology Embedding Protocol

Secure Virtual Topology Embedding Protocol (SVTEP) uses a concept of node and link stresses which represent the loads on physical router and link respectively. The concept of node and link stresses is borrowed from a survey of virtual network embedding [3]. The node stresses are updated in local databases of physical routers when a virtual routers is placed/removed on/from physical router. Similarly, the link stresses are updated in local databases of physical routers when a virtual links are added/removed on/from physical link. SVTEP has two alternative scheme to update node and link stresses. First, SVTEP may use centralized algorithm to update the node and link stresses of physical routers and links. Second, it may use a distributed wave traversal algorithm which does the function of propagating node and link stress to all physical routers in the data center. Once the propagation of node and link stresses is finished, SVTEP can securely find the optimal physical router and path for the placement of virtual router and link respectively.

4.3 Access Control Enforcement Engine

Access Control Enforcement Engine (ACEE) carries out two functions. First, ACEE manages virtual resources through a combined enforcement of Mandatory Access Control (MAC) and Discretionary Access Control (DAC) policies. ACEE also updates security labels of federated subjects and virtual resources in SLIB of whenever management operations are executed. Second, ACEE uses VLIB and SLIB to multiplex and de-multiplex traffic of local virtual routers. The non-local traffic is forwarded using FVLIB and FSLIB.

4.4 Label Information Bases

Label Information Base (LIB) consists of two types of data structures for storing virtual labels and security labels of virtual routers and virtual links respectively. Virtual Label Information Base (VLIB) stores virtual labels of collocated virtual routers and links which have the forms of ⟨pnode,vnode⟩ and ⟨source pnode,source vnode, destination pnode,destination vnode⟩ respectively where pnode and vnode stand for local virtual and physical labels respectively. The formation of virtual labels is cooperatively carried out by SVTEP, NPLDP and ACEE. Security Label Information Base (SLIB) stores security labels of virtual resources and it is automatically updated by FACM. To perform fast forwarding of incoming non-local packets, the data plane uses FVLIB and FSLIB. FVLIB and FSLIB consist of subsets of VLIB and SLIB respectively. FVLIB is frequently updated at some regular interval.

4.5 Working Principle and Theoretical Evaluation

The design of proposed router framework for secure network virtualization must address the placement of ACEE, SVTEP and NPLDP at various locations. There are two alternatives for the placement, namely, distributed and centralized. The placement of ACEE must be distributed because the decision about packet forwarding must be very fast and on-line. The placement of SVTEP and NPLDP can either be centralized or distributed. The centralized placement may results into two dis-advantages (i) single point of failure and (ii) delay in load balancing and virtual network management. But it has two advantages (i) less computation and (ii) less message complexity which result into low network-centric traffic. The distributed placement results into least delay-time for network management with more message complexity as compared to centralized version.

All the recent virtual router architecture support security provision only for data-centric traffic and no provision for network-centric traffic except AVR. AVR concentrates on confidentiality of all types of traffic whereas the proposed router framework provides a full-fledged solution for secure management of collocated virtual resources, efficient load balancing among physical resources and transparent mapping between physical and virtual networks.

5 Conclusion

This paper proposes a router framework by adding a virtual plane without significantly changing the existing standard router architecture. The paper also proposes two abstract protocols namely *Secure Virtual Topology Embedding Protocol* (SVTEP) and *Node and Path Label Allocation Protocol* (NPLDP) to efficiently and securely manage the placement of virtual resources on physical network through *Access Control Enforcement Engine* (ACEE). Further, the proposed router framework is the first step towards the design of router which supports secured multi-tenant network collocation. The future work includes evaluation of NPLDP, SVTEP and the proposed router framework.

References

1. Chowdhury, N.M.K., Boutaba, R.: A survey of network virtualization. Comput. Netw. **54**, 862–876 (2010)
2. Nimkar, A.V., Ghosh, S. K.: Towards full network virtualization in horizontal iaas federation: security issues. J. Cloud Comput.: Adv. Syst. Appl., SpringerOpen **2**(19), 19:1–19:13 (2013)
3. Fischer, A., Botero, J., Till Beck, M., de Meer, H., Hesselbach, X.: Virtual network embedding: a survey. Commun. Surv. Tutorials IEEE **15**, 1888–1906 (2013)
4. Fan, P., Chen, Z., Wang, J., Zheng, Z., Lyu, M.: Topology-aware deployment of scientific applications in cloud computing. In: 2012 IEEE 5th International Conference on Cloud Computing (CLOUD), pp. 319–326, June 2012
5. Egi, N., Greenhalgh, A., Handley, M., Hoerdt, M., Huici, F., Mathy, L.: Fairness issues in software virtual routers. In: Proceedings of the ACM Workshop on Programmable Routers for Extensible Services of Tomorrow, PRESTO '08. ACM, New York, NY, USA, pp. 33–38 (2008)
6. Rathore, M., Hidell, M., Sjödin, P.: Data plane optimization in open virtual routers. In: Networking 2011, Lecture Notes in Computer Science, vol. 6640, pp. 379–392, Springer Berlin Heidelberg (2011)
7. Mattos, D.M.F., Ferraz, L.H.G., Costa, L.H.M.K., Duarte, O.C.M.B.: Evaluating virtual router performance for a pluralist future internet. In: Proceedings of the 3rd International Conference on Information and Communication Systems, ICICS '12. ACM, New York, NY, USA, pp. 4:1–4:7 (2012)
8. Lakshman, T.V., Nandagopal, T., Ramjee, R., Sabnani, K., Woo, T.: The softrouter architecture. In: Third Workshop on Hot Topics in Networks HotNets-III, ACM, San Diego, CA, USA, Nov 2004
9. Bozakov, Z., Papadimitriou, P.: Openvroute: an open architecture for high-performance programmable virtual routers. In: IEEE 14th International Conference on High Performance Switching and Routing (HPSR), pp. 191–196 (2013)
10. Sarrar, N., Feldmann, A., Uhlig, S., Sherwood, R., Huang, X.: Fibium-towards hardware accelerated software routers. EuroView 2010 (poster session) **9**, 1–17 (2010)
11. Louati, W., Houidi, I., Zeghlache, D.: Autonomic virtual routers for the future internet. In: Proceedings of the 9th IEEE International Workshop on IP Operations and Management, IPOM '09, Springer-Verlag, Heidelberg, pp. 104–115 (2009)
12. Nimkar, A.V., Ghosh, S.K.: A theoretical study on access control model in federated systems. In: Communications in Computer and Information Science, Recent Trends in Computer Networks and Distributed Systems Security, vol. 420, pp. 310–321, Springer Berlin Heidelberg (2014)
13. Bozakov, Z.: An open router virtualization framework using a programmable forwarding plane. SIGCOMM Comput. Commun. Rev. **40**(4), 439–440 (2010)
14. Bozakov, Z.: Architecture and algorithms for virtual routers as a service. In: 2011 IEEE 19th International Workshop on Quality of Service (IWQoS), pp. 1–3 (2011)
15. Wang, Y., Keller, E., Biskeborn, B., van der Merwe, J., Rexford, J.: Virtual routers on the move: live router migration as a network-management primitive. SIGCOMM Comput. Commun. Rev. **38**, 231–242 (2008)

Analysis of Machine Learning Techniques Based Intrusion Detection Systems

Rupam Kr. Sharma, Hemanta Kumar Kalita and Parashjyoti Borah

Abstract Attacks on Computer Networks are one of the major threats on using Internet these days. Intrusion Detection Systems (IDS) are one of the security tools available to detect possible intrusions in a Network or in a Host. Research showed that application of machine learning techniques in intrusion detection could achieve high detection rate as well as low false positive rate. This paper discusses some commonly used machine learning techniques in Intrusion Detection System and also reviews some of the existing machine learning IDS proposed by authors at different times.

Keywords Intrusion detection system · Supervised learning · Unsupervised learning · KDD'99 · Anomaly detection · Host intrusion system

1 Introduction

Intrusion Detection System (IDS)s are security tools that detect intrusions to a network or a host computer. An IDS is either host based or network based. A host based IDS detects attacks on a host computer, whereas, a network based IDS, also called Network Intrusion Detection System (NIDS), detects intrusions into a network by analyzing network traffic and are generally installed in network gateway or server. Host based intrusion detection systems can be divided into four types, namely (a) File System Monitors, (b) Log file analyzers, (c) Connection analyzers,

R.Kr. Sharma (✉) · H.K. Kalita
North Eastern Hills University, Shillong, India
e-mail: sun1_rupam1@yahoo.com

H.K. Kalita
e-mail: hemata91@yahoo.co.in

P. Borah
Assam Don Bosco University, Guwahati, India
e-mail: parashjyoti@hotmail.com

© Springer India 2016
A. Nagar et al. (eds.), *Proceedings of 3rd International Conference on Advanced Computing, Networking and Informatics*, Smart Innovation, Systems and Technologies 44, DOI 10.1007/978-81-322-2529-4_51

485

(d) Kernel-based IDS [1, 28]. Furthermore, based on the data analyzing technique there are mainly two categories of IDSs, signature-based and anomaly based. A signature-based system detects attacks by analyzing network data for attack signatures stored in its database. This type of IDS detects previously known attacks, whose signatures are stored in its database. On the other hand, an anomaly-based IDS looks for deviations from normal behavior of the subject. Anomaly-based systems are capable of detecting novel attacks [2]. Machine learning techniques could be effective for detecting intrusions. Several Intrusion Detection Systems have been modeled based on machine learning techniques. Learning algorithms are built either on offline dataset or real data collected from university or organizational networks. This paper portrays some commonly used machine learning techniques in the field of intrusion detection and reviews the results of the implementation. The reviewed models consider training on offline data. A tabular summarization of the reviewed models is also presented in Table 1. The remaining of this paper is organized as follows: Sect. 2 briefs on the definition of Machine Learning. Section 3 presents a theoretical overview of commonly used machine learning techniques. Section 4, summarizes some of the machine learning based Intrusion Detection Systems. Finally, a conclusion is derived and presented in Sect. 5.

2 Machine Learning

Machine learning is concerned with teaching a machine how to perform certain tasks by themselves in presence or absence of training data or example(s) [3]. In context to Intrusion detection, Machine Learning is a discipline concerned with the learning methodology of a computer system to classify or cluster future unseen data from the experience and knowledge gained from the past data. These algorithms or techniques help computers learn from their computational environment. Generally machine learning techniques can be categorized into the following types.

2.1 Supervised Learning

In supervised learning the training dataset is readily available along with its target vector. The learner learns from available data taking guidance of the output vector.

2.2 Unsupervised Learning

In contrast to supervised learning, unsupervised learning systems learn from its environment. Systems learn from training data, but there is no target vector available.

Table 1 Summarization of some existing machine learning based IDSs

Author(s)	Soft computing technique (s)	Dataset	Compared with (techniques)	Average detection rate or accuracy	Year
Dewan Md. Farid et al.	ABA	KDD'99	ANN, SVM, NBC	99.52 %	2008
Gang Wang et al.	FC, ANN	KDD'99	DT, NBC and BPANN	96.71 %	2010
Sufyan T. Faraj Al-Janabi et al.	BPANN	KDD'99	–	93 % (on testing)	2011
Yu-Xin Meng	ANN, SVM and DT separately	KDD'99	Among ANN, SVM and DT	Higher than 99 %	2011
Zahra Jadidi et al.	ANN and GSA	Sperotto et al. [27]	PSO	Higher than 99 %	2013
WenJie Tian et al.	ANN and PSO	KDD'99	BPANN	–	2010
Srinivas Mukkamala et al.	ANN and SVM	KDD'99	Between ANN and SVM	Higher than 99 %	2002
Kamarularifin Abd Jalil et al.	Decision Tree	KDD'99	ANN, SVM	99 %	2010
Dong Seong Kim et al.	GA and SVM	KDD'99	SVM	Higher than 99 %	2005
Adriana-Cristina Enache et al.	ACO/PSO and SVM	NSL-KDD	–	98.89 %	2014
Wenying Feng et al.	SOACN and SVM	KDD'99	SVM, SOACN and KDD99 Winner	94.86 %	2014
Guanghui Song et al.	SVM	KDD'99	–	95.31 %	2010
Shih-Wei Lin et al.	DT, SVM, SA	KDD'99	DT, SVM, DT with SA, SVM with PSO, SVM with SA	99.96 %	2012
Yinhui Li et al.	ACO and SVM	KDD'99	–	98.62 %	2012

3 Commonly Used Machine Learning Techniques

In this section some commonly used machine learning techniques in the field of intrusion detection are briefed.

3.1 Artificial Neural Network

Artificial Neural Network (ANN)s are the computational models of neural structure of human brain. Neurons are the basic building blocks of human brain. An ANN is a layered network of artificial neurons. An ANN may consist of an input layer, one or more hidden layer(s) and an output layer. The artificial neurons of one layer are fully or partially connected to the artificial neurons of the next layer. Each of these connections are associated with a weight, and feedback connections to the previous layers are also possible [4].

3.2 Decision Tree

Decision tree is one of the simplest machine learning techniques. A decision tree can easily be represented as a set of if-then rules. The classification starts from root node, traversing down the tree till the suitable leaf node. Each node of the tree represents a solution. Each node tests on an attribute of the instance and each descending branch of that node corresponds to one of the values of that attribute. Starting from the root node, each node tests the attribute specified by that node and moves down the tree through the branch matching the value, till it reaches a leaf node [4].

3.3 Support Vector Machine

Support Vector Machine (SVM) maps the input vector into a higher dimensional feature space. It is a binary classification technique that classifies input instances into two classes. Only the Support Vectors determine the optimal separating hyper-plane to classify input instance into one of the two classes. Support Vectors are the points closest to the separating hyper-plane. During classification, mapped input vectors placed on one side of the separating hyper-plane in the feature space falls into one class and placed on the other side of the plane falls into the other class. In case the data points are not linearly separable, SVM uses suitable kernel function to map them into higher dimensional space, so that, in that higher dimensional space they become separable [5].

3.4 Bayesian Classification

Bayesian learning is a statistical learning method based on probabilities of hypotheses. A prior probability is assigned to each candidate hypothesis based on prior knowledge. Training examples may increase or decrease the probability of a hypothesis to be correct. This probability can be calculated using Bayes' theorem. Classification is done by combining the predictions of multiple hypotheses, weighted by their probabilities. These probabilities in Bayesian method could be calculated using Bayes' theorem. Requirement of initial knowledge of many probabilities make practical application of Bayesian methods difficult [4].

3.5 Self-organizing Map

SOM is a special class of unsupervised learning Artificial Neural Network that consists of units located on a regular low-dimensional grid. Initially, each unit is assigned an associated initial weight vector. An input vector is compared with the weight vector of every unit of the SOM. The weights of the closest unit and its neighbors are updated after each iteration during the training process. Once the training process is over, each input vector has a corresponding output vector and the Euclidean distance between the input and each unit. The unit with the smallest distance is called the Best Matching Unit (BMU) [6].

4 Available Offline Machine Learning Based Intrusion Detection Systems

Farid et al. [7] proposed Adaptive Bayesian Algorithm (ABA) for anomaly intrusion detection and found that training and testing time of ABA is almost half of the time needed to train and test Naïve Bayesian Algorithm. The ABA achieved much higher performance with average of 99.52 % detection rate.

In the algorithm, proposed by Lin et al. [8], Simulated Annealing (SA) [9] was used with SVM for feature selection and SA with Decision Tree (DT) was used to increase the testing accuracy and to build the decision rules. Simulation results of the proposed algorithm were compared with the hybrid process of DT, SA, and feature selection; the hybrid process of SVM, PSO, and feature Selection; the hybrid process of SVM, SA, and feature Selection; SVM only; and DT only. It showed that the proposed algorithm with 23 features outperforms the others.

Song et al. [10] have proposed a model that utilizes the advantages of Multiple Kernel Learning combined with SVM by tackling feature selection problems and classification accuracy problem in NIDS. They have compared SVM method based on single kernel without feature selection; SVM method based on single kernel with feature selection; l_1-norm Multiple Kernel Learning method; and l_2-norm Multiple Kernel Learning method. Multiple Kernel Learning (MKL) method is used to train the SVM classifier. By MKL, it is generally meant for sparse solutions. Though, sparse MKLs appear to be beneficial, sometimes improvements are required to improve the performance of SVM. Some of these improvements could be higher accuracy, lesser training time etc. l_p-norm MKL ($1 \leq p \leq \infty$) method incorporates these improvements by covering both sparse as well as non-sparse MKLs [11]. Authors had experimented with l_p-norm MKL for p = 1 and 2. Experimental results showed that l_2-norm MKL method achieves the highest Detection Rate (DR) and lowest False Positive Rate (FPR).

Li et al. [12] had considered the preprocessing of the training data by deleting the redundant instances of KDD'99 dataset and by applying k-means clustering to compact the dataset into 5 clusters. Ant Colony Optimization (ACO) algorithm was then used to select a small representative subset of the whole dataset. They had proposed a method called Gradually Feature Removal (GFR) to reduce the size of the feature set. Experimental results proved this method advantageous. At the final step, SVM classified the attack instances from benign data. The SVM classifier achieved highest performance on selecting 19 features based on their importance order derived from GFR method.

ANN-based IDSs have drawbacks of (1) lower detection precision, especially for low-frequent attacks, e.g., Remote to Local (R2L), User to Root (U2R), and (2) weaker detection stability. To overcome these two drawbacks of ANN-based IDS, Wang et al. [13] proposed a new approach based on ANN and fuzzy clustering that claims to achieve higher detection rate, less false positive rate and stronger stability as compared to other ANN-based only IDS. The proposed technique divides the KDD'99 dataset into a training set and a test set. The training set is divided into further k smaller subsets using Fuzzy Clustering (FC). Each of the k subset is used for training k different ANNs. Simulation to the k ANNs were performed using the whole training set to reduce the error for every ANN. The membership grades generated by fuzzy clustering module were then used to combine the results. Subsequently, another new ANN was trained using the combined results. In the testing phase, test set data were input into the k different ANNs and received outputs. Based on these outputs, the final results were then achieved by the last fuzzy aggregation module. The new method achieved better results as compared to Decision Tree, Naïve Bayes and Back Propagation Neural Network.

Al-Janabi and Saeed [14] proposed a four stage/module model for anomaly-based NIDS. The four stages are (1) Monitoring Module, (2) Detection Module, (3) Classification Module, and (4) Alert Module. The model used ANN for classification of anomalous network traffic from normal traffic. Authors had calculated detection rate and false positive rate for three scenarios—(1) Detection only scenario, (2) Detection and classification, (3) Detection and detailed classification [14]. In their experiment, the system performance degraded as the system had been subjected to detect more specific attack, or attack types or sub-types.

Meng et al. [15] compared ANN, SVM and DT schemes for anomaly detection in an uniform environment and concluded that J48 algorithm of DT gives better performance than the other two schemes. The detection rate of low frequent attack types (U2R, R2L) was also high in all three implemented machine learning techniques.

ANN is proven to be effective in detecting intrusions, but it also suffers from numerous drawbacks, such as, deciding the number of neurons and number of layers to be selected, problem of over fitting, convergence to local minima, etc. [16]. Particle Swarm Optimization (PSO) algorithm was used by Tian et al. [16] to optimize these parameters of ANN. The experiment was carried out on pre-processed KDD'99 dataset with only 8 features selected. Mean Square Error of the proposed algorithm was compared with Back Propagation Neural Network. The proposed algorithm generates smaller Mean Square Error than Back Propagation Neural Network.

In a high-speed network, analysis of the whole packet is time consuming. Subject to real-time detection in a high speed network environment a flow-based anomaly detection technique was proposed by Jadidi et al. [17]. Analysis of only the packet header speeds up the packet processing task for real-time detection. The machine learning technique used here is ANN optimized with Gravitational Search Algorithm (GSA). The GSA optimizes the weights of the two-layer perceptron network. GSA is compared with PSO for weight adjustment. Experimental results showed that GSA trained Multi Layer Perceptron (MLP) network gives higher accuracy of 99.43 % with 0.56 % error and miss rate and 0.64 % false alarm rate.

Intrusion detection approaches using ANN and SVM were described by Mukkamala et al. [18] and were compared one against the other. In developing the IDS on ANN the author used three different Feed Forward Neural Network (FFNN) architectures were used with number of layers 4, 3 and 3 respectively. Feed Forward Back propagation algorithm using scaled conjugate gradient decent or SCG for learning and was used to learn the ANN that achieved 99.25 % detection accuracy. On comparison, SVM based IDS was found to provide a slightly higher detection accuracy over the ANN based IDS.

Abd Jalil et al. [19] evaluated the performance of Decision Tree (J48) algorithm in anomaly detection in comparison to ANN and SVM detection methods conducted by Osareh and Shadgar [20]. The KDD Cup dataset was used for training and testing of the models. The experimental results showed that J48 algorithm outperforms the other two methods.

The popular machine learning technique, SVM, was used by Kim et al. [22] to detect anomaly in fusion with Genetic Algorithm (GA). Authors have used the KDD Cup 1999 data [21], and Stolfo et al. [23] defined higher-level features. GA was executed up to 20 generations. The system could achieve more than 99 % detection rate which is higher than the general SVM classification performed by them.

Feng et al. [24] introduced a new classification technique and utilized the advantages of SVM and Clustering based on Self-Organized Ant Colony Network (CSOACN) by combining them with little modifications to fit with each other. Author named the new algorithm as Combining Support Vector with Ant Colony (CSVAC). One of the advantages of the developed system was its flexibility for modification, extension, maintenance and re-usability. Another advantage was the small volume of training data. The system performance was tested and compared in three different modes (SVM only, CSOACN only and CSVAC) on the same test dataset. The proposed algorithm outperformed pure SVM in terms of average detection time, training time, false positive rate and false negative rate; and it outperformed CSOACN in terms of training time with comparable detection rate and false alarm rates. Also, the experimental results showed comparable performance of the proposed model as compared to KDD99 winner [25].

Enache and Patriciu [26] presented an IDS model based on Information Gain (IG) for feature selection combined with the SVM classifier. The parameters for SVM are selected using Swarm Intelligence algorithm (ACO or PSO). In their experiment, authors had tested the detection accuracy for sole SVM classifier, SVM classifier with feature selection using Information Gain, PSO for SVM parameters and Information Gain for feature selection, and ACO for SVM parameters and Information Gain for feature selection. Experimental results showed that ACO is more suitable than PSO for selecting SVM parameters.

Table 1 summarizes different machine learning techniques and their comparative output results.

5 Conclusion

Machine learning techniques are proved to be efficient for intrusion detection. High accuracy in intrusion detection can be achieved using machine learning techniques even though the detection accuracy depends on some other factors too. Some of them are selection of correct feature set, selection of appropriate training and testing data, etc. With the selection of the appropriate attributes for these factors, a higher performance could be achieved.

However, machine learning algorithms may exhibit some vulnerabilities, such as, misclassification of network data due to poison learning. In future, we would undertake analysis of such vulnerabilities of the learning algorithm and respective response mechanism of the learning algorithm to such attacks would be undertaken.

References

1. De Boer, P., Pels, M.: Host-Based Intrusion Detection Systems. Amsterdam University, Amsterdam (2005)
2. Garcia-Teodoro, P., et al.: Anomaly-based network intrusion detection: techniques, systems and challenges. Comput. Secur. 28.1, 18–28 (2009)
3. Richert, W.: Building Machine Learning Systems with Python. Packt Publishing Ltd, UK (2013)
4. Mitchell, T.M.: Machine Learning. McGraw-Hill Science/Engineering/Math, (March 1, 1997), ISBN: 0070428077
5. Hu, W., Liao, Y., Vemuri, V.R.: Robust Support Vector Machines for Anomaly Detection in Computer Security. ICMLA (2003)
6. Kohonen, T.: The self-organizing map. Proc. IEEE 78(9), 1464–1480 (1990)

7. Farid, D.M., Rahman, M.Z.: Learning intrusion detection based on adaptive bayesian algorithm. In: 11th International Conference on Computer and Information Technology, 2008. ICCIT 2008, IEEE (2008)

8. Lin, S.-W., et al.: An intelligent algorithm with feature selection and decision rules applied to anomaly intrusion detection. Appl. Soft Comput. **12**(10), 3285–3290 (2012)

9. Bertsimas, D., Tsitsiklis, J.: Simulated annealing. Stat. Sci. **8**(1), 10–15 (1993)

10. Song, G., et al.: Multiple kernel learning method for network anomaly detection. In: 2010 International Conference on Intelligent Systems and Knowledge Engineering (ISKE), IEEE (2010)

11. Kloft, M., et al.: Lp-norm multiple kernel learning. J. Mach. Learn. Res. **12**, 953–997 (2011)

12. Li, Y., et al.: An efficient intrusion detection system based on support vector machines and gradually feature removal method. Expert Syst. Appl. **39**(1), 424–430 (2012)

13. Wang, G., et al.: A new approach to intrusion detection using artificial neural networks and fuzzy clustering. Expert Syst. Appl. **37**(9), 6225–6232 (2010)

14. Al-Janabi, S.T.F., Saeed, H.A.: A neural network based anomaly intrusion detection system. Dev. E-syst. Eng. IEEE (2011)

15. Meng, Y.-X.: The practice on using machine learning for network anomaly intrusion detection. In; 2011 International Conference on Machine Learning and Cybernetics (ICMLC), vol. 2, IEEE (2011)

16. Tian, W.J., Liu, J.C.: A new network intrusion detection identification model research. In: 2010 2nd International Asia Conference on Informatics in Control, Automation and Robotics (CAR), vol. 2. IEEE (2010)

17. Jadidi, Z., et al.: Flow-based anomaly detection using neural network optimized with GSA algorithm. In: 2013 IEEE 33rd International Conference on Distributed Computing Systems Workshops (ICDCSW), IEEE (2013)

18. Mukkamala, S., Janoski, G., Sung, A.: Intrusion detection using neural networks and support vector machines. In: Proceedings of the 2002 International Joint Conference on Neural Networks, 2002. IJCNN'02, vol. 2. IEEE (2002)

19. Abd Jalil, K., Kamarudin, M.H., Masrek, M.N.: Comparison of machine learning algorithms performance in detecting network intrusion. In: 2010 International Conference on Networking and Information Technology (ICNIT), IEEE (2010)

20. Osareh, A., Shadgar, B.: Intrusion detection in computer networks based on machine learning algorithms. Int. J. Comput. Sci. Netw. Secur. **8**(11), 15–23 (2008)

21. KDD-CUP-99 Task Description: http://kdd.ics.uci.edu/databases/kddcup99/task.html

22. Kim, D.S., Nguyen, H.-N., Park, J.S.: Genetic algorithm to improve SVM based network intrusion detection system. In: 19th International Conference on Advanced Information Networking and Applications, 2005. AINA 2005, vol. 2. IEEE (2005)

23. KDD Cup 1999 Data: http://kdd.ics.uci.edu/databases/kddcup99/kddcup99.html

24. Feng, W., et al.: Mining network data for intrusion detection through combining SVMs with ant colony networks. Future Gener. Comput. Syst. **37**, 127–140 (2014)

25. Pfahringer, B.: Wining the KDD99 classification cup: bagged boosting. SIGKDD Explor. Newsl. **1**(2), 65–66 (2000)

26. Enache, A.-C., Patriciu, V.V.: Intrusions detection based on Support Vector Machine optimized with swarm intelligence. In: 9th International Symposium on Applied Computational Intelligence and Informatics (SACI), 2014 IEEE (2014)

27. Sperotto, A., et al.: A labeled data set for flow-based intrusion detection. IP Operations and Management. Springer, Berlin, pp. 39–50 (2009)

28. Sharma, R.K., Kalita, H.K., Issac, B.: Different firewall techniques: a survey. In: 2014 International Conference on Computing, Communication and Networking Technologies (ICCCNT), IEEE (2014)

Compression and Optimization of Web-Contents

Suraj Thapar, Sunil Kumar Chowdhary and Devashish Bahri

Abstract Various image formats used online have different characteristics and can only be used in an appropriate manner. The paper creates a boundary around each format, for it to be used efficiently. It also discusses neat tricks and tips for optimizing rich content websites in mobile devices, which generally have low processing power. To save bandwidth usage and provide speedy access to content, compression is required. And hence this paper will do more in increasing efficiency of websites and greater effect on the performance and overall load time in slow internet connections. Reducing network latency has become very important so that many devices can connect to the server at the same time. However, it is even more important to save the bandwidth of the users who pay to view the websites. This research paper aims at outlining techniques to reduce the size of a website and encourage developers to code efficiently.

Keywords Image · Compression · Optimization · Web

1 Introduction

Ever since internet has become an integral tool for business and education, compression of web content has become more important than ever. In just 60 s, nearly 640 terabytes of IP data is transferred across the globe [1]. Compared to the amount

S. Thapar (✉) · S.K. Chowdhary
Department of Information Technology, Amity University, Noida,
Uttar Pradesh, India
e-mail: surajthapar1@gmail.com

S.K. Chowdhary
e-mail: skchowdhary@amity.edu

D. Bahri
University Institute of Engineering and Technology, Chandigarh, India
e-mail: tranclix@gmail.com

© Springer India 2016
A. Nagar et al. (eds.), *Proceedings of 3rd International Conference on Advanced Computing, Networking and Informatics*, Smart Innovation, Systems and Technologies 44, DOI 10.1007/978-81-322-2529-4_52

495

of data transferred for images, HTML documents, scripts and fonts, it'd be in multiples of IP data. An average internet user consumes 17 gigabytes of bandwidth per month [2], accounting for considerable internet usage charges. Loading heavy websites is expensive not only for desktop users but mobile users are the worst affected. Even 1 MB of web-page costs Rs 10 on 3G on popular networks like Airtel and Vodafone. The increasing reliance of people on digital methods to ingress contents that matter in everyday life, be it online shopping, reading news, or socializing, there is an increasing need to provide a method for faster, easier and simultaneously economical access to these online services. As an obvious fact, the use of images and other multimedia contents in a web-page directly affects its size and performance. This requires the developers to use less resource extensive methods of designing the website which in turn affects the website's appeal and overall user experience. A solution to this problem is to have or to develop compression techniques so that the designers and the developers can focus on the functionality and user experience of the website rather than focusing on website's size or its performance. The findings reveal that the impact of a website on a mobile device's hardware is directly proportional to the website's size. The paper will discuss on how compression of websites will have a greater effect on the performance and overall load time in slow internet connections.

2 Images

Images take up larger percentage of the page size. It is important to study what format is best for your image, browser and platform. Every format in use has its pros and cons. JPEG, PNG and GIF are most popular formats being adopted by web browsers. Currently, there is no format for the web that has all the ideal features. Depending on the type of image, browser support and the requirements of the page, the image may be encoded in different formats. Before deciding the format of the image, a typical developer must keep three major criteria in mind:

- Is animation required?
- Is transparency required?
- Is high quality data required?

The developer has the choice to select either GIF or PNG if transparency is required. PNG being a higher quality encoding format and GIF being highly compressible format. If a developer is looking for high quality data then they may choose either PNG or JPEG. Latter being better in quality and compression. This is not to say that PNG is lacking in its reign.

2.1 Animated Images

GIF or Graphics Interchange Format Currently, for animation GIF is the only format supported by major browsers. However it is a compromise on quality and thus, an alternative is an urgent necessity. There was a time when Flash took over GIFs but as a matter of fact, Flash is not an image and is also not supported on many devices.

GIF limits its image to 255 colors and 1 alpha channel in one image. There is work around to use more colors but GIF is almost never used for True Color images. For including more than 256 colors, a GIF image has to be divided into matrix blocks, each of which can have its own 256-color palette. The blocks can then be rendered into a tiled stripe image. However, each image block requires its own separate color table i.e., a GIF image with a lot of image blocks will be very large and thus defeat the whole purpose of compression. Layered and tiled GIF images are not rendered or supported by many browsers. Most browsers interpret these tiled or layered GIF images as animated image and display the layers one after another with an automatic time gap of 0.1 s. Technical limitations are observed while creating compressed GIF files because many GIF generators use single global palette for colors of all frames of an animation.

Animated PNG or APNG APNG is a high quality animated image format which is rendered as PNG if there is no support for it and shows only the first frame. The MIME data of PNG and APNG are same, so there's actually no easy way to determine if the PNG you're looking at is animated or still. To actually identify the hidden frames from APNG image, one can search the image's source code for "aCTL" (Animation Control Chunk) string [3].

The problem with less popularity of APNG is that tools for creating, editing and rendering are not widely available. It is a high quality format with full alpha channel and is supported by Firefox, Opera and Google Chrome (via a plugin). The problem is that Google is holding back with using this high-potential animated format. Google has its own team working on creating a new format for the internet—WebP. WebP technology is based on On2 Technologies' V8 project which was acquired by Google on On2's purchase [4].

WebP A GIF image may provide better compression but it is always a compromise on quality. GIFs support 8-bit RGB color channel and 1-bit alpha channel whereas WebP supports 24-bit color channel and 8-bit alpha channel. WebP supports both, lossy and lossless compression whereas GIF only supports lossy compression. In WebP, lossless and lossy frames can be combined into a single animation. In presence of seeking, WebP takes less time to decode than GIF. Upon scrolling or changing tabs, the animated images have to drop and re-pickup frames which results in more CPU usage and utilization spikes. In such cases, WebP takes about half of the time required than the time required to decode GIF. WebP is a newly introduced format and technology with extensive support on various devices, browsers and websites.

Alternate Techniques Images created with HTML5 canvas Object can also be animated. An alternative method for animations in web sites is to use standard static images or SVGs and animate them using JavaScript, Flash, SVG, Canvas, WebGL or other plugin based technologies. Most of these plugin-based technologies are device/OS centric and are not available on all devices, rendering them ineffective in most cases. Moreover, these are just some work around and not an actual "image". These techniques make it hard to modify and interfere with portability.

2.2 Transparent Images

GIF or Graphics Interchange Format GIF images have a 1-bit alpha channel which means that it has binary transparency. It can either have a 100 % transparent pixel or 0 % transparent pixel. This means that there is no way to have transparency values anywhere in between. In other words, partial transparency is not supported in GIF. The effect of partial transparency can be studied in a linear gradient image with one end being opaque and the other end being transparent, as seen in Figs. 1 and 2.

Fig. 1 PNG image

Fig. 2 GIF image

Fig. 3 GIF image

Fig. 4 PNG image

PNG or Portable Network Graphics PNG offers a variety of options when it comes to transparency and supports partial transparency. Two ways to add transparency to the image in this format are either by declaring a single pixel value as transparent or by enabling any percentage of transparency with the help of an alpha channel. Images that contain the pixel color value in palettes, can have their alpha values added to the palette entries. In this method, when the values for transparency are less than the total number of palette entries, then the remaining entries are considered fully opaque by default. Unlike GIF images, PNG images handle full transparency very well. Figures 3 and 4 compare the transparency in both formats.

WebP PNG only supports lossless compression in order to provide high quality and transparency. However WebP, for the first time, has brought lossy compression along with transparency. This compression was not available in any format until the WebP team used the techniques of lossy compression from VP8 engine. When compared to the amount of compression between PNG and lossless WebP, WebP is about 30 % less in the size and keeps the same visual quality.

2.3 High Quality Images

JPEG or Joint Photographic Experts Group JPEG is an image format which is primarily used because of its capability of retaining high quality and 16 million colors which makes it an excellent format to be used for photography and colorful illustrations. It also supports lossy and lossless compression, where the former reduces the image quality and the latter one does not. It uses Huffman encoding for compression. In lossy compression there is a dramatic reduction in size by about 81.9 % at the compression ratio of 15:1. At 15:1 there is very less loss of quality, almost not distinguishable visually. However, at (Fig. 4) higher ratios of compression there is greater loss of image quality. Under lossy compression (at low quality), the JPEG compressor creates radial patches of pixcls, it can be clearly seen inside the text "JPEG" in Fig. 5.

PNG or Portable Network Graphics PNG is also a high quality image format. It only supports lossless compression and there is no degradation of the image quality. The phenomenon of radial patches by JPEG encoder as seen in Fig. 5 is not seen in PNG image (Fig. 6).

Sharp edged PNG images are smaller in size when compressed. PNG features interlacing for the web which allows quicker loading of image. Instead of loading the pixels of image serially, the deflator loads the first pixel of each small block. This way the user gets an idea of what image is loading before 100 % of the image is downloaded. This phenomenon is best described in the above Fig. 7. The image loads as step 1, 2, 3 and 4. In JPEG images progressive feature, an alternative to interlacing, is used which has customizable number of passes. While pass 1 of JPEG image is more distorted than pass 1 of PNG images but it turns out that passes that follow the first pass are much clearer than that of PNG.

WebP WebP is the only format which supports animation, transparency and high quality. It is still in a development stage and is not widely used yet. Being developed by Google, it has got lot of support and recognition by major companies, websites and developers. Image editors like ImageMagick, Photoshop and GIMP

Fig. 5 JPEG Image with Lossy Compression

Fig. 6 PNG Image with Lossless Compression

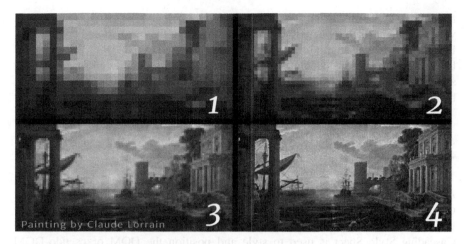

Fig. 7 Interlacing in PNG Image

Table 1 Comparative study of image formats

	APNG	PNG	GIF	JPEG	WebP
Compression	Good	Good	Ok	Good	Better
Lossless	Yes	Yes	Yes	Yes	Yes
Lossy	No	No	Yes	Yes	Yes
Transparency	Yes	Yes	Binary	No	Yes
Animation	Yes	No	Yes	No	Yes
Quality	Good	Good	Ok	Good	Good
Browser support	Firefox, safari	All	All	All	Chrome, opera and android

have already received a plugin to export WebP images, which was a not in the case of APNG.

WebP's lossy compression encoding has been derived from WebM video format [5], which in itself is a very popular video format online. It offers about 30 % more compression than JPEG and PNG and retains an excellent image quality. Under high compression it doesn't create a radial patch as seen in Fig. 5 in JPEG. The only downside of using this format is that it is not yet fully supported in most web-browsers and devices. Google has already introduced this technology from Android 4.x but the developers are yet to exploit its full potential (Table 1).

3 Overview

WebP is a major competitor as it has got everything a modern image format for the web requires. Android being the most popular OS for mobile devices has provided support for WebP on its browser starting from Android 4.x. WebP is a patented technology by Google and has been introduced to the public as "Royalty Free" [6], which means it may get better support in future from the open-source community. Recently Google also introduced a webkit forked engine called "Blink" [7] which is being used by latest versions of Google Chrome and Opera. Blink itself doesn't support APNG and has an extensive support for WebP which gives WebP an upper hand.

4 Code

A basic web page uses multiple languages like HTML, CSS and JavaScript among others. HTML generates DOM elements, JavaScript is a programming language that communicates with web-browser to add functionality to the page and Cascading Style Sheet is used to style and position the DOM or pseudo-DOM elements. Proper positioning and compression of code optimizes the web page.

4.1 HTML or Hyper Text Markup Language

Rich content web pages often have larger size. For a responsive website one might opt for a less resource consuming website. This section will discuss about reducing document size by eliminating redundant structures, and using minification techniques to obtain a rich, responsive web page.

The redundancy in HTML document is inclusion of comments <!– –..– –> in script blocks. The browsers that required this error-prevention measure (such as IE6 and Netscape) are just about dead. Comments in scripts are excess bytes and should be removed from the document. Another uncalled-for method is to include CDATA in script block. CDATA prevents the parser to detect < and & as a markup inside the script, but it is only for documents with content-type as "application/xhtml + xml" and not the widely used "text/html". While comments and CDATA take less part of the web page, the style attribute style = '...', takes the most of a page. As the developer desires to style a non-redundant element or style by placing the specific style inside the style = '...' tag. The styles of the element should be rather placed in an external css file which enables the mechanism of caching on the client side.

Script tags <script type = 'text/javascript'> ... </script> are used throughout the HTML page. It has been noted that the JavaScript files are often heavier in size and make separate server requests. The browser won't further process the content until the download(s) of the JavaScript file(s) is/are completed. It is recommended to place the JavaScript just before ending the </body> tag. This will display the text and image content before the JavaScript has even loaded, which decreases the initial load time. For most websites, the content such as images and texts is more important than loading JavaScript at first. Some websites heavily depend on JavaScript, such as Canvas based websites and hence require the JavaScript to load first. But in reality only a negligible percentage of websites completely rely on Canvas/JavaScript.

The most redundant element of a web page is the *doctype*. Without assigning a *doctype* to the page, the page is automatically displayed in quirks mode, a feature that made old websites live in modern browsers. But the quirks mode is old in itself as modern browsers seem to have made the HTML world just about error-free. From HTML 4.1, the *doctype* formats seemed to be scary but Dustin Diaz has found a way to skin [8] it down. Normally the *doctype* is a special tag with long unnecessary values. <!DOCTYPE html PUBLIC "-//W3C//DTD XHTML 1.0 Strict//EN", "http://www.w3.org/TR/xhtml1/DTD/xhtml1-strict.dtd"> into <! *DOCTYPE html>* is obviously some reduction for every page on the web. The HTML5 doctype <!DOCTYPE html> is fully backwards compatible which means it is supported in all older web-browsers.

4.2 CSS or Cascading Style Sheet

All the styles are written in CSS language which are placed internally inside *<style>* … *</style>* tag or externally in a separate CSS file. The latter is recommended. External CSS file reduces the load time after the files are cached on the first load. Furthermore, developers should use CSS shorthands which are explained in the example code below (Fig. 8).

Shorthands reduce the number of lines of code, hence reducing file size. The CSS file should be minified/compressed in a single line of code while removing unnecessary spaces and line breaks. At times in a website there are multiple CSS files which increase the download requests hence increasing the page load time. For the best performance, all the CSS files should be combined into a single file. This will diminish the network latency. Use CSS minifiers for further compression. A CSS minifier removes all the unnecessary spaces from the CSS file at once.

```
Original Code
div {
    padding-left: 10px;
    padding-right: 20px;
    padding-top:30px;
}
```

```
Shorthand Code
div {
    padding: 30px 20px auto 10px;
}
```

Fig. 8 CSS Shorthands

4.3 JS or JavaScript

JavaScript is a client-side programming language which is more powerful in terms of control. JavaScript is downloaded and processed by the parser, taking more time to load unlike CSS or HTML. The situation worsens in some web-browsers which do not allow the download of any other file while the JavaScript is being processed. The only solution to this under-weighted problem is to bring down the JavaScript files to the bottom of the page. This forces the browser to load the images, text and other content even before it reaches the <script type="text/JavaScript" src="..."> ... </script> tag, whence the scripts are loaded. When possible, JavaScript files belonging to the same page should be placed in a single file to decrease network latency. JavaScript minifiers or compressors remove the unnecessary spaces in the script. A well formatted JavaScript file (jQuery in this case) with 10,000 lines of code (288 Kb) gets reduced to 1 line of code (96 Kb). There is a considerable decrease in size—almost by a fraction of 3, while retaining the functionality (Fig. 9).

4.4 GZip

GZip is a popular and efficient server-side compression utility which has freedom from patented algorithms [9]. Most other compression techniques are restricted by patents. Most browsers have support for client side deflation of GZip files. It was noted that a 100 Kb JavaScript file after GZip compression reduces to just 36 Kb. This is a 65 % reduction in size. It is recommended to GZip all the HTML, CSS, and JavaScript files on the server. The Gzip version of the file will be downloaded and deflated on the client's web-browser.

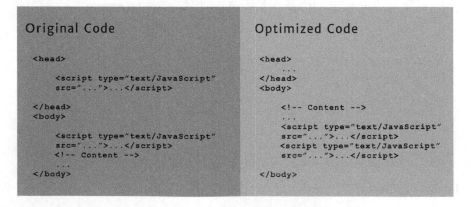

Fig. 9 Optimizing webpage by proper positioning of Javascript

GZip can be used to compress images as well but it may just be waste of CPU for the server. If images on the web are already compressed, further compression through GZip may sometimes increase the file size. Though GZip deflation is very quick, it isn't the case with low-processor mobile devices. Deflation of GZip image is slower than downloading the uncompressed version on low-processor mobile devices.

5 Conclusion

The compression and optimization of web pages is a necessity as websites are actively being used in low-processor mobile devices by billions of people. Optimization has two major bifurcations—images and code. Choosing the best image format for the web will become easier as WebP technology reaches its peak. Keeping a single format makes it easier to understand compression. It is obvious that the upcoming format will support high levels of compression which might just out pass the latest compression levels. From the findings of this paper we conclude that **WebP** format might just become the most powerful format.

When it comes to coding, optimization plays an equally important role. While various in-depth JavaScript programming techniques are used to optimize the web, the most common involves declaration and deferring of the scripts. CSS has its own tricks but they vary from browser to browser. While HTML can't be left behind in optimization as it is the root of the DOM.

References

1. Burgess Rick, One minute on the Internet: 640 TB data transferred, 100 k tweets, 204 million e-mails sent, http://www.techspot.com/news/52011-one-minute-on-the-internet-640tb-data-transferred-100k-tweets-204/million-e-mails-sent.html, 8th June 2014
2. Arthur Charles, Average home broadband user downloads 17 gigabytes a month, http://www.theguardian.com/technology/2011/nov/01/home-broadband-download-17-gigabytes, 8th June 2014
3. A PNG showing differently in Firefox and Chrome, http://onox.com.br/2014/02/07/a-png-showing-differently-in-firefox-and-chrome.html, 7 February 2014
4. Claburn Thomas, Google Acquires On2 Technologies, InformationWeek, 29th June 2014
5. WebP—Faster Web with smaller images—The WebM Project, http://downloads.webmproject.org/ngov2012/pdf/09-ngov-webp-still-image.pdf, 29 June 2014
6. VP8—Wikipedia, the free encyclopedia, http://en.wikipedia.org/wiki/VP8, 2014
7. Google Forks WebKit And Launches Blink, A New Rendering Engine That Will Soon Power Chrome And Chrome OS, http://techcrunch.com/2013/04/03/google-forks-webkit-and-launches-blink-its-own-rendering-engine-that/will-soon-power-chrome-and-chromeos/, 2013
8. The Skinny on Doctypes—Dustin Diaz, http://www.dustindiaz.com/skinny-on-doctypes/, 2008
9. The gzip home page, http://www.gzip.org/

Harnessing Twitter for Automatic Sentiment Identification Using Machine Learning Techniques

Amiya Kumar Dash, Jitendra Kumar Rout and Sanjay Kumar Jena

Abstract User generated content on twitter gives an ample source to gathering individuals' opinion. Because of the huge number of tweets in the form of unstructured text, it is impossible to summarize the information manually. Accordingly, efficient computational methods are needed for mining and summarizing the tweets from corpuses which, requires knowledge of sentiment bearing words. Many computational techniques, models and algorithms are there for identifying sentiment from unstructured text. Most of them rely on machine-learning techniques, using bag-of-words (BoW) representation as their basis. In this paper, we have applied three different machine learning algorithm (Naive Bayes (NB), Maximum Entropy (ME) and Support Vector Machines (SVM)) for sentiment identification of tweets, to study the effectiveness of various feature combination. Our experiments demonstrate that NB with Laplace smoothing considering unigram, Part-of-Speech (POS) as feature and SVM with unigram as feature are effective in classifying the tweets.

Keywords Bag-of-words (BoW) · Machine learning algorithms · Laplace smoothing · Part-of-Speech (POS)

A.K. Dash (✉) · J.K. Rout · S.K. Jena
Department of Computer Science & Engineering, National Institute
of Technology Rourkela, Rourkela, India
e-mail: 213CS1141@nitrkl.ac.in

J.K. Rout
e-mail: 513cs1002@nitrkl.ac.in

S.K. Jena
e-mail: skjena@nitrkl.ac.in

© Springer India 2016
A. Nagar et al. (eds.), *Proceedings of 3rd International Conference
on Advanced Computing, Networking and Informatics*, Smart Innovation,
Systems and Technologies 44, DOI 10.1007/978-81-322-2529-4_53

507

1 Introduction

Sentiment analysis (SA) is concerned with automatically extracting sentiment related information from a text and aims to categorize text as positive or negative on the premise of the positive or negative sentiment (opinion) expressed in the document/sentence towards a topic. A document/sentence with positive or negative sentiment is also said to be of positive or negative polarity respectively [1]. The granularity of polarity can be up to the level of words. That is textual information can be classified as either objective or subjective. Objective (non-polar) sentences and words represent facts, while subjective (polar) sentences and words represent perceptions, perspectives or opinions. It is important to make distinction between subjectivity detection and sentiment analysis as they are two separate task in natural language processing. Sentiment analysis can be dependently or independently done from subjectivity detection. Pang and Lee [2] state that to get better result subjectivity detection performed prior to sentiment analysis.

The task of sentiment analysis is very challenging, not only due to the syntactic and semantic variability of language, but also because it involves the extraction of indirect or implicit assessments of objects, by means of emotions or attitudes. That is why automatic identification of sentiment requires fine grained linguistic analysis techniques and substantial efforts to extract features for machine learning or rule-based approaches.

In this paper, we have used three different machine learning algorithm on tweets for automatic sentiment identification and compare the results with Movie reviews obtained by Pang et al. [3]. We investigated a mixture of features like unigram, bigram, POS and adjectives to search out the effective feature for sentiment analysis. It was observed from our experiment that NB classifier with Laplace smoothing and SVM taking unigram and POS as feature gives better result than other features that we have employed.

2 Related Work

The business potential of sentiment analysis has resulted in an exceedingly important quantity of analysis and Pang [4] provides an overview. Ibrahim et al. [5] presents an in depth survey about different techniques used for opinion mining and sentiment analysis. During this section, we restrict our discussion to the work that is most relevant to our approach. Pang Lee et al. use Naive Bayes, Maximum Entropy and Support Vector Machines for SA of movie reviews considering distinctive features like unigrams, bigrams, combination of both, including parts of speech and position information with unigram, adjectives etc. [2, 3]. It was seen from their experiment that Feature presence is more important than feature frequency. It was also observed that for small feature space Naive Bayes performs better than SVM but when feature space is increased SVM perform better than Naive Bayes classifier.

Our work is to perform sentence-level sentiment identification, where we classified tweets using above three machine learning classifier and compare the results with movie reviews.

3 Machine Learning Methods

Our aim in this work is to find out the effective feature for sentiment classification of tweets being positive sentiment or negative sentiment. We used the standard bag-of-feature frame work for implementing these machine learning algorithms. Let $\{w_1, \ldots, w_m\}$ be the m words that can appear in a tweet/sentence; examples include the unigram word "silent" or the bigram "low price". Let $n_i(t)$ be the number of times w_i occurs in tweet t. Then, each tweet t is represented by the tweet vector $t: = (n_1(t), n_2(t), \ldots, n_m(t))$.

3.1 Naive Bayes

Naive Bayes classifier is a simple probabilistic classifier that relies on Bayes theorem. The most likely class according to the Naive Bayes classifier is the class among all classes which maximizes the product of two probabilities prior and likelihood, the word in a tweet given the class i.e. how often that word is expressed in a positive tweets or in a negative tweets

$$C_{NB} = \underset{c_j \in C}{\arg \max}\, p(c_j) \underset{i \in positions}{\Pi} p(w_i | c_j) \tag{1}$$

Research on sentiment analysis tells that word occurrence may matter more than word frequency. As tweets are 140 character length occurrence of a word tell us a lot, but the fact that if it occurs more than once may not tell us much more. So we need to clip all the word count in each tweet at one and remove duplicate words in each tweet to retain a single instance of the word. So for our work we have used another variant of Naive Bayes classifier i.e. binarized (Boolean feature) Multinomial Naive Bayes classifier which assumes the features to be occurrence of count.

Laplace Smoothing. Here we used Laplace smoothing assuming that even if we have not seen a given word in the whole corpus, there is still a chance that our sample of tweets happened to not include that word.

$$\hat{p}(w|c) = \frac{count(w, c) + 1}{count(c) + |V|} \tag{2}$$

3.2 Maximum Entropy

The maximum entropy classifier is a probabilistic classifier which belongs to the class of exponential models that has proven effective during a variety of language process applications. It does not assume that the features are conditionally independent of each other. Here our target is to use the contextual information of the tweets (unigram, bigram, and other characteristics) within the text in order to categorize it to a given class (positive or negative). Maximum entropy estimates takes the following exponential form:

$$P_{ME} = \frac{1}{Z(t)} \exp\left(\sum_i \lambda_{i,c} F_{i,c}(t, c) \right) \tag{3}$$

where $Z(t)$ is the size of the training dataset used as a normalization function. $F_{i,c}$ is a indicator function for feature f_i and class c, defined as follows,

$$F_{i,c} = \begin{cases} 1, & n_i > 0 \text{ and } c' = c \\ 0, & \text{otherwise} \end{cases}$$

For estimating the λ parameters we use ten iteration of IIS (improved iterative scaling) algorithm, together with a Gaussian prior to counteract over fitting.

3.3 Support Vector Machine

Support vector machines (SVMs) are widely used for various text categorization in past, usually outperforming Naive Bayes classifier. In case of two-class problem with d dimension, the basic idea is to search a hyperplane, represented by vector $\overleftarrow{w}$, that not just differentiates the tweet vectors in one category from those in alternative, yet for which the separation, or margin, is as large as attainable. Let positive and negative be the correct class of tweet t_j and $c_j \in \{1, -1\}$ refers to the class labels positive and negative, then searching a hyperplane corresponds to a constrained optimization problem; where the solution is described as,

$$\overrightarrow{w} = \sum \alpha_j c_j \overrightarrow{t_j}, \quad \alpha_j \geq 0 \tag{4}$$

where the α_j's are derived by solving a dual optimization problem. The tweet vectors t_j are called support vectors for which α_j is greater than zero, as these tweet vectors contribute to the hyperplane. Classification of tweets includes primarily deciding that facet of $\overleftarrow{w}$'s hyperplane they fall on.

4 Experimental Set-up

4.1 Dataset

For implementation we have used *Niek j. Sanders* data set and *Polarity* data set. *Niek j. Sanders* data set contains 5513 hand classified tweets. The corpus contains tweets about apple, goggle, Microsoft and twitter. Tweets are classified into four classes positive, negative, neutral and irrelevant. Irrelevant tweets are those tweets that are not in English language or not related to the topic. In our experiment we have consider three classes positive, negative and neutral. So we converted all irrelevant class to neutral class. The polarity data set is a set of film review documents available for research in sentiment analysis and opinion mining. The most recent available data set is version 2.0, and it comprises 1000 positive labeled and 1000 negative labeled film reviews extracted from the Internet Movie Database Archive.

4.2 Data Pre-processing

Data pre-processing is done to eliminate the incomplete, noisy and inconsistent data [6]. Data must be pre-processed to apply any of the data mining functionality. We have employed the following pre-processing task before applying Machine learning algorithms.

Replace all URLs with a tag AT_USER, replace targets (e.g. @John) with tag USER; replace all the emoticons with a their sentiment polarity by looking up the emoticon dictionary; replace all negations (e.g. not, no, never, cannot) by tag NOT; replace a sequence of repeated characters by two characters, for example, convert coooooooool to cool; replace the words like what, which, how etc., are not going to contribute to polarity (called stop words);special character like, [],{},() ... should be removed in order to remove discrepancies during the assignment of polarity; stripped hash symbol (#*tomorrow* → *tomorrow*). We have used python regular expression for data pre-processing. We employed python Natural Language Toolkit (*NLTK*3.0) to get unigram, bigram, POS features of tweets.

4.3 Evaluation Metrics

The overall performance of individual classifier is measured by:

$$accuracy = \frac{\#of\ correctly\ labeled\ tweets}{\#of\ all\ the\ tweets\ in\ the\ test\ dataset}$$

Table 1 Accuracy of tweets using different features

	Features	No. of features	Frequency or presence	Naive Bayes		Maximum entropy		Support vector machine	
				Reviews (%)	Tweets (%)	Reviews (%)	Tweets (%)	Reviews (%)	Tweets (%)
1	Unigram	5989	Presence	81.0	81.5	80.4	78.36	82.9	82.5
2	Bigram	19,148	Presence	77.3	78.60	77.4	78.0	77.1	77.8
3	Unigram + bigram	25,748	Presence	80.6	80.92	80.8	79.78	82.7	81.6
4	Unigram + POS	19,061	Presence	81.5	82.0	81.2	80.3	81.9	81.99
5	Adjectives	1197	Presence	77.0	69.48	77.7	76.4	75.1	76.4

5 Results and Discussion

We explore a variety of features that are potent for sentiment analysis. We have used N-gram features like unigrams (n = 1), bigrams (n = 2) that are widely used in different of text classification, including sentiment analysis. In our study we experimented with unigrams and bigrams with boolean features. Each n-gram feature is associated with a boolean value, which is set true if and only if the n-gram is present in the tweet [3]. Table 1 represents the different features we have used and the accuracy results of individual classifier. Here we have performed a comparison between the movie review data set used by Pang Lee et al. and our dataset. From Table 1, it has been observed that when we used NB classifier with Laplace smoothing, the classification accuracies resulting from using unigram as features gives better result in case of tweets than movie reviews, but when we used MaxEnt classifier the accuracy result of Movie reviews are more than the tweets.

We additionally considered usage of bigrams to capture negation words for handling negation and phrases for dealing with Word Sense Disambiguation (WSD). *Line*(2) of results table demonstrates that using bigram as feature does not improve performance of the classifier as that of unigram presence. In our experiment we observed that, although bigram presence does not improve the classification accuracy it is as equally useful a feature as unigram; in reality bigrams are found to be effective features for handling word sense disambiguation. We also experimented considering bigram as single feature but the results were not as good, but combination of unigram and bigram features (*Line*(3) of results table) produces results competitive with those obtained by using unigram.

POS features are verified effective in sentiment analysis. Since adjectives are good indicators of sentiment, they are usually considered as effective feature for sentiment analysis. Our experiment shows (*Line*(5) of results table) that considering only adjectives produces results competitive with those obtained by using unigram and bigram. *Line*(4) of results table shows that all the three classifier produces better result considering unigram and POS as feature. *Line*(1) of results table shows that SVM with unigram as feature produces best result out of all the features we have considered.

6 Conclusion

We have studied the sentiment analysis result for tweets collected from twitter public domain. The results table shows that classification accuracies using unigram presence and POS as feature turned out to be most effective as compared to other alternative features we employed. Though there are many machine learning techniques are available, however no single technique has proven to consistently outperform the other across many domains.

References

1. Liu, B.: Sentiment analysis and opinion mining. Synth. Lect. Human Lang. Technol. **5**(1), 1–167 (2012)
2. Pang, B., Lee, L.: A sentimental education: sentiment analysis using subjectivity summarization based on minimum cuts. In: Proceedings of the 42nd Annual Meeting on Association for Computational Linguistics, p. 271. Association for Computational Linguistics (2004)
3. Pang, B., Lee, L., Vaithyanathan, S.: Thumbs up?: sentiment classification using machine learning techniques. In: Proceedings of the ACL-02 Conference on Empirical Methods in Natural Language Processing, vol. 10, pp. 79–86. Association for Computational Linguistics (2002)
4. Pang, B., Lee, L.: Opinion mining and sentiment analysis. Found. Trends Inf. Retrieval **2**(1–2), 1–135 (2008)
5. Sadegh, M., Ibrahim, R., Othman, Z.A.: Opinion mining and sentiment analysis: a survey. Int. J. Comput. Technol. **2**(3), 171–178 (2012)
6. Wang, W., Chen, L., Thirunarayan, K., Sheth, A.P.: Harnessing twitter "big data" for automatic emotion identification. In: Privacy, Security, Risk and Trust (PASSAT), 2012 International Conference on and 2012 International Conference on Social Computing (SocialCom), pp. 587–592. IEEE (2012)
7. Nigam, K., Lafferty, J., McCallum, A.: Using maximum entropy for text classification. In: IJCAI-99 Workshop on Machine Learning for Information Filtering. vol. 1, pp. 61–67 (1999)
8. Turney, P., Littman, M.L.: Unsupervised Learning of Semantic Orientation from a Hundred-Billion-Word Corpus (2002)
9. Turney, P.D.: Thumbs up or thumbs down?: semantic orientation applied to unsupervised classification of reviews. In: Proceedings of the 40th Annual Meeting on Association for Computational Linguistics, pp. 417–424. Association for Computational Linguistics (2002)
10. Vipul Pandey, C.I.: Sentiment analysis of microblogs. In: Diploma Thesis, CS 229 Project Report, Stanford University

Part VII
Big Data and Recommendation Systems

Analysis and Synthesis for Archaeological Database Development in Nakhon Si Thammarat

Kanitsorn Suriyapaiboonwattana

Abstract Nakhon Si Thammarat, a province in Southern Thailand, was the heartland of ancient cultures and kingdoms, developed over several centuries, which had left behind tremendously valuable archaeological finds, including artifacts, structures, and sites. Unfortunately, these precious archaeological treasures are disturbed and destroyed daily by looting and modern land development. To keep a record of this irreplaceable archaeological data, we decided to develop an archaeological database. This database is managed within an RDBMS aimed at describing the relationship between artifacts, structures, and historical, stratigraphic and environmental complexity of archaeological sites, and is directly connected to a digital archive containing the raw data. In order to analyze and synthesize the archaeological database, we surveyed currently popular standard metadata for archaeological data VRA and other existing solutions. In this paper, we propose an extension metadata structure of VRA and a logical database for specific archaeological data in the area.

Keywords Archeological database · Archeology · Metadata · Database design · VRA

1 Introduction

Archaeology is the study of human activities from the past to present, primarily through the recovery and analysis of the material remains and environmental data, which includes artifacts, architecture, ecofacts, biofacts, and cultural landscapes. Archaeology is the study of material culture people have left behind and analyze it to understand the lives, behaviors, societies and cultures of the people from past to present. Indeed, archaeology is very important for all of us because it helps us to

K. Suriyapaiboonwattana (✉)
222 Walailak University, ThaiBuri, ThaSala, Nakhon Sri Thammarat 80161, Thailand
e-mail: kanitsorn.su@wu.ac.th

© Springer India 2016   517
A. Nagar et al. (eds.), *Proceedings of 3rd International Conference on Advanced Computing, Networking and Informatics*, Smart Innovation, Systems and Technologies 44, DOI 10.1007/978-81-322-2529-4_54

understand ourselves and others who have shared the same world. Southern Thailand is another region which has had a long-term civilization related to the fact that it was a crossroad of the world. However, no database, designed specifically for archaeological data has been created in Southern Thailand.

This research, therefore, aims to analyze and synthesize the specific metadata for developing archaeological database in Nakhon Si Thammarat, a province in Southern Thailand, which was the heartland of ancient cultures and kingdoms, developed over several centuries and had left behind tremendously valuable archaeological finds. Unlike other forms of archeological databases, the design of our database relies specifically on the analyses of unique archaeological finds found in this area. This database describes the contextual relationship between sites, structures, artifacts, and their historical, stratigraphic and environmental complexity, and is directly connected to a digital archive containing the raw data that can be used by archeologists to do archaeological data mining and to work with ontologies for knowledge management. This paper is organized into 6 sections. Section 1 is the introduction. Section 2 summarizes related previous literature. In Sect. 3, we explain our research methods. In Sect. 4, we introduce our proposed metadata and implementing database. Section 5 discusses the results, while Sect. 6 concludes this paper.

2 Related Work

The most difficulty encountered during this research was synthesizing the archaeological data gradually acquired. We can gather tons of archaeological information from museums, documents, archaeological reports, and the internet, for example, but not to understand what exactly the archaeologists need for their work. We believe it is archaeologist's duty to decide what kind of information and processing should be performed. To obtain that information, Research Unit for Archaeology (RUA) has become our case study. RUA is one of Walailak University research units, which was founded in 2013 to initiate and support research projects related to archaeology, history, art history, and cultural geography in southern Thailand. Directed by Dr. Wannasarn Noonsuk, who received his Ph.D. in archaeology and art history from Cornell University, the unit has strong research team of professors, specialists, and graduate students from various fields in the university. It focuses on the interdisciplinary approaches of research and collaboration across the academic fields to enrich our understanding of human socio-cultural development. We gather most of important information from RUA. However, in order to analyze and synthesize metadata for archaeological database, we surveyed current existing database and standard metadata that are related to our work, as follows:

2.1 Database for Archaeological Data Recording and Analysis by Anichini et al.

This database is designed based on 4 different logical levels that gradually manage the information through an interpretative synthesis process: from definition of the material trace to classification of data into typological and chronologically-divided macro categories which have archiving and analysis functions, as well as categorization purposes for the predictive calculation of archaeological potential [1]. The proposed logical structure, gathered from archaeological excavation data from MAPPA archaeological research team, is implemented with Microsoft Access because of the compatibility with other software, especially GIS software, and the greater knowledge of this software by research team. However, this particular research is an internal work tool and public disclosure is not expected.

2.2 Developing Archaeological Database in Finland

This research in Finland proposed the development of database for archaeological excavations on the Lahti market square by using Microsoft Access. This research aimed to figure out how to improve the database and share it for further usage. The results showed that the improvement process is very case-specific and depended on earlier experiences and current database of other museums [2].

2.3 IADB (Integrated Archaeological Database System)

The Integrated Archaeological Database system or IADB [3], is designed for data management which is important throughout the archaeological excavations. IADB can help manage the excavation record to be used during the post-excavation analyses and research. The IADB is implemented based on web application technology. There is no need to installed IADB software on the client's computer. Users can access IADB via web browser such as Internet Explorer, Google Chrome, and etc. IADB consists of relationships between data resources. There are various kinds of resources such as finds, contexts, sets, groups, phases, objects, and images (also see [3]).

2.4 VRA Core 4

As seen on the VRA CORE official website, "The VRA Core is a data standard for the description of works of visual culture as well as the images of that works. The standard is hosted by the Network Development and MARC Standards Office of the Library of Congress (LC) in partnership with the Visual Resources Association" [4]. The Core 4 is an effective standard metadata for the images and the cultural objects but still lacks some important elements for recording components related to archaeological field research, such as site, structure, trench, and etc.

3 Research Framework

In this section, we explain our research framework and methods. This research is constructed according to the book written by Havner et al. [5]. The framework is divided into 3 parts: environment, IS research and knowledge base. The environment defines the problem space in which reside the phenomena of interest. The knowledge base provides the foundation and methodologies. Both of the environment and the knowledge base will be used to create the data that leads to research's goal. In this research, the goal is to perform the analysis and synthesis for archaeological database development in Nakhon Si Thammarat.

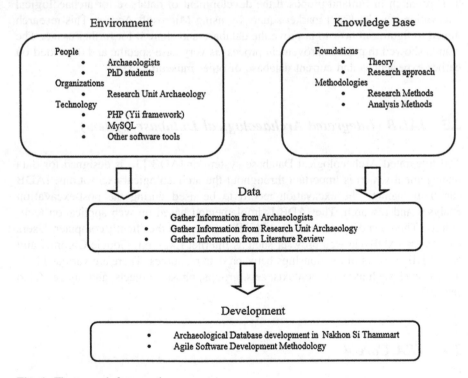

Fig. 1 The research framework

In Fig. 1, we can explain for the details in the following:

- "People" in the environment section are the archaeologists who need to use archaeological database and providing valuable information such as questionnaires, feedback and personal opinion for the research.
- "Organizations" is Research Unit for Archaeology at Walailak University who provide most important data and also currently work on archaeological projects in Nakhon Si Thammart.
- "Technology" is tools that we used for the research which includes MySQL, PHP (Yii framework) and other software tools (MySQL Workbench, CentOS, JQuery, Blueprint framework).
- "Foundations" in knowledge section consist of theory and research approach. The theory is theoretical information about database design, VRA core metadata, MVC programming and archaeology. For research approach, we used Inductive approach as our research approach to analysis of data and examination of archaeological database practice problems.
- "Methodologies" are divided into research method and analysis method. In this study, the research method is qualitative research. This research focuses on analysis and synthesis of current archaeological metadata standard, and based on the requirement of RUA and local archaeologists. To design specific metadata that will be used for developing a real case study of archaeological database in Nakhon Si Thammarat. Hence, a qualitative method is selected. Finally, analysis method is done using thematic analysis.
- The data is the result of the processing of the environment and the knowledge base.
- The development is the goal of the research. We use agile software development methodology to implement archaeological database in Nakhon Si Thammarat.

4 Metadata Design and Database Development

In this section, we discuss how we gathered user requirements to design our archaeological metadata and to implement the metadata to database.

4.1 Analysis and Synthesis of Metadata

First of all, we surveyed all of previous works [1–3, 6, 7]. We found some problems of previous works such as

- The metadata was designed for internal usage.
- The metadata do not follow VRA Standard.
- The metadata was not designed for the field research in archaeology.

Because of these problems, we used the information to design and create our questionnaires. Next, we used questionnaires to collect user requirements from RUA research team. After we gathered all the feedbacks, we picked up some interesting points to discuss with the interviewees. We gained several important advices and archaeology-specific requirements that were missing from VRA Core 4.0, which are the significant elements for explaining the relationship between the data related to environment and material remains. Following several face-to-face interviews with RUA research team, we discovered their actual needs for the database. In Table 1, we describe our extended metadata from VRA core 4.0, namely RUA Metadata Core.

The distribution and development of sites are of significance in archaeology. They profoundly reflect the socio-cultural and environmental changes. Hence, RUA Metadata Core proposes 9 newly elements to describe the specific characteristics of the field research in archaeology. These elements can collect the detailed distribution and chronology of the sites, structures, and specific artifacts. For example, we provide an example of RUA Metadata Core in Table 2.

4.2 Database Design and Development

Most archaeological database developments, including existing metadata, do not concern with the relationship between different kinds of information found in archaeological excavations. Instead, they were designed to describe each artifact independently. However, it is the most important part of the design of any archaeological database because it can tell us about the distribution and chronology of archaeological finds. According to RUA Metadata Core as shown in Table 2, we classified the types of information into 4 groups: site, structure, trench and artifact.

Table 1 RUA Metadata Core

No	Metadata elements	
	XML element	Description
1	aliasName	alternative name of the artifact
2	story	belief or rumor about the artifact
3	trench	excavation pit
4	phase	General time period
5	basket	Unit of soil and artifacts within it
6	scientificDating	Dates acquired from scientific dating methods
7	surfaceElevation	Elevation of the soil surface above sea level
8	depthBelowSurface	Depth of artifacts found below surface
9	register	Person discovering the artifact

Table 2 Example of RUA metadata core

No	Metadata elements	
	XML element	Data example (a pot)
1	agent	Unknown
2	culturalContext	Tambralinga
3	date	circa 800 (creation)
4	description	It is an earthenware pot with short foot, wide body, and flaring mouth. It also have cord-mark decoration on its body
5	inscription	–
6	location	Ban Hua Ton archaeological site, Sichon District, Nakhon Si Thammarat Province, Thailand
7	material	Pottery
8	measurements	20 cm (height) × 16 cm (width)
9	relation	It was found near artifact no. 572
10	rights	–
11	source	–
12	stateEdition	–
13	stylePeriod	Tambralinga 1
14	subject	Ritual pottery; Hinduism
15	technique	Wheel-turned pottery making
16	textref	
17	title	HT-2014-2-Pot 2
18	worktype	Pottery, pot
19	aliasName	Ban Hua Ton Pot, Mo Ban Hua Ton
20	Story	Local people believed that this kind of pot was used for containing sacred water in ancient Hindu shrine
21	Trench	HT-2014-2
22	Phase	Early Tambralinga
23	Basket	4
24	scientificDating	TL 750 ± 850 CE
25	surfaceElevation	10 m above sea level
26	depthBelowSurface	105 cm. below surface
27	Register	Wannasarn Noonsuk

In Fig. 2, we designed the database that synthetically describes the levels and relations among the RUA metadata core elements. The result of this synthesis is important to the main structure of the database, which has "site" as the primary data. Site primarily contains geographical data of each archaeological site. To create the other groups, including structure, trench and artifact, we need to create "site" first. Structure refers to architecture and features on the ground. In terms of relationship, structure is subset of site. Trench is excavation pit. Trench can be in a site

Fig. 2 Classification of data related to environment and material remains

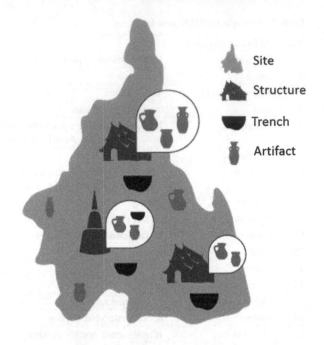

or in structure. Artifact is a man-made object, such as tools, statues, and pottery sherds, sometime characteristic to specific time period.

To develop archaeological database, LAMP is chosen. LAMP is a generic software stack model that consists of Linux, Apache, MySQL and PHP. Our archaeological database is implemented on CentOS, a Linux distribution derived entirely from the Red Hat Enterprise Linux (RHEL). According to our survey of related work, we found limitations of previous archaeological database, such as:

- No scalability
- No public access (designed for internal usage)
- Do not support advance database access control such as RBAC, TRBAC
- No software development methodologies such as Agile.

Our research aims to develop archaeological database for everyone, including RUA research team, other archaeologists, students, and the public in general, and to develop high scalability web application for archaeological database. Therefore, we decided to use Yii framework, an open-source Web application development framework, to implement the project. In order to handle the complexity of the database and the considerable variability of users, we need strong database access control policies. In this research, we implemented RBAC database access control policy, a method of regulating access to database based on the roles of individual users (Figs. 3 and 4).

Fig. 3 RUA archaeological database—website for public usage

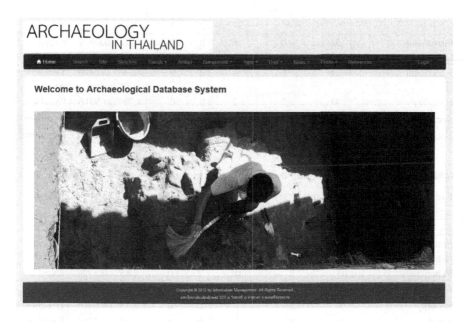

Fig. 4 RUA archaeological database—website for internal usage

5 Evaluation and Discussion

To evaluate archaeological database, we use SUMI (Software Usability Measurement Inventory), questionnaire for the assessment of the usability of software which has been developed, validated, and standardized on an international basis. The evaluation is divided into 4 parts: Functional Requirement, Functional, Usability and Security. The archaeological database was evaluated by 2 groups of users. The first group is 50 public users using our public website to search for the information. The second group is 12 members of RUA research team using CRUD in all database functions. In Tables 3 and 4, shows the summary of all evaluation results.

According to Table 3, the level of the public user's satisfaction with the database is excellent in 3 parts. In the functional testing, 18 participants required mobile optimized website to serve the browsing across the platforms using any mobile devices and 4 other participants wished for a mobile application version.

Table 4 indicates the level of RUA research team's satisfaction with the database, which is excellent in all parts. The result shows that the archaeological database function is perfectly fit for field research requirements. The archaeologists can record more contextual relationship between sites, structures, trenches and artifacts that can lead to establish ontology knowledge base for the distribution and chronology of archaeological finds.

Table 3 Summary evaluation result by public users

Evaluation	Excellent	Very good	Good	Fair	Poor	$\bar{x}$	S.D.	Result
	5	4	3	2	1			
Functional requirement	34	10	6	0	0	4.56	0.6974	Excellent
Functional	26	18	4	2	0	4.36	0.7940	Very good
Usability	41	7	2	0	0	4.78	0.5016	Excellent
Security	33	14	3	0	0	4.60	0.6000	Excellent
Total	134	49	15	2	0	4.58	0.6741	Excellent

Table 4 Summary evaluation result by RUA research team

Evaluation	Excellent	Very good	Good	Fair	Poor	$\bar{x}$	S.D.	Result
	5	4	3	2	1			
Functional requirement	9	2	1	0	0	4.67	0.6236	Excellent
Functional	10	2	0	2	0	4.83	0.3727	Excellent
Usability	11	1	0	0	0	4.92	0.2764	Excellent
Security	8	3	1	0	0	4.58	0.6401	Excellent
Total	38	8	2	0	0	4.75	0.5204	Excellent

6 Conclusion

The primary objective of our research is to design and develop archaeological database for describing the contextual relationship between different kinds of information in the archaeological field research. In this research, we propose RUA Metadata Core, extended from VRA Core 4.0, to collect important archaeological data from Nakhon Si Thammarat, because some elements have been previously missing from other databases. The RUA metadata core was implemented to archaeological database for RUA research team and provided data to public usage. The evaluation showed demands from public users. We learned from the public users that we should improve our archaeological database to support mobile devices, using responsive web design. Also, there are still various classifications that need more metadata designs to support different types of artifacts. At the end, this research is an important invention in Thailand, and perhaps in the field of archaeology in general as well, which would contribute to the progress of the research in this field tremendously in the future.

References

1. Anichini, F., Fabianni, F., Gattiglia, G., Gualandi, M.L.: A database for archaeological data recording and analysis. MapPapers. 1en-II 21–38 (2012)
2. Luo, W., Hartonen, V.: Developing archaeological database in Finland. Lahti University of Applied Sciences (2014)
3. Integrated Archaeological Data. IADB. http://www.iadb.org.uk/. (2014)
4. VRA Core. VRA Core Schemas and Documentation. http://loc.gov/standards/vracore/. (2014)
5. Hevner, A., March, S., Park, J., Ram, S.: Design science in information system research. MIS Q. 28(1), 75–105 (2004)
6. Bhurimpai, P., Lamphun, N.R.: Development of the organization of information resources for buddha images museum. J. Inf. Sci. 28(3), 21–32 (2010)
7. Denzin, N.K., Lincoln, Y.S.: The Sage Handbook of Qualitative Research 3rd edn. Thousand Oaks. (2005)
8. Huvila, I.: The Ecology of Information Work: A case study of bridging archaeological work and virtual reality base knowledge organization. Abo Academi University Press, Turku (2006)
9. Mocanu, A.-N., Velicanu, M.: Building a spatial database for romanian archaeological sites. Database Syst. J. 2(1), 3–12 (2011)
10. Hua-guang, G., Yue-ping, N.: Study of Archeology Spatial Database in Xinjiang Based on ArcGIS Engine. In: Computer Science and Network Technology. 232–235 (2011)
11. Makridis, M., Daras, P.: Automatic classification of archaeological pottery sherds. J. Comput. Cult. Heritage. 5(4), 15:1–15:21 (2012)
12. Wu, L., Zhu, C.: Design and Realization on the Object-oriented Remote Sensing Archaeological Information System, pp. 3050–3053. Environment and Transportation Engineering, Remote Sensing (2011)
13. Hong-dan, Z., Qin-xia, W.: Research and Design on the Data Model Based on Archaeology GIS. Multimedia Information Networking and Security. (2010) 961–963

Evaluation of Data Mining Strategies Using Fuzzy Clustering in Dynamic Environment

Chatti Subbalakshmi, G. Ramakrishna and S. Krishna Mohan Rao

Abstract The recent applications of data mining such as biological, scientific, financial and others are changing data regularly, which is uncertain and incomplete. For finding tendency in these data up-to-date, we need to modify existing data mining algorithms with dynamic characteristics. Soft computing methods are suitable for finding changes in uncertain data. In order to adopt change in data we can apply any of two approaches, update algorithm by ignoring earlier state or update with respect to earlier state. In this paper, we have framed two fuzzy clustering methods based on these approaches and implementation done using R software with comparison.

Keywords Data mining strategy · Changing data · Fuzzy c-means · Silhouette index

1 Introduction

In many data mining applications such as business intelligence, image processing, scientific and online applications often apply technique is cluster analysis, is process of grouping a set of similar data points into subsets [1]. In recent years all these applications are generating dynamic data, i.e. changing over time. For example, in business intelligence we group the customers based on their buying behavior to develop the business strategies to enhance the customer relation management. However, the behavior of customer is always changes over time and they have to

C. Subbalakshmi (✉)
Guru Nanak Institutions Technical Campus, Hyderabad, Telangana, India
e-mail: subbalakshmichatti@gmail.com

G. Ramakrishna
K L University, Vijayawada, Andhra Pradesh, India

S. Krishna Mohan Rao
Siddhartha Engineering College, Hyderabad, India

© Springer India 2016
A. Nagar et al. (eds.), *Proceedings of 3rd International Conference on Advanced Computing, Networking and Informatics*, Smart Innovation, Systems and Technologies 44, DOI 10.1007/978-81-322-2529-4_55

529

alter their strategies. Therefore, such applications we need to change in the process of clustering to find new trends. The traditional data mining algorithms are not suitable for finding new trends on these application data, due to algorithms takes static inputs. Consider a case, where we have to input number of clusters (k) to partition clustering algorithms (K-means and K-medoids) which is decide before [2–4]. But k value depends on characteristics of data set (the size of data) and updates it as data changes. Hence, we need a new approach for mining on dynamic environment. In this direction many clustering methods are proposed to perform on dynamic data.

The aim of dynamic model is to find changes in data and adjust the input parameters accordingly. Soft Computing methods are suitable for finding changes in uncertain and vagueness data. In this field Crespo and Webber introduced data mining strategy in changing environment [5]. According to, when user wants to adopt the changes and he can apply any of two approaches, perform complete data mining task from the base or follow the updating of present system according arrival of new data. In this paper, we evaluate these two strategies on changing data using fuzzy clustering. The paper is structured as follows. We discussed the related work in Sect. 2, proposed dynamic clustering methods are presented in Sect. 3, and results are presented in Sect. 4 and finally conclusion in Sect. 5.

2 Related Work

From literature, the focus of dynamic clustering algorithm is to identify the changes in the data and adopt these by updating clustering parameters. The change in data is uncertain and incomplete, hence soft computing approaches (Fuzzy sets, Rough sets, and Evaluation computing and neural networks) are suitable for clustering on these data. Crespo and Webber proposed a methodology on dynamic data using fuzzy clustering and rough k-means clustering algorithms [5–7]. We motivated to their methodology and evaluate proposed strategies using fuzzy clustering.

2.1 Fuzzy Clustering

Unlike in hard clustering, fuzzy clustering assign a data point to more than one cluster using degree of membership values as defined in fuzzy set theory [8, 9]. For all data point, algorithm calculates degree of membership value with each cluster and assigns to cluster with high membership value. The basic fuzzy clustering algorithm is Fuzzy C Means (FCM) and achieves using minimization of objective function (J) as in Eq. (1).

$$J = \sum_{i=1}^{n} \sum_{k=1}^{c} \mu_{ik}^{m} |p_i - v_k|^2 \tag{1}$$

where, n: number of data objects; c: number of clusters; μ: fuzzy membership value; m: fuzziness factor (>1); p_i: data point; v_k: center of kth cluster.

The center of the kth cluster is calculated using Eq. (2) as,

$$v_k = \frac{\sum_{i=1}^{n} \mu_{ik}^{m} p_i}{\sum_{i=1}^{n} \mu_{ik}^{m}} \tag{2}$$

The fuzzy membership can be calculated using Eq. (3) as,

$$\mu_{ik} = \frac{1}{\sum_{i=1}^{c} (|p_i - v_k| \, ||p_i - v_l|)^{\frac{2}{m-1}}} \tag{3}$$

2.2 Silhouette Index

Silhouette index is for evaluating internal cluster results and used for finding optimal number of clusters [10]. For each data point (i), the silhouette width s(i) is defined as,

$$s(i) = \frac{b(i) - a(i)}{\max\{a(i), b(i)\}} \tag{4}$$

where $a(i)$ is average dissimilarity between data point (i) *and* all other data within the similar cluster and $b(i)$ is the minimum average dissimilarity of i to all other cluster. The data point with positive $s(i)$ is correctly clustered and with negative value indicates wrong clustering.

2.3 Data Mining Strategies on the Changing Data

The process of mine the knowledge from data warehouse is cycle of extract data, set input parameters of mining algorithm and execute algorithm. As database adds new data into data warehouse, this cycle is repeated to get accurate results. In [5] described three approaches for data mining system on changing databases:

1. Ignore changes in data and keep on apply initial parameters.
2. For each new incoming data entire cycle is repeated by ignoring previous state.
3. For each new incoming data, identify need of change and update based on existing clusters with new data.

The first strategy does not adopt the changes, thus no need of updating the data mining system and reduces the computational cost. But it does not give the correct results as data change. In strategy two, adopts the changes and gives the accurate results, for that, it repeats entire cycle for every instance and performs mining on the entire data. However, it requires more computation cost. Strategy three, also adopts the changes in data, but it does not do the data mining process from the scratch and it identifies need for update based on the new data and performs the update with respect to previous system. It is computationally cheap and identifies the changes in environment based on previous state of system.

3   Dynamic Clustering Algorithms

Here we consider second and third data mining strategies to adopt the changes in data and framed a fuzzy clustering method for each approaches.

3.1   Dynamic Clustering Using First Strategy

We proposed a method to cluster on dynamic data in two phases. 1. Find optimal number of clusters using Silhouette width on given data set. 2. Execute fuzzy c-means clustering. For each cycle of new incoming data, combine new data with existing data and repeat two phases. The algorithm steps are giving below:

Phase 1: Find right number of clusters on initial data set $D_{initial}$ with data size of n.
Step 1: Repeat for each cluster number (c) for $c = 2$ *to* ($n/2 - 1$) Calculate the silhouette width (S) using Eq. 1;
Step 3: find c which has maximum average cluster silhouette width;
Phase 2: Execute fuzzy c-means on data set (D)

Repeat (phase 1&2) for every new incoming data by combining new and old data as,

$$D = D_{initial} + D_{new}$$

3.2   Dynamic Clustering Using Second Strategy

In this section, we present a method for finding soft changes in data and update cluster structure as proposed in [5]. The algorithm steps are given in two phases: 1. Initial clustering 2. Iteration. Iteration step is executed for each new incoming data, it can perform any of three actions (create new cluster, move cluster centers and delete cluster) as for each change.

Initial clustering:

Step 1: calculate average cluster silhouette width for each c.
Step 2: Find cluster (c) which gives maximum average silhouette width;
Step 3: Execute fuzzy-c means clustering with c on initial data set $D_{initial}$.
Iteration: repeat for every new incoming data

Step 1: Let D_{new} has m number of new data points added.
Step 2: Identify the new points which represent the changes in cluster structures; To find the new objects which are not fit in existing cluster centers, apply two properties. For each new data point i with current clusters center,
Property 1: If all membership values near $1/c$ cannot be classified correctly.

$$\left| \mu_{ik} - \frac{1}{c} \right| \le \alpha, \forall k \in \{n+1, n+2, \ldots n+m\}, \forall i \in \{1, 2, \ldots c\} \qquad (5)$$

Property 2: If the distance between data point (i) to all cluster centers is more than the minimum distance among any two cluster centers.

$$\overline{d_{ik}} > \frac{1}{2} \min\{d(v_i, v_j)\} \qquad (6)$$

If any data point satisfies both the properties represent need of change.
Step 3: Identify structural changes.
Case 1: Create new cluster.
Property: If average number of new objects requires changes is beyond threshold (β) then create new cluster. Otherwise, go for next case.

$$\sum_{k=n+1}^{n+m} \frac{IC(x_k)}{m} \ge \beta \ with \ a \ parameter \ \beta, \ 0 \le \beta \le 1 \qquad (7)$$

Case 2: Move cluster centers.
Combine new objects with old data and perform fuzzy c means clustering with same number of cluster. $D = D_{initial} + D_{new}$

4 Results

We implemented proposed methods in R software language on dynamic customer segmentation. In customer relationship management, tracking customer behavior is significant and it changes always over time. We use customer wholesale data set collected from UCI repository to show effectiveness of these methods. It refers to 440 customers of wholesale data with two channel and three regions. To show dynamic behavior of customers, in each cycle we added randomly generated subsets and executed methods to track behavior of customers.

4.1 Results of First Strategy

As defined in Sect. 3.1, we executed dynamic clustering method in three cycles and the results of first phase are given in Table 1. After finding right number of clusters, we executed fuzzy c-means.

4.2 Results of Second Strategy

As defined in Sect. 3.2, in cycle-1 initial clustering executed on 20 objects. From the next cycle iterations are started. In cycle-2, we added 20 new objects and identified two objects satisfy both conditions. It shows that need of change in cluster structure and then applied condition three which results to move of cluster centers. In cycle-3, we added 20 objects and apply the condition 1, 2 and 3 on new data, number of data objects requires changes in clusters and indicates to create new cluster. The results of cycles are given in Table 2.

4.3 Evaluation

From the results of both methods, we can observe that in first method does not give internal changes in cluster structures. For every cycle old results are refreshed with new results. Hence, we cannot track customer behavior as it not maintaining

Table 1 Results of first phase to decide the right number of clusters

Cycle	Data size	Optimal no of clusters	Max. avgsil width
Cycle-1	20	2	0.676344
Cycle-2	40	2	0.545914
Cycle-3	60	3	0.5362692

Table 2 Results of three cycles

Cycle	Cluster	No. of objects	Changes	Avgsil width
Cycle-1	1	15	–	0.6773940
Initial data	2	5	–	0.6731938
20 objects			Avgsil width	0.676344
Cycle-2	1	23	Move	0.7436776
20 new	2	17	Move	0.1786387
Data added			Avgsil width	0.545914
Cycle-3	1	35	Move	0.7833169
20 new	2	15	Move	0.3769178
Data added	3	20	Create	−0.0893708
			Avgsil width	0.5362692

previous results. But in second method, it shows the moving of objects between clusters, changes in cluster centers and arrival of new groups. It indicates changes in buying behavior of customer over time. Therefore, the applications which require the internal changes of data with respect to previous cluster structure can offer second method.

5 Conclusion

We considered the problem of clustering on dynamic data set as most of applications are generating the changing data over time and discussed the merits and demerits in changing environment. We proposed and executed two dynamic clustering algorithms based on fuzzy clustering, to show the evaluations of two strategies on wholesale customer data. From the results, we identify that first method is simple and it does not give changes in behavior of data, but second method shows the changing behavior in data and as for that we can modify the clusters. Most of present applications are required adopt dynamic model and further we can modify these methods by considering the issues noise, complexity and size of data to get more accurate results.

References

1. Hartigan, J.A. Clustering Algorithms. Wiley, New York (1975)
2. Hartigan, J.A., Wong, M.A.: Algorithm AS 136: a K-Means clustering algorithm. J. Roy. Stat. Soc. Ser. C **28**(1), 100–108 (1979)
3. Kaufman, L., Rousseau, P.J.: Clustering by means of medoids. In: Dodge, Y. (ed.) Statistical Data Analysis Based on the L_1–Norm and Related Methods, pp. 405–416. North-Holland, Amsterdam (1987)

4. Park, H.S., Jun, C.H.: A simple and fast algorithm for K-medoids clustering. Expert Syst. Appl. **36**(2), 3336–3341 (2009)
5. Crespo, F., Weber, R.: A methodology for dynamic data mining based on fuzzy clustering. Fuzzy Sets Syst. **150**(2), 1 (2005)
6. Peters, G., Weber, R., Nowatzke, R.: Dynamic rough clustering and its applications. J. Appl. Soft Comput. **12**(2012), 3193–3207 (2012)
7. Peters, G., Weber, R.: Intelligent cluster algorithms for changing data structures. Int. J. Intell. Def. Syst. **2**(2), 105–119 (2009)
8. Nock, R., Nielsen, F.: On weighting clustering. IEEE Trans. Pattern Anal. Mach. Intell. **28**(8), 1–13 (2006)
9. Bezdek, J.C.: Pattern Recognition with Fuzzy Objective Function Algorithms. Plenum Press, New York (1981). ISBN 0-306-40671-3
10. Visvanathan, M., Adagarla, B.S., Gerald, H.L., Smith, P.: Cluster validation: an integrative method for cluster analysis. In: IEEE International Conference BIBMW, pp. 238–242 (2009)

A Survey of Different Technologies and Recent Challenges of Big Data

Dipayan Dev and Ripon Patgiri

Abstract Big Data, the buzz around the globe in recent days is used for large-scale data which have huge volume, variety and with some genuinely difficult complex structure. The last few years of internet technology as well as computer world has seen a lot of growth and popularity in the field of cloud computing. As a consequence, these cloud applications are continually generating this big data. There are various burning problems associated with big data in the research field, like how to store, analysis and visualize these for generating further outcomes. This paper initially points out the recent developed information technologies in the field of big data. Later on, the paper outlines the major key problems like, proper load balancing, storage and processing of small files and de-duplication regarding the big data.

Keywords Big data · Key technologies · Hadoop · Load balancing · Storage

1 Introduction

Big data analytics has become the focal point of current computer world, cloud computing [1] and similar business world. Big Data is generated from various manners like, social networking, various smart phones and its applications, huge number of online transactions per day, the emails, photos, logs of stock exchanges, health records of governments as well as numerous other online statistical records [2, 3]. So, it could be easily understood that, this amount of data is prone to grow

D. Dev (✉) · R. Patgiri
Department of Computer Science and Engineering,
National Institute of Technology Silchar, Assam 788010, India
e-mail: dev.dipayan16@gmail.com

R. Patgiri
e-mail: ripon@cse.nits.ac.in

© Springer India 2016
A. Nagar et al. (eds.), *Proceedings of 3rd International Conference on Advanced Computing, Networking and Informatics*, Smart Innovation, Systems and Technologies 44, DOI 10.1007/978-81-322-2529-4_56

massively with the course of time and therefore it becomes a burdensome to store, process, visualize and analyze those data using traditional database management systems.

It might be hard to believe that, 5 EB (10^{18}) of data was produced from the dawn of civilization to 2003, whereas, we generate this amount of data in every 2 days. Up to 2012, the online world had a size of nearly 2.72 ZB (10^{21}). As predicted, this data got nearly doubled every 2 year which reaches up to 8 ZB by 2015 [4, 5]. If on an average a PC holds about 500 GB data, then nearly 20 billion computers are needed to accommodate all the data around the globe. Human genome decryption, which usually took nearly 10 years in past, now-a-days does not take more than 5–6 days [6]. The multimedia data which seems to be the mainstay of internet world grew by 70 % till 2013 [7] and might by 90 % by 2015. In current days, there are more than 8 billion mobile subscribers and more than 20 billion text messages are sent every day. As per the report, more than 30 billion devices will get connected to internet by 2015 [8].

In 2015, the Big Data which is termed as a universal project, focusing mainly in analytic, visualization and real time data collection of massive scale data. Many statistics have come out from the media project. Facebook currently has more than 1 billion active users, 200 billion photos gets uploaded, nearly 150 billion friend connections, more than 50 billion segments of content and 3 billion comments and likes are getting posted. More than 300 h of videos are posted in YouTube in every minute by the users [9] and almost 1000 new sites are developed in every minute of a day [10]. So, in can be easily inferred that, in the coming decade, the amount the information will be increased by at least 50 % from the current size. But the number of IT specialists who maintain all those data will also get increased by 1.5 times [11].

This shows how important the Big Data is in the current research world! But there is no in-depth study on it because of restricted by condition in the past. But now, with the hardware costs come down and more and more development in the science and technology, people can focus into the big data more.

The article is worded as follows: Sect. 2 deals with the different technologies that have been evolved so far for Big Data. Section 3 deals with the storage challenges. Section 4 of the paper gives a detail description of the challenges in three key techniques of Big Data. Section 5 concludes the work.

2 Different Technologies Used for Big Data

The data are rich in sources and possess large variety in the environment of big data. The availability and efficiency of the large-scale data processing is necessary, moreover the storage and analysis of data provide high quality to the big data. The data acquisition in the past was single and needed only tiny amount of data for storage, analysis and management. Most of those could be managed with parallel and relational databases. In those traditional database technologies, fault tolerance as well as degree of consistency was relatively high so as to improve the speed up the data processing. The cost of big data processing should be abated, and for that we cannot use the traditional method. This is mainly because, big data needs data-centric mode whereas the traditional approach of data processing is typically based on a centralized one. Therefore, it is quite impractical to implement the traditional centralized method on different operations of big data.

The processing of big data is actually homogeneous to that of conventional data. The basic difference in processing is that, big data performs parallel processing in each of the stages, such as Map-Reduce [12] that deal with the huge amount of unstructured data. In Map-Reduce, the data processing is done in three stages viz. splitting, sorting and lastly merging the output. So the Map-Reduce speed up the data processing of big data using its parallel processing capability. Initially, Map-Reduce achieve the parallel processing of big data using a huge number of low-commodity hardware such that it doesn't need the data consistency at a high level. The main advantage of Map-Reduce is its availability and flexibility, because of its capacity to process unstructured, structured and semi-structured data. Map-Reduce, due to its distributed processing capacity, possess high level of parallel processing ability.

2.1 Data Acquisition

Conversion of data format calls for a huge cost. Due to which the traditional approach of data collection doesn't work easily and fails to match the need of big data collection. Following are the new methods that are used in the data acquisition:

A. **Collecting method of system log**: A lot of industries use their own framework to collect data, but majority of those are used to the system log, viz. Chukwa of Hadoop [13], Flumen of Clodera, Scribe of Facebook etc. These all follow distributed approach that process hundreds MB of log data acquisition per seconds and fulfil the transmission needs.
B. **Method to collect the unstructured data from network**: Data information gathering using web crawler or any API is known the network data collection. The method refers to the extraction of the unstructured data and saving it to some local file. The method supports video documents, audio, images and collection of other accessories too. It does not only mean to collect the content of the network. We can use different bandwidth management technology, like DFI or DPI to collect the network flow.
C. **Other data acquisition methods**: Data which are of higher security, a user can consult any company or research institute for collection of that data by any system specific interface.

2.2 Data Preprocessing

Big data pre-processing involves three basic steps: viz. Data extraction, Transformation and Loading. Extract the data from various data sources to a transitional layer for conversion, integration and finally loading into the ultimate database or a particular file in the local storage, this is the basic process for data mining. ETS is made responsible for the whole phenomenon.

A. **Data extraction**: Data can be extracted from various ways like heterogeneous and distributed data source with the help of interfaces like ODBC etc. The metadata is a crucial factor to perform the data extraction operation.
B. **Data transformation**: After the first phase of extraction, the raw data is transformed into target data structure so as to meet the business needs such as data conversion, abatement, data cleaning, integration and other processing to attain summary.
C. **Data loading**: In the data loading phase, the converted and summarized data is stored in the desired database or some local file system.

2.3 Data Storage

What is the key to efficient big data technology? It is surely the way of storing and improving the fault tolerance of the distributed system. As discussed earlier, the amount of data is growing hugely as compared to the storage capacity. It is nearly impossible to match the data storage of any enterprise's using tens or rather hundreds of servers. The solution to it is how to use thousand of low-commodity hardware so as to reduce the costs of large servers. As the low-commodity hardware to prone to failures, the need of replication of same data into few (generally three) other servers is necessary to increase reliability of the system. These schemes provide high handing capacity when implemented in large distributed frameworks.

A. **File storage technology for big data**: In big data environment, two widely accepted open source file storage technologies are used for the storage management, viz. Google File System (GFS) [14] and HDFS [15, 16]. The data reliability is ensured by redundant storage and no data loss is provided by the distributed storage. HDFS and GFS, both have master-slave architecture. A special master server, called metadata server keeps all the metadata and manages any R/W requests from the clients. The slaves in turn manage the task of application data storage.
B. **Data management technology of big data**: There are two main storage technologies that are used to represent and store the mass data.

   (a) **Column storage technology**: To represent data structures such as image, video, URL, geographic positions etc. we should use some multi dimensional data management system to manage and organize them efficiently. We save the data of same attribute in the same column. According to the increasing demand, different column fields possess different properties. The real time column distribution unifies the data management and thus avoids the traditional storage mode and hence ameliorates the processing speed.

(b) **NoSQL database**: NoSQL [17], which is widely accepted by most of the industries to deal with the big data, sorts out the problem of complex database structure by eliminating the relational database characteristics. Therefore, it is quite convenient to enlarge the volume of system architecture and data. On the other hand, NoSQL is extremely flexible with the data storage format.

2.4 Data Mining

In the generation of big data, data mining has become the hardest challenges of all. The big data mining in the distributed systems are performed in parallel. That means, distributing the data across the cluster and then operating mining by parallel processing and ultimately generating the output. The process is almost similar to the traditional parallel processing of data. Both the methods break the task into multiple chunks and then assign each of them to different nodes. But in Map-Reduce, after each stages of processing, the slaves inform the results to the master node, whereas in traditional approach, there is some interaction of data between the processing nodes [18]. In big data environment, it is essential to analyze unstructured, semi-structured and structured data together which is not the case with traditional approaches. The typical approaches of data mining algorithm are as follows:

A. **Hive**: Hive is a data ware-house, placed on the top of Hadoop, for various data query and data mining operations by the Hadoop management. It has its own query language; HiveQL which transforms users' SQL query so that it can be used by Map-Reduce task.

B. **Mahout**: It presents data mining and a machine learning algorithms, which help to realize some elegant algorithms like classification, cluster etc. It also can bestow numerous data mining operations when combined with Hadoop.

3 Big Data Storage Challenges

As explained earlier, the traditional data storage reaches a serious bottleneck when dealing with explosive growth of data. The Table 1 shows such a scenario taking Social Networks as example.

Table 1 New challenges on storage

Characteristics	Social networks	Challenges
Large amount of data	Large amount of data that are rapidly growing	Security, integrity, real-time intelligence, reliability, high efficiency low consumption, high concurrency
Various forms of data	Structured, semi-structured, un-structured	
High potential value	Mining, social relationship, customized recommendation	
Huge fluctuations of data	7–8 times	

- As the load fluctuation is unpredictable and quite high in nature, the big data storage system should dynamically match all the load characteristics.
- The length and delay of I/O path in data processing should be abated, due to the need of real time big data applications.
- High I/O, large bandwidth and huge system storage capacity are keys to maintain high efficiently of big data applications. Otherwise, large concurrent user access, speedy data growth and storing huge amount of data cannot be handled.

Although there are various such technologies for matching the need of large-scale data, but still huge scope of improvement is left to fulfill the actual demand of big data storage.

4 Challenges for Big Data in Distributed File Systems

File system is the foundation of upper layer application [19]. Storing the massive-scale data in the file system has become an onerous task for all the big data industries and research world. It is impossible for a single hard disk to store all the data and therefore distribution of the data among hard disks of multiple nodes is necessary. Grouping all the hard disk into a global storage, the DFS provides increased scalability, better I/O bandwidth and huge storage capacity. The most common file system like GFS, Lustre, and HDFS keep their metadata and application data separately in two different types of server because of the difference in storage and access of the same. Despite of all those numerous advantages, still many loopholes are found in DFS when it comes to growth of data, complexity in the storage etc. These shortcomings are given more and more attention and it has become the main focus of research in computer world.

4.1 Small Files Problem

A major portion of the Internet data is represented by large number of small files. The famous distributed file system, like HDFS and GFS are designed to deal with the large files and thus they lack efficiency in accessing the smaller files.

Table 2 depicts the drawbacks associated with small file management for traditional DFS.

- Small files possess very poor disk I/O performance due to its high frequency access.
- As the files are relatively smaller in size, the fragmentation of files results in huge disk space wastage.
- One of the crucial drawbacks of small files is with their metadata management. They generates large amount of metadata, which becomes almost impossible task for the metadata server and engenders overall performance degradation.
- When creating links for each of the files, it leads to network delays. Table 2 shows some general method of optimization[20–27] for creation of link for each file. The paper, [28] shows that; if all the small files of a same directory are written to the same block, then the task speed of Map-Reduce can be increased by a large manner.

Table 2 Classification and analysis of algorithms for small files problems

Method	Illustration	Advantage	Disadvantage
Metadata Management	Metadata compression reduces metadata size	Enhance space utilization	The metadata lookup performance is poor due to extra steps involved
Performance Optimization	Utilizing caching technologies and pre-fetching to ameliorate the access efficiency e.g. Hot Files Caching, Metadata Caching	The betterment of cache hit ratio	Just for particular application
Small file merging	A set of correlated files is combined into a single large file to reduce the file count	Reduce the metadata content	Extra indexes, affect the speed
Sequence files	Form by a series of binary, where key is the name of the file, the value of the file content	Free access for small files, but did not restrict how much users and files	Platform dependent
Way to store	Small files stored separately in separate areas	Reduced disk fragmentation	Complexity of the movement

4.2 Load Balancing

Numerous load balancing algorithms have been proposed to get rid of unbalancing load, long reaction time and data traffic congestion among multiple nodes of a cluster.

Two ways of load balancing is proposed in [29–32]. One of which is through prevent, other is by migration. All I/O requests are uniformly distributed throughout the cluster such that every node gets almost equal responsibility. With proper load balancing, no nodes get overloaded or under loaded under any condition.

Another process is to regulate after the imbalance happened. After the load imbalance occurs, it is removed effectively by copy or migrate data among other servers. Generally speaking, after many years of research in the field of load balancing, its content has become relatively rich for big data. There are two types of algorithms: dynamic load balancing and static load balancing.

In dynamic load balancing, the loads among the nodes are adjusted based on the current load of the overall system. Few definitive dynamic load balancing algorithms are source address hash algorithm, weighted least connection algorithm, minimum connection algorithm and destination address hash algorithm.

On the other hand, static load balancing algorithms don't track the current situation of the system. Rather, it allocates the tasks based on some pre-established strategy or experience. The algorithm has blindness, as it does not consider the dynamic changes of the cluster. These kinds of algorithms are generally suitable for homogeneous and small service systems. Few classic static load balancing algorithms are, priority algorithm, polling algorithm and ratio algorithm.

The algorithms of dynamic load balancing can be categorized into distributed and centralized type. Table 3 portraits the advantages and disadvantages between them.

In distributed file system, the load balancing algorithms again can be categorized into recipient starter method and sender starter method. The recipient starting method starts from the under loaded nodes, which take away the excess loads from overloaded nodes. This method is worthwhile when the whole cluster is in a state of overload, so overloaded nodes are easy to found and hence frequent migration of data doesn't happen. Whereas, in sender starting method, a part of load from the overloaded nodes is transferred to the under loaded one. This phenomenon is suitable when the whole is in under load condition and follows a similar reason stated above.

Above mentioned load balancing approaches have both advantages and disadvantages, which lead to some mix procedures of all. The approach in [29] uses three queues (overload, optimal load and light load) in the master server. It uses priority algorithms and weighting polling algorithms on the queues. An important characteristic of centralized load balancing is its low communication overhead, which is certainly important for massive data storage. There after lots of researches, increases its reliability by altering the system structure. A general way of

Table 3 Comparison between centralized and distributed strategy

	Centralized strategy	Distributed strategy
Principle	In dynamic load balancing, all the operations are centralized to a single special server. It becomes its sole responsibility to collect load information and how to maintain the whole system	No such special load management servers, so each node need to collect, analyze and manage the load information of the neighbor nodes
Advantages	Communication costs are relatively low and simple. It's very much preferable for large-system model	High scalability; high reliability; failure of any node doesn't hamper the working condition of the whole system
Disadvantages	Tough to increase scalability, poor in reliability; once the single server fails, the complete load balancing system becomes offline. With the increase in size of the system, the complexity in management also increases	Here, the communication cost of load exponentially increases. Each server has a tiny amount of load; the load regulation is local whose result may not be satisfactory; so, a chance to engender the load to move back

decomposing complex load balancing is done by grading or layering and accomplished by several servers.

In [30], a dynamic load balancing of hierarchical strategy was introduced. A layer was designed to accumulate load status information, between the clients and metadata server. The purpose of this is to reduce the cost of load information acquisition. In the load management, an addition of backup server also increases the reliability of the system.

In [31], an algorithm that involves two stage metadata server filesystem was proposed for proper load balancing. The task processing and task allocation function of the metadata server were decentralized to improve the extensibility and parallel processing capability of the system.

In dynamic environment, the load balancing processes are quite complex. There are many other factors associated with it, other than the scheduling policy. Various questions like, which type of information should be collected, how should the information be collected, the time of task migration, finding the node where the task to be migrated etc. are to be answered. HDFS's uses its equilibrium strategy for single evaluation index of disk usage that has multiple loopholes. The paper in [32] uses multiple attribute to calculate the evaluation function and double threshold value for evaluation of server's load condition.

All the above mentioned methods could not properly solve the bottlenecks of threshold. When the system load is lighter on average and if the pre-determined set threshold is too high, then a part of the nodes will become busy while the other will turn idle. If the threshold is set very low, but the average load is on the higher side, then frequent load migration will appear a lot. So, a pre-determined threshold is dangerous and leaves the system in highly unstable condition. A threshold value based on the current load condition is appropriate and arises as a significant research problem.

4.3   De-duplication

Various types of redundant data might exist in a distributed file system. Those are, exact file headers of same file types, non-identical versions of a unified file, different users uploading same files and so on. De-duplication is really effective technology mainly for optimizing storage capacity. It removes redundant data in the data set, leaving a single copy of it, hence eliminating all duplicate data from the storage.

The conventional procedure to determine the duplicate data block is the way how the hash value is computed. If the hash-code of any data blocks matches with another existing value in the hash table, it's an indication that the block is redundant and not needed to be stored. As the indices increases, the memory efficiency declines drastically. So, the current research focus is to find efficient algorithms to reduce the redundant data and perform data-duplication at higher speed.

Other key technology is the partitioning of data. While the detection of duplicate in the whole file might be very fast, but if same data remains in many different files, then the redundant data of the files can't be removed. So, a file should be divided into block units before they are written into the file system. In [33], five crucial chunking algorithms of data de-duplication are proposed and a comparison of their efficiently on real data set is shown.

De-duplication helps to control the drastic growth of data effectively. It's also helps in increasing effecting storage space, improves storage efficiency and so on. However, with de-duplication, the reliability of data gets affected. If several files rely on a particular data block, then removing that block with eventually damage those files together. So, performing data-duplication at first, and then creating a backup for the dataset with a low access rate, will ensure reliability and no wastage of storage space.

5   Conclusion

The paper provides an overview of different technologies, approaches, advantages and disadvantages of various aspects of big data. Our research shows that, though many techniques, tools and data are available in the world of data science, still there are various points that are needed to be analyzed, developed and discussed for the betterment.

Despite the huge popularity of big data in recent computer world, the study of its still not yet completed and many more things are needed to be explored. There should be lot more of countermeasures in the related field to overcome the bottlenecks of big data. This literature paper explores the key technologies of big data in distributed file system and provides a platform for the further study.

References

1. Dev, D., Baishnab, K.L.: A review and research towards mobile cloud computing. In: 2014 2nd IEEE International Conference on Mobile Cloud Computing, Services, and Engineering (MobileCloud), pp. 252, 256, 8–11 Apr 2014
2. Eaton, C., Deroos, D., Deutsch, T., Lapis, G., Zikopoulos, P.C.: Understanding Big Data: Analytics for Enterprise Class Hadoop and Streaming Data. Mc Graw-Hill Companies, New York (2012). ISBN 978-0-07-179053-6
3. Schneider, R.D.: Hadoop for Dummies, Special Edition. Wiley, Canada (2012). ISBN 978-1-118-25051-8
4. Intel IT Center.: Planning Guide: Getting Started with Hadoop. Steps IT Managers Can Take to Move Forward with Big Data Analytics (2012). http://www.intel.com/content/dam/www/public/us/en/documents/guides/getting-started-with-hadoop-planning-guide.pdf
5. Singh, S., Singh, N.: Big data analytics. In: 2012 International Conference on Communication, Information & Computing Technology Mumbai India, IEEE (2011) http://hpccsystems.com/. Accessed 11 Mar 2013
6. http://hpccsystems.com/. Access 11 Mar 2013
7. Manyika, J., Chui, M., Brown, B., Bughin, J., Dobbs, R., Roxburgh, C., Byers, A.H.: Big data: the next frontier for innovation, competition, and productivity. McKinsey Global Institute (2011). http://www.mckinsey.com/~/media/McKinsey/dotcom/Insights%20and%20pubs/MGI/Research/Technology%20and%20Innovation/Big%20Data/MGI_big_data_full_report.ashx
8. Gerhardt, B., Griffin, K., Klemann, R.: Unlocking value in the fragmented world of big data analytics. Cisco Internet Business Solutions Group (2012). http://www.cisco.com/web/about/ac79/docs/sp/Information-Infomediaries.pdf
9. https://www.youtube.com/yt/press/en-GB/statistics.html
10. http://www.humanfaceofbigdata.com/. Accessed 11 Mar 2013
11. Tankard, C.: Big data security. Network Security Newsletter, Elsevier (2012). ISSN 1353-4858
12. Dean, J., Ghemawat, S.: MapReduce: simplified data processing on large clusters. CACM **51** (1), 107–113 (2008)
13. Apache, Hadoop.: Open source implementation of MapReduce. http://hadoop.apache.org
14. Ghemawat, S., Gobioff, H., Leung, S.T.: The Google file system. In: Proceedings of ACM SOSP (2003)
15. Apache.: HDFS Architecture Guide. Apache Software Foundation, Canada (2008)
16. Dev, D., Patgiri, R.: Performance evaluation of HDFS in big data management. In: 2014 International Conference on High Performance Computing and Applications (ICHPCA), pp. 1, 7, 22–24 Dec 2014
17. Cattell, R.: Scalable SQL and NoSQL data stores. ACM SIGMOD Rec. **39**(4), 12–27 (2010)
18. Lee, K.H., Lee, Y.J., Choi, H., Chung, Y.D., Moon, B.: Parallel data processing with MapReduce: a survey. ACM SIGMOD Rec. **40**(4), 11–20 (2011)
19. Ci, X., Meng, X.: Big data management: concepts, techniques and challenges. J. Comput. Res. Dev. **50**, 146–169 (2013)
20. Li, X., Dong, B., Xiao, L. Ruan, L., Ding, Y.: Small files problem in parallel file system. In: 2011 International Conference on Network Computing and Information Security, NCIS 2011, pp. 227–232. Guilin, Guangxi, China, 14–15 May 2011
21. Dong, B., Zheng, Q., Tian, F., Chao, K., Ma, R., Anane, R.: An optimized approach for storing and accessing small files on cloud storage. J. Netw. Comput. Appl. **35**, 1847–1862 (2012)
22. Dong, B., Qiu, J., Zheng, Q., Zhong, X., Li, J., Li, Y.: A novel approach to improving the efficiency of storing and accessing small files on Hadoop: a case study by PowerPoint files. In: 2010 IEEE 7th International Conference on Services Computing, SCC 2010, pp. 65–72. Miami, FL, United States, 5–10 July 2010

23. MacKey, G., Sehrish, S., Wang, J.: Improving metadata management for small files in HDFS. In: 2009 IEEE International Conference on Cluster Computing and Workshops, CLUSTER '09. New Orleans, LA, United States, 31 Aug–4 Sept 2009
24. Chandrasekar, S., Dakshinamurthy, R., Seshakumar, P.G., Prabavathy, B., Babu, C.: A novel indexing scheme for efficient handling of small files in Hadoop distributed file system. In: 2013 3rd International Conference on Computer Communication and Informatics, ICCCI 2013. Government of India, Department of Science and Technology, Council for Scientific and Industrial Research (CSIR), Coimbatore, India, 4–6 Jan 2013
25. Zhang, Y., Liu, D.: Improving the efficiency of storing for small files in HDFS. In: 2012 International Conference on Computer Science and Service System, CSSS 2012, pp. 2239–2242. Nanjing, China, 11–13 Aug 2012
26. Li, X., Dong, B., Xiao, L., Ruan, L.: Performance optimization of small file I/O with adaptive migration strategy in cluster file system. In: 2nd International Conference on High-Performance Computing and Applications, HPCA 2009, pp. 242–249. Shanghai, China (2010), 10–12 Aug 2009
27. Mohandas, N., Thampi, S.M.: Improving Hadoop performance in handling small files. In: 1st International Conference on Advances in Computing and Communications, ACC 2011, pp. 187–194. Kochi, India, 22–24 July 2011
28. Liu, J., Bing, L., Meina, S.: The optimization of HDFS based on small files. In: 2010 3rd IEEE International Conference on Broadband Network and Multimedia Technology, IC-BNMT2010, pp. 912–915. Beijing, China, 26–28 Oct 2010
29. Zhang, C., Yin, J.: Dynamic load balancing algorithm of distributed file system. J. Chin. Comput. Syst. 32, 1424–1426 (2011)
30. Wu, W.: Research on Mass Storage Metadata Management, vol. D. Huazhong University of Science and Technology, Wuhan (2010)
31. Tian, J., Song, W., Yu, H.: Load-balance policy in two level cluster file system. Comput. Eng. 33, 77–79, 82 (2007)
32. Gu, F.: Research on Distributed File System Load Balancing in Cloud Environment, vol. D. Jiaotong University, Beijing (2011)
33. Cai, B., Zhang, F.L., Wang, C.: Research on chunking algorithms of data de-duplication. In: International Conference on Communication, Electronics, and Automation Engineering, 2012, pp. 1019–1028. Xi'an, China (2013)

Privacy Preserving Association Rule Mining in Horizontally Partitioned Databases Without Involving Trusted Third Party (TTP)

Chirag N. Modi and Ashwini R. Patil

Abstract In this research work, we design a security protocol that derives association rules securely from the horizontally distributed databases without a trusted third party (TTP), even communication channel is unsecured between involving sites. It ensures privacy and security of owner's with the help of elliptic curve based Diffie-Hellman and Digital Signature Algorithm.

Keywords Distributed association rule · Privacy · Elliptic curve cryptography

1 Introduction

Data mining played a vital role for extracting potentially hidden information from the large collection of datasets. It has been widely demonstrated over centralized and distributed data with numerous significant applications (refer Fig. 1). The global mining result over distributed data is derived by exchanging data among all sites. Data distribution among sites may have following types. The scenario of horizontal distribution of data is that each of the involving has different number of data records but same set of attributes, which in case of vertical distribution of data, each site has same number of data record but different set of attributes. However, these techniques pose a threat to privacy and security of individual site's data as each of the involving site exchanges its data among other sites. For better motivation, consider following real time scenario: Assume that the data of one corporation is distributed among their sites. For deriving business statistics, owner of that corporation needs to perform mining over the global data from all sites. Here, each

C.N. Modi (✉)
National Institute of Technology Goa, Farmagudi, Goa, India
e-mail: cnmodi@nitgoa.ac.in

A.R. Patil
National Institute of Technology, Surat, India
e-mail: ashwinipatil29@gmail.com

© Springer India 2016   549
A. Nagar et al. (eds.), *Proceedings of 3rd International Conference on Advanced Computing, Networking and Informatics*, Smart Innovation, Systems and Technologies 44, DOI 10.1007/978-81-322-2529-4_57

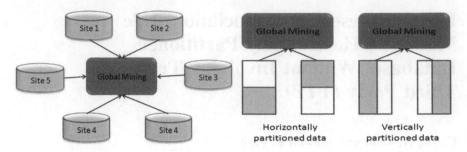

Fig. 1 Distributed data mining scenarios

site exchange its data with data miner. However, such data exchange may pose a privacy threat to data of sites.

To address such privacy problem in distributed data mining, we propose a security protocol that derives privacy preserving association rules (a data mining technique) over horizontally distribution of data without involving a trusted third party (TTP). It ensures data privacy and security with the help of elliptic curve based Diffie-Hellman (ECDH) and Digital Signature Algorithm (ECDSA).

The organization of this paper is as follows: Sect. 2 discusses the paradigm of distributed mining and investigate the existing approaches. The proposed security protocol is presented in Sect. 3. Section 4 analyzes the security and privacy of the proposed protocol. Section 5 concludes our research work with references at the end.

2 Literature Survey

Deriving Association Rule in Horizontally Distributed Databases [1]. Let $A = \{a_1, \ldots, a_n\}$ is n attributes of items in Dataset D that is horizontally distributed in m different sites like $Site_1$, $Site_2$, ..., $Site_m$. Each site has D_i ($1 \leq i \leq m$), where each D_i has same attributes $\{a_1, \ldots, a_n\}$ but different number of records. Now, the itemset $I \subseteq A$ has support count $L_i(I)$ at $site_i$ if I is included in $L_i(I)$ of the records at $site_i$. The overall support count of I is given as $G_s(I) = \sum_{i=1}^{m} L_i(I)$. An itemset $I \subseteq A$ is overall supported if $G_s(I) \geq MST \times \sum_{i=1}^{m} |D_i|$, where MST is the minimum support threshold and $|D_i|$ is number of records at $site_i$. Association rule is an implication formula like $I \Rightarrow R$, where $I \subseteq A$, $R \subseteq A$. The rule has overall support and confidence, if $G_s(I \Rightarrow R) \geq MST \times \sum_{i=1}^{m} |D_i|$ and $G_s(I \Rightarrow R) \geq MCT \times G_s(I)$, where MCT is the minimum confidence threshold.

Privacy Problem. In unsecured distributed association rule mining environment, the privacy problem is to collect data from the involving sites and derive global association rules that must satisfy the given MST and MCT with following objectives: The privacy and security of any site's data should not be affected either by any other sites or by an external adversary through monitoring communication channel among sites. In addition, this global mining process should have low communication and computational cost.

Existing Privacy Approaches. For centralized data, many rule hiding approaches have been proposed till date [2, 3]. However, for distributed data, Vaidya and Clifton [4] introduced in order to preserve privacy of association rules, while mining them over a vertically distributed datasets. Kantarcloglu and Clifton [1] have presented an approach to derive overall association rules in horizontally distributed datasets by deriving the union of candidate association rules securely. Liu *et al.* [5] have used hashing of candidate itemsets and filter out the itemset whose support exceeds the minimum thresholds. In [6], extended Distributed Rk-Secure Sum Protocol that is used for deriving association rule securely, while in [7] Chaturvedi *et al.* have used multi-party computation approach in order to derive the combinations of subsets securely, and it is checked whether an item available at a site is also available within a subset at another site. In [8], the authors have used Sign based secure sum concept and scalar product technique for mining privacy preserving association rules over mixed partitioned data. Borhade and Shinde [9] have proposed a condensation based approach which uses secure multi party computation to protect data privacy. DARM [10] finds the global strong association rules confidentially, and satisfies ε-differential privacy.

Existing approaches find global association rules by offering privacy. They assume that communication channel for data exchange is secure. However, in real life, practically the communication channel is unsecured and thus, an external adversary can monitor communication channel among sites and may threat to the global mining result or information privacy. Thus, there is a need of protecting data privacy against unauthorized access. Moreover, TTP may pose privacy threat to global mining result.

3 Privacy Preserving Distributed Mining of Association Rules: Our Proposal

3.1 Elliptic Curve Cryptography: Basic Concepts

Elliptic curve based cryptosystem has security strength of discrete logarithm problem over integer modulo a prime. It offers comparative security as RSA with shorter key length [11].

ECDH protocol for key sharing. Let, there are two sites ($Site_1$ and $Site_2$) which are agreed on elliptic curve $Ez(z, a, b)$, z is a large prime number and base point g with an order of n such that $ng = O$. $Site_1$ and $Site_2$ generate random numbers (their private key) R_1 and R_2 respectively, where $R_1, R_2 < n$. Now $Site_1$ generates a public key (Y) by adding base point (g), R_1 times, where $Site_2$ generates a public key (Y') by adding base point (g), R_2 times. Then both sites send their public key to each other and compute shared key by their private key times adding received public key. As ECDH uses addition instead of exponentiation it is fast for key exchange.

ECDSA for authentication and verification: Let $Site_1$ selects an elliptic curve $E(p, a, b)$. Now a large integer i in the interval $[1, n - 1]$ is selected as private key of $Site_1$ and the point P of order n is selected to compute public key $Q = P_{ri}P$ of $Site_1$. Thus, $Site_1$ contains a public key (E, z, n, Q) and private key P_{ri}. For authentication, it generates a signature for the message (e.g. M) using following algorithm:

Signature generation at $Site_1$: First a random number k is selected in $[1, n - 1]$ and $kP = (x1, y1)$ is calculated. Then taking $r = x1 \bmod n$, $s = k - 1$ (*hash* $(M) + P_{ri}r$) mod n is derived. The (r, s) is taken as a digital signature of message M, which is to be sent to $Site_2$ for verification.

Digital signature verification at $Site_2$: $Site_2$ obtains signed message M and public key (E, z, n, Q) from $Site_1$. For signature verification, first $w = s^{-1} \bmod n$ and *hash* (M) is derived. Then $U_1P + U_2Q = (x_2, y_2)$ is found, where $U_1 = hash(M)w \bmod n$ and $U_2 = rw \bmod n$. Finally, it finds $v = x_2 \bmod n$ and agree to signature, if $v = r$.

In proposed protocol, we use ECDH for key sharing and to encrypt support count of itemsets (before sending to other site), where ECDSA is used for authentication and verification between involving parties.

3.2 Proposed Protocol

Let there are five semi honest sites, (one master site e.g. mining expert, and other local sites) having homogeneous databases D_1, D_2, D_3, D_4 and D_4 respectively (refer Fig. 2). For finding global association rules, master site needs global support of each itemset (e.g. X). At the same time, any transaction data or support count of any item should not be revealed to any involving site. To address this, the proposed communication protocol works in following phases:

Phase 1: Key sharing between *master site* and local sites: Each site S_i ($1 \le i \le n$) shares a key Key_i ($1 \le i \le n$) with master site using ECDH.

Phase 2: Finding global transaction count with privacy preservation: Here, number of global transactions are counted (step 1–4 in Fig. 2) securely. $Site_1$ encrypts its transaction count $|D_1|$ by adding Key_1 and sends ($|D1| + Key_1$) to $Site_2$ after digitally signing it with the help of ECDSA. $Site_2$ verifies signature and adds in its encrypted count of transactions ($|D2| + Key_2$) and sends ($|D1| + Key_1 + |D2| + Key_2$) to $Site_3$. $Site_3$ and $Site_4$ follow same procedure as $Site_2$, and sends final digitally signed ($|D1| + Key_1 + |D2| + Key_2 + |D3| + Key_3 + |D4| + Key_4$) to *master site*. *Master site* verifies signature by using public key of $Site_4$ and calculates global count of transaction by subtracting summation ($Key_1 + Key_2 + Key_3 + Key_4$) and adding its transaction count $|D_5|$. Thus, *master site* gets global $Gs(|D|) = (|D1| + Key_1 + |D2| + Key_2 + |D3| + Key_3 + |D4| + Key_4) - (Key_1 + Key_2 + Key_3 + Key_4) + |D_5|$.

Phase 3: Finding frequent itemset with privacy preservation: Here, global support count G_s is counted for each itemset (Steps 5–8 in Fig. 2). $Site_1$ sends

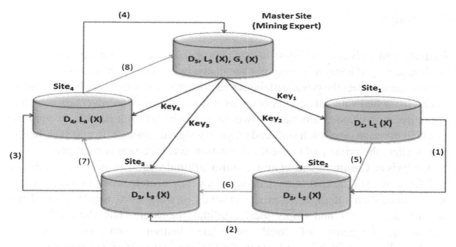

Fig. 2 Proposed protocol

encrypted local support count of item X ($L_1(X) + Key_1$) to $Site_2$, after signing it. $Site_2$ verifies signature and adds in its encrypted local support count of X ($L_2(X) + Key_2$) and sends ($L_1(X) + Key_1 + L_2(X) + Key_2$) to $Site_3$ with digital signature. $Site_3$ and $Site_4$ follow same procedure as $Site_2$, and sends final digitally signed ($L_1(X) + Key_1 + L_2(X) + Key_2 + L_3(X) + Key_3 + L_4(X) + Key_4$) to *master site*. *Master site* verifies signature and calculates global support count of X ($G_s(X)$) by subtracting summation ($Key_1 + Key_2 + Key_3 + Key_4$), and adding its local support count of X, $L_5(X)$. Thus, *master site* gets global support count of X, $G_s(X) = (L_1(X) + Key_1 + L_2(X) + Key_2 + L_3(X) + Key_3 + L_4(X) + Key_4) - (Key_1 + Key_2 + Key_3 + Key_4) + L_5(X)$. Then, *master site* calculates frequency of X by $G_s(X)/Gs(|D|)$. This procedure will be followed to get frequency of each itemset.

Phase 4: Finding global association rules: After finding the frequency of each itemset, *master site* generates association rules with the help of association rule mining technique (e.g. Apriori Algorithm) [12]. An algorithm is shown below:

```
Input: n homogeneous databases Dᵢ(1 ≤ i ≤ n) belong to each sites Sᵢ
(n ≥ 3) respectively, MST and MCT
Output: Global Association rules over D, where D = Σⁿᵢ₌₁|Dᵢ|
Algorithm:
    1) Master site shares key with all the local sites
    2) Master site calculates global count of transactions (Gₛ(|D|))
    3) Master site calculates global support count of all itemsets
    4) Master site finds global frequent itemsets whose global
       support Gₛ ≥ MST × Gₛ(|D|)
    5) Finally, Master site finds global association rules like X⇒Y
       whose global support Gₛ ≥ MST × Gₛ(|D|) and global confidence
       Gₑ≥ MCT × Gₛ(Y), where X, Y are globally frequent itemsets.
```

4 Analysis

Security and Privacy Analysis. In our approach, each site exchanges encrypted local frequency of items with neighbor site instead of original transaction data. Thus, any site cannot affect the privacy of the actual contents of transactions. It is difficult to predict the actual value of support count as it is encrypted. If an adversary monitors communication channel among any two sites, he/she will not be able to send the modified message since each site sends digitally signed message. In addition, he/she cannot affect the privacy of the sent information since the sent is encrypted.

Analysis of Communication and Computational Cost. Let there are n sites and m different items. The cost of exchanging frequency of transaction to master site is $(n - 1)$ since each site (apart from *site* 1) sends message to its neighbor site. Each site contains $2^m - 1$ non empty local candidate itemsset. Thus, the total cost of exchanging frequency of local candidate itemset with master site is $(n - 1) \times (2^m - 1)$, which is resulted into $O(2^m)$ since n is negligible compared to 2^m.

Let computation cost for generating shared key and digital signature are C_1 and C_2 respectively. The cost for deriving non-empty candidate itemsets at each site is m^2, and total cost for encrypting and signature generating all non-empty candidate itemsets is $2^m(C1 + C2)$. Thus, the total cost for generating encrypted and digitally signed candidate itemsets are $O(n(2^m + m^2))$ since $(C_1 + C_2) \ll 2^m$. Thus, the total cost for generating association rules is $O(2^m)$.

5 Conclusions

We have introduced a security protocol that derives privacy preserving association rules over horizontally partitioned data without using TTP, where communication channel between involving sites is unsecured. Here, ECDH helps to protect confidentiality of each site's data, while ECDSA helps to improve security of data exchanged between sites. Proposed algorithm offers security and privacy of each involving site's data and mines global association rules securely. In addition, it has affordable communication and computational cost.

References

1. Kantarcioglu, M., Clifton, C.: Privacy-preserving distributed mining of association rules on horizontally partitioned data. Trans. Knowl. Data Eng. **16**, 1026–1037 (2004)
2. Modi, C., Rao, U.P., Patel, D.R.: A survey on preserving for sensitive association rules in databases. In: Recent Trends in Business Administration and Information Processing, pp. 538–544 (2010)
3. Modi, C.N., Rao, U.P., Patel, D.R.: Maintaining and in privacy preserving association rule. In: International Conference on Computing, Communication and Networking Technologies, pp. 1–6 (2010)

4. Vaidya, J., Clifton, C.: Privacy preserving association rule mining in vertically partitioned data. In: ACM SIGKDD Conference on Knowledge Discovery and Data Mining, pp. 639–644 (2002)
5. Liu, J., Piao, X., Huang, S.: A privacy-preserving mining algorithm of association rules in distributed databases. In: IMSCCS **2**, 746–750 (2006)
6. Mathews, M.T., Manju, E.V.: Extended distributed rk—secure sum protocol in apriori algorithm for privacy preserving. Int. J. Res. Eng. Adv. Technol. **2**(1), 1–5 (2014)
7. Chaturvedi, G.K., Gawande, R.M.: Privacy preserving association rules mining in horizontally distributed databases using FDM and K and C algorithm. Int. J. Eng. Dev. Res. **2**(3), 3334–3337 (2014)
8. Muthu Lakshmi, N.V., Rani, K.S.: Privacy preserving association rule mining without trusted party for horizontally partitioned databases. Int. J. Data Min. Knowl. Manage. Process **2**(2), 17–29 (2012)
9. Borhade, S.S., Shinde, B.B.: Privacy preserving data mining using association rule with condensation approach. Int. J. Comput. Sci. Inf. Technol. **5**(2), 1560–1563 (2014)
10. Wahab, O.A., Hachami, M.O., Zaffari, A., Vivas, M., Dagher, G.G.: DARM: a privacy-preserving approach for distributed association rules mining on horizontally-partitioned data. In: 18th International Database Engineering and Applications Symposium, pp. 1–8 (2014)
11. Koblitz, N.: Elliptic curve cryptosystems. Math. Comput. **48**, 203–209 (1987)
12. Han, J., Kamber, M.: Data Mining: Concepts and Techniques. Morgan Kaufmann Publishers Inc., San Francisco (2001)

Performance Efficiency and Effectiveness of Clustering Methods for Microarray Datasets

Smita Chormunge and Sudarson Jena

Abstract Numerous of clustering methods are proficiently work for low dimensional data. However Clustering High dimensional data is still challenging related to time complexity and accuracy. This paper presents the performance efficiency and effectiveness of K-means and Agglomerative hierarchical clustering methods based on Euclidean distance function and quality measures Precision, Recall and F-measure for Microarray datasets by varying cluster values. Efficiency concerns about computational time required to build up dataset and effectiveness concerns about accuracy to cluster the data. Experimental results on Microarray datasets reveal that K-means clustering algorithm is favorable in terms of effectiveness where as efficiency of clustering algorithms depends on dataset used for empirical study.

Keywords Clustering · K-means · Agglomerative hierarchical · F-measure · Precision · Recall

1 Introduction

To extract useful information from large sum of data is a main purpose of data mining. Classification and Clustering is a basic task of data mining. Clustering is unsupervised learning and Classification is supervised learning method. Now a days clustering is used in number of fields such as Clustering high-dimensional data like genes data [1], for mining text data, pattern recognition, geodatabase applications, heterogeneous data analysis, mining Web data, and image analysis, also in

S. Chormunge (✉)
Department of Computer Science, GITAM University, Hyderabad, India
e-mail: smita2728@rediffmail.com

S. Jena
Department of Information Technology, GITAM University, Hyderabad, India
e-mail: sudarsonjena@gitam.edu

© Springer India 2016
A. Nagar et al. (eds.), *Proceedings of 3rd International Conference on Advanced Computing, Networking and Informatics*, Smart Innovation, Systems and Technologies 44, DOI 10.1007/978-81-322-2529-4_58

information retrieval field, statistical data analysis and biomedical data. Main task of Cluster analysis is to group set of objects based on similarity between objects. It groups the objects in one group which are similar but different than other group of objects and the objects which are not fit any group is outliers. Better clustering achieved if the similarity within a group and the difference between groups are superior [2].

Traditional clustering algorithms find challenging to cluster large dataset which is computationally expensive. Datasets can be large in different ways; when huge number of objects in the dataset, each object can have many attributes, and there can be many clusters to discover. Clustering algorithms are generally classified into partitioning and Hierarchical method. Partitioning method subdivided into K-means and K-medoids algorithms. Some partitioning methods CLARA and CLARANS work well for large datasets. Hierarchical method classified into Agglomerative and Divisive. Several hierarchical methods BIRCH, Chameleon, ROCK and CURE work efficiently for clustering huge amount of data. Hierarchical clustering group's data objects either from number of clusters to one cluster or singleton cluster divide into number of clusters without the hierarchical structure [3]. Some more categories of clustering methods are Grid-based method, Model based clustering, Constrained based clustering and Density based clustering method. There are certain issues to cluster the high dimensional data such as to work with data which have many attributes, handling outliers, storing and visualizing data and computational time complexity is also a big issue.

In past year research reported that for document clustering, hierarchical clustering is best quality clustering approach but there is limitation because of its Computational time. Time complexity of K-means method is linear as per the number of documents, although it produces inferior clusters. Combining K-means and agglomerative hierarchical approaches get better results [4]. Most of work has been done in Clustering low dimensional data, when considering genes data which has high dimensional data these clustering methods may fails to handle such kind of data. To predict value from such data is very difficult task because it affects the accuracy and complexity cost of clustering algorithms.

This paper evaluates the computational time and accuracy to cluster the data of K-means and Agglomerative hierarchical clustering methods for Microarray datasets. We used quality metrics f-measure, precision and recall and Euclidean distance function for measuring computational time. For empirical study we use well known Microarray datasets. Experimentally we evaluated these two clustering methods for computational time to form a clusters and accuracy to cluster the gene data. We analyzed datasets for different number of cluster values for each dataset for both clustering methods.

Section 2 describes the review of clustering methods and Evaluation metrics. In Sect. 3 we present performance evaluation of clustering methods on Microarray datasets. Experimental results discuss in Sect. 4. Conclusion of work discuss in Sect. 5.

2 Clustering Methods

In this section we describe a review of K-means and Agglomerative hierarchical clustering methods.

2.1 K-Means Clustering Method

The k-means algorithm is one of the popular and easiest clustering method for implementation. It is partition based clustering method and used in different applications. K-means clustering method form groups for cluster objects without any prior knowledge of those objects relationships.

The k-means Algorithm [5]

(1) Arbitrarily selects m as initial centers $C = \mathbf{c}_1,..., \mathbf{c}_m$,
(2) Set the cluster $\mathbf{c}_i$ for each $i \in \{1, ..., m\}$, to be the set of points in χ which is closer to $\mathbf{c}_i$ than the cluster $\mathbf{c}_j$ for all $j \neq i$.
(3) For each $i \in \{1,..., m\}$, set center point in a set C_i of cluster i to $\mathbf{c}_j$. cluster center ci is recomputed as $ci = \frac{1}{|Ci|}\sum_{x \in Ci} x$
(4) A step (2) and (3) repeats until there is no further change in $\mathbf{c}_i$.

K-means algorithm is popular because of its linear complexity. The complexity of k-means is $O(T * s * m * N)$ where T iterations performed on a sample size of s instances, for N attributes. Its adaptability to sparse data and proficient speed of convergence is also one reason of popularity of K-means [6]. But this method has some drawbacks, main drawback is that user have to give cluster values, the algorithm is unable to determine the appropriate number of clusters. To provide input values to the algorithm, user has to define the cluster values in advanced. For better results, user has to experiments with different values of clusters and finds the value which is suitable to their data. It handles nominal data and numerical data efficiently but inefficient to work with categorical data.

2.2 Hierarchical Clustering

Hierarchical clustering constructs a hierarchy like a tree which can be represented by graphical way, called dendrogram. The branches of tree keep track of clusters and also show the similarity between the clusters. By cutting the dendrogram at some level, we get specified number of clusters, it arrange the similar objects together. Hierarchical clustering mainly classified into two types [1]:

- *Agglomerative*: It is also called as bottom up approach: It start with each object forming its individual group. For every pair of clusters the similarity distance are calculated and based on this criterion, clusters are merged until termination condition reached. Merging of clusters is based on Euclidean distance between any two objects from different clusters.

- *Divisive*: This is also called as a "top down" approach: It initiates with considering all objects in one cluster then this cluster is splits into smaller cluster by calculating Euclidean distance between objects until termination condition reached.

Simple Agglomerative Clustering Algorithm

1. Initialize the cluster set considering each data object is an individual cluster.
2. Calculate the similarity between all pairs of clusters, i.e. calculate the similarity distance between the b and d clusters.
3. Merge the two clusters which are most similar.
4. Revise the similarity matrix to reproduce the pair-wise similarity between the old clusters and new cluster.
5. Steps 3 and 4 repeat until cluster criterion meets.

As per the similarity measure the hierarchical clustering methods could be further classified into [7] *Single-link clustering* also called the nearest neighbor method, the link between two clusters whose two elements are closest to each other is made by a single element pair. *Complete-link clustering* also known as diameter, it considers the distance between two clusters whose elements are similar in same cluster but different from other cluster elements. *Average-link clustering* also known as minimum variance method, it considers the mean distance between elements of each cluster.

High-computational complexity is one drawback of hierarchical clustering algorithm. Another is lack of robustness where small alteration in data changes a structure of the hierarchical dendrogram. In additional, the greedy nature of this method not allows the modification for previous clustering in both approaches agglomerative and divisive [1]. Initial step is important while taking decision of merging and splitting the cluster, once forwarded then it can never be corrected.

2.3   Evaluation Metrics

This section we describes the metrics for evaluating performance of clustering methods. Here we used F-measure as a quality measure and Euclidean distance function for efficiency point of view. Performance of different clustering method is depends on the measures to be used. Based on these measures we can consider the best clustering algorithm for our dataset which is being evaluated. For measuring the quality of clustering method two measures are used. One is internal quality measure which are not referring external knowledge and another is external quality measure clustering, it calculate the effectiveness of clustering methods to known classes, by comparing the groups created by the clustering algorithms. F-measure is one of the external metric which measures the effectiveness of clustering algorithms.

There are three categories for comparing clustering [8] the first is based on entropy measure or information based measures, it measures the information shared by two clustering. The second is computing recall, precision or other measures for the clusters which are most similar clusters. Third category is based on pair counting.

2.3.1 F Measure

F-measure is an external measure for accuracy. For computing F-score it depend on two factors Precision and recall. F-score is calculated by weighted average of recall and precision.

$$\text{Recall } (i, \ j) = N_{ij}/N_i$$
$$\text{Precision } (i, \ j) = N_{ij}/N_j$$

Define class i and cluster j, where N_j is elements of cluster j and N_i is the number of elements of class i. N_{ij} is the numbers of elements of class i in cluster j [9].

The F measure is calculated for class i and cluster j as follows

$$F(i, \ j) = \frac{(2 * \text{ Recall } (i, \ j) * \text{ Precision}(i, \ j \ ))}{(\text{Precision } (i, \ j \ ) + \text{Recall}(i, \ j \ ))}$$

F measure is computed as a result of weighted average of Precision and recall for each class i, as shown in Eq. (1).

$$Fc = \frac{\sum_i(|i| * F(i))}{\sum_i |i|} \tag{1}$$

where $|i|$ is the size of class i.

2.3.2 Euclidean Distance Function

Euclidean distance is distance between two points in Euclidean space. It is calculated by squared length of a vector $x = [x_1 \ x_2]$ shown in Eqs. (2) and (3), which is a coordinates summation of the squares. The two coordinates squared distance for example $x = [x_1 \ x_2]$ and $y = [y_1 \ y_2]$ is the sum of squared differences in their coordinates given in Eq. (4). Notation x, y is the vectors x and y, where as d refer the distance between two vectors x and y, it can be represented as [10]:

$$d_{x,y}^2 = (x_1 - y_1)^2 + (x_2 - y_2)^2 \qquad (2)$$

The distance between two vectors is the square root

$$d_{x,y} = \sqrt{(x_1 - y_1)^2 + (x_2 - y_2)^2} \qquad (3)$$

All coordinates of the vector are zero then it will be a zero vector $0 = [0\ 0]$. In such case the distance between the vector $x = [x_1\ x_2]$ and zero vector is given by

$$d_{x,0} = \sqrt{x_1^2 + y_1^2} \qquad (4)$$

The zero vector is also called as *origin* of the space and finally we can write $d_{x,0}$ as d_x.

3 Performance Evaluation of Clustering Methods

Here we discussed details of datasets. The summary of dataset used for empirical study shown in Table 1. We collected publically available well-known Microarray datasets. Our study is on four data sets such as SRBCT, Lymphoma, CNS and Leukemia for evaluation of clustering methods.

Gene selection problem is one of the typical application domains, clustering such high dimensional data is challenging task. We analyzed average time required to build up datasets and quality of two clustering methods; K-means and Agglomerative clustering method. These clustering methods evaluated on collected datasets which have different features, instances and classes shown in Table 1.

SRBCT is a Gene's data which contains 2308 features and 83 samples. It is taken from the microarray experiments of Small Round Blue Cell Tumors (SRBCT) [11]. Out of 83 samples 63 is training samples and 25 test samples. Lymphoma is a broad term encompassing a variety of cancers of the lymphatic system [11]. It contains total 4026 genes and the samples are 62. There are all together three types of lymphomas. The first category, Chronic Lymphocytic Lymphoma, the second type Follicular Lymphoma and the third type Diffuse Large B-cell Lymphoma. CNS [11] represents a heterogeneous group of tumors about which little is known biologically. It contains 7129 genes and 42 numbers of samples. The Leukemia data

Table 1 Summary of datasets used for empirical study

Datasets	Features	Instances	Class
SRBCT	2308	83	4
Lymphoma	4026	62	3
CNS	7129	60	2
Leukemia	7129	72	2

set [11] contains 7129 genes on 72 samples. Two sample variants of leukemia is (AML, 25 samples, or ALL, 47 samples).

For evaluation above datasets we use Weka software as data mining tool [12]. We uploaded datasets in software and calculated the computational time of each dataset, varying number of clusters as 10, 15 and 20 based on Euclidean distance function for both clustering method k-means and Hierarchical method. We calculated the Precision, Recall and F-measure for collected microarray datasets to observe quality of these clustering methods.

4 Results and Analysis

In this section we present the experimental results obtained for evaluation of clustering methods varying cluster values 10, 15 and 20 on Microarray data based on quality metrics and distance function. Table 2 shows the tabular representation of results obtain for K-means and Agglomerative hierarchical clustering method to evaluate 10 clusters based on quality measures Precision, Recall, F-measure and Euclidean distance function to calculate time. Here 'K' denoted as number of clusters. Tables 3 and 4 present the results of evaluation of clustering method for 15 and 20 cluster values respectively.

From Table 2 results we observed that K-means is good for quality and takes less time for evaluation of 10 numbers of clusters for each dataset. Whereas Agglomerative clustering method behaves poor in accuracy and takes more time to compute all datasets. Table 3 shows the results where k-means takes more time to evaluate SRBCT and CNS datasets than Agglomerative; in terms of accuracy

Table 2 Evaluation of clustering methods for clusters K = 10

Datasets	K-means				Agglomerative			
	Time	Precision	Recall	Fmeasure	Time	Precision	Recall	Fmeasure
SRBCT	1.88	0.609	0.614	0.607	3.05	0.374	0.36	0.238
Lymphoma	1.63	0.867	0.844	0.847	2.44	0.466	0.683	0.554
CNS	3.09	0.711	0.682	0.684	3.81	0.771	0.635	0.511
Leukemia	2.36	0.81	0.795	0.778	5.03	0.78	0.667	0.548

Table 3 Evaluation of clustering methods for clusters K = 15

Datasets	K-means				Agglomerative			
	Time	Precision	Recall	Fmeasure	Time	Precision	Recall	Fmeasure
SRBCT	4.89	0.857	0.816	0.811	3.28	0.436	0.379	0.246
Lymphoma	3.5	0.971	0.967	0.967	3.74	0.474	0.689	0.562
CNS	6.45	0.691	0.565	0.545	5.27	0.767	0.617	0.491
Leukemia	5.63	0.8	0.793	0.777	7.77	0.796	0.712	0.608

Table 4 Evaluation of clustering methods for clusters K = 20

Datasets	K-means				Agglomerative			
	Time	Precision	Recall	Fmeasure	Time	Precision	Recall	Fmeasure
SRBCT	2.7	0.889	0.8	0.796	2.69	0.413	0.387	0.256
Lymphoma	4.16	1	1	1	3.69	0.46	0.679	0.549
CNS	5.92	0.55	0.522	0.527	4.19	0.774	0.643	0.525
Leukemia	5.78	0.853	0.808	0.789	8.55	0.793	0.704	0.598

K-means is better. Table 4 represent the results of clusters 20, here also time and accuracy varies based on datasets. Both clustering method shows variation when we increase the cluster values.

Figures 1, 2 and 3 is a comparison graph of evaluation of clustering methods based on quality measure; F-measure for Microarray datasets for cluster values 10, 15 and 20 respectively. K-means method works effectively for cluster value 10. Even after increasing cluster values 15 and 20 K-means shows better performance than Agglomerative method.

Figures 4, 5 and 6 is a comparison graph of evaluation of clustering methods based on Time for Microarray datasets for clusters 10, 15 and 20 respectively. We observed some variation in results after increasing cluster values. For cluster values 15 and 20 there is variation in time for both clustering methods. From these observations we found that the K-means is better in accuracy point of view than Agglomerative clustering method. Efficiency point of view there is variation in results for both clustering methods.

Fig. 1 Comparison graph of clustering methods based on F-measure for K = 10

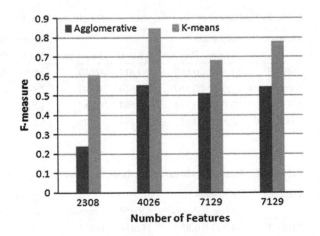

Fig. 2 Comparison graph of clustering methods based on F-measure for K = 15

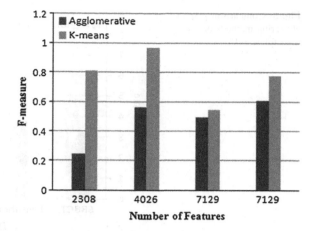

Fig. 3 Comparison graph of clustering methods based on F-measure for K = 20

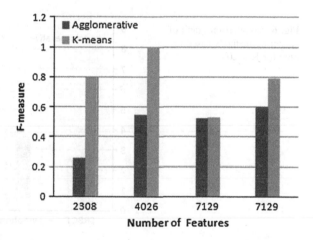

Fig. 4 Comparison graph of clustering methods based on time for K = 10

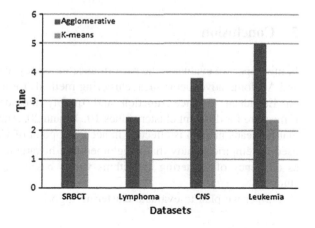

Fig. 5 Comparison graph of
clustering methods based on
time for K = 15

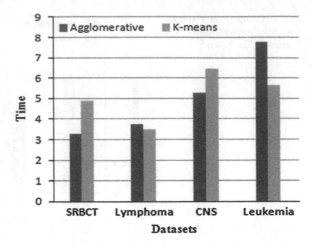

Fig. 6 Comparison graph of
clustering methods based on
time for K = 20

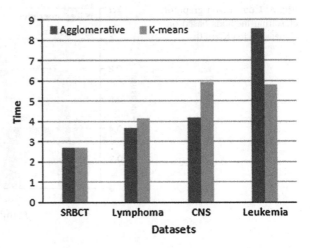

5   Conclusion

In this paper we evaluated the performance efficiency and effectiveness of K-means
and Agglomerative hierarchical clustering methods for high dimensional data based
on Euclidean distance function and quality measures Precision, Recall and
F-measure for different cluster values 10, 15 and 20. Analyzing all results we found
that K-means method is effective in accuracy point of view for Microarray datasets
used for empirical study than Agglomerative hierarchical clustering method where
as efficiency of clustering algorithms varies based on dataset used for empirical
study.

Further we plan to evaluate clustering algorithms for different quality metrics.

References

1. Jiang, D., Tang, C., Zhang, A.: Cluster analysis for gene expression data: a survey. IEEE Trans. Knowl. Data Eng. **16**(11), 1370–1386 (November 2004). doi:ieeecomputersociety.org/10.1109/TKDE
2. Steinbach, M., Ertöz, L., Kumar, V.: The challenges of clustering high dimensional data. In: New Vistas in Statistical Physics—Applications in Econophysics, Bioinformatics, and Pattern Recognition. Springer (2004)
3. Xu, R., Wunsch, D., Survey of clustering algorithms. IEEE Trans. Neural Netw. **16**(3), 645–678 (May 2005)
4. Steinbach, M., Karypis, G., Kumar, V.: A comparison of document clustering techniques. In: KDD Workshop on Text Mining (2000)
5. Onoda, T., Sakai, M., Independent component analysis based seeding method for k-means clustering. In: IEEE/WIC/ACM International Conferences on Web Intelligence and Intelligent Agent Technology (2011). doi:10.1109/WI-IAT.2011.29
6. Dhillon, I., Modha, D.: Concept decompositi-on for large sparse text data using clustering. Mach. Learn. **42**, 143–175 (2001)
7. Rokach, L., Maimon, O.: Clustering Methods Data Mining and Knowledge Discovery Handbook. Springer (2005)
8. Achtert, E., Goldhofer, S., Kriegel,H.-P., Schubert, E., Zimek, A.: Evaluation of clusterings—metrics and visual support. In: Proceedings of the 28th International Conference on Data Engineering (ICDE), Washington, DC (2012)
9. Bourennani, F., Pu, K.Q., Zhu, Y.: Visualization and integration of databases using self-organizing map. IEEE Int. Conf. Adv. Databases Knowl. Data Appl. 155–160 (2009). doi:10.1109/DBKDA.2009.30
10. Greenacre, M., Primicerio, R.: Measures of Distance between Samples: Euclidean, pp. 47–59. Fundacion BBVA Publication (December 2013). ISBN:978-84-92937-50-9
11. http://csse.szu.edu.cn/staff/zhuzx/Datasets.html
12. Bouckaert, R.R., Frank, E., Hall, M., Kirkby, R., Reutemann, P., Seewald, A., Scuse, D.: WEKA Manual for Version 3-7-10, July 31, 2013

Data Anonymization Through Slicing Based on Graph-Based Vertical Partitioning

Kushagra Sharma, Aditi Jayashankar, K. Sharmila Banu
and B.K. Tripathy

Abstract Data anonymization is a technique that uses data distortion to preserve privacy of public data to be published. Several data anonymization techniques and principles have been proposed in the past such as k-anonymity, l-diversity, and slicing. Slicing promises to address the drawbacks of the other two anonymization models. Our proposition is the use of a graph-based vertical partitioning algorithm (GBVP) in the process of Slicing instead of the originally proposed Partition Around Medoids (PAM). We will present several arguments that favor GBVP against PAM as a choice for clustering algorithm.

Keywords Data anonymization · Slicing · Attribute partitioning · k-medoids · Clustering · Graph-based vertical partitioning

1 Introduction

With advancements in internet, database systems and data storage systems, it has become increasingly easier to collect, curate and serve a large variety of public data at a large scale. Public data is often used for analytics to gain useful insights and trends. However, it also presents the risk of exposing the privacy of an individual. Thus, it is ethical to ensure that the data being publicly released for analysis is anonymized, i.e. the identity of the individual is preserved. In this paper, we take a

K. Sharma (✉) · A. Jayashankar · K.S. Banu · B.K. Tripathy
School of Computing Science & Engineering, VIT University, Vellore 632 014, India
e-mail: kushagra8888@gmail.com

A. Jayashankar
e-mail: aditi.ja14@gmail.com

K.S. Banu
e-mail: sharmilabanu.k@vit.ac.in

B.K. Tripathy
e-mail: tripathybk@vit.ac.in

© Springer India 2016   569
A. Nagar et al. (eds.), *Proceedings of 3rd International Conference
on Advanced Computing, Networking and Informatics*, Smart Innovation,
Systems and Technologies 44, DOI 10.1007/978-81-322-2529-4_59

look at some of the existing data anonymization techniques. Slicing is a data anonymization technique that promises to address the drawbacks of other widely used data anonymization models. We propose that Graph Based Vertical Partitioning Algorithm (GBVP) can be involved in the process of Slicing. The original Partition Around Medoids (PAM) algorithm, involved in Slicing, is substituted and we present our arguments favoring the modification. GBVP is a recent approach for vertical partitioning in distributed databases.

2 Data Anonymization Techniques

In the past two decades, several anonymization principles and techniques have been proposed. In this section, we briefly introduce some of the important developments in the field of data anonymization.

Some of the general terminologies used with respect to the attributes in the dataset are:

(1) *Identifiers*: uniquely identifies an individual, such as Social Security Number.
(2) *Quasi-Identifiers (QI)*: information that do not uniquely identify an individual, however, combined with other attributes, can lead to possible identification, such as Zip code, Age and Birthday put together.
(3) *Sensitive Attributes (SA)*: this information is unknown to the opposite end and considered sensitive due to its highly personal nature, such as Salary.

2.1 k-Anonymity

L. Sweeney proposed a model for data anonymization called k-anonymity. The k-anonymity is achieved when every record in a set of quasi-identifiers (QI) is indistinguishable from at least $k - 1$ other records in the set. The k-anonymity model is prone to homogeneity attacks when the values of the sensitive attributes (SA) in a group are the same. This method works well against direct linking attacks but fails against homogeneity attacks [1].

2.2 l-Diversity

The l-diversity model was introduced to address the shortcomings of k-anonymity model. It achieves this by diversifying the sensitive attribute (SA) values within each bucket. This diversification can be achieved in different ways. The l-diversity model is susceptible to background knowledge attacks because the values within an attribute are treated in the same manner, regardless of the value's distribution in the data [2].

2.3 Slicing

Slicing is one of the recent approaches to data anonymization. The dataset is partitioned such that the data belonging to a partition semantically corresponds with each other but not with the data in other partitions. This approach splits the dataset along the horizontal and vertical directions. Vertical fragment will be a subset of attributes that have a strong correlation with each other and, the set of records in each horizontal fragment are grouped to form buckets. Within each group, the values of each attribute are permutated to remove any semantic relation with other attributes [3].

2.4 Graph-Based Vertical Partitioning (GBVP)

Vertical partitioning is the process of splitting the attributes in a relational table into groups or fragments. An attribute affinity matrix is the input of the GBVP algorithm. The attribute affinity matrix is a complete graph with each attribute as a node and each edge having the value of attribute affinity of the corresponding nodes. It starts with an arbitrary node and constructs a linearly growing tree by adding edges of largest weights. Candidate partitions are identified by formation of cycles. These cycles can be extended to improve the partition decision. The process goes on until all the nodes are exhausted. It was developed primarily to address vertical fragmentation in database design. However, the algorithm operates on a set of data points where the distance between each pair of data points is known thus making it suitable as a generic clustering algorithm [4].

3 Attribute Partitioning in Slicing

Slicing involves several steps. Attribute partitioning involves the creation of subsets of the attributes present in the dataset. Tuple partitioning is the creation of subsets. In [3] the mean-square contingency coefficient is used. The reason behind this is the categorical nature of the attributes.

Let there be two attributes:

Attribute: A_1, Domain: $\{v_{11}, v_{12}... v_{1d1}\}$, Domain size: d_1
Attribute: A_2, Domain: $\{v_{21}, v_{22}... v_{2d2}\}$, Domain size: d_2

The mean-square contingency coefficient between the attributes A_1 and A_2, follows the formula:

$$\emptyset^2(A1, A2) = \frac{1}{\min\{d1, d2\} - 1} \sum_{i=1}^{d1} \sum_{j=1}^{d2} \frac{(f_{ij} - f_i f_j)^2}{f_i f_j}. \tag{1}$$

f_i fractional occurrence of v_{1i}
f_j fractional occurrence of v_{2j}
f_{ij} fractional co-occurrences of v_{1i} and v_{2j}

Formal definition of f_i and f_j:

$$f_{ij} : f_{i.} = \sum_{j=1}^{d_2} f_{ij} \text{ and } f_j = \sum_{i=1}^{d_1} f_{ij}. \tag{2}$$

Here,

$$0 \leq \emptyset^2(A1, A2) \leq 1.$$
$$\text{Distance } d(A1, A2) = 1 - \emptyset^2(A1, A2).$$

Attributes pairs with low distance value are strongly correlated.

We wrote a python script to compute the distance matrix for a given dataset using the above method. Table 2 shows the distance matrix computed from Table 1.

3.1 Attribute Partitioning in Slicing

The PAM algorithm [5] is an application of the k-medoid algorithm. The algorithm makes a random initialization of k cluster centers (called medoids). All data points are assigned to their closest medoid. A point other than the medoid is tentatively considered as a medoid and the costs are computed. This is repeated for all possible candidate medoids. The point with the least cost is selected as the new medoid and

Table 1 Original Table: A sample dataset consisting of Quasi-Identifiers (QI) and Sensitive Attributes (SA)

Age	Sex	Zipcode	Race	Salary ($/year)
23	M	632014	Caucasian	35,000
23	F	632014	Hispanic	23,000
25	M	632011	African American	25,000
30	M	632011	Caucasian	44,000
30	F	732013	Native American	23,000
27	F	732013	Hispanic	35,000

Table 2 Distance Matrix: 2-D array describing attribute pairs and the distance between them

	Age	Sex	Zipcode	Race	Salary
Age	1.000000	0.666667	0.250000	0.416667	0.416667
Sex	0.666667	1.000000	0.333333	0.000000	0.333333
Zipcode	0.250000	0.333333	1.000000	0.500000	0.500000
Race	0.416667	0.000000	0.500000	1.000000	0.416667
Salary	0.416667	0.333333	0.500000	0.416667	1.000000

the data points are reassigned accordingly. The process is repeated until no changes are observed in the configuration (Fig. 2).

3.2 Attribute Partitioning Using GBVP

GBVP algorithm starts from an affinity matrix which has as its elements $aff(A_1, A_2)$; that is the affinity between any two attributes A_1 and A_2. This affinity matrix is transformed into a graph format in which the attributes are the nodes and the edge values are the corresponding affinity values between the attribute pairs. The algorithm starts by considering an initial starting node. It then successively adds linearly connected maximal edges keeping the degree of each node less than or equal to 2. If a cycle is formed then it is checked if the nodes in the cycle can be considered as a candidate partition based on the weights of edges involved in the cycle. Further expansions of the candidate partition are explored and when no such expansions are possible, the candidate partition is saved as a partition (Fig. 3).

Fig. 1 Distance graph:
Table 2 as a complete graph

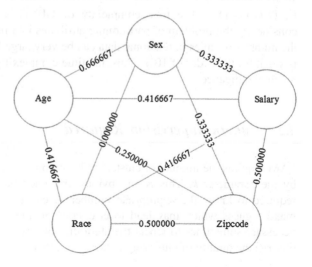

Fig. 2 Result of PAM
performed on Fig. 1

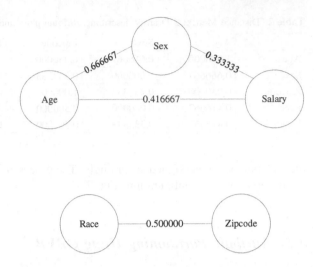

GBVP was proposed in the context of partitioning relational tables vertically in the process of database design. However, the goals of database design match with those of Slicing i.e. grouping correlated attributes and keeping uncorrelated ones separate. GBVP thus intuitively fits into the problem context.

4 Comparison Between PAM and GBVP

4.1 Time Complexity

Both PAM and GBVP have high computational complexity for large datasets. Given a dataset with n attributes and k initial medoids, the time complexity of PAM is: $O(k(n-1)^2)$. The time complexity of GBVP is: $O(n)^2$. However, we are considering the problem of partitioning attributes in a relational dataset. Generally, the number of tuples in relational data can be very large but the number of attributes is small (of the order of 10^4). Thus, the time complexity comparison does not have much significance.

4.2 Human Supervision Required

PAM requires the number of clusters to be supplied in advance. This is represented by the parameter k. This is an obvious drawback because the user is somehow required to know the appropriate number of clusters beforehand or be forced to make a guess which may lead to erroneous or misleading results. This makes it necessary sometimes to repeat the algorithm by choosing different values of k and observe the quality of clustering either manually or by using some automation tool.

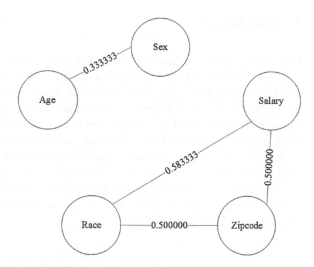

Fig. 3 Result of GBVP performed on Fig. 1

This adds to the complexity of the process. The parameter k does need not be provided while using the GBVP algorithm. Instead, the algorithm generates the number of partitions or clusters required in a single run. The overhead in using PAM is, therefore, avoided.

Moreover, PAM is an iterative algorithm. Since the initial choice of k medoids is random, the resulting clusters can vary when the process is repeated on the same dataset. Thus, in practice the PAM algorithm is made to run on a problem i times where i is the number of iterations. The quality of clusters obtained from these iterations is compared and the best one is chosen as the result. Thus the larger the value of i, the better is the expected value of clustering in PAM. The user thus has to specify a value for i. If it is too small the quality of clustering can suffer. On the other hand if it is too large, the time taken by the algorithm to complete will be large. GBVP on the other hand makes no such random initialization. The choice of initial starting node for GBVP process does not affect the end result. Thus for a given dataset the end result will be the same for any number of iterations with any choice of starting node.

4.3 Relevance to Problem

The requirement of attribute partitioning in slicing is that the attributes within a partition are highly correlated and those in different partitions be uncorrelated. This is required to preserve data utility while preserving privacy. Both GBVP and PAM attempt to achieve this objective. However, the clusters identified by PAM are generally spherical in shape which restricts the variety of clusters that could be identified. GBVP on the other hand can identify clusters of arbitrary shape as long as they are sufficiently close, making it more versatile.

5 Conclusion

Our analysis indicates that on all grounds of comparison GBVP is either better than or as good as PAM when the process of attribute partitioning is concerned. The time complexity of GBVP is comparable to PAM. The amount of supervision required is negligible for GBVP and there is no restriction on the shape of clusters that can be identified by GBVP making it more robust. This makes GBVP a viable and in fact a more appropriate choice for the clustering algorithm in the process of Slicing.

Future work on these grounds includes experimental comparison of the running times of the two implementations over datasets of varying size. The quality of clusters produced can be compared by using a suitable metric. Since distance matrices are used, mean-square error computation can be a feasible choice for the determination of cluster quality.

$$SSW(C, k) = \sum_{i=1}^{N} \left\| x_i - c_{p(i)} \right\|^2. \tag{3}$$

SSW represents sum of squares within the clusters; where C is the clustering space with k clusters. This is the sum of squares of each of the data points to its corresponding cluster center.

Further, it needs to be verified that the anonymized datasets produced by using GBVP in the attribute partitioning step do preserve privacy. This can be accomplished by considering the Sensitive Attribute(s) as the outcome of a classification model (e.g. Random Forest, Support Vector Machines). The model can be trained on the anonymized dataset with the sensitive attribute removed. The trained classifier can then be used to predict the SA's for the un-anonymized dataset for which the values are already known. The predictions can then be compared against the known values to estimate the accuracy of prediction. For anonymization to be effective it would be desirable to have a low accuracy of prediction.

References

1. Sweeney, L.: k-anonymity: a model for protecting privacy. Int. J. Uncertainty Fuzziness Knowl. Based Syst. **10**, 557–570 (2002)
2. Machanavajjhala, A., Kifer, D., Gehrke, J., Venkitasubramaniam, M.: l-diversity: privacy beyond k-anonymity. ACM Trans. Knowl. Discov. Data **1**(1), 3 (2007)
3. Li, T., Li, N., Zhang, J., Molloy, I.: Slicing: a new approach to privacy preserving data publishing. Arxiv: 0909.2290v1, 12 September (2009)
4. Shamkant, M.R., Navathe, B.: Vertical partitioning for database design. In: ACM, p. 11 (1989)
5. Kaufman, L., Rousueeuw, P.: Finding Groups in Data: an Introduction to Cluster Analysis. Wiley (1990)

An Efficient Approach for the Prediction of G-Protein Coupled Receptors and Their Subfamilies

Arvind Kumar Tiwari, Rajeev Srivastava, Subodh Srivastava and Shailendra Tiwari

Abstract G-protein coupled receptors are responsible for many physiochemical processes such as neurotransmission, metabolism, cellular growth and immune response. So it necessary to design a robust and efficient approach for the prediction of G-protein coupled receptors their subfamilies. To address the issue of efficient classification G-protein coupled receptors and their subfamilies, here in this paper we propose to use a weighted k-nearest neighbor classifier with UNION of best 50 features selected by Fisher score based feature selection, ReliefF, fast correlation based filter, minimum redundancy maximum relevancy and support vector machine based recursive feature elimination feature selection methods. The proposed method achieved an overall accuracy of 99.9, 98.3 % MCC values of 1.00, 0.98 ROC area values of 1.00, 0.998 and precision of 99.9 and 98.3 % using 10-fold cross validation to predict the G-protein coupled receptors and their subfamilies respectively.

Keywords G-protein coupled receptors · Weighted k-nearest neighbor · Minimum redundancy maximum relevance · Sequence derived properties · Matthew's correlation coefficient

A.K. Tiwari (✉) · R. Srivastava · S. Tiwari
Department of Computer Science & Engineering, Indian Institute
of Technology (BHU), Varanasi 221005, U.P, India
e-mail: arvind.rs.cse12@itbhu.ac.in

R. Srivastava
e-mail: rajeev.cse@iitbhu.ac.in

S. Tiwari
e-mail: stiwari.rs.cse@itbhu.ac.in

S. Srivastava
School of Biomedical Engineering, Indian Institute of Technology (BHU),
Varanasi 221005, U.P, India
e-mail: subodhresonance@gmail.com

© Springer India 2016
A. Nagar et al. (eds.), *Proceedings of 3rd International Conference
on Advanced Computing, Networking and Informatics*, Smart Innovation,
Systems and Technologies 44, DOI 10.1007/978-81-322-2529-4_60

577

1 Introduction

G-protein coupled receptors (GPCRs) are seven-transmembrane domain receptors that sense molecules outside the cell and activate inside signal transduction pathways for cellular responses. They are called seven-transmembrane receptors because they pass through the cell membrane seven times. GPCRs can be grouped into six classes based on sequence homology and functional similarity these are Rhodopsin-like, Secretin like, Metabotropic glutamate, cyclic AMP, Taste and Vomeronasal receptors. There are a larger number of G-protein coupled receptors are available in human in these some have been identified their function like growth factors, light, hormones, amines, neurotransmitters, and lipids etc. However a large number of the GPCRs found in the human genome have unknown functions and so it is necessary to design an efficient approach to predict families and subfamilies of G-protein coupled receptors for the new drug discovery.

Initially Bhasin et al. [1] proposed a SVM based method by using amino acid composition and dipeptide of amino acids for the prediction of G-protein coupled receptor. Later, Bhasin and Raghava [2] proposed a SVM based method for the classification of amine type of G-protein-coupled receptors by using of amino acid composition and dipeptide composition of proteins. Gao et al. [3] proposed a nearest neighbor method to discriminate GPCRs from non-GPCRs and subsequently classify GPCRs at four levels on the basis of amino acid composition and dipeptide composition of proteins. Gu and Ding [4] have proposed binary particle swarm optimization algorithm to extract effective feature for amino acids pair compositions of GPCRs protein sequence. Then they have used ensemble fuzzy k-nearest neighbor classifier to predict GPCRs families. Gu et al. [5] proposed an Adaboost classifier to predict G-protein-coupled receptors by pseudo amino acid composition with approximate entropy and hydrophobicity patterns. Peng et al. [6] proposed a principal component analysis based method for the prediction of GPCRs, family and their subfamilies by using sequence derived features.

This paper proposes a weighted k-nearest neighbor in which inverse kernel function are applied to calculate weighted distance to improve the prediction performance of G-protein coupled receptors families and their subfamilies by using sequence derived properties. In this paper 1497 sequence derived features are used to predict G-protein coupled receptors and their subfamilies. In this paper five supervised filter based methods, Fisher score based feature selection [7], ReliefF [8], fast correlation based filter [9], minimum redundancy and maximum relevancy (mRMR) [10] and support vector machine based recursive feature elimination (SVM-RFE) [11] methods are used to obtain optimal number of features. Further, to incorporate the advantages of these feature selection algorithms apply UNION for the fusion of these selected features and apply weighted k nearest neighbor to G-protein coupled receptors families and their subfamilies. This paper used hold one out cross validation to find the best k between 1 and 30 the value specified to the weighted kNN parameter.

2 Materials and Methods

In this paper the sequence of the G-protein coupled receptors are extracted from GPCRDB (http://www.gpcr.org/7tm/). Here, 2962 number of GPCR sequences including 974 number of Rhodopsin like, 621 number of Secretin like, 454 number of Metabotropic glutamate, 08 number of cAMP receptors, 480 number of taste and 425 number of vomeronasal receptors with 576 non-G-protein coupled receptors. Here, all the 576 non-G-protein coupled receptors proteins are selected from Uniport database with the keyword NOT G-protein coupled receptors. To avoid the homology bias the CD-HIT [12] server is used to remove the homologous sequence using 70 % sequence identity as the cutoff, because when we decrease the cutoff as 0.5 and 0.4 respectively then very small sequences are left for the evaluation of classifier which is associated with lower performance values in comparison to 70 % cutoff.

2.1 Features Extraction of Protein Sequences

In this paper, to fully characterize protein sequence eight feature vectors are extracted from PROFEAT server [13] to represent the protein sample, including amino acid composition, dipeptide composition, correlation, composition, transition, distribution of physiochemical properties, sequence order descriptors, and pseudo amino acid composition with total of 1497 features being calculated for the prediction of G protein coupled receptors and their subfamilies.

2.2 Feature Subset Selection

In this paper five supervised filter based methods, Fisher score based feature selection [7], ReliefF [8], fast correlation based filter [9], minimum redundancy and maximum relevancy (mRMR) [10] and support vector machine based recursive feature elimination (SVM-RFE) [11] feature selection methods are used to obtain optimal number of features. If we apply, these feature selection methods on the same dataset then each of them results in different feature subset where features are ranked according to their rank. Also the performance of a classifier for each feature subset selected by different method may be different. Therefore, here we address this problem by proposing a method for optimal feature selection by the fusion of feature subsets produced by these methods using union of the selected features by different feature selection algorithms.

2.3 Classification of G-Protein Coupled Receptor and Their Subfamilies

For the classification of G-protein coupled receptor and their subfamilies the weighted k neighbor classifier has been used. This paper proposes a weighted k-nearest neighbor in which inverse kernel function are applied to calculate weighted distance to improve the prediction performance of G-protein coupled receptors families and their subfamilies. Basically the traditional kNN classifier is used each time a different k, starting from k = 1 to k = square root of the training set. This paper used hold one out cross validation to find the best k between 1 and 30 the value specified to the kNN parameter. The weighted k-nearest neighbor classifier has performed the following steps.

1. Let $L = \{(y_i, x_i), i = 1, \ldots \ldots \ldots, n_L\}$ be a training datasets where $y_i \in \{1 \ldots c\}$ represents the class and $x_i' = (x_{i1}, \ldots \ldots \ldots, x_{ip})$ represents the predictor values. Let x be the test sample whose class level y has to be predicted.
2. Obtain k + 1 nearest neighbor to x by using Manhattan distance function d(x, x_i). Here

$$Manhattan\ distance = d(x, x_i) = \sum_{s=1}^{p} |x_s - x_{is}|$$

3. The (k + 1)th neighbor is used for the standardization of the k smallest distance by using following equation

$$D(x, x_i) = \frac{d(x, x_i)}{d(x, x_{k+1})}$$

Transform and normalize distance by using inverse kernel function $K = \frac{1}{|d|}$ to obtain the weight $w_i = \frac{1}{D(x, x_i)}$.

4. Assign a class, y of test sample, x which shows a weighted majority of the k-nearest neighbor.

In this paper the proposed method used two level strategies to predict G-protein coupled receptors and their subfamilies. The complete procedure of the proposed method for the prediction of G-protein coupled receptors and their subfamilies are as follows:

1. Produce seven feature vectors with 1497 features that represent a protein sequence.
2. Select best 50 number of features with Fisher score based, ReliefF, FCBF, mRMR, and SVM-RFE feature selection methods.
3. Fusion of feature subsets produced by these methods using UNION of the selected features by different feature selection algorithms.

4. Apply weighted k-nearest neighbor classifier for the prediction of G-protein coupled receptors and their subfamilies are as follows:

First, it is determined that protein sequence is G-protein coupled receptors or non-G-protein coupled receptors and then, if protein is classified as G-protein coupled receptors then the method classify the subfamilies of G-protein coupled receptors.

3 Performance Measures

In this paper, 10-fold cross validation is used to measure the performance of weighted k-nearest neighbor classifier. In K-fold cross validation the dataset of all proteins is partitioned into K subsets where one subset is used for validation and remaining $K - 1$ subsets is used for training. This process is repeated for K times so that every subset is used once as a test data. In this paper accuracy (ACC), Precision, Receiver Operating Characteristics (ROC) and Matthew's correlation coefficient (MCC) is used to measure the performance.

4 Results and Comparative Analysis

In this paper, a weighted k-nearest neighbor is proposed to be used with inverse kernel function to calculate weighted distance to improve the prediction performance of G-protein coupled receptors and their subfamilies by using sequence derived properties. This paper used hold one out cross validation to find the best k between 1 and 30 the value specified to the kNN parameter. For partitioning of the datasets into train and test sets and evaluating the performance of the proposed model the 10-fold cross-validations are used. In subsequent subsections the results and performance analysis of the proposed model for the prediction of G-protein coupled receptors and their subfamilies are presented and discussed. The performance analysis of the proposed model is showed for the best 50 number of features selected among 1497 total number of features by using Fisher score based feature selection algorithms, ReliefF, FCBF, mRMR, and SVM-RFE feature selection methods, for each cases as well as for the UNION of these selected features.

4.1 Prediction of GPC Receptors and Non-GPC Receptors

To predict the GPC receptors and Non-GPC receptors, a weighted k-nearest neighbor is evaluated with for the best 50 number of features selected among 1497 total number of features by using Fisher score based feature selection algorithms,

ReliefF, FCBF, mRMR, SVM-RFE feature selection methods and the UNION of these selected features. It is observed that the performance of the classifier is improved by using the UNION of the best 50 features selected by five different feature selection algorithms (see Fig. 1).

From the analysis of Table 1 it is observed that the performance of a weighted k-nearest neighbor methods provides overall accuracy of 99.9 %, MCC of 1.00, ROC Area of 1.00 and Precision of 99.9 % and Support vector machine based method with Kernel = RBF, γ = 200, C = 100 provides overall accuracy of 99.9 %, MCC of 0.996, ROC Area of 0.997 and Precision of 99.9 % for the prediction of GPC receptor and Non-GPC receptor with the UNION of the best 50 features (187 features) selected by five different feature selection algorithms. Here, it is also observed that weighted k-nearest neighbor based classifier performed better in comparison to SVM classifier to predict the G-protein coupled receptors and non G-protein coupled receptors (see Table 1).

4.2 Prediction of Subfamilies of G-Protein Coupled Receptors

To predict the subfamilies of G-protein coupled receptors, a weighted k-nearest neighbor is evaluated with for the best 50 number of features selected among 1497 total number of features by using Fisher score based feature selection algorithms, ReliefF, FCBF, mRMR, SVM-RFE feature selection methods and the UNION of these selected features. It is observed that the performance of the classifier is

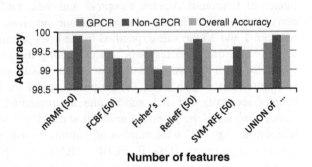

Fig. 1 Accuracy for classification of GPC receptor and Non-GPC receptor with different data sets

Table 1 Result analysis for the classification of GPCR and Non-GPCR

Family	Proposed feature selection + w-kNN				Proposed feature selection + SVM			
	ACC	MCC	ROC area	Precision	ACC	MCC	ROC area	Precision
Non-GPCR	99.7	1.00	1.00	99.5	99.5	0.996	0.997	99.8
GPCR	99.9	1.00	1.00	99.9	100	0.996	0.997	99.9
Overall	99.9	1.00	1.00	99.9	99.9	0.996	0.997	99.9

Fig. 2 Accuracy for classification of families of GPCRs with different data sets

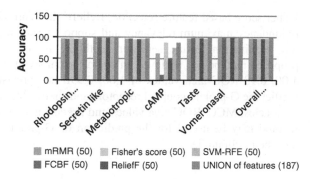

■ mRMR (50)   ▨ Fisher's score (50)   ■ SVM-RFE (50)
■ FCBF (50)   ■ ReliefF (50)   ■ UNION of features (187)

Table 2 Result analysis for the classification of subfamilies of GPC receptors

Subfamilies of GPCR	Proposed feature selection + wk-NN				Proposed feature selection + SVM			
	ACC	MCC	ROC area	Precision	ACC	MCC	ROC area	Precision
Rhodopsin like	97.7	0.96	0.997	97.3	96.7	0.942	0.973	95.5
Secretin like	99	0.99	0.999	99.8	98.6	0.987	0.992	99.4
Metabotropic	97.1	0.98	0.999	98.7	96.9	0.975	0.984	98.9
cAMP	87.5	0.78	0.977	70	37.5	0.612	0.688	100
Taste	98.5	0.98	0.998	98.3	97.1	0.963	0.982	96.7
Vomeronasal	99.8	0.99	1.00	98.6	98.4	0.974	0.989	97.2
Overall	98.3	0.98	0.998	98.3	97.3	0.964	0.981	97.3

improved by using the UNION of the best 50 features selected by five different feature selection algorithms (see Fig. 2).

From the analysis of Table 2 it is observed that the performance of a weighted k-nearest neighbor methods provides overall accuracy of 98.3 %, MCC of 0.98, ROC Area of 0.998 and Precision of 98.3 % and for Support vector machine based method with Kernel = RBF, $\gamma = 200$, C = 100 provides overall accuracy of 97.3 %, MCC of 0.964, ROC Area of 0.981 and Precision of 97.3 % for the classification of subfamilies of GPC receptors with the UNION of the best 50 features (187 features) selected by five different feature selection algorithms. Here, it is also observed that weighted k-nearest neighbor based classifier performed better in comparison to SVM classifier to predict the subfamilies of G-protein coupled receptors (see Table 2).

5 Conclusions

The G-protein coupled receptors are the largest superfamilies of membrane proteins and important targets for the drug design In this paper a weighted k-nearest neighbor classifier has been proposed with UNION of best 50 features selected by

Fisher score based feature selection, ReliefF, fast correlation based filter, minimum redundancy maximum relevancy and support vector machine based recursive feature elimination feature selection methods. In the 10-fold cross validation the proposed method has achieved an overall accuracy of 99.9, 98.3 % MCC values of 1.00, 0.98 ROC area values of 1.00, 0.998 and precision of 99.9 and 98.3 % to predict the G-protein coupled receptors and their subfamilies respectively. The high accuracies, MCC, ROC area values and precision values indicate that the proposed method may be useful for the prediction of G-protein coupled receptors families and their subfamilies.

References

1. Bhasin, M., Raghava, G.P.S.: GPCRpred: an SVM-based method for prediction of families and subfamilies of G-protein coupled receptors. Nucleic Acids Res. **32**(Suppl. 2), W383–W389 (2004)
2. Bhasin, M., Raghava, G.P.S.: GPCRsclass: a web tool for the classification of amine type of G-protein-coupled receptors. Nucleic Acids Res. **33**(Suppl. 2), W143–W147 (2005)
3. Gao, Q.B., Wang, Z.Z.: Classification of G-protein coupled receptors at four levels. Protein Eng. Des. Sel. **19**, 511–516 (2006)
4. Gu, Q., Ding, Y. Binary particle swarm optimization based prediction of G-protein-coupled receptor families with feature selection. In: Proceedings of the First ACM/SIGEVO Summit on Genetic and Evolutionary Computation, pp. 171–176. ACM (June 2009)
5. Gu, Q., Ding, Y.S., Zhang, T.L.: Prediction of G-protein-coupled receptor classes in low homology using chous pseudo amino acid composition with approximate entropy and hydrophobicity patterns. Protein Pept. Lett. **17**(5), 559–567 (2010)
6. Peng, Z.L., Yang, J.Y., Chen, X.: An improved classification of G-protein-coupled receptors using sequence-derived features. BMC Bioinform. **11**(1), 420 (2010)
7. Guyon, I., Elisseeff, A.: An introduction to variable and feature selection. J. Mach. Learn. Res. **3**, 1157–1182 (2003)
8. Kira, K., Rendell, L.A.: The feature selection problem: traditional methods and a new algorithm. In: AAAI, pp. 129–134 (July 1992)
9. Yu, L., Liu, H.: Feature selection for high-dimensional data: a fast correlation-based filter solution. In: ICML, vol. 3, pp. 856–863 (August 2003)
10. Ding, C., Peng, H.: Minimum redundancy feature selection from microarray gene expression data. J. Bioinform. Comput. Biol. **3**(02), 185–205 (2005)
11. Yu, Y. SVM-RFE Algorithm for Gene Feature Selection. Computer Engineering (2008)
12. Li, W., Godzik, A.: Cd-hit: a fast program for clustering and comparing large sets of protein or nucleotide sequences. Bioinformatics **22**, 1658–1659 (2006)
13. Rao, H.B., Zhu, F., Yang, G.B., Li, Z.R., Chen, Y.Z.: Update of PROFEAT: a web server for computing structural and physicochemical features of proteins and peptides from amino acid sequence. Nucleic Acids Res. **39**(suppl 2), W385–W390 (2011)

Part VIII
Fault and Delay Tolerant Systems

Max-Util: A Utility-Based Routing Algorithm for a Vehicular Delay Tolerant Network Using Historical Information

Milind R. Penurkar and Umesh A. Deshpande

Abstract A vehicular delay tolerant network (VDTN) has emerged as special type of delay tolerant network (DTN), characterized by the non-existence of end-to-end connectivity between different source nodes and destination nodes thereby diminishing the possibility of delivering messages to the destination node. Hence, routing becomes a vital issue in VDTN. In this paper, a utility-based routing algorithm (Max-Util) for a VDTN has been proposed. This algorithm exploits the historical information of a node for computing the utility of a node to take a decision to forward a message among nodes. This history comprises the past encounters of a node with destination and relay nodes, the contact duration between nodes and the remaining buffer size of a node in contact. Our evaluations show that, Max-Util performs better in terms of message delivery ratio, overhead ratio and average delivery latency as compared to some existing DTN routing approaches like PRoPHET and Spray-and-Wait.

Keywords Vehicular delay tolerant networks · Intermittent connectivity · Routing · Utility · Forwarding · Flooding

1 Introduction

Delay tolerant networks (DTN) are mobile ad hoc networks where end-to-end connectivity does not exist most of the time because of short radio transmission range, limited power of evolving wireless communication devices, disruptions caused in the networks and network partitioning. VDTN is an example of DTN where vehicles act as moving nodes to carry devices to transmit messages in the

M.R. Penurkar (✉) · U.A. Deshpande
Department of Computer Science and Engineering, VNIT, Nagpur, India
e-mail: milindpenurkar@gmail.com

U.A. Deshpande
e-mail: uadeshpande@cse.vnit.ac.in

© Springer India 2016
A. Nagar et al. (eds.), *Proceedings of 3rd International Conference on Advanced Computing, Networking and Informatics*, Smart Innovation, Systems and Technologies 44, DOI 10.1007/978-81-322-2529-4_61

587

network. VDTNs seize help of DTN capabilities to support disruptions for providing the network connectivity. VDTNs are more useful in rural or mountainous areas because of the existence of intermittent connectivity.

Traditional routing protocols such as DSDV [1], DSR [2], AODV [3], etc. in the MANET/VANET establish an end-to-end path between a source and a destination node before sending any packets. Routing in such a network is all about finding the shortest path in the network and to forward packets on this path. These protocols naturally fail when applied to a VDTN because an instantaneous end-to-end path would be lacking to route the packets. To make communication possible in such a network, one must facilitate the routing of messages without establishing an end-to-end path in advance. Thus, routing becomes a challenging issue in this class of a network. Routing in these networks is not to devise the shortest path among the nodes but to ensure that messages reach the destination in one way or another. VDTNs employ store-carry-forward mechanism to route messages to the destination.

Among the number of routing algorithms discussed in the literature, an important type of routing algorithms discussed in a DTN literature focuses on the utility of nodes. Utility of a node can be defined as its fitness to forward a message to the destination. A utility based algorithm brings into play the utility of nodes for taking a decision in the routing process. In this paper, we propose Max-Util, a hybrid utility based routing algorithm that exploits both destination dependent as well as destination independent utility functions to route messages in a VDTN.

We show that Max-Util performs better by increased delivery ratio, reduced overhead ratio and decreased average delivery latency as compared to PRoPHET [4] (the utility based algorithm), except Spray-and-Wait [5] (a controlled flooding algorithm) that succeeds Max-Util in terms of reducing the overhead ratio.

The remainder of this paper is organized as follows. Section 2 describes the related work and contributions in this area. Section 3 presents the Max-Util algorithm in detail. Section 4 shows simulation scenarios and analyzes the obtained results, Sect. 5 concludes the paper and Sect. 6 provides guidelines for future work.

2 Related Work

A number of routing approaches are designed, developed and simulated in a VDTN. In the following section, we discuss basic approaches for routing algorithms followed by the utility based routing algorithms.

2.1 Basic Approaches

2.1.1 Forwarding Based Routing Algorithms

In these types of based First is the forwarding technique, in that only one copy of a message exists at any time in the network. First Contact [6], Spray-and-Focus [7] and FRESH [8] are some examples of the forwarding approach.

2.1.2 Flooding and Controlled Flooding Based Routing Algorithms

In the flooding routing technique, every node floods messages to every other node that it meets and therefore a number of copies of a message exist at any time in the network. Most familiar form of flooding is Epidemic [9].

To prevail over the operating cost of a fully flooding approach, researchers have arrived at a number of approaches [6–8]. Controlled flooding algorithms have moderate message delivery ratio compared to their fully flooding based counterparts with less overhead ratio.

2.2 Utility-Based Routing Algorithms

In this kind of algorithms, a forwarding decision depends on the context of the current node that holds messages and the node that is encountered. A set of parameters associated with the nodes in contact are obtained for estimating the nodes' utility for a given message targeted to a specific destination. This utility is evaluated based on utility functions that are categorized in two types in the literature as destination dependent (DD) utility functions and destination independent (DI) utility functions [6]. Examples of parameters for building DD utility functions comprise of history of past encounters, pattern of locations visited and social networks whereas examples of parameters for building DI utility functions include amount of mobility, node resources and cooperative behavior [6]. There are two types of utility-based replication: uncontrolled and controlled utility-based replication as discussed in [3]. An example of a controlled utility-based replication is SMART replication [6] and that of an uncontrolled utility-based replication is PRoPHET [4].

SMART replication integrates DD and DI utility functions for computing utility, i.e. it is hybrid in nature (It makes use of "History of past encounters" as destination dependent utility function whereas "mobility and node resources" are destination independent utility functions used). Encounter-Based Routing (EBR) [10] is another example of a controlled utility-based replication. In this algorithm, future rate of node encounters is predicted using nodes past encounters. EBR makes use of DD utility function (history of past encounters) to compute utility.

Uncontrolled utility-based replication has featured in PRoPHET. In this algorithm, each node maintains a probability vector that holds nodes' probability of future encounters using the history of past encounters. Based on this probability, a message copy is forwarded to a new node only if that node is highly probable of encountering the destination in the future. PRoPHET makes use of DD utility function (history of past encounters) to compute utility.

3 Max-Util: The Utility-Based Algorithm

3.1 Background

In a VDTN, the following issues are yet to be addressed at length.

1. Nodes remain in contact for a short duration of time because of the speeds with which they travel. In [11], researchers show that the duration of contacts between cars using IEEE 802.11 g crossing at 20 km/h is about 40 s, at 40 km/h is about 15 s and at 60 km/h is about 11 s. When Transmission Control Protocol (TCP) is used at 60 km/h, on average 80 Kb data could be transferred and in most of the instances no data was transferred at all. UDP yielded better results, with about 2 Mb transferred in a contact at 60 km/h. Thus, the number of messages that can be transmitted during the contact opportunity is dependent on the contact duration.
2. Since a mobile node may have limited storage, it may not be able to accommodate all the messages that it receives. Thus remaining buffer storage of nodes in contact has a great impact on message transmission.
3. Node's history of encounters with a destination node or with different road side units (RSUs) also termed as relay nodes plays a significant role for making a decision of forwarding messages to a node in question. This seems realistic because if a node has encountered the destination node or RSU in the past, then it is more likely to meet them again in the future.

Considering all these parameters that outline our history information, we propose Max-Util—a hybrid utility based routing algorithm using nodes' history information to route messages in a VDTN. It computes the overall utility by applying a weighted average of parameters at hand in history information and then forwards a message to a candidate node that has the maximum utility. Max-Util is an uncontrolled utility-based algorithm and it is hybrid in nature because it combines both destination dependent as well as destination independent functions. The destination dependent functions considered are destination meeting count and relay meeting count where as contact duration and remaining buffer size have been included as destination independent functions.

3.2   Assumptions

Following are the assumptions for the Max-Util algorithm.

3.2.1   General

All the nodes in the network have unique identifications (IDs). Two types of terminal nodes are present in the network: source terminal nodes and destination terminal nodes. They act as stationary nodes and are deployed at two ends of the vehicular network (E.g. some devices having storage and internet connectivity). Mobile nodes are vehicles that carry homogeneous or heterogeneous devices (E.g. mobile phones, PDAs, laptops having Bluetooth or Wi-Fi) for transmitting messages. Mobile nodes considered here are cars that follow map based movements i.e. they do follow different paths every time from one location to another during their travel as opposed to buses that follow predefined paths using map route based movements, therefore this appears to be a realistic assumption. Road Side Units (E. g. some devices having storage and wireless connectivity) deployed in the vehicular network vary and they are placed at different intersections to improve the delivery ratio. They are also termed as relay nodes as per VDTN terminology.

3.2.2   Message Creation and Replication Policy

Only source nodes create messages. All nodes (source nodes, relay nodes and mobile nodes) except destination nodes employ replication of messages.

3.2.3   Scheduling and Dropping Policy

a. Scheduling of messages is based on the random policy in that any random message from the queue would be selected for scheduling. Dropping of messages is based on time-to-live (TTL) of the network, i.e. when TTL of a message reaches, it would be dropped from a queue.

3.2.4   Forwarding/Flooding Policy

Source nodes, relay nodes and mobile nodes employ selective flooding (This would be based on the algorithm described in Sect. 3.4).

3.3 Notations

Following notations are used in the phase-II of the algorithm.

(i) U_d (Destination utility)—Utility value for destination node meeting count—
      This is the count stored by every mobile node when it meets the destination
      node or destination nodes.
(ii) U_r (Relay utility)—Utility value for relay node meeting count—This is the
      count stored by every mobile node when it meets any of the relay nodes.
(iii) U_b (Buffer utility)—Utility value for remaining buffer size—This is the value
      stored for the percentage of remaining size of mobile node's buffer (This is the
      buffer percentage of a mobile node that is currently in contact).
(iv) U_c (Contact utility)—Utility value for the contact duration—This is the value
      stored for the inter-contact duration among the mobile nodes (Amount of time
      a mobile node is in contact with another mobile node).
(v) U_o (Overall utility)—This is the weighted average of the all the utility values
      mentioned above. We use three constants as weights—α for U_d, or U_r, β for U_b
      and γ for U_c such that $\alpha + \beta + \gamma = 1$. The detailed computation is explained in
      the *Phase–II* of the algorithm.

Node Ids: Source Nodes–*S*, Mobile Nodes–*M, M1 & M2*, Relay nodes—*R*,
Destination Nodes—*D*, Node meeting count (Any type of a node meets any other
type of a node)—*U*.

3.4 The Algorithm

The algorithm runs in two phases—first phase does contact history creation and
updating the history if needed, whereas second phase marks the actual message
transmission. The history used by the nodes is shown in the Table 1 named as
"Contact History Table".

Table 1 Contact history table

Node types	History stored at nodes
Source nodes and mobile nodes	Node ID, destination meeting count (*DMC*), relay meeting count (*RMC*), contact duration (*CD*), remaining buffer size (*RMB*)
Road side units or relay nodes	Node ID, destination meeting count (*DMC*), contact duration (*CD*), remaining buffer Size (*RMB*)
Destination nodes	Nil

The algorithm emphasizes on *DMC* or RMC primarily, hence the value of α carries more weight whereas β and γ set to have less weights. The threshold value of α is decided based on the experimentation. Based on the value of α, β and γ values are decided and have been set to $(1 - α)/2$. The thresholds for *RMB* and *CD* have also been decided empirically.

Initially, Node Ids = NULL, *RMC* = 0, *DMC* = 0, *CD* = 0, *RMB* = 0, *U* = 0; Minimum value for $RMB-RMB_t$ and Minimum value for $CD-CD_t$.

Phase–I (Contact history creation or updating)

For all nodes
>   If *M* comes in contact with *R* {Update *RMC* at *M* (*RMC*++) }
>   If *M* comes in contact with *D* {Update *DMC* at *M* (*DMC*++) }

For all nodes *S* do:
>   If *M* comes into contact with *S* & *U=0* {
>   *S* generates history information of *M* and record the Node id, *RMC, DMC,*
>   *RMB* and compute the contact duration *(CD)*.
>   *U++;* }
>   Else If *M* comes into contact with *S* & *U>1*
>   { *S* updates the history information of *M*. }

For all nodes *R* do:
>   If *M* comes into contact with *R* & *U=0*
>   { *R* generates history information of *M* and record the Node id,
>   *DMC, RMB* and compute the contact duration *(CD)*.
>   *U++;*
>   }
>   Else If *M* comes into contact with *R* & *U>1*
>   { *R* updates the history information of *M* (Without *RMC*).   }

For all mobile nodes *M1 and M2* do:
>   If one mobile node *M1* comes into contact of other mobile node *M2* & *U=0*
>   {
>   *M1* generates history information of *M2* and record the Node id,
>   *RMC, DMC, RMB* and Compute the contact duration *(CD)*.
>   *U++;*}
>   If a mobile node *M1* comes into contact of other mobile node *M2* & *U>1*
>   {*M1* and *M2* exchange and update the history information of each other. }

Phase–II (Message Transmission)

If a mobile node $M1$ comes in contact with at least one of the nodes $M2$
If $(DMC$ at $M2 > = 1$ $\&\&$ RMC at $M2 = 0)$
{

$$U_o = (\alpha \times U_d + \beta \times U_c + \gamma \times U_b) \qquad (1)$$

}
Else If a mobile node $M1$ comes in contact with at least one of the nodes $M2$
    If(DMC at $M2 > = 0$ $\&\&$ RMC at $M2 >=1)$
    {

$$U_o = (\alpha \times U_r + \beta \times U_c + \gamma \times U_b) \qquad (2)$$

    }
Else If a mobile node $M1$ comes in contact with at least one of the nodes $M2$
    If(DMC at $M2 = 0$ $\&\&$ RMC at $M2 =0)$
    {

$$U_o = (\beta \times U_c + \gamma \times U_b) \qquad (3)$$

    }
For all nodes $(S, M1, R)$
// RMB is less than the maximum size of a message and CD is not enough to transmit
a maximum sized message.
If $(RMB < RMB_t$ and $CD < CD_t)$
        { No message is forwarded..}
Else
{ If $M2$ has highest overall utility value (U_o) among the mobile nodes in contact
        { Forward a message to $M2$. }
Else If there are mobile nodes with equal overall utility values (U_o)
        { Forward a message to any node randomly. }   }

4   Experimentation and Results

4.1   Experimentation

The experimentation has been carried out in the Opportunistic Network
Environment (ONE) [12] simulator. Following assumptions have been made for all
scenarios. The size of the network termed as *world size* is set to 4000 × 3500 m^2.
Terminal nodes are placed on the extreme ends of the vehicular network whereas
relay nodes are placed on the intersections of the roads on the network. A number of
source and destination terminal nodes in the network are fixed to *five* each. All
source nodes are placed near each other but are not in the range of each other. The
same analogy has been marked for the destination nodes. The number of relay
nodes in the network is fixed to *ten*. These are placed on different intersections of

the roads in the network to increase in message delivery ratio as mentioned in the case 2 of the algorithm. The number of mobile nodes is varied from 20 to 140 in the network, leading to the variations in the node density.

ONE simulator is assumed to have no MAC and physical layers, interfaces specifications have done the abstraction of these layers [12]. The Bluetooth interface is assumed with the transmission range of all the nodes as 100 m and transmission speed is 3 Mbps. The random way point model is employed for node mobility. The size of a message is chosen randomly from 512 KB to 1 MB. Messages creation interval has been set randomly between 40 and 50 s. TTL of a message is set to 300 min. Each simulation run is for 43,200 s (i.e. for 12 h).

The following scenario has been experimented. In this scenario, the number of mobile nodes is varied from 20 to 140 and the buffer size of all nodes is fixed at 50 MB. After experimentation, we have come to a conclusion that the typical value of that α should be 0.7 and so values β and γ become 0.15.

Following performance measures are used for comparing the results.

1. *Delivery ratio*: This is defined as the ratio of the number of messages delivered to the destination to a number of all messages created.
2. *Average delivery latency*: This is defined as the ratio of total latency of delivered messages to the destination to the total number of delivered messages. Note that the delivery latency is calculated only for delivered messages and not for all created messages.
3. *Overhead ratio*: This is defined as the ratio of difference between number of relayed messages and number of delayed messages to the number of delayed messages.

4.2 Results

We compare Max-Util algorithm with PRoPHET (uncontrolled replication based utility algorithm), Spray-and-Wait (controlled flooding algorithm). For the Spray-and-Wait scheme, we set the maximum number of copies to *six* (to ensure high delivery ratio).

As shown in Fig. 1, when the number of mobile nodes is increased from 20 to 160, the delivery ratio increases for all the algorithms in almost all situations. The Max-Util algorithm outperforms all other algorithms in terms of delivery ratio. This can be explained as follows. When the node density increases, the number of contacts made by a node with destination and relay nodes increases, thereby increasing the destination meeting count (DMC) as well as relay node meeting count (RMC). As per Max-Util algorithm, DMC and RMC have a higher contribution to the overall utility and a message is forwarded to a node with the highest overall utility value.

As shown in Fig. 2, as the number of nodes increases, the message delivery latency at times decreases and increases at times. The decrease in latency is because

Fig. 1 Delivery ratio versus node density

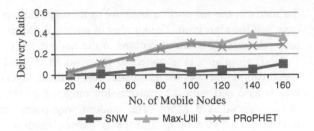

Fig. 2 Average delivery latency versus node density

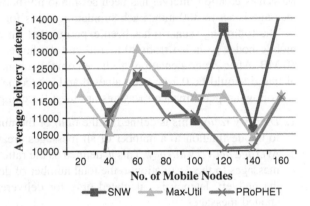

Fig. 3 Overhead ratio versus node density

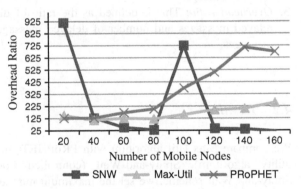

more nodes would be involved in message forwarding, delivery ratio increases as well and thereby the amount of time required for messages reaching destination nodes reduces. The increase in the delivery latency is because of the combination of underlying routing algorithm and the mobility. This is in general true for all the algorithms. For Max-Util algorithm, the same analogy is applied as well. For Max-Util, as nodes store history about other nodes decision making becomes faster and thereby reduction in time delivered messages, hence the delivery latency is low despite good message delivery ratio.

Figure 3 shows the overhead ratio for all the routers against node density. It is observed that the Max-Util has less overhead ratio as compared to PRoPHET algorithm but has higher overhead ratio as compared to Spray-and-Wait algorithm.

This is because Max-Util not only relays fewer messages compared to PRoPHET algorithm, but also has a good delivery ratio. In Spray-and-Wait algorithm, nodes spray very limited number of messages to the nodes (In our case, Spray-and-Wait assumes *six* copies) until they meet the destination node directly.

Also, as shown in Fig. 3, as the node density increases from 20 to 160, overhead ratio for Max-Util decreases in almost all cases of varying density with exception of initial stages. Initially (from a node density of 20 to 80) as node density increases, the overhead ratio is fluctuating because the delivered messages are very much less compared to relayed messages. Decrease in overhead ratio is because of the increase in no of nodes as well as the increase in number of relayed messages. For a node density above 80, the overhead ratio increases and is almost increasing constantly as the relative increase in delivery ratio and relayed messages.

5 Conclusion

In this paper, we have devised a selective utility based forwarding strategy for a VDTN called Max-Util. Instead of making use of forwarding messages without applying any intelligence, Max-Util forward messages to those nodes which have more chances of delivering the messages. This choice is based on the history information stored in the nodes. Simulation results show that Max-Util performs much better for two performance measures of delivery ratio and overhead ratio when compared to PRoPHET algorithm. It outperforms Spray-and-Wait algorithm for the delivery ratio but has a higher overhead ratio. Thus, it is fair to say that Max-Util succeeds in its goal of providing efficient routing in an intermittently connected network with a low routing overhead.

6 Future Work

In our algorithm, we have assumed that the nodes manage their queues in a random fashion, i.e., whenever a new message arrives to a full queue, a randomly chosen message from the queue is dropped. It might be better to use some other strategy here. For example, we could assign priorities to messages and based on the chances of their delivery. Finally, to reduce the buffer requirement at the nodes for further increase in the performance, injection of acknowledgement messages by a destination node to mobile nodes (indicating the receipt of certain messages) might be helpful. In our future work, we propose to include these improvements to the proposed algorithm and perform extensive experimentation for performance evaluation.

References

1. Johnson, D., Maltz, d.: Dynamic source routing in ad-hoc wireless networks. In: ACM SIGCOMM, Aug 1996
2. Perkins, C., Royer, E.: Ad hoc on-demand distance vector routing. In: 2nd IEEE workshop on mobile computing systems and applications, Feb 1999
3. Spyropoulos, T., Rais, R.N.B., Turletti, T., Obraczka, K., Vasilakos, T.: Routing for Disruption Tolerant Networks: Taxonomy and Design. In: ACM/Kluwer Wireless, Networks (WINET), vol. 16. No. 8, (2010)
4. Lindgren, A., Doria, A., Schelen, O.: Probabilistic routing in intermittently connected networks. Lect. Notes Comput. Sci. **3126**, 239–254 (2004)
5. Spyropoulos, T., Psounis, K., Raghavendra, C.S.: Spray and Wait: Efficient routing in intermittently connected mobile networks. In Proceedings of ACM SIGCOMM workshop on delay tolerant networking (WDTN-2005)
6. Spyropoulos, T., Turletti, T., Obraczka, K.: Routing in delay tolerant networks comprising heterogeneous node populations. IEEE Trans. Mob. Comput. (TMC) **8**(8), 1132–1147 (2009)
7. Jain, S., Fall, k., Patra, R.: Routing in a delay tolerant network. In: Annual international conference of the special interest group on data communication, ACM SIGCOMM'04, Aug 30-Sept 3: Portland. Oregon, USA (2004)
8. Spyropoulos, T., Psounis, K., Raghavendra, C.: Spray and focus: Efficient mobility-assisted routing for heterogeneous and correlated mobility. In Proceedings of IEEE PERCOM, on the international workshop on intermittently connected mobile ad hoc networks (ICMAN) (2007)
9. Vahdat, A., Becker, D.: Epidemic routing for partially connected ad hoc networks. Technical Report, CS-06, Duke University (2000)
10. Nelson, S.C., Bakht, M., Kravets, R.: Encounter-based routing in DTNs. In Proceedings of the IEEE INFOCOM. Rio de Janerio, Brazil (2009)
11. Rubinstein, M., et al.: Measuring the capacity of in-car to in-car vehicular networks. IEEE Commun. Mag. **47**(11), 128–136 (2009)
12. Ari Keränen, Jörg Ott, Teemu Kärkkäinen, The ONE Simulator for DTN Protocol Evaluation. In: Proceedings of the 2nd international conference on simulation tools and techniques, SIMU Tools'09, Rome, Italy (2009)

A Novel Approach for Real-Time Data Management in Wireless Sensor Networks

Joy Lal Sarkar, Chhabi Rani Panigrahi, Bibudhendu Pati
and Himansu Das

Abstract Wireless sensor networks (WSNs) become an emerging area of research now-a-days. It plays an important role to support a wide range of applications. In this work, we propose an architecture for data management in WSNs where in each cluster, sensor nodes transfer data to their cluster-head (CH) and among the CHs one CH is selected based on the highest energy label. The CH aggregates data and reduces the size of the data to be transmitted to the base station. We have also proposed an algorithm which determines how fast the data is transferred from the source node to the base station.

Keywords Wireless sensor networks · Distributed data management · Real-Time data management · Assembly line scheduling

1 Introduction

A WSN is composed of a large number of self-contained devices called sensor nodes [1, 2]. Sensor nodes have a lot of functionalities such as sensing, processing and transmitting information about the environment on which they are deployed.

J.L. Sarkar (✉) · C.R. Panigrahi · B. Pati
Department of Computer Science and Engineering, C.V. Raman College of Engineering,
Bhubaneswar 752054, India
e-mail: joy35032@rediffmail.com

C.R. Panigrahi
e-mail: panigrahichhabi@gmail.com

B. Pati
e-mail: bpati@iitkgp.ac.in

H. Das
School of Computer Engineering, KIIT University, Bhubaneswar, India
e-mail: das.himansu2007@gmail.com

© Springer India 2016
A. Nagar et al. (eds.), *Proceedings of 3rd International Conference
on Advanced Computing, Networking and Informatics*, Smart Innovation,
Systems and Technologies 44, DOI 10.1007/978-81-322-2529-4_62

599

WSNs collect information from the target environment by the sensor nodes and sensor nodes are usually distributed in a geographical area [3]. In addition to one or more sensors, each node in a sensor network is typically equipped with a radio transceiver (or other communication devices), a small microcontroller and an energy source. WSNs have a number of applications which are generally connected to the databases [4]. But for certain applications processing time becomes increasingly critical. There are two main approaches to data storage and query processing in WSNs. These approaches are warehousing approach and distributed approach [1]. In case of warehousing approach, there are central databases and in distributed approach each sensor node is considered as a data source.

In this work, we propose an algorithm in which a source node selects a path where total time to reach the destination is minimum. In WSNs, there may be more number of nodes deployed in a distributed manner. The size of memory and storage on board are often limited primarily by economic consideration and is improving day by day. In distributed approach, the sensors act as local databases. In case of real-time databases, data must represent the current state of the environment on which they are captured and is very important and are especially the risk areas. The real time applications are subjected to process a very huge amount of data and are having temporal constraints. For that many transactions may miss their deadlines which result in the degradation of the performance [5].

1.1 Motivation

It is a challenging task to work on real- time data management system in wireless environment as wireless environment is always vulnerable. Dave and Gupta proposed an architecture for real -time data management [6]. In their architecture, a CH collects the data from the sensor nodes and aggregates the data and reduces the size of the data that need to be transmitted to the base station. But to collect instant data from the environment is a challenging task. So, there should be some algorithm which can collect the instant data in an energy efficient way.

1.2 Contributions

In this section, we present our major contributions. We first propose an architecture for real-time data management. Then we design an algorithm to collect instant data in a real-time system and also we have designed an algorithm to find the shortest path to increase the lifetime of the network.

2 Related Works

In distributed data management systems, there are several methods such as in-network processing, acquisition query processing, cross-layer optimization, data-centric data or query dissemination has been proposed [7]. In case of in-network processing, Tiny DB and COUGAR [1] were the first optimization techniques for reducing the data transmission in WSNs. Noel et al. [8] proposed a real time spatiotemporal indexing scheme for agile sensors, named as *pasTree* [7]. Dave and Gupta proposed an architecture where in each cluster, the sensor nodes communicate data to their cluster heads. Cluster heads have some protocol to reduce the data and the data is transmitted to the base station [9]. In [1], authors developed an approach for in-network processing based on clustering. In this approach, a network is composed of clusters and each cluster is managed by a CH. A child node sends its reading periodically to the CH. In this architecture, each node embeds a query layer and acts as a query proxy between network and the application layer. In [10], the authors proposed an algorithm for reducing the data transmission in WSNs. In [6], Dave and Gupta proposed a set of algorithm to minimize the energy consumption of the sensor nodes by considering a real-time scenario where data are gathered from the source nodes and transmitted to the sink.

3 Proposed Architecture

We have proposed an architecture based on clusters in order to handle a reliable and real-time data management. In distributed networks, the clusters and the CHs are formed in arbitrary manner [7]. In each cluster, a sensor node communicates data to the corresponding CH that aggregates data and reduces the size of the data. But, in our architecture the CHs of different clusters form another cluster and based on their energy label a particular CH is selected and then the selected CH aggregates data and reduces the size of the data which is transmitted to the base station. The proposed architecture is shown in Fig. 1. In Fig. 1, CH_1, CH_2,..., CH_n represent the cluster-heads. CH_{new} is the newly selected CH from CH_1, $CH2$,...,CH_n. Action Relay Station (ARS) is used to prevent the excessive dissipation of energy. Also to timely perform actions, the ARS sends data to the base stations via another ARS.

3.1 Algorithm to Select a Cluster-Head

We propose an algorithm to select a cluster-head. In our algorithm, from all clusters, we select a CH. The CH aggregates and reduces the size of the data. Among all the CHs, a new cluster-head (called CH_{new}) is selected based on their highest energy labels. The CH_{new} is again aggregates the data and reduces the size of the

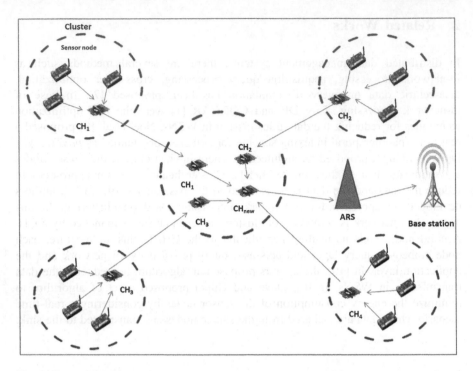

Fig. 1 Proposed architecture

data and finally data is sent to the base station via ARS. The algorithm to select a cluster-head is given in Algorithm 1 and the steps involved in Algorithm 1 are as follows:

Algorithm 1: To select a cluster-head

Step1: Find the cluster-head (CH_i) in each cluster (C_i), where i = 1, 2, 3, ..., n.
Step2: CH_i aggregates the data and reduces the size of the data.
Step3: Among all the CH_i, a new cluster-head (CH_{new}) is selected based on their highest energy label.
Step4: CH_{new} aggregates the data and reduces the size of the data.
Step5: The data from the CH_{new} is sent to the base station via ARS.

3.2 Algorithm to Determine the Shortest Path

We propose another algorithm, where sensor nodes send the necessary information to the sink via other sensor nodes by selecting the path where time consumption is minimum with respect to all other paths. In this algorithm, a sensor node sends the data to its neighbouring nodes using assembly line scheduling algorithm. There may be many options to select a path but it can select only those paths where time consumption is minimum. For example, suppose an information X is coming from the source node S and if there are two paths to reach the second node where total time of the first path is 50 ms and the other is 52 ms, then information X is transferred via first path because it takes less time. Let there are i number of lines and j number of nodes. Let fastest possible time to get a source node from the starting point through $node_{i,j}$ be $fi[j]$. The entry time to enter $node_{1,1}$ and $node_{2,1}$ be $e_{i,j}$. The time to transfer a nodes away from line i after passing through $node_{i,j}$ is $t_{I,j}$ and let X_i be the exit time. The algorithm to determine the shortest path is given in Algorithm 2.

Algorithm 2. To determine the shortest path

1. f1 [1] $\leftarrow e_1 + a_{1,1}$	10. then f2[j]$\leftarrow$f2[j-1] + $a_{2,j}$
2. f2 [1]$\leftarrow e_2 + a_{2,1}$	11. l2[j]$\leftarrow$2
3. for j$\leftarrow$2 to n	12. else f2[j]$\leftarrow$f1[j-1] + $t_{1,j-1}$ + $a_{2,j}$
4.do if f1[j-1] + $a_{1,j}$<=f2[j-1] + $t_{2,j-1}$ + $a_{1,j}$	13. l2[j]$\leftarrow$1
5. then f1[j]$\leftarrow$f1[j-1] + $a_{1,j}$	14. if f1[n]+x1<=f2[n] + x_2
6. l1[j]$\leftarrow$1	15. then f*=f1[n]+ x_1
7.else f1[j]$\leftarrow$f2[j-1]+$t_{2,j-1}$ + $a_{1,j}$	16. l*=1
8. l1[j]$\leftarrow$2	17.else f*=f2[n]+x
9. if f2[j-1]+$a_{2,j}$<=f1[j-1] + $t_{1,j-1}$+ $a_{2,j}$	18.l*=2

In an assembly line [11] there may be any number of lines and the data move from one node to another. The information processing system in an assembly line is shown in Fig. 2. In Fig. 2, $node_{1,1}, node_{1,2,...}, node_{1,n}$ represent the nodes in line1. Similarly, $node_{2,1}, node_{2,2}, ..., node_{2,n}$ represent the nodes in line2 and so on. In Fig. 2, e_1 represents the entry time to reach $node_{1,1}$ and similarly e_2 represents the entry time to reach node $node_{2,1}$. Now, let us consider the first line where, identified node A sends the required information to the $node_{1,1}$ [time required is $e_1 + a_{1,1}$] then $node_{1,2}$ and so on. Ultimately, when information is reached at the last node ($node_{1,n}$), then it exits (total time = time to reach $node_{1,n} + x_1$). The same process is followed also in case of second line. If information is passed from the first line (suppose from $node_{1,1}$) to the second line then total time required is $(a_{1,1} + t_{1,1} + a_{2,2})$. Here we can see that extra time $(t_{1,1})$ is needed. Let the fastest possible way for the data to get from the starting point through sensor $nodes_{1,j-1}$ (where $j >= 2$). Let the time from A to $node_{1,1}$ is $(e_1 + a_{1,1})$, A to $node_{2,1}$ is $(e_2 + a_{2,1})$, A to $node_{1,2}$ $(e_2 + a_{2,1} + t_{2,1} + a_{1,2})$ and also A to $node_{2,2}$ is $(e_1 + a_{1,1} + t_{1,1} + a_{2,2})$. Here, $e_1 + a_{1,1} <= e_2 + a_{2,1} + t_{2,1} + a_{1,2}$. So we take

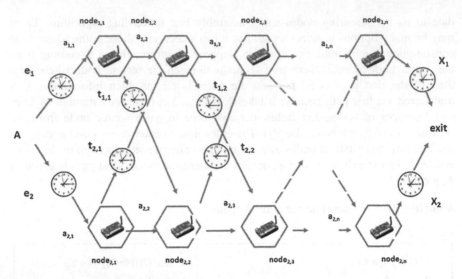

Fig. 2 Information/data processing system

$(e_1 + a_{1,1})$ and the same process is continued for other nodes. When data is reached at the last node ($node_{1,n}$ or $node_{2,n}$) then extra time is also added and finally data is transmitted to the base station.

4 Experimental Set up

For implementation of our proposed architecture, we use MATLAB (R2013a) and we take different values at a particular moment which may change from time to time. We take 60 sensor nodes for each line and then we take the average time. We compare our proposed architecture with Dave and Gupta's approach [6], where in each cluster, CH aggregates the data and reduces the size of the data and then data is transmitted to the sink via ARS but to get instant data we use another CH which aggregates more data and reduces the size of the data.

5 Analysis of Results

The time to reach the data from the source node to the neighbouring nodes varies from node to node. The minimum time to reach the first line nodes is shown in Fig. 3 and minimum time to reach the second line nodes is shown in Fig. 4. The maximum time to reach nodes of first and second line is shown in Figs. 5 and 6

Fig. 3 Minimum time to
reach the first line nodes

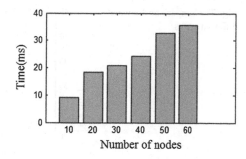

Fig. 4 Minimum time to
reach the second line nodes

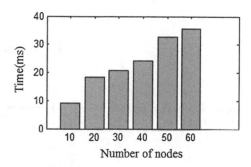

Fig. 5 Maximum time to
reach the first linenodes

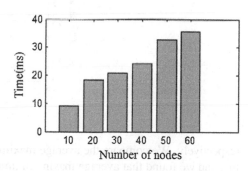

Fig. 6 Maximum time to
reach the second line nodes

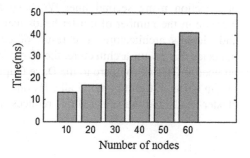

Fig. 7 Comparison between
different paths

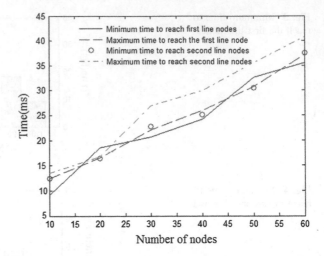

Fig. 8 Comparison between
two architectures

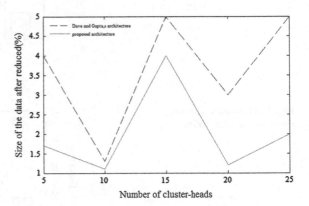

respectively. We compare the average maximum as well as minimum time for each
line and we found that average maximum time to reach the second line nodes is less
as compared to first line and is shown in Fig. 7. So information will be sent to the
base station using second line. We have also compared the size of data with
increase in the number of cluster heads using our proposed architecture and Dave
and Gupta's architecture. The result of comparison is shown in Fig. 8 and it
indicates that in our architecture, the size of the data decreases with increase in the
number of CHs as compare to the Dave Gupta's architecture. This is due to the fact
that in our architecture among all CHs one is selected based on the energy level of
cluster heads and the selected CH reduces the size of data.

6 Conclusion

In this work, we propose an architecture where information of the first node is sent via other neighbouring nodes to the sink and each sensor node collects different kinds of information from the environment. This work has reviewed the solution for managing sensor data in real-time as it is important to response a particular query within its deadline. In future, we will consider the frequency changes of the neighbouring nodes after receiving the data from the identified node and we will also try to incorporate the different query processing at the same time from different nodes.

References

1. Diallo, O., Rodrigues, J., Sene, M.: Real-time date management on wireless sensor networks: a survey. J. Netw. Comput. Appl. (Elsevier), **35**(3), 1013–1021 (2012)
2. Das, H., Naik, B., Pati, B., Panigrahi, C.: A survey on virtual sensor networks framework. Int. J. Grid Distrib. Comput. **7**(5), 121–130 (2014)
3. Pati, B., Misra, S., Saha, B.K.: Advanced network technologies virtual lab: a human-computer interface for performing experiments on wireless sensor networks. In: 4th International Conference on Intelligent Human Computer Interaction (IHCI), 1–5 (2012)
4. Mokashi1, M., Alvi, A.S.: Data management in wireless sensor network: a survey, Int. J. Adv. Res. Comput. Commun. Eng. **2**(3), (2013)
5. Lee, W.Y.: Energy-efficient scheduling of periodic real-time tasks lightly loaded multicore processors, IEEE Trans. Parallel Distrib. Syst. **23**(3), (2012)
6. Gupta, S., Dave, M.: Real-time approach for data placement in wireless sensor networks. Int. J. Electron. Circuits Syst. **2**(3), 132–139 (2008)
7. Diallo, O., Rodrigues, J., Sene, M.: Distributed data management techniques for wireless sensor networks. IEEE Trans. Parallel Distrib. Syst. **26**(2), 604–623 (2013)
8. Noel, G., Servigne, S., Laurini, R.: The Po-tree: a real-time spatiotemporal data indexing structure. J. Develop. Spat. Data Handling (Springer), 259–27 (2005)
9. Callaway, E.H.: Wireless Sensor Networks: Architectures and Protocols. CRC press, (2004)
10. Rooshenas, A., Rabiee, H.R., Movaghar, A., Naderi, M.Y.: Reducing the data transmission in wireless sensor networks using the principal component analysis, In Proceeding of Sixth International Conference on Intelligent Sensors, Sensor Networks and Information Processing (ISSNIP), (2010) 133–138
11. Cormen, Thomash H., Leiserson, Charles E., Rivest, Ronald L., Stein, Clifford: Introduction to Algorithms, 3rd edn. The MIT PressCambridge, Massachusetts London, England (2009)

Contact Frequency and Contact Duration Based Relay Selection Approach Inside the Local Community in Social Delay Tolerant Network

Nikhil N. Gondaliya, Dhaval Kathiriya and Mehul Shah

Abstract Social Delay Tolerant Network (DTN) allows mobile devices to exchange the data opportunistically without end-to-end path when they come in contact. As the connectivity is opportunistic so, the routing task becomes very challenging. By investigating the real social network traces, it exhibits that a node tends to meet a certain group of nodes more frequently and regularly compared to other nodes outside their group which forms the local community. There are many social characteristics (e.g. centrality, similarity) of human beings which are exploited to select the appropriate relay node. Many community based routing protocols proposed in the literature which select the relay node as one of the community members or the most central node inside the community. In this paper, we propose two different approaches of the relay selection inside the local community: contact frequency based approach and contact duration based approach. When message carrier and encountered node belongs to the same community of the message's destination, the relay selection is done based on contact frequency and contact duration of the node with the message's destination. It is usually inside the social community that people in the same community meet very often and spend more time with only a few members of their community than all other members. To evaluate the performance of our approaches, we choose the real traces from the campus and the conference environment. The simulation result on the real traces

N.N. Gondaliya (✉) · M. Shah
G H Patel College of Engineering and Technology, Vallabh Vidyanagar, Gujarat, India
e-mail: nikhilgondlaiya@gcet.ac.in

M. Shah
e-mail: mehulshah@gcet.ac.in

D. Kathiriya
Information Technology Department, Anand Agriculture University, Anand, Gujarat, India
e-mail: drkathiriya@gmail.com

© Springer India 2016
A. Nagar et al. (eds.), *Proceedings of 3rd International Conference on Advanced Computing, Networking and Informatics*, Smart Innovation, Systems and Technologies 44, DOI 10.1007/978-81-322-2529-4_63

shows that the proposed approaches of selecting a relay node outperform better in terms of delivery ratio for the campus environment where as they perform quite similar as an existing scheme in the conference environment.

Keywords Community detection · Social delay tolerant networks · Contact frequency · Contact duration

1 Introduction

The Delay tolerant networking has been proposed to support wireless network application where end to end path is not available due to frequent changing network topology and high node mobility. These properties make the routing in DTN a very challenging problem. Message transmission in DTN follows store-carry-and-forward paradigm. If the source node encounters any other node then a message is forwarded and this process is repeated until it is reached to the destination via intermediate nodes. When a node has forwarding opportunity then all encountered nodes can be candidate to relay the message. Relay selection and forwarding decision made by current node is based on several routing strategies. Communication in social DTN is made between human being who carries a mobile device with them so; performance depends on human mobility patterns and their social behavior. It has been proved that human mobility is strongly correlated with social relationship between individuals [1]. Individuals meet family and friends frequently and regularly, to colleague less frequently and with stranger accidentally.

In the recent years, social relationship based approaches [2–5] have been used to help forwarding in social delay tolerant network. They use one or more metric like similarity, degree and betweenness centrality and community information to select the relay node. It is revealed that the community based methods perform well under human mobility environment than other metrics [3]. Many centralized community detection methods are proposed in literature to detect community on the graphs, but it is very hard to detect it in DTN due to its distributed nature. Hui [6] first defines the idea of distributed community detection with three different methods (Simple, K-Clique and Modularity) to detect the community structure based on aggregated contact duration in DTN environment. K-Clique method is very efficient method to detect the community structure than other two. The community based routing algorithm forwards the message to intermediate nodes on the path to the destination until it reaches to the member of a community of the message's destination which has higher chances to deliver. It also uses the strategy of the message forwarding inside the local community (e.g. Local centrality) means the node which is the most central in the community has more chances to meet all other nodes in its community

[3]. In this paper, we propose two different approaches of the relay node selection in the local community: Contact Frequency Based Approach (CFBA) and Contact Duration Based Approach (CDBA). When a message carrier node meets to another node and if it belongs to the local community of the message's destination then the message is forwarded to that node. But if both the nodes are in the community of the message's destination, then CFBA selects the node which has the highest contact frequency and CDBA selects the node which has more contact duration with the message's destination. Our work is motivated by Greedy-Total [7] in which the node x_0 forwards the message to the node x_1 if it has contacted the destination more number of times than x_0.

The rest of the paper is organized as follows: Sect. 2 presents the related work, and Sect. 3 describes the forwarding approach based on contact frequency and contact duration. Section 4 evaluates the performance of protocol on real traces which is followed by conclusions and future work in Sect. 5.

2 Related Work

Different routing protocols have been proposed in the recent years to optimize the performance of DTN. These protocols select the relay node either without any knowledge or based on some local information at the node. Author categorized routing protocols in flooding based approach, forwarding based approach and social based approach [8]. Flooding based approaches are very simple [9, 10] and deliver the message to all encountered nodes. These methods are more efficient, but incur a heavy cost. Forwarding based methods use some strategies for selection of the relay node like encounter count [11, 12] contact time, most recent encounter time. Social based methods are on the basis of social network analysis study which tries to discover the different characteristics like contact frequency, contact duration which remains stable over the time between mobile devices carried by the human beings. Our focus of study is social based approaches hence we explore only those methods.

Social based routings are influenced by study of the social network analysis and discovered the different social characteristics from the contact between mobile devices. The majority of the routing methods which is based on the social characteristics build the community structures dynamically from the local information collected at each node like contract frequency or contact duration or both. Label routing [2] works based on the social community concept with the community are labeled and select the relay node which is in the community of destination of the message. Bubble Rap [3] applies the idea of the community and the centrality to relay the message. It uses a distributed community detection method [6] to detect the community from the aggregated contact duration between the node pairs.

Friendship routing [4] construct friendship community using the social pressure metric which measures the quality of friendship based on their long lasting and regular contacts. In [13] the community detection is done based on contact frequency and contact duration between the node pairs. [14] Defines similarity metrics from nodes' encounter history to describe the neighboring relationship between node pairs and detect the community based on strong intra-community connections. Very few [3] community based routing protocols proposed the strategy to select the relay node inside the local community. Our main focus of this work is to study the social relationship between the member of the same community based on contact duration and contact frequency which is discussed in the next section.

3 Contact Duration and Contact Frequency Based Relay Selection Approach

Community based message forwarding would be carried out in two phases: inter community forwarding and intra community forwarding. In the first phase, the message is forwarded inside the local community to one of its member or destination directly if it is encountered. In the later phase, the message is forwarded among the members of the community based on the higher value of some utility (e.g. local centrality). We use K-Clique distributed community detection method [6] to detect the local community at each node based on the aggregated contact duration for node pair.

Our approaches use inter community forwarding scheme which is the same as found in [3]. If encountered nodes do not belong to the local community of the message's destination, then the node which is the most central (global centrality) in the network is chosen as the relay node. But if both the nodes: carrier node and encountered node are in the same community of the message's destination, then how to select the appropriate relay node. This tie is broken by the local centrality metric [3] means the node which is the most central in its community, has more chances to deliver the message. After investigating the contact duration and contact frequency between the node pairs inside the same community, we come to know that nodes meet with only few members of their community very frequently and for a longer period of time. Based on this intuition, we propose the relay selection approaches which select the relay node inside the local community which has the highest contact duration (CDBA) and contact frequency (CFBA) with the message's destination. The following algorithm shows the steps for the message forwarding using both the approaches.

Algorithm 1. CFBA and CDBA algorithm

$CF_{n0}(n1)$: Contact frequency of node n_o with node n_1
$CD_{n0}(n1)$: Total contact duration of node n_o with node n_1
GC: Global centrality of node
When node n_0 with message m encounter node n_1
if n_1 is the destination of message m **then**
   forward message m to n_1
end if
if $dest(m)$ is in the local community of n_1 **then**
   forward the message to n_1
             //inter community forwarding
else if n_0 and n_1 both are not in local community of destination **then**
   if $(GC(n_0) < GC(n_1))$ **then**
      forward the message to n_1
   else
      keep message with n_0
   end if
end if
                 // CFBA
else if n_0 and n_1 both have destination in their local community **then**
   if $(CF_{no}(dest(m)) < CF_{n1}(dest(m)))$ **then**
      forward the message to n_1
   else
      keep the message m with n_0
   end if
                 // CDBA
else if n_0 and n_1 both have destination in their local community **then**
   if $(CD_{no}(dest(m)) < CD_{n1}(dest(m)))$ **then**
      forward the message to n_1
   else
      keep the message m with n_0
   end if

---

4 Simulation Results

In order to evaluate the proposed relay selection approaches, we have used ONE simulator [15] which is discrete event driven simulator and used real time datasets as input to the simulator.

4.1 Dataset and Simulation Configuration

We have chosen five different datasets in which three are from campus and two are from conference environment. The characteristic of dataset is shown in Table 1. In

Table 1 Characteristics of five real trace dataset

Dataset	Device	Network type	Number of nodes	Duration (days)	Environment
Cambridge	iMote	Bluetooth	36	11	Campus
Pmtr	Pmtr	–	49	19	
MIT Reality	Phone	Bluetooth	97	21	
Infocom05	iMote	Bluetooth	41	3	Conference
Infocom06	iMote	Bluetooth	98	3	

the simulation, we set a familiar threshold and k value in K-Clique method in such a way that the number of the communities is as per the experiment group. The messages are generated at intervals of 30–40 s with the size of the message is 25 k and buffer size of each node is set to 10 MB. Time To Live (TTL) values vary from 10 min to 3 weeks depending upon the duration of each trace file.

4.2 Results

We evaluate the performance of protocols using three metrics: (a) delivery ratio: the ratio of the number of delivered message to the number of generated messages (b) average delivery delay: the average time for delivering all the messages to its destination (c) number of forwarding: the total number of the messages forwarded in the network.

Both the forwarding approaches are compared with the local centrality metric [3] of the BubbleRap. Figure 1 shows that delivery ratio of CFBA and CDBA is on the average 15 and 25 % higher than BubbleRap from TTL value of 3 h in Cambridge and Pmtr dataset respectively. They also achieve on average 20 % more delivery ratio from TTL of 6 h in MIT dataset. Figure 1d shows that delivery ratio is the almost similar for InfoComm05 trace. We have not shown the results for InfoComm06 due to space constraint, but it is similar to InfoComm05. The reason for performing well in the campus environment is that people come in contact with each other very frequently and spend more time due to strong social tie between them. Even the graph density and modularity is higher in the campus than the conference environment and due to that reason it is not possible to exploit contact duration and contact frequency between the node pairs. Moreover, in the conference environment, there is more diversity of the people who came from different regions and they might interact very less frequently and for shorter time duration. It is also concluded from the simulation that CDBA performs 2–5 % better than CFBA. Contact frequency and contract duration are the best metrics than the local centrality in the selection of the relay node in the local community for achieving the higher delivery ratio.

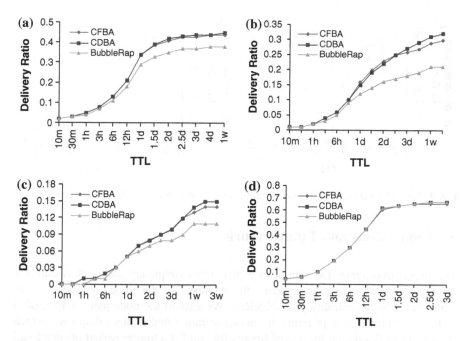

Fig. 1 Delivery ratio versus time to live. **a** Cambridge. **b** Pmtr. **c** MIT. **d** InfoComm05

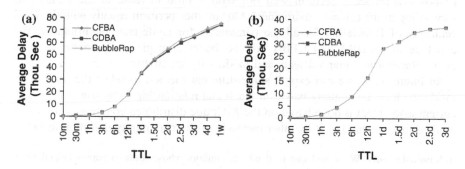

Fig. 2 Average delay (In Thousands Sec.) versus time to live. **a** Cambridge. **b** InfoComm05

Figure 2 demonstrates average delivery delay comparison. We can see that the delivery delay is increased by maximum 5 % for CDBA and CFBA than BubbleRap. The same result is achieved with other datasets of the campus environment. The reason is that the message is forwarded only when the message carrier node encounters the appropriate relay node in case of CDBA and CFBA.

A number of the message forwardings are increased in CDBA and CFBA than BubbleRap in all datasets for campus environment in search of the suitable relay node as shown in Fig. 3. The performance of all the parameters is the similar for a proposed approaches with the local centrality in the conference environment.

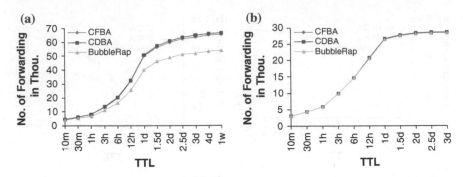

Fig. 3 No. of forwarding (in Thousands) versus time to live. **a** Cambridge. **b** InfoComm05

5 Conclusions and Future Work

The paper investigates the significance of contact frequency and contact duration utility to forward the message among the members of the same community compared to the local centrality of BubbleRap. We exploit the node pair's relationship in the local community in terms of contact duration and contact frequency which shows that nodes do not meet very frequently and for a longer period of time to all the members of their community. The improvement in delivery ratio shows that our proposed approaches perform well but with a little increase in the number of forwarding in the campus environment. Our utilities perform equally with the local centrality of BubbleRap for all the parameters for conference environment. We conclude that CDBA and CFBA are the better strategies only in the campus environment where graph density and modularity are higher.

In future work, we can exploit some different method to select the relay node inside the local community based on the social relationship in human beings. The current assessment is done based on the K-Clique distributed community detection method, but we can test it on other methods of the community detection in DTN.

Acknowledgments We would like to thank the authors whose research papers helped us in making this research.

References

1. Zhu, T., Wang, C., Liu., D.: Commuity roamer: a social based routing algorithm in opportunistic mobile networks. In: Proceedings of 14th International Conference, (ICA3PP 2014), Dalian, China, 24–27 August 2014
2. Hui, P., Crowcroft, J.: How small labels create big improvements. In: Proceedings of the Fifth IEEE International Conference on Pervasive Computing and Communications Workshops (PERCOMW 2007), pp. 65–70 (2007)

3. Hui, P., Crowcroft, J., Yoneki, E.: Bubble rap: Social-based forwarding in delay tolerant networks. IEEE Trans. Mob. Comput. **10**(11), 1576–1589 (2010)
4. Bulut, E., Szymanski, B.K.: Exploiting friendship relations for efficient routing in mobile social networks. IEEE Trans. Parallel Distrib. Syst. **23**(12), 2254–2265 (2012)
5. Mtibaa, A., May, M., Diot, C., Ammar, M.: Peoplerank: social opportunistic forwarding. In: Proceedings of the 29th Conference on Information Communications (INFOCOM 2010), pp. 111–115 (2010)
6. Hui, P., Yoneki, E., Chan, S.Y., Crowcroft, J.: Distributed community detection in delay tolerant networks. In: Proceedings of 2nd ACM/IEEE International Workshop on Mobility in the Evolving Internet Architecture (MobiArch 2007), pp. 1–8 (2007)
7. Erramilli, V., Crovella, M.: Forwarding in opportunistic networks with resource constraints. In: Proceedings of the Fourth ACM Workshop on Challenged Networks (CHANTS 08) (2008)
8. Shen, J., Moh, S., Chung, I.: Routing protocols in delay tolerant networks: a comparative survey. In: Proceedings of the 23rd International Technical Conference on Circuits/Systems, Computers and Communications, p. 1577 (2008)
9. Vahdat, A., Becker, D.: Epidemic Routing for Partially-Connected Ad Hoc Networks. Duke University, Durham (2000)
10. Spyropoulos, T., Psounis, K., Raghavendra, C.S.: Spray and wait: an efficient routing scheme for intermittently connected mobile networks. In: Proceedings of the, ACM SIGCOMM Workshop on Delay-Tolerant Networking (WDTN 2005), pp. 252–259 (2005)
11. Lindgren, A., Doria, A., Schelén, O.: Probabilistic routing in intermittently connected networks. SIGMOBILE Mob. Comput. Commun. Rev. **7**(3), 19–20 (2003)
12. Nelson, M.B., Kravets, R., Encounter-based routing in DTNs. In: Proceedings of the IEEE INFOCOM, pp. 846–854 (2009)
13. Wei, K., et al.: On social delay-tolerant networking: aggregation, tie detection, and routing. IEEE Trans. Parallel Distrib. Syst. **99**, 1–10 (2013)
14. Li, F., Wu, J.: LocalCom: a community-based epidemic forwarding scheme in disruption-tolerant networks. In: IEEE Secon (2009)
15. Keranen, A., Ott, J., Karkkainen, T.: The one simulator for dtn protocol evaluation. In: Proceedings of the 2nd International Conference on Simulation Tools and Techniques (Simutools 2009), pp. 1–10 (2009)

Part IX
Satellite Communication, Antenna Research, and Cognitive Radio

Part IX
Satellite Communication, Antenna Research, and Cognitive Radio

Optimal Structure Determination of Microstrip Patch Antenna for Satellite Communication

A. Sahaya Anselin Nisha

Abstract Microstrip Antenna design has experienced more development in the past period of years and still is subjected to more development. There are different kind of microstrip antenna which can be used for many handheld devices and in the communication devices like satellite link, radar system, radio and cellular mobiles. In this paper the behavior of microstrip antenna is analyzed with two types of electromagnetic band-gap structure such as frequency selective Structure (FSS), and photonic band-gap (PBG) structure. The major characteristics of EBG structure are to show the band gap features in the suppression of surface-wave propagation. This aspect helps to give better antenna performance such as increasing the gain and reducing back radiation. In particular, the distance between EBG cells and the patch is free from the outside control of the cell period, which can be arbitrarily selected, and the final setup offers footprint reduction. With these factors the Gain, bandwidth, and the Voltage standing wave ratio (VSWR) are analyzed using the FEM based EM simulator HFSS.

Keywords Duroid materials · EBG structure · FSS structure · HFSS · PBG structure · Probe feed

1 Introduction

Antennas represent a good role in the area of wireless communication. They can be patch antenna, folded dipole antenna, parabolic antenna and slot antennas each of which having separate their own habituations and properties. It is a main field in the research concerning the data transmission and reception. Microstrip antennas are being identical to the parallel plate capacitors. Since both have metal layer and a sandwiched dielectric substrate as a parallel plate between them. In microstrip

A. Sahaya Anselin Nisha (✉)
Sathyabama University, Chennai 600119, India
e-mail: nishalawrence.nisha@gmail.com

© Springer India 2016
A. Nagar et al. (eds.), *Proceedings of 3rd International Conference on Advanced Computing, Networking and Informatics*, Smart Innovation, Systems and Technologies 44, DOI 10.1007/978-81-322-2529-4_64

antenna, metal plate's one layer is extended in a great amount on comparing with the other layers. The main aspect of EBG structure is their power to make a difference in the radiative dynamics within the structure so there are no electromagnetic modes obtained in the dielectric. Alam and Islam [1] 1realized the EBG structure by inserting split-ring slots inside two reversely connected rectangular patches and achieved band gaps have widths of 4.3 GHz (59.31 %) and 5.16 GHz (38.88 %), which are centered at 7 and 13 GHz, respectively. The properties of different dielectric is analyzed [2, 3] using EBG structure patch antenna and [4] low profile dipole antenna is matched with optimal reflection phase of an electromagnetic band gap by relating theoretically the return loss by discrete frequencies in the range of frequency of interest. Comparison between normal mushroom structure EBG and folk like structure EBG shows the folk like structure is providing the degree of freedom in adjusting the band gaps and used to develop multi band structure [5] and this antenna producing steerable array with a linearly discrete beamsteering at 2.468 GHz. The electromagnetic Band Gap Structure using metamaterials for new applications in developed and analysed by many researchers [6–8]. The review of one, two and three dimensional EBG structure is analysed by Alam [9]. Here comparison between the types of Electromagnetic band gap structure such as photonic bandgap structure and frequency selective structure is analysed using the software High Frequency Structure Simulation (HFSS).

2 Photonic Bandgap (PBG) Structure

The photonic bandgap structure is the regular periodic arrangement of structure which controls the electromagnetic radiation in a similar way to semiconductor devices which controls the electrons. Semiconductor material exhibits an electronic band-gap with no electrons in it. In the similar way in the photonic crystal structure that has photonic band-gap which does not allow the electromagnetic radiation to propagate for a specific frequency within the band-gap. The PBG model is constructed on the crystal structure by mechanically drilling holes a millimeter in diameter into the substrate of high dielectric constant material which is shown in Figs. 1 and 2. The periodicity in an EBG structure is unsettled by either removing or adding a material with a different dielectric constant, size, or shape, a "defect" state is created in the forbidden gap, where an electromagnetic mode is allowed, and localization of the energy occurs. The feed line and the radiating element is generally photo etched on the substrate. The fringing field between the ground plane and the patch edge cause the microstrip patch antenna to radiate mainly. Here the contacting method is being employed where in which the RF power can be directly fed to the radiating element patch with the use of a connecting line such as a probe feed. The other method non-contacting in which the aperture feeding and proximity feeding is employed in which a electromagnetic field coupling is done for

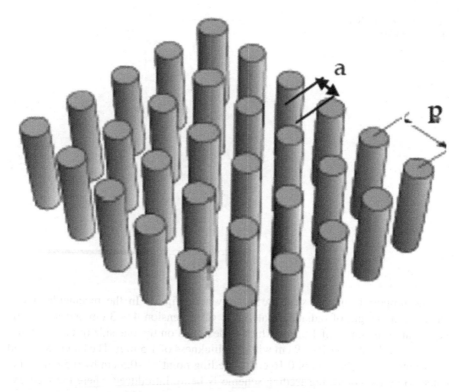

Fig. 1 Holes of PBG structure

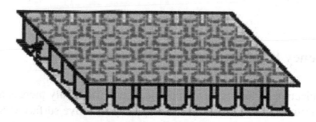

Fig. 2 Substrate material with holes

transferring the power between the patch and microstrip line. Rectangular micro-strip patch antenna is described by more physical parameters than on comparing with conventional patch antenna. They are used in application in which it is required to function for frequency range from 2 to 2.4 GHz. The substrate material Rogers RT/Duroid 5880 is used to design an antenna is pasted and return loss, VSWR, and gain are its performance characteristics which are obtained.

Fig. 3 Proposed PBG
structure antenna

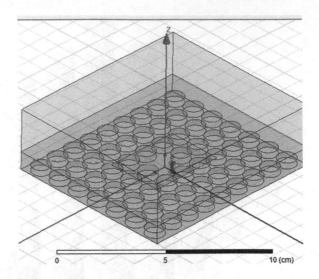

The proposed antenna structure is shown in Fig. 3. In the rectangular patch antenna, the length of radiating material is of dimension 4 × 3 cm and it is made very thin of thickness of 1.5 mm which is designed on the one side of the dielectric substrate of dimension 10 × 9 cm with the thickness of 1.5 mm. The antenna is feed power through probe feed of 0.16 cm at feeding point of −0.5 cm by 0.5 cm on the patch. Here the contacting feeding scheme is being introduced where the feed can be directly given on any desired location on the patch of the antenna to match the input impedance and have less spurious radiation. The frequency of operation is selected at 2.4 GHz.

3 Frequency Selective Surface (FSS) Structure

Frequency selective surface is a conducting sheet periodically pierces and makes a hole with apertures, which constitutes a frequency selective surface (FSS) structure to electromagnetic waves. There are different types of frequency selective surface structure there which can be used for modeling the high directive antenna, since to achieve better impedance matching other than directivity enhancement. Frequency selective structure for microstrip antenna for dual frequency is developed and it is resonating at the frequency of 5.2 and 8.0 GHz [10]. It shows when the incident angle is increasing from 0° to 40°; the filter response is stable in frequency selective structure. Frequency selective structure with resonators and multimode cavities for finer filtering characteristics is demonstrated using the introduction of poles and zeros [11–13]. The FSS structure encompasses a development with high electrical phenomenon surface that reflects the plane wave in-phase and suppresses surface wave [14]. Jerusalem cross-shaped frequency selective surfaces used an artificial

Fig. 4 Proposed FSS
structure

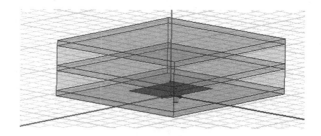

magnetic ground plane and invasive weed optimization algorithm is engaged to get best possible dimensions of the patch antenna. This antenna is resonating at 5.8 GHz having a broad bandwidth of about 10.44 % [15]. The improvement in the bandwidth planar antennas is done by placing conventional PEC (perfect electrical conductor) ground plane, on top of the surface [16].

The FSS superstrate layer is made of three substrate layers, in this top layer is made with bakelite below that second dielectric material is polyamide followed by rogers duroid dielectric and its image in the ground plane are in regular interval with two dimensions to analyze using periodic boundary condition. The FSSs are desired candidates as alternatives to dielectric substrate EBGs for directivity improvement. The Superstrate layer above a printed patch antenna and the ground plane is replaced by the image of superstrate layer.

In the substrate material the dielectric constant is very important in case of design parameters. For better efficiency, low quality factor and higher bandwidth the dielectric substrate with low dielectric constant is chosen which increases the fringing field at the radiating element patch and there by increasing the radiation. Where Rogers RT/Duroid 5880 is very good in this case regarding the radiation pattern and the loss tangent is been given less attention during the simulation. The thickness of the substrate plays a vital role in increasing the fringing field thereby causing the increased radiation. The thickness i.e. the height of the substrate (h) of the rectangular patch with probe feed is used in s band. The height of the substrate material of the antenna employed in the proposed antenna is h = 1.5 mm. Figure 4 shows the designed frequency selective structure antenna using HFSS.

4 Results and Discussions

The Return loss for PBG obtained is −14.6530 dB at the center frequency of 2.543 GHz as shown in Fig. 5. The return loss obtained for FSS is −29.4407 dB this shows in Fig. 6 that 9.6 % of the energy is reflected and 91 % of energy is transferred. By comparing it is found that the return loss for PBG structure is more than the FSS structure. Further the VSWR is a measurement for perfect match of

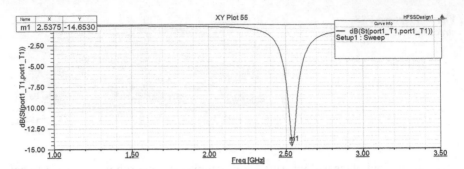

Fig. 5 Return loss curve for PBG antenna

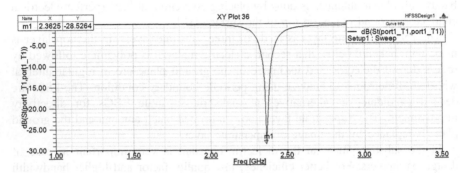

Fig. 6 Return loss curve for FSS antenna

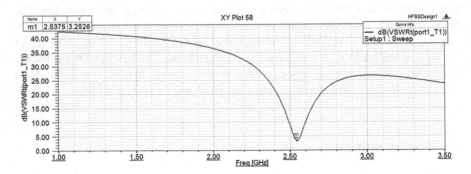

Fig. 7 VSWR curve for PBG antenna

VSWR would be of 1:1 this show how much energy reflected or transferred into the cable. Figures 7 and 8 shows the VSWR obtained from the graph is 3.2526 dB for PBG and 0.5861 dB for FSS antenna which is like 1.1:1 which is a desirable value

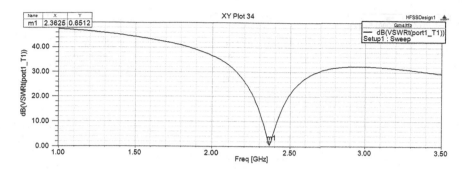

Fig. 8 VSWR curve for FSS antenna

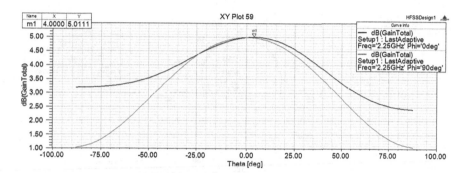

Fig. 9 Gain curve for PBG antenna

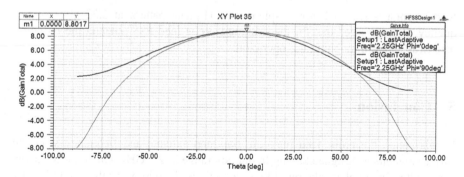

Fig. 10 Gain curve for FSS antenna

with low level of mismatch. Figures 9 and 10 shows the gain of the designed probe feed PBG antenna is 5.0111 dB and the gain increased for FSS antenna structure to be 8.8068 dB. Figures 11 and 12 shows the radiation pattern for the designed probe

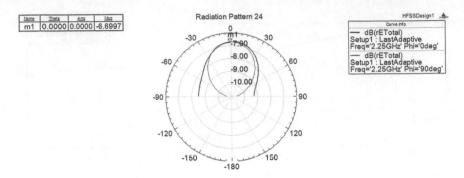

Fig. 11 Radiation pattern for PBG antenna

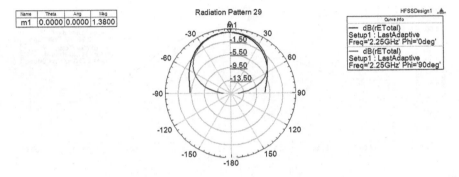

Fig. 12 Radiation pattern for FSS structure

feed antenna is shown below. E-plane and H-plane pattern at 2.25 GHz of frequency. It is observed that for the designed probe feed FSS antenna the radiation pattern is stable throughout the operation.

5 Conclusion

In this paper, the probe feed photonic bandgap antenna and frequency selective surface antenna is designed having frequency range 2–2.5 GHz which has a narrow bandwidth. The implementation of these techniques is to improve the bandwidth by increasing substrate height i.e. thickness, patch height is been increased and by reducing the substrate material permittivity. The result is found that the optimal EBG structure among the two different antenna structure is the frequency selective surface with very less return loss −29.4407 dB and the gain is increased to 8.8068 dB which can be used for satellite application.

References

1. Islam, M.T., Alam, M.S.: Compact EBG structure for alleviating mutual coupling between patch antenna array elements. Prog. Electromagnet. Res. **137**, 425–438 (2013). doi:10.2528/PIER12121205
2. Shaban, H.F., Elmikaty, H.A., Shaalan, A.A.: Study the effects of electromagnetic band-gap (EBG) substrate on two patches microstrip antenna. Prog. Electromagnet. Res. B **10**, 55–74 (2008). doi:10.2528/PIERB08081901
3. Biancotto, C., Record, P.: Triangular lattice dielectric EBG antenna. Antennas Wirel. Propag. Lett. **9**, 95–98 (2010). doi:10.1109/LAWP.2010.2043494
4. McMichael, I.T., Mirotznik, M., Zaghloul, A.L., A method for determining optimal EBG reflection phase for low profile antennas. Antennas Propag. Soc. Int. Symp. (APSURSI), (2012). doi:10.1109/APS.2012.6348945
5. Yang, L., Fan, M., Chen, F., She, J., Feng, Z.: A novel Compact electromagnetic-bandgap structure and its applications for microwave circuits. IEEE Trans. Microw. Theory Tech. **53** (1), 183–190 (2005). doi:10.1109/TMTT.2004.839322
6. Loh, T.H., Mias, C.: Photonic bandgap surfaces with inter digitated corrugations. IEEE Electron. Lett. **40**, 1123–1125 (2004). doi:10.1049/el:20045756
7. Lee, Y.L.R., et al.: Dipole and tripole metallodielectric photonic bandgap (MPBG) structures for microwave filter and antenna applications. IEEE Proc. Optoelectron. **147**, 395–400 (2000). doi:10.1049/ip-opt:20000892
8. Rahmat-Samii, Y., Yang , F.: Electromagnetic Band Gap Structures in Antenna Engineering. Cambridge University Press, Cambridge (2009). doi:10.1109/APMC.2008.4958195
9. Alam, M.S., Misran, N., Yatim, B., Islam, M.T.: Development of electromagnetic band gap structures in the perspective of microstrip antenna design. Int. J. Antennas Propag. 2013, Article ID 507158, 22 p. doi:dx.doi.org/10.1155/2013/507158
10. Li, B., Shen, Z.: Three-dimensional Dual-band Frequency Selective Structure Using Microstrip Lines. Progress in Electromagnetics Research Symposium Abstracts, Stockholm, Sweden, 12–15 Aug 2013
11. Rashid, A.K., Shen, Z.: A novel band-reject frequency selective surface with pseudo-elliptic response. IEEE Trans. Antennas Propag. **58**(4), 1220–1226 (2010). doi:10.1109/TAP.2010.2041167
12. Rashid, A.K., Shen, Z.: Scattering by a two-dimensional periodic array of vertically placed microstrip lines. IEEE Trans Antennas Propag. **59**(7), 2599–2606 (2011). doi:10.1109/TAP.2011.2152332
13. Li, B., Shen, Z.: Miniaturized bandstop frequency-selective structure using stepped impedance resonators. IEEE Antennas Wirel. Propag. Lett. **11**, 1112–1115 (2012). doi:10.1109/LAWP.2012.2219571
14. Chen, H.-Y., Tao, Y.: Bandwidth enhancement of a U-Slot patch antenna using dual-band frequency selective surface with double rectangular ring elements. MOTL **53**(7), 1547–1553 (2011)
15. Monavar, F.M., Komjani, N.: Bandwidth enhancement of microstrip patch antenna using jerusalem cross-shaped frequency selective surfaces by invasive weed optimization approach. Progress Electromagnet. Res. **121**, 103–120 (2011). doi:10.2528/PIER11051305
16. Yeo, J., Mittra, R., Chakravarthy, S.: A GA-based design of electromagnetic bandgap (EBG) structures utilizing frequency selective surfaces for bandwidth enhancement of microstrip antennas. Antennas Propag. Soc. Int. Symp. **2**, 400–403 (2002). doi:10.1109/APS.2002.1016108

Multi-metric Routing Protocol for Multi-radio Wireless Mesh Networks

D.G. Narayan, Jyoti Amboji, T. Umadevi and Uma Mudenagudi

Abstract Wireless Mesh Networks (WMNs) are emerging as low-cost technology to build broadband access networks. With the demand for the real-time multimedia applications like voice over IP and video on demand, providing better quality of service (QoS) in WMNs has become as important research issue. To address this, many cross layer routing metrics are proposed in the literature. But the usage of single link quality metric for different types of traffic within the network is insufficient to guarantee fine grained QoS. Towards this, we propose a multi-metric routing protocol which finds the optimal path based on two routing metrics depending on the type of traffic. We use two metrics namely Metric for Interference, Load and Delay (MILD) and Contention Aware Transmission Time (CATT) for TCP and CBR traffic respectively. MILD is throughput optimizing routing metric where as CATT is delay sensitive metric suitable for multimedia traffic. We implemented these metrics in well-known proactive routing protocol Optimized Link State Routing (OLSR) in ns-2.34. The results indicate that the multi-metric protocol perform better in terms of throughput and average end-to-end delay compared to single metric.

Keywords WMN · MILD · CATT · Multi-metric · OLSR

D.G. Narayan (✉)
Department of Information Science and Engineering, B.V.B College
of Engineering & Technology, Hubli 580031, India
e-mail: narayan_dg@bvb.edu

J. Amboji · T. Umadevi
Department of Computer Science and Engineering, B.V.B College
of Engineering & Technology, Hubli 580031, India
e-mail: ambojij@yahoo.com

T. Umadevi
e-mail: pearlpapu.21@gmail.com

U. Mudenagudi
Department of Electronics and Communication Engineering, B.V.B College
of Engineering & Technology, Hubli 580031, India
e-mail: uma@bvb.edu

© Springer India 2016
A. Nagar et al. (eds.), *Proceedings of 3rd International Conference on Advanced Computing, Networking and Informatics*, Smart Innovation, Systems and Technologies 44, DOI 10.1007/978-81-322-2529-4_65

631

1 Introduction

Wireless Mesh Networks [1] is a communication network which consists of set of nodes structured in a mesh topology. These networks are becoming popular to connect wired and wireless networks to internet in a cost effective way. It is comprised of mesh routers, mesh clients and gateways. In mesh topology one or certain number of nodes can be selected as mesh gateway to enable communication with the internet and some of the nodes can be acting as access point to provide communication to client stations. These mesh routers and gateways have multi-radio capabilities which introduce more interference in the network. These nodes have less mobility and energy consumption is not a major issue. Some of the applications of WMN include metropolitan area, transportation systems, surveillance system, security systems and city network. The key issue for all these applications is QoS. Thus designing routing protocol which can provide better QoS is an important research issue.

The routes in the multi-radio mesh network changes periodically because quality of wireless links fluctuates dynamically. This is due to the several reasons such as interference, packet losses, and dynamic traffic load patterns. To address these issues, routing algorithms and cross layer routing metrics are significant to improve the performance of applications [2]. Cross layer routing metrics are used to estimate the quality of the link in terms of various parameters obtained from PHY and MAC layers. In addition to this, there is degradation in the performance of wireless mesh network when a single routing metric is used to route different types of packets. TCP traffic may require different parameters to be considered to route a packet to get best results and UDP traffic may require some other parameters to be considered while routing the packets. Therefore it is necessary to analyze the performance of routing metrics under different types of traffic and application. Multi-metric routing approach requires careful study of existing metrics, their merits, demerits and suitability for the type of traffic.

The multi-metric routing protocols are proposed in literature [3, 4]. However these are designed for single radio mesh network. Thus we propose a multi-metric routing protocol which integrates two multi-radio routing metrics namely MILD and CATT. The task of multi-metric is to dynamically select a metric to route the packets based on type of traffic. We use delay sensitive CATT for UDP traffic and MILD for TCP traffic. We choose OLSR as routing protocol on which these routing metrics are implemented. Optimized Link State Protocol (OLSR) [5] is a table driven routing protocol which is recommended for 802.11s mesh standard. It is the most suitable protocol for infrastructure mesh network as the mesh routers are static. The results reveal that the multi-metric approach improves the QoS of the considered traffic types compared to single routing metric.

The organization of this paper is as follows. Section 2 deals with the related work i.e. review on cross layer routing metrics, and also discusses about multi-metric routing and OLSR protocol. Section 3 discusses the proposed multi-metric protocol along with cross layer implementation. Section 4 presents the simulation environment and the analysis of simulation results. Finally, Sect. 5 concludes the work.

2 Related Work

This section discusses few cross layer routing metrics, multi-metric routing and functionality of OLSR routing protocol.

2.1 Routing Metrics for WMN

The routing metrics for wireless mesh network are discussed below.

2.1.1 Expected Transmission Count (ETX)

ETX [6] is stated as the number of transmissions and retransmissions entailed to deliver a packet to intended sink node over a wireless link. This metric is an amalgamation of path length and packet loss ratio. Link quality using this metric is computed using Eq. (1).

$$ETX_i = \frac{1}{d_{fwd} * d_{rvs}} \tag{1}$$

d_{fwd} is the probability that the packet is received successfully at destination node and d_{rvs} is the probability that the acknowledgement is successfully received at source node. ETX does not provide information about the traffic in the network and hence fails in load balancing.

2.1.2 Expected Transmission Time (ETT)

ETT [7] was designed to overcome the problems of ETX. Transmission rate of each link is considered in this metric. It is calculated using Eq. (2):

$$ETT_i = \sum_i^n ETT_i = \sum_i^n (1/(1 - P_i)) * (S/B_i) \tag{2}$$

where S indicates the size of the packet, P_i indicates the packet loss rate and B_i indicates rate of transmission of link i. This metric considers both packet loss rate and transmission rate of each link i. This metric does not consider interference parameters, traffic load and channel diversity.

2.1.3 Weighted Cumulative ETT (WCETT)

WCETT [8] is an extended version of ETT which captures the transmission rate and packet loss ratio of each link. WCETT is designed to work in multi-hop and multi-radio wireless networks. It is calculated using Eq. (3).

$$WCETT_i = (1 - \beta) \sum_{i=1}^{m} ETT_i + \beta \max_{1 \leq j \leq ch} X_k \tag{3}$$

where m is the total number of hops on the path, ch is the total number of channels, β is a smoothing parameter ranging between 0 and 1 and X_k is summation of transmission times by each hop on the channel k. Inter-flow interference and traffic loads are not considered in this metric.

2.1.4 Interferer Neighbors Count Routing Metric (INX)

INX [9] is the extended version of ETX which reduces the radio utilization cost by considering interference parameters experienced by nodes in the network. It is calculated as follows:

$$INX_i = ETX_i * \sum_{k \in N_i} r_k \tag{4}$$

where N_i is total number of interfering links caused from the transmission taking place on link i, r_k is the transmission rate of the link k. Asymmetric links are considered when defining set of interfering neighbours because of which interference is considered in better way then MIC metric.

2.1.5 Metric for Interference and Channel Diversity (MIND)

MIND [10] considers load aware and interference factors by passive supervision mechanism hence reduces overhead caused by active probing technique. It can be defined as

$$MIND = \sum_{link_i \in p}^{n} INTERLOAD_i + \sum_{node_j \in p}^{m} CSC_j \tag{5}$$

INTERLOAD factor is for inter-flow interference and load, CSC considers intra-flow interference, n estimates the number of connections, m is the number of nodes over route p.

2.2 Multi-metric Routing

In [3] authors propose OLSR-DC (Dynamic Choice) protocol, it is a multi-metric approach which is a combination of ETX and Minimum Delay (MD) Metrics. Here Routing is done based on type of traffic. It uses ETX metric to route TCP packets and MD metric to route UDP packets. This approach gives the acceptable QoS support because packets are routed as per the metrics that best reveal their requirements.

In [4] authors extend the work using fuzzy logic in which problem associated with multiple metrics for routing is solved. But both the metrics are implemented for single radio networks. We extend the work which overcomes the drawbacks by considering multi-radio metrics.

2.3 Optimized Link State Routing Protocol (OLSR)

The OLSR [5] protocol is a popular proactive protocol which is recommended in IEEE 802.11s mesh standard. It has three messages which are used to compute the routes. The three messages are namely Hello, Topological control (TC) and Multiple interface declaration (MID). The Hello messages are used to sense the link over interfaces by which connectivity is checked and they help in construction of MPRs. TC messages are spread regularly throughout network. Every node in a network maintains topology information and this information is obtained from TC messages. MID messages are used to broadcast the information about multiple interfaces present in node. OLSR has four important components namely neighbour detection, Multi Point Relay (MPR) selection, TC message flooding and routing table calculation. Neighbour detection is done using Hello messages. TC message flooding is carried out using MPR which minimizes the control packets. The route calculation by each node is done using Dijkstras algorithm. As OLSR is table driven protocol no route detection delay is associated to find new route but scalability issues arise when network size grows. It can sustain loss of few packets as control packets do not need reliable transmission because every node sends control messages regularly.

3 Proposed System

Multi-metric chooses the suitable metric to route the diverse packets like if there is a TCP traffic it chooses a suitable routing metric to route the packets and if there exists a CBR traffic (multimedia application) then a multi-metric chooses a suitable metric to route the packets such that it results in overall throughput improvement. In the proposed System we are using metric MILD to route TCP packets which takes into account interference, load and delay parameters. CBR packets are routed using metric CATT which is delay sensitive.

3.1 Implementation of Routing Metric

Cross layer routing metrics used in the implementation of multi-metric protocol are explained here.

3.1.1 Metric for Interference, Load and Delay (MILD)

MILD [11] assigns cost to individual link that takes into account link delay, link load and interferences. MILD for link l is defined as follows

$$MILD_l = \frac{1}{c_n}[CW_{avg} + iETT_l] * N_l \qquad (6)$$

C_n is channel utilization of node n, CW_{avg} is average contention window on link l and N_l is the interfering link count of node n. It considers mainly three factors i.e. load, delay and interference measurement. Where CW_{avg} is defined as:

$$CW_{avg} = \frac{(1 - PER_l)(1 - (2PER_l)^{r+1})}{(1 - 2PER_l)(1 - (PER_l)^{r+1})} \times CW_0 \qquad (7)$$

and $iETT_l$ is defined as:

$$iETT_l = ETT \times \frac{Packet\_overhead}{B_{aval}} \qquad (8)$$

Existing metric MILD is for single Radio. In our work we have enhanced this metric for multi-Radio.

3.1.2 Contention Aware Transmission Time (CATT)

CATT [8] metric is the combination of contention aware routing metric and iAWARE routing metric. It can be used in link state protocols, CATT helps in load balancing on the links and to obtain path that optimizes total packet transmission time. It can be calculated as follows:

$$CATT_i = ETX_i \cdot \sum_{j \varepsilon N_i}\left(\left(\sum_{k \varepsilon N_j} \frac{R_j}{L_k}\right) \cdot T_j \cdot \frac{R_k}{L_j}\right) \qquad (9)$$

where N_i is the number of connections that interfere with the transmission on connection i, N_j is number of connections that interfere with transmission on the connection j. R_j and R_k indicate the packet size of the connections containing 1 and

2 hop neighbours respectively. T_j is defined as data transmission attempt rate on connection j.

3.2 Cross Layer Implementation

Parameter acquisition and path computation using cross layer mechanism is as shown in the Fig. 1. Steps are as follows:

- Fetch RSSI parameter from Physical layer and calculate IR value in MAC layer.
- Calculate average CW and Cn parameters in MAC layer.
- Take ETX and CSC parameters from network layer.
- Based on traffic choose the appropriate Metric.
- Compute the path by considering above parameters using OLSR protocol in network layer.

The multi-metric algorithm checks for the type of the traffic and chooses suitable metric to route the packets as shown in Algorithm 1:

Algorithm 1 Multimetric

1: **if** (Traffic Type == TCP) **then**
2:     Select Metric MILD
3: **else**
4:     Select Metric CATT
5: **end if**
---

4 Simulation Model

A detailed simulation study is carried out in order to analyze the performance of multi-metric with the existing metric.

4.1 Simulation Environment

We conduct the simulation using NS-2.34 [12] to obtain the results for multi-metric with the help of CMU tool to create the network topology with which flows are generated randomly. We have used static scenario for comparing the performance of proposed multi-metric with metric MILD. The different parameters used are illustrated in Table 1.

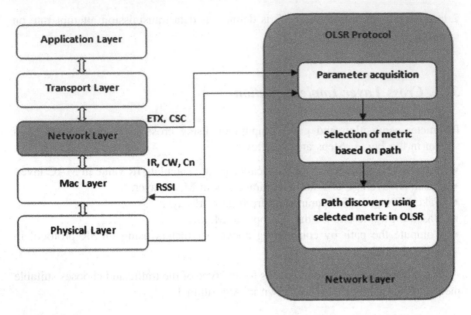

Fig. 1 Cross Layer Architecture

Table 1 Simulation Parameters

Parameters	Values
Number of nodes	30
Number of flows	15
Simulation time	50 s
Propagation model	Shadowing
Traffic type	TCP and CBR
Packet size	512
Transmission range	250 m
Environment size	750 m * 750 m
Packet rate	50, 55, 60, 65, 70 (packets/s)
IEEE MAC	802.11b

4.2 Performance Parameters

- *Average throughput*: It is the rate at which packets are received successfully at destination.
- *End-to-end delay*: It is the average end-to-end delay of all the packets that are received successfully at destination.

4.3   Simulation Results

This part presents the comparison of the performance of multi-metric with the metric MILD based on QoS parameters like throughput and delay. For a system to perform better throughput should be high and delay should be minimum. We create topology using Ns-2.34 tool. This topology creation involves specifying node positions, total number of nodes and generating random traffic flows of CBR and TCP between all sources and destinations. The simulation comprised of 7 CBR flows and 8 TCP flows each of which starts at different simulation time. There are five source nodes and six destination nodes for CBR flows, five source nodes and eight destination nodes for the transmission of TCP flows.

Figure 2a, b gives the overall throughput and delay respectively where throughput of multi-metric is high compared to metric MILD. As shown in Fig. 2 throughput of multi metric is improved compared to MILD as it is based on MILD and CATT. At 50 and 65 traffic rate throughput is same as MILD also considers channel busy time. The delay of multi metric is also improved compared to MILD with similar observation. From Fig. 3a it is clear that CBR throughput is high compared to TCP as we are emphasizing more on multi-media applications. From Fig. 2a, b it is observed that delay is inversely proportional to throughput and delay obtained from multi-metric is less compared to metric MILD.

Figure 3a, b shows packet differentiation for throughput and delay performance of TCP and CBR traffic with regard to multi metric and MILD routing. The results reveal that TCP and CBR throughput and delay is better with regard to multi-metric routing; this is because multi-metric protocol route the packets based on type of traffic. In our work we use CATT to route CBR packets as it considers the delay and rate diversity of link which results in minimum delay. We use MILD metric to route TCP packets as it optimizes throughput by considering traffic load on interfering

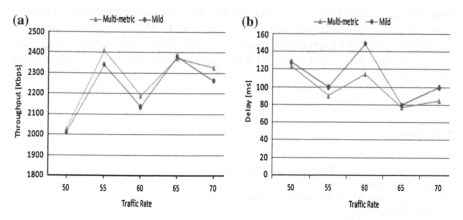

Fig. 2 **a** Network throughput versus traffic rate. **b** Network delay versus traffic rate

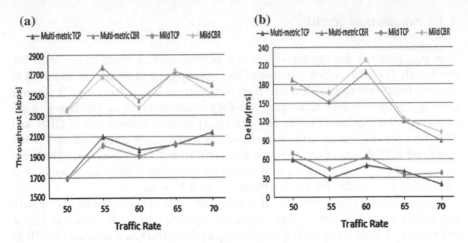

Fig. 3 **a** CBR, TCP throughput versus traffic rate. **b** CBR, TCP delay versus traffic rate

nodes. Thus we are able to differentiate CBR and TCP traffic based on these two metrics as discussed earlier the results are same at packet rate 50 and 65 as topology is random.

5 Conclusion

Wireless Mesh Network is a communication network which consist of set of nodes structured in a mesh topology. The two factors that influence the performance in WMNs are routing metrics and routing algorithms. A single routing metric fails to recognize characteristics such as different transmission rates, packet loss rates, and interference required by the type of the traffic. Thus we design multi-metric routing protocol using OLSR based on MILD and CATT. MILD improves the TCP throughput and CATT improves average packet delay. The results using NS2 simulations reveal that multi-metric OLSR performs better than the single routing metric MILD.

References

1. Akyildiz, I.F., Wang, X.: A survey on wireless mesh networks. IEEE Commun. Mag **43**(9), S23–S30 (2005)
2. Campista, M.E.M., et al.: Routing metrics and protocols for wireless mesh networks. IEEE Netw. **22**(1), 6–12 (2008)
3. Gomes, R., et al.: Dynamic metric choice routing for mesh networks. In: 7th International Information and Telecommunication Technologies Symposium, I2TS (2008)

4. Lopes Gomes, R., Moreira Junior, W., Cerqueira, E., Jorge AbelÃl'©m, A.: Using fuzzy link cost and dynamic choice of link quality metrics to achieve QoS and QoE in wireless mesh networks. J. Netw. Comput. Appl. **34**(2), 506–516 (2011)
5. Clausen, T., et al.: Optimized Link State Routing Protocol (OLSR). (2003)
6. De Couto, D.S., Aguayo, D., Bicket, J., Morris, R.: A high-throughput path metric for multi-hop wireless routing. Wireless Netw. **11**(4), 419–434 (2005)
7. Draves, R., Padhye, J., Zill, B.: Routing in multi-radio, multi-hop wireless mesh networks. In: Proceedings of the 10th Annual International Conference on Mobile Computing and Networking. ACM (2004)
8. Genetzakis, M., Siris, V.A.: A contention-aware routing metric for multi-rate multi-radio mesh networks. Sensor, Mesh and Ad Hoc Communications and Networks. In: SECON'08. 5th Annual IEEE Communications Society Conference on. IEEE (2008)
9. Borges, V.C.M., et al.: Routing metric for interference and channel diversity in multi-radio wireless mesh networks. In: Ad-Hoc, Mobile and Wireless Networks, pp. 55–68. Springer, Berlin Heidelberg (2009)
10. Borges, V.C., Pereira, D., Curado, M., Monteiro, E.: Routing metric for interference and channel diversity in multi-radio wireless mesh networks. In: Ad-Hoc, Mobile and Wireless Networks, pp. 55–68. Springer, Berlin Heidelberg (2009)
11. Narayan, D.G., et al.: A cross layer Interference and delay aware routing metric for Wireless Mesh Networks. Computational Intelligence and Information Technology. In: CIIT 2013. Third International Conference on. IET (2013)
12. The Network Simulator: http://www.isi.edu/nsnam/ns

A Novel Dual Band Antenna for RADAR Application

Kousik Roy, Debika Chaudhuri, Sukanta Bose and Atanu Nag

Abstract We have proposed the design, performance and analysis of a novel dual band antenna for radar communication. The working frequency of the antenna is 1.55 and 6 GHz and it is designed on a Rogers RT/Duroid 6202 laminate substrate (dielectric constant = 2.2). The length and width of the patch are respectively 15 and 20 mm, while the length (l) and width (w) of the slots are respectively 14 and 1 mm with a 10 mm feed length. We have used IE3D Zealand (Ver-12) for obtaining the simulation result and analysis of our proposed antenna. The results are analyzed by investigating current distribution, S(1,1), elevation pattern, directivity pattern and voltage standing wave ratio (VSWR). Our antenna finds both L band (1–2 GHz) and C band (4–8 GHz) applications where it can be effectively used for RADAR application.

Keywords Dual band antenna · Radar communication · L and C band · S(1,1) · Elevation pattern · IE3D

1 Introduction

Low-profile antennas are preferred because of certain factors such as subsidized cost, good performance, size and so on for outstanding performances of RADAR application. Now-a-days, there are numerous usage like mobile, radio and satellite

K. Roy (✉)
Asansol Engineering College, Asansol, India
e-mail: kousikroy002@gmail.com

D. Chaudhuri · S. Bose · A. Nag
Modern Institute of Engineering & Technology, Bandel, India
e-mail: debika.chaudhuri@gmail.com

S. Bose
e-mail: sukanta.vlsi@gmail.com

A. Nag
e-mail: tnnag79@gmail.com

© Springer India 2016
A. Nagar et al. (eds.), *Proceedings of 3rd International Conference on Advanced Computing, Networking and Informatics*, Smart Innovation, Systems and Technologies 44, DOI 10.1007/978-81-322-2529-4_66

643

communications that require similar specifications. Due to these requirements, microstrip antennas can be used effectively for its low profile and subsidized manufacturing cost. To make our analysis simpler we have considered the patch, which is very thin in comparison to the free-space wavelength λ, to be of common geometrical shape [1–7].

2 Design Procedure

The practical design of the proposed dual band antenna designed for radar application is based on two models viz. transmission line model and cavity model [8]. The design procedure is accomplished with the knowledge of different specific information of the proposed antenna.

Length (l) is found out by the equation,

$$l = \frac{\lambda}{2} - 2\Delta l \tag{1}$$

where λ = free-space wavelength.

The extended length (Δl) is given by,

$$\Delta l = 0.412 \frac{\left(\varepsilon_{reff} + 0.3\right)\left(\frac{w}{h} + 0.264\right)}{\left(\varepsilon_{reff} - 0.258\right)\left(\frac{w}{h} + 0.8\right)} \tag{2}$$

The effective dielectric constant (ε_{reff}) is given by,

$$\varepsilon_{reff} = \frac{\varepsilon_r + 1}{2} + \frac{\varepsilon_r - 1}{2}\left[1 + 12\frac{h}{w}\right]^{-1/2} \tag{3}$$

ε_r = relative dielectric constant, h = height of the substrate and w = width of the substrate.

$$w = \frac{v_0}{2f_r}\sqrt{\frac{2}{\varepsilon_{r+1}}} \tag{4}$$

where f_r is the resonant frequency and v_0 = velocity of light in free space. Furthermore, the effective length (L_{eff}) is,

$$L_{eff} = L + 2\Delta l \tag{5}$$

3 Antenna Configurations and Characteristics

Figure 1 shows the proposed geometry of the antenna, designed on a Rogers RT/Duroid 6202 laminate substrate with dielectric constant = 2.2, composed of four slots and a feed line. The feed length is 10 mm and the selected substrate height is set as 1.6 mm. The length and width of the patch are respectively 15 and 20 mm while the length of slot 1 and slot 2 are 14 mm, the width of slot 1 and slot 2 are 1 mm. We have used IE3D Zealand (Ver-12) for obtaining the simulation result and analysis of our proposed antenna [9].

3.1 Current Distribution

Figure 2 shows the current distribution at 1.55 and 6 GHz respectively and can be observed with universal dB color measure scale.

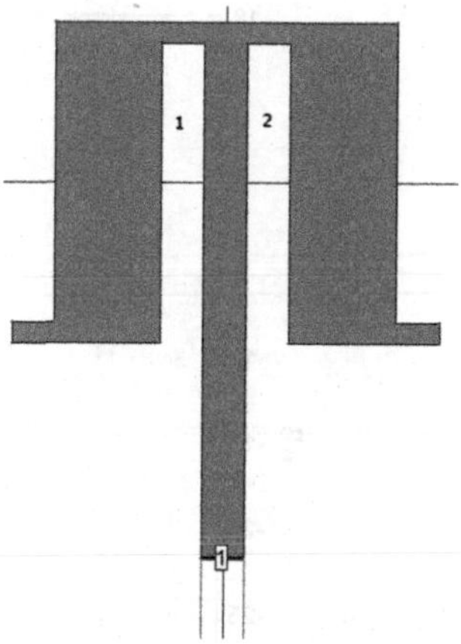

Fig. 1 Antenna geometry

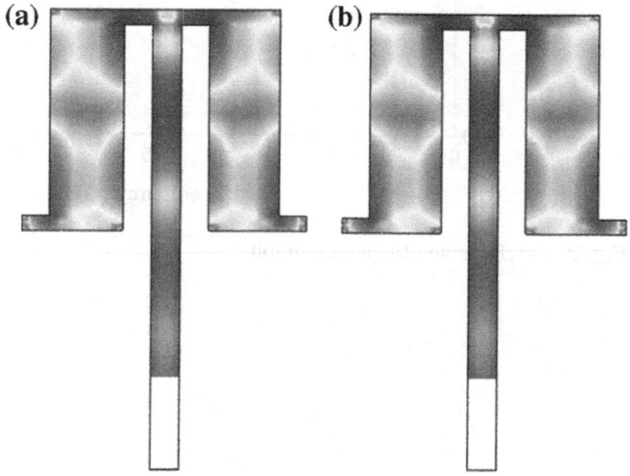

Fig. 2 Current distribution at **a** 1.55 GHz and **b** 6 GHz

3.2   S(1,1)

In Fig. 3 the S(1,1) obtained at 1.55 GHz is about −38 dB and the S(1,1) obtained at 6 GHz is about −39.5 dB.

3.3   *Elevation Pattern and Directivity Pattern*

Elevation pattern of gain at $\phi = 0°$ and $\phi = 90°$ is shown in Fig. 4, while that of directivity is shown in Fig. 5 for (a) f = 1.55 GHz and (b) f = 6 GHz.

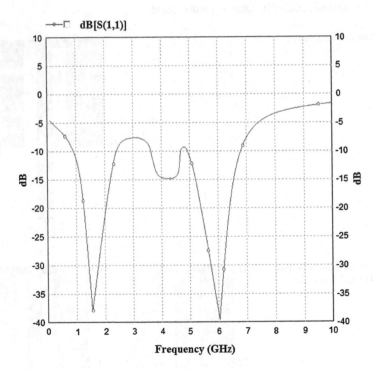

Fig. 3 S(1,1) versus frequency graph

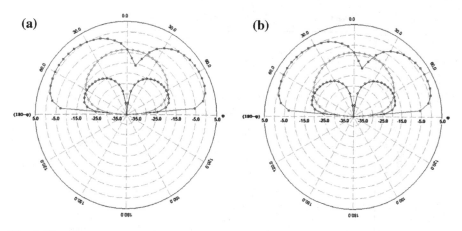

Fig. 4 Elevation pattern gain display at **a** 1.55 GHz and **b** 6 GHz

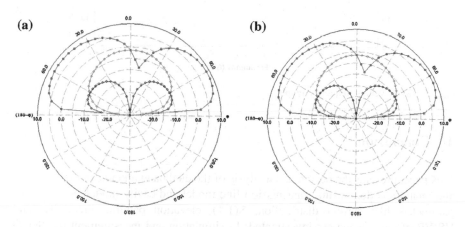

Fig. 5 Elevation pattern directivity display at **a** 1.55 GHz and **b** 6 GHz

3.4 VSWR

In Fig. 6 the variation of VSWR with frequency is depicted. The VSWR obtained at 1.55 GHz is about 1.1 and the VSWR obtained at 6 GHz is about 1.2.

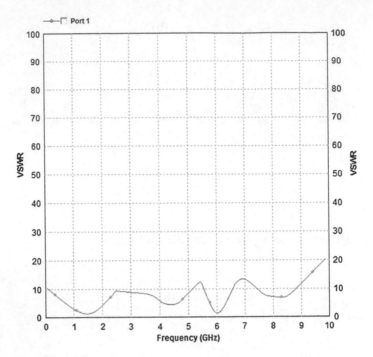

Fig. 6 VSWR versus frequency graph

4 Conclusion

This paper presents the design and analysis of the dual band antenna using IE3D [9] and analysis is done using transmission line model and cavity model. Some major parameters like current distribution, S(1,1), elevation pattern, directivity, and VSWR of the antenna are investigated. In simulation and measurement the S(1,1) has a negative value indicating that the losses are minimum during the transmission. As for a dual band antenna, S(1,1) is acceptable at two frequencies, i.e. 1.55 and 6 GHz and its value are −38 and −39.5 dB. The observed VSWR at 1.55 and 6 GHz are 1.1 and 1.2 indicating that the level of mismatch is not so high. The simulated antenna can be effectively used for radar communication in L band (1–2 GHz) and C band (4–8 GHz).

References

1. Kraus, J.D.: Antennas for All Applications. 3rd ed. McGraw Hill Inc., New York (2002)
2. Yang, F., Zhang, X.X., Ye, X., Rahmat-Samii, Y.: Wide-Band E shaped patch antennas for wireless communications. IEEE Trans. Antennas Propag. **49**, 1094–1100 (2001)
3. Chen, H.Y., Tao, Y.: Performance improvement of a U-slot patch antenna using a dual-band frequency selective surface with modified jerusalem cross elements. IEEE Trans. Antennas Propag. **59**, 3482–3486 (2011)
4. Balanis, C.A.: Antenna Theory Analysis and Design, 2nd edn. Wiley, New York (2005)
5. Pandey, V.K., Vishvakarma, B.R.: Analysis of an E-shaped patch antenna. Microwave Opt. Technol. Lett. **49** 4–7 (2007)
6. Kim, B.: Novel single-feed circular microstrip antenna with reconfigurable polarization capability. IEEE Trans. Antenna Propag. **56** 630–638 (2008)
7. Yang, W., Zhou, J.: A single layer wideband low profile tooth-like-slot microstrip patch antenna fed by inset microstrip line. In: International Workshop on Antenna Technology (iWAT), pp. 23–26 (2013)
8. Jayanthy, T., Sugadev, M., Ismaeel, J.M., Jegan, G.: Design and simulation of microstrip M-patch antenna with double layer. IEEE Trans. Boomed. Eng. **54**, 2057–2063 (2008)
9. Zeland Software Inc. IE3D: MoM-Based EM Simulator.Web. http://www.zeland.com

Author Index

A

Abubakar, Babangida, 125
Akhtar, Md. Amir Khusru, 169
Akhtar, Nadeem, 281
Alisha, 133
Amboji, Jyoti, 631
Anitha, R., 327
Annappa, B., 39, 51
Apoorva Chandra, S., 63
Arya, Rajeev, 223
Asharani, M., 117
Aswini, S., 439
Awasthi, Amit K., 143

B

Bahri, Devashish, 495
Banka, Haider, 85
Banu, K. Sharmila, 569
Bhattacharjee, Shrutilipi, 407
Bhattacharyya, Aditya, 337
Biswas, Soumojit, 355
Bopche, Ghanshyam S., 315
Borah, Parashjyoti, 485
Bose, Sukanta, 643

C

Chaitanya, J.N.V.K., 63
Chandrasekaran, K., 17
Chandrasekhar, P., 117
Chaudhuri, Debika, 643
Chhotaray, Animesh, 355
Chhotaray, Sukant Kumar, 355
Chirumamilla, Narendra, 389
Chitti, Sridevi, 117
Chormunge, Smita, 557
Chowdhary, Sunil Kumar, 495

D

Dadhich, Reena, 373
Dani, Ajay, 457
Das, Himansu, 599
Dash, Amiya Kumar, 507
Dash, Ratnakar, 363, 427
Datta, Somjit, 289
Deshmukh, Madhukar, 197
Deshpande, Umesh A., 587
Dev, Dipayan, 537
Dharani, M.K., 109
Dhivya, R., 109
Dinesh, Sushant, 17
Divya, M., 39
Dora, Durga Prasada, 161
Drira, Ghofrane, 103
Dueck, Marcel, 29

F

Faiz, Sami, 103

G

Gaikwad, Arun N., 151
Gangadharan, G.R., 447
Gawali, Dnyaneshwar, 197
Gayathri Devi, B., 109
Ghogare, Shraddha, 457
Ghosh, Soumya K., 407, 475
Goel, Jai Narayan, 299
Gondaliya, Nikhil N., 609

H

Harikrishnan, R., 209
Husain, Akhtar, 215

I

Iyappan, P., 439

© Springer India 2016
A. Nagar et al. (eds.), *Proceedings of 3rd International Conference
on Advanced Computing, Networking and Informatics*, Smart Innovation,
Systems and Technologies 44, DOI 10.1007/978-81-322-2529-4

J

Jakhar, Amit Kumar, 397
Jawahar Senthil Kumar, V., 209
Jayashankar, Aditi, 569
Jena, Sanjay Kumar, 363, 507
Jena, Sudarson, 557
Jevitha, K.P., 417
Joseph, Swathy, 417

K

Kaiwartya, Omprakash, 161
Kalita, Hemanta Kumar, 485
Kamath, S. Sowmya, 63
Kathiriya, Dhaval, 609
Kaushik, Baijnath, 85
Kaushik, R., 63
Krichen, Saoussen, 103
Krishna Mohan Rao, S., 529
Krishnamurthy, G., 117
Kumar, Sujit, 239
Kumar, Sushil, 161, 239
Kushwah, Virendra Singh, 185

M

Mahto, Dindayal, 347
Majhi, Banshidhar, 427
Majumder, Subhashis, 289
Mallya, Dilip, 63
Mary Saira Bhanu, S., 259
Meghanathan, Natarajan, 95
Meher, Lingaraj, 271
Mehtre, B.M., 299
Mehtre, Babu M., 315
Mishra, Tapas Kumar, 231
Mitra, Anirban, 3
Modi, Chirag N., 549
Mohanta, Taranisen, 389
Mohapatra, Durga Prasad, 381
Muddi, Leena, 389
Mudenagudi, Uma, 631

N

Nag, Atanu, 643
Naik, Nenavath Srinivas, 465
Narayan, D.G., 631
Narwariya, Priusha, 185
Nayak, Deepak Ranjan, 427
Negi, Atul, 465
Nimkar, Anant V., 475

P

Panigrahi, Chhabi Rani, 599
Patgiri, Ripon, 537

Pati, Bibudhendu, 599
Patil, Ashwini R., 549
Paul, Aakash, 289
Pawar, Ambika, 457
Penurkar, Milind R., 587
Poonacha, P.G., 133
Prakash, Shiv, 161
Prasad, Munaga V.N.K., 249, 271
Praveen, Kanakkath, 309

R

Raheja, Supriya, 373
Rai, Gopal N., 447
Raja Sree, T., 259
Rajnish, Kumar, 397
Rajpal, Smita, 373
Ramakrishna, G., 529
Rao, Bapuji, 3
Rao, Sriram, 17
Rath, Girija Sankar, 355
Revuri, Aravinda Babu, 389
Rout, Jitendra Kumar, 507
Roy, Kousik, 643

S

Sahaya Anselin Nisha, A., 621
Sahoo, G., 169
Sahu, Madhusmita, 381
Sahu, Santosh Kumar, 363
Sangeetha, S., 109
Sarkar, Joy Lal, 599
Sarma, Sayan Sen, 71
Sastry, V.N., 465
Sathiyanarayanan, Mithileysh, 125
Sathya, P., 109
Schiek, Michael, 29
Schloesser, Mario, 29
Sethumadhavan, M., 309
Setua, S.K., 71, 337
Shah, Mehul, 609
ShanmugaSundaram, G., 439
Sharma, Kushagra, 569
Sharma, Rupam Kr., 485
Sharma, S.C., 215, 223
Sharma, Saransh, 289
Shendre, Kanchan, 363
Singh, Akansha, 143
Singh, Karan, 143
Sinha, Nikita, 51
Siva Reddy, L., 249
Soni, Rishi, 185
Sridevi Ponmalar, P., 209
Srivastava, Rajeev, 577

Srivastava, Subodh, 577
Subbalakshmi, Chatti, 529
Suriyapaiboonwattana, Kanitsorn, 517

T
Thanalakshmi, P., 327
Thapar, Suraj, 495
Tiwari, Arvind Kumar, 577
Tiwari, Shailendra, 577
Tlili, Takwa, 103
Tripathi, Sachin, 231
Tripathy, B.K., 569

U
Umadevi, T., 631
van Waasen, Stefan, 29

V
Vatti, Rambabu A., 151

Y
Yadav, Dilip Kumar, 347

Printed in the United States
By Bookmasters

Printed in the United States
By Bookmasters